# TRANSBOUNDARY WATER RESOURCES IN AFGHANISTAN

# TRANSBOUNDARY WATER RESOURCES IN AFGHANISTAN

## Climate Change and Land-Use Implications

JOHN SHRODER and SHER JAN AHMADZAI

*Center for Afghanistan Studies, University of Nebraska at Omaha, Omaha, NE*

ELSEVIER

AMSTERDAM • BOSTON • HEIDELBERG • LONDON • NEW YORK • OXFORD
PARIS • SAN DIEGO • SAN FRANCISCO • SINGAPORE • SYDNEY • TOKYO

Elsevier
Radarweg 29, PO Box 211, 1000 AE Amsterdam, Netherlands
The Boulevard, Langford Lane, Kidlington, Oxford OX5 1GB, UK
50 Hampshire Street, 5th Floor, Cambridge, MA 02139, USA



Front cover legend: Kajakai Dam, Helmand province, is utilized for hydro power, September 9, 2009 in Helmand, southern Afghanistan. Provides electricity to Helmand and Kandahar

Back cover legend: Afghanistan's first national park, Band-e-Amir in Bamyan



**Library of Congress Cataloging-in-Publication Data**
A catalog record for this book is available from the Library of Congress

**British Library Cataloguing-in-Publication Data**
A catalogue record for this book is available from the British Library

ISBN: 978-0-12-801886-6



*Publisher:* Candice Janco
*Acquisition Editor:* Louisa Hutchins
*Editorial Project Manager:* Emily Thomson
*Production Project Manager:* Maria Bernard
*Cover Designer:* Matthew Limbert

Typeset by SPi Global, India

# Table of Contents

## 6. Characteristics and Implications of Climate Change in Afghanistan and Surrounding Regions

J. SHRODER



# II

# WATER-MANAGEMENT ISSUES IN CENTRAL AND SOUTH ASIA

## 7. Modern Water Management Issues in Central and Southwest Asia

J. SHRODER



## 8. Introduction to the Indigenous Water-Management System in Afghanistan

S.J. AHMADZAI



## 9. Dams in Afghanistan

A.K. STEWART



## 10. Water Quality Matters

P. McNAMARA



## 11. Issues of Hydrologic Data Collection by Remote Sensing in Afghanistan and Surrounding Countries

J. SHRODER

## 12. Afghanistan Water Treaties and Relevant International Water Law

J. SHRODER



## 13. Hydro-Cognizant Capacity-Building in Afghanistan

J. SHRODER



## 14. Hydro-Hegemony in Afghanistan and Surrounding Countries

J. SHRODER



## 15. Water in Islam

J. SHRODER



## 16. Water and the Opium Economy in Afghanistan

J. SHRODER



## 17. Hydropolitics of Afghanistan and Its Future

S.J. AHMADZAI



## 18. Future of Water Management in Afghanistan: Capacity Building, Risk Assessment, Cooperation, and Good Water Governance

J. SHRODER

# III

# DISTANCE LEARNING IN CENTRAL AND SOUTH ASIA

# Contributors

**N. Eqrar** University of Kabul, Kabul, Afghanistan

**P. McNamara** University of Nebraska at Omaha, Omaha, NE, United States

**L. Sinfield** US Navy, San Diego, CA, United States

**A.K. Stewart** St. Lawrence University, Canton, NY, United States

# About the Authors

**John (Jack) F. Shroder** is a Senior Research Scholar in the Center for Afghanistan Studies (CAS), as well as an Emeritus Professor of Geography and Geology at the University of Nebraska at Omaha, where he has worked since 1969. Co-founder with Chris Jung of the CAS in 1972, Dr. Shroder has studied Afghanistan's natural environment in detail using ground surveys and satellite-image analysis for much of the past four decades. Author of many scientific papers and opinion editorials about Afghanistan, and most recently a book entitled *Natural Resources in Afghanistan*, Dr. Shroder has offered factual information and observations about his geographic and geologic perspectives on centuries of conflict in the region.

Dr. Shroder has traveled extensively throughout Afghanistan, mainly in the 1970s when life in the country was considerably safer than presently. During the time that Shroder was Director of the National Atlas of Afghanistan in 1977–78, four-wheel drive and ambulatory field work, as well as low-altitude aircraft provided by the American embassy, allowed extensive survey of the country. Once the wars started in 1979 and field work was no longer possible, Dr. Shroder used multiple satellite sensors to investigate ground conditions nationwide for the next 30 years. Co-directing the GLIMS Project in Afghanistan after 2000, Dr. Shroder helped to assess the solid and liquid water resources in the Hindu Kush and Western Himalaya as part of the efforts in the Global Land Ice Measurement from Space, which was begun and managed by the U.S. Geological Survey and NASA (National Aeronautics and Space Administration). After the 9/11 attacks in the U.S.A. by Al Qaeda, Dr. Shroder returned to Kabul numerous times to renew professional associations with Afghan professors and to assist in educational development of Kabul Polytechnic University and Kabul University.

Now retired in Crested Butte in the mountains of Colorado, Dr. Shroder maintains an avid interest in all things Afghanistan, and continues to publish extensively on that country.

**Sher Jan Ahmadzai** is the Director for the Center for Afghanistan Studies (CAS) at the University of Nebraska at Omaha. He has been working at the center since 2007 in various positions. Before joining the CAS, he served in the Office of the President of the Islamic Republic of Afghanistan. Prior to leaving Afghanistan, he served as Director of the Presidential Schedule at the Office of the President of Afghanistan where he was responsible for managing President Hamid Karzai's day-to-day work schedule. During Mr. Ahmadzai's efforts at the palace, he was able to foster and maintain professional relations with government, community, and political leaders across Afghanistan and with the representatives of the international community in Kabul.

Mr. Ahmadzai's areas of specialty are Afghan governance, Afghan hydropolitics, tribal dynamics, and their impacts on the public and their relationship with the government as an institution providing public services based on appropriate public policies. Other specialty areas of Mr. Ahmadzai are Afghanistan–Pakistan relations, U.S. involvement in Afghanistan, and regional security in South-Central Asia.

Mr. Ahmadzai has written op-eds and articles for the BBC, CNN, and *Omaha World-Herald*, and a number of papers on various political issues concerning Afghanistan, the United States, and the region of South-Central Asia. He is also a regular contributor/commentator on BBC Pashto and Dari radio programs and the Voice of America (VOA) Pashto and Dari services on various social, political, and current affairs regarding Afghanistan and the U.S.A.

**Professor Naim Eqrar** has held teaching positions for more than 20 years in the Department of Geology of the Faculty of Geosciences, University of Kabul as a Dean, Head of the Department of Geology, and Professor of Water Resources. He has developed several course materials for a national training of experts for the national and international agencies at various levels and supervised over 50 training courses. Professor Eqrar has recently (2010-2015) developed his doctoral dissertation on the Hydrogeology and Hydrochemistry of Kabul City at Bonn University-Germany to the final stage and will defend his dissertation this year (2016). Professor Eqrar has participated as a guest professor of water resources, as a speaker, presenter, and Afghanistan country representative at several conferences and workshops abroad, including the USA, Norway, India, Germany, Belgium, Iran, Nepal, China, Japan, Sri Lanka, Russia, and the Czech Republic. Professor Eqrar has published a large number of scientific reports, journal articles, and conference papers. He has been an editorial board member of five national journals and a technical reviewer for over 10 scientific papers. He has an extensive teaching experience in the fields of hydrology, hydrogeology, water resources, and hydrochemistry, particularly concerning water resources of the area of Afghanistan. Professor Eqrar is a member of the Supreme Council of the Water and Land of Afghanistan (SCoWLA), a member of the National Hydrological Committee of Afghanistan (NHCA), a member of the steering committee of the Small Grand Program (SGP) OP6 Country Program Strategy Development of UNDP, and a member of the Strengthening the Resilience of Rural Livelihood Options of Climate Change Projects for Afghan Communities, promoted by UNDP/MAIL.

**Patrick McNamara** serves as Director of International Studies at the University of Nebraska Omaha. His expertise is in facilitating multi-party stakeholder processes, with a growing focus on the intersection between public policy and natural-resources management. Dr. McNamara has been a Fulbright-Nehru Senior Scholar in India studying water conflict, a scholar-in-residence at the International Water Management Institute in Sri Lanka, a participant in the Nebraska Water Leaders Academy, and most recently with Dr. Shroder and Mr. Ahmadzai, Dr. McNamara is helping to facilitate an ongoing transboundary water-cooperation project in Afghanistan, Tajikistan, and Pakistan that is funded by the U.S. Institute of Peace, and the Lounsbery Foundation from Washington, D.C.

**Mr. Len Sinfield** has over 30 years of experience working as a hydrogeologist and is currently a water compliance manager with the U.S. Navy in San Diego, California. He worked as a civilian with the U.S. Army Corps of Engineers on water-development projects throughout southern Afghanistan in 2012 and early 2013, with an emphasis on developing water sources for the City of Kandahar along with Afghan governmental partners.

**Dr. Alexander K. Stewart**, U.S. Army (retired), is a veteran of the Cold War and three foreign wars; most recently (2009), he spent a year as a soldier-geologist with the Texas Agricultural Development Team (143rd Long-Range Surveillance Detachment, 36th Infantry and 82nd Airborne Divisions) in Afghanistan. His time with this unique counterinsurgency team provided the opportunity to help the people of Afghanistan, not by means of his military training, but through his civilian skills as a professor and scientist. It was during his many dam-assessment missions in Afghanistan that Dr. Stewart became interested in water-resources in the region. Dr. Stewart is currently an Associate Professor of Geology at St. Lawrence University in Canton, New York State.

# Foreword to Transboundary Water Resources in Afghanistan

One of the most urgent challenges of the 21st century is to ensure sustainable food and water security in the face of population and income increases, a changing climate, and growing demands for scarce water resources. This enormous challenge, which has significant political, environmental, social and economic implications, plays itself out in different ways in different contexts.

In the almost 150 countries whose rivers, lakes and/or aquifers cross international boundaries, addressing this challenge requires effective transboundary water management, especially when these water bodies constitute the major or only source of renewable water. In such contexts, no one country can fully control its water resources, and policy makers need to work through international cooperation and transboundary water agreements. To do so successfully, they need to understand the intricacies of the hydrological interrelationships among countries. They need to be able to analyze the way in which dams and irrigation projects in upstream countries impact water availability in downstream countries, and how that might affect their ability to attain water, food, and energy security. They also need to be aware of the impacts that climate change, population growth, and other factors might have on future supply and demand for water.

This is clearly the situation in Afghanistan and the Southwest Asian countries that surround it: Pakistan, Iran, Uzbekistan, Turkmenistan, and Tajikistan. A map of the Amu Darya, Kabul, Helmand, Harirud, and Murghab river basins shows how the waters that emanate from Afghanistan are intertwined with Afghanistan's development and that of its downstream neighbors. Importantly, all these countries have arid and semi-arid climates, and depend heavily on irrigated agriculture for their food security. Afghanistan and its neighbors irrigate over 38 million hectares of land, almost twice as much as is irrigated in all of the United States and constituting in some of these countries as much as 80% or 90% of the land available for agricultural production. In a region where food, water, and energy supplies are already vulnerable and subject to increasing pressures from political and social conflicts, population growth, and climate change, the rivers that originate in Afghanistan have a major impact on the water, food, and energy security of their neighbors downstream. And while Afghanistan's water resources are at present relatively underdeveloped, over time the country will inevitably need to make a greater use of its waters, with significant downstream impacts.

Policy makers in Afghanistan and in the region require a broad understanding of the hydrological interconnections of the region's shared water resources to use and manage them successfully to meet their development goals. The success of negotiations between countries in the region over the allocation of shared water resources and their use for irrigation, flood/drought management and energy generation will require increasing the capacity of decision-makers in Afghanistan and the region to understand and analyze the hydrology of Afghanistan and its neighboring countries and the management of these shared water resources.

This is why *Transboundary Water Resources in Afghanistan*, by John Shroder and Sher Jan Ahmadzai, fills an important niche. By providing a comprehensive and science-based view of the interconnected water resources of Afghanistan and its neighbors, the book will equip policy makers and technical advisors in the region with the tools they need to develop effective policies and interact cooperatively with the countries that share its water resources. The book provides a solid introduction to the physical characteristics of water, both in general and specifically in Afghanistan and its neighboring countries, as well as in-depth discussion of water management issues in Central and South Asia and a look at the future of water management in the region. The book also includes chapters on such vital subjects as the characteristics and implications of climate change, hydrological data collection, and international water law.

John Shroder, Emeritus Professor of Geography and Geology at the University of Nebraska at Omaha, has dedicated his academic career to studying and analyzing transboundary water negotiation issues between Afghanistan and Pakistan, and has brought together in one volume his extensive knowledge and experience of the water resources of the region, as well as the views and perspectives of various other specialists in the field. By helping policy makers in this critically important region to advance water and food security, the book is a significant contribution to the work and mission of the Robert B. Daugherty Water for Food Institute at the University of Nebraska, where Shroder and chapter author Patrick McNamara are Faculty Fellows. The institute was established in 2010 to bring the best of the University's expertise in water and food to address the challenges of improving water and food security across the globe. Working with Faculty Fellows like Shroder and McNamara and a global network of partners, our mission is to have a lasting and significant impact on food and water security through research, engagement, communication, and education programs.

*Roberto L. Lenton*
Founding Executive Director
Robert B. Daugherty Water for Food
Institute at the University of Nebraska

# Preface

The 21st century is generally regarded by knowledgeable scientific and intelligence analysts as a century of looming water issues because of worldwide diminution of access to that essential resource as a result of burgeoning human populations and resultant climate change. Afghanistan is at the top of the watershed, or the place from where several major rivers (Amu Darya, Kabul, Helmand, Harirud, and Murghab rivers), emanate from their headwaters. These rivers flow over the international borders into the surrounding countries in the north into Tajikistan, Uzbekistan, and Turkmenistan, in the northwest into Iran and Turkmenistan, in the southwest into Iran, and in the southeast and east into Pakistan. These rivers have been the foundations of millennia of human development in Asia.

In the ever-evolving world of international water law, it has been recognized in recent years that the lowland countries of the world were the places where humankind first began its earliest irrigation schemes thousands of years ago. This resulted in the development of extensive civilizations and expanding human populations in the lowlands, who came to depend upon regular delivery of waters unhindered and undammed from the undeveloped uplands of places such as Afghanistan.

The importance of water in the modern or ancient worlds cannot be overstressed; if enough of the treasured resource was not available back in antiquity, then that particular portion of civilization must ultimately collapse. Such has happened many times before (Fagan, 2008; and many others). But now we know this, and can better predict the future, rather than allowing such an event of disappearing water to be as much of a surprise as it was to the rulers in antiquity, when clearly they thought their gods must have deserted them as the rains stopped and the waters dried up. Of course, it is possible to live in a 9th-century fantasy world of antiquity as many of the Taliban and ISIS jihadis would love to try to do as they promote their particular warped barbarism, but the alternative in the modern world is the use of good science to try to overcome human-caused difficulties, rather than give in to some supposed "will of God."

In the situations outlined in this book, learning about the many variables in Central and Southwest Asia that affect, deliver, and control water in all its forms and conditions, including human machinations, would seem to have a high benefit to those who indulge in calculating all of its possible variations. It is obvious that the study and understanding of water delivery in Afghanistan and its neighboring countries of Pakistan, Iran, Turkmenistan, Uzbekistan, and Tajikistan have much to recommend them. This will help to bring up the capacity levels of understandings in the government and among academic water cognoscenti in Afghanistan to head off any possible confrontations later on with surrounding countries who may try to take issue with Afghanistan on its use of its waters. Local water law is always made by acquiescent governments who seek reciprocity when it is convenient to them, but adequate forewarning of coming changes does help to protect against less desired climatic outcomes, and can influence water legalities in the international eye. Training of Afghan government officials and bringing them up to international levels of hydro-understanding will enable them to profit in any sort of water treaty negotiations with hydro-experts from other countries as they all negotiate the international distributions of water allocations and seek additional funding to build dams and irrigation infrastructures.

Recently, much has been made of rivers that dry up before they reach the sea because of human withdrawals upstream (neo-arheism) (Falkenmark and Lannerstad, 2005; Lannerstad, 2002; L'vovich and White, 1990; Meybeck, 2003). In fact, however, all of the large rivers (Kabul, Helmand, and Amu Darya) originating in Afghanistan have been drying up repeatedly as a result of overuse, likely coupled with climate change before they reach their historical destination. These facts are the strongest evidence possible that water issues in Afghanistan are most serious and serve as a strong harbinger of things to come in Central and South Asia, as well as many other semiarid and arid areas of the world. Because of this forecast of such serious problems in a world of coming climate change, study of these aspects of water in Central and South Asia is a highly valuable exercise.

The study of water in all its forms runs the gamut from the stark physics of an unusual material that behaves quite differently from almost all other compounds in nature (expanding when cold, decreasing its density when solid, etc.), and is so essential to life on this planet. This critical necessity extends out to exert control on the bewildering plethora of all the different levels of water delivery and usage for humans. Everyone has some little understanding of water in their own experience with the material, but few can understand all of its attributes unless they study the character of water and the physical and human manipulations thereof. This means that not only is water an unusual substance essential to life, but its core uses to humanity and all other life transcend simplistic understandings so that great analytical expertise must be applied to manage it properly. Because all too many people can take water for granted, and quite misunderstand some of its apparently mysterious and diverse attributes, they can be rather surprised when the water somehow diminishes or disappears. One of the purposes of this book is to explain some of these mysteries of water in Central and Southwest Asia so that no one in government who has been paying attention will have to wonder too much about what is happening to the hydrologic character of their region.

Building hydro-capacity in any region is thus an issue of education; teaching about the basic physics and chemistry of this vital $H_2O$ substance, even while the basic geoscience (surfical physical geography and hydrology of rivers, coupled with the hydrogeology of water underground) is added in to give an essential background context. This is all the basic science, which is not all that complex; it is just that there can be a great deal of it, or even nearly an overwhelming amount, and people may be loath to take on understanding of too much of this information lest they feel that they do not need to understand all of it. Then because so much cultural or sociological baggage concerning water also exists in any given society to surmount as well, the amount of information required can feel fairly overwhelming. Nonetheless, because this information is so essential to wise management of this critical resource, this book is designed to serve as a foundation document to assist water-cognizant agents in Central and Southwest Asia.

Many people in Afghanistan now recognize that most of the illiterate Taliban insurgency against their government does not know that the lowland neighbors who support the insurgency actually are most interested in unrestricted access to the water that comes from Afghanistan. Keeping that water under-utilized in an under-developed Afghanistan is thus obviously of advantage to the neighboring countries. For lowland countries to foreclose on the rights of upland countries to also develop is obviously not fair. Some solutions to this problem are water-capacity building in Afghanistan so that the government officials will comprehend the situation and be more able to deal with hydro-issues more effectively in future. The age of hydro-diplomacy (Eliasson, 2015) is now firmly upon us, and it is seen as an essential aspect of the human condition as the rising pressures of global water shortages loom largely everywhere, but especially in Central and Southwest Asia.

## References


Eliasson, J., 2015. The rising pressure of global water shortages. Nature 517 (7532), 6.

Fagan, B., 2008. The Great Warming: Climate Change and the Rise and Fall of Civilizations. Bloomsbury Press, New York.

Falkenmark, M., Lannerstad, M., 2005. Consumptive water use to feed humanity—curing a blind spot. Hydrol. Earth Syst. Sci. 9, 15–28.

Lannerstad, M., 2002. Consumptive Water Use Feeds the World and Makes Rivers Run Dry, MSc Thesis. Royal Instutue of technology (KTH), Sweden, Stockholm.

L'vovich, M.I., White, G.F., 1990. Use and transformations of terrestrial water systems. In: Turner, B.L. II (Eds.), The Earth as Transformed by Human Action: Global and Regional Changes in the Biosphere Over the Past 300 Years. Cambridge University Press, Cambridge.

Meybeck, M., 2003. Global analysis of river systems: from Earth system controls to Anthropocene syndromes. Phil. Trans. R. Soc. Lond. B 358, 1935–1955.

# Introduction

*Transboundary Water Resources in Afghanistan: Climate Change and Land-Use Implications* is a new book by John (Jack) F. Shroder, Sher Jan Ahmadzai, and several of their colleagues, taking a look at a vital resource of increasing scarcity in the world. Afghanistan, being at the top of the Hindu Kush watershed, is blessed with precipitation at the higher altitudes but starved for water in its lowland steppes and deserts. Inequality of water availability has long been a contentious issue in a region now caught in the throes of climate-change stressors that add to the volatile mix of ethnicities and religious ideologies contributing to much violence to the area.

This book is divided into three parts: an introduction to the physical characteristics of water in the region; extensive assessment of issues of water management in Afghanistan and neighboring countries; and distance learning where necessary information and understanding about water can be transmitted electronically across borders most easily.

The book begins with an assessment of the regional hydrologic cycle and hydrogeography of Afghanistan and surrounding countries through which the water from the highlands flows. The underground water of Afghanistan is considered as a whole, with detail in the vital Kabul and Kandahar basins. Water as a hazard (flood, drought, snow avalanches, and wet landslides) is a major problem, adding additional climate-change threats and disasters. Water management in Afghanistan is considered in both traditional and more modern forms. Adding to the comprehensive nature of this book are issues of old and new dams, decreasing water quality and quantity, and lack of historic water-data collection in the face of new opportunities for remote-sensing assessments of water above and below ground. Limited-to-absent water treaties with neighboring countries are addressed in detail, with new ideas of international water law that can be applied through diplomatic negotiations. Educational limitations caused by long wars have necessitated hydro-cognizant capacity building in Afghanistan to deal with new ideas about gaining various hydro-hegemonic controls over transboundary water resources. Some means to help with these issues can be obtained by attention to the beneficial treatment of water in Islam, as well as recognition of climate-change reasons for the rise of drought-resistant opium poppies. Hydropolitics in Afghanistan is fraught with complexity and political machinations due to an uncertain future of water management in Asia. Some diverse water issues can be alleviated through concerted efforts to raise educational levels by electronic distance learning and use of the hydro-information that this book provides.

PART I

# INTRODUCTION TO PHYSICAL CHARACTERISTICS OF WATER



The focus of this work is on the water that arises in the mountains of Afghanistan and flows more or less radially outward across the borders of Afghanistan and into the neighboring countries of Pakistan, Iran, Turkmenistan, Uzbekistan, and Tajikistan. China also borders Afghanistan in the far northeast, but no water is transmitted across this border, so that area is not included. Because Afghanistan is the center of our focus and the pivot point of our expertise, we provide the most information about transboundary water deriving from that country. By necessity, however, some discussion of the hydrological problems of the neighboring countries must be included as well. Less important then, for the neighboring countries bordering Afghanistan, is any great detail about their hydrological situation, because such is neither directly relevant nor deeply bedded in our expertise. Instead the discussions herein can serve as a basis for more advanced, or detailed assessments of the water in bordering countries.

The result of the focus being described here is that in any exposition of transboundary waters, the hydrological geography and geology of the source region provides an essential set of attributes upon which to focus. Thus the energy sources that run the hydrological cycle are relevant, as are the processes that move gaseous, liquid, and solid $H_2O$ between locations. The basic principles and characteristics of the regional hydrological cycle must figure in the initial discussion, which also includes the regional and local climate controls. Aspects such as the winds that bring precipitation-bearing storms are essential, as are discussions of the snow and ice in the mountains, which later melt seasonally or, after much delay as glacier storage, also melt to eventually become river water downstream. On hillsides, the water runs off and initiates the soil-destroying sheet and gulley erosion that is so detrimental to long-term agricultural productivity and fills up essential water impoundments with sediment so that water availability for irrigation is reduced, and hydroelectric power generation is curtailed. Finally, the configuration of the drainage basins and the major surface rivers is essential to comprehend the main aspects of water availability.

Hidden below the ground surface, of course, are all the unusual aspects of underground water, which, because it has been hidden from view for so long everywhere in the world, tends to be poorly understood and improperly used or under-utilized. Nonetheless, in areas where the environment has long been grossly abused by the hand of man—and Afghanistan is known as emblematic of such a place—great over-utilization or contamination of this vital resource can become a severe problem. In fact, as the climate of the world changes, warming and drying in Central and Southwest Asia, these problems will only get worse, so it is essential for agents of the governments of the region to understand these issues and attempt to seek solutions where possible. In addition, the hazards, risks, and disasters associated with all the forms of $H_2O$ in action in Central and Southwest Asia are critical to understand as well, particularly because some of the associated hazard problems are among the worst in the world. Water, of course, constitutes one of the chief natural hazards at work, mainly in the cases of either too much or too little of the compound so that flood and drought disasters are greatly feared and little understood in Afghanistan or throughout the region.

Finally, various aspects of climate change have become ever more obvious to many people of the world as mega-droughts and mega-storms rage across the planet. Among the high mountain water towers of Asia, more and more recognition of glaciers wasting away, increasing meltwater lakes in breakout floods, formerly dependable perennial rivers drying up in certain seasons, increasing dust storms—all appear to be short-term evidence akin to weather variability that might be harbingers of the new realities of common coming climate change. This book is, at least in part, an attempt to bring recognition to the new problems, and by exposing them to greater light, perhaps to help alleviate some societal problems.

CHAPTER

# 1

# Characteristics of the Regional Hydrological Cycle

*J. Shroder*

**University of Nebraska at Omaha, Omaha, NE, United States**

OUTLINE



## Abstract

The hydrological cycle in Central and Southwest Asia, of course, operates essentially the same as it does in the rest of the world, but it does have regional variations in character and timing of its phases, energy sources, winds, distributions, climate and topographic influences, and other controlling factors that need to be understood. The high mountains of the region serve as the water-tower catchments for the elusive moisture that passes over the dry lowlands, but fortunately for the people who live there, that moisture precipitates orographically in the mountains above them. Not so fortunately, however, the common devegetation and soil erosion that also occur so commonly in the region, end up despoiling the surficial environments and reducing water infiltration into the surficial soils so that the runoff is accelerated into flashfloods and is thereby wasted. In any case, multiple drainage basins have resulted, the development and use of which are the focus of this book.

**Keywords:** Water physics, $H_2O$ phases, Precipitation, Climate controls, Mountain topography, Glaciers, Devegetation, Rapid runoff, Liquid and solid flow, Drainage systems



http://dx.doi.org/10.1016/B978-0-12-801886-6.00001-X

## 1.1 INTRODUCTION

Water in Central and Southwest Asia is such a precious resource in the dominantly semi-arid and arid environments in which people live in the region that this book is designed to answer a number of essential questions. First and foremost, of course, is a basic exposition of the science behind this aqueous resource before we investigate the human aspects of water management in the region. Thus in this case, our intent is to present the basic facts about how hydrological cycles (Fig. 1.1; in multiple languages) work in general, and progressively work into the more arcane details of local water situations in the region.

The basic physics of the system $H_2O$ that everyone knows more or less is that the liquid phase is *water*, the solid phase is frozen *ice*, and the gaseous phase is water *vapor* (Fig. 1.2). The basic principles of the hydrological cycle are that these different aspects or phases of the $H_2O$ system cycle endlessly back and forth to the many different locations in the oceans, atmosphere, glaciers, soil moisture, freshwater and salt lakes, and underground water. The processes that move the $H_2O$ between its many locations include:

- evaporation (liquid water to water vapor);
- precipitation (vapor phase to liquid rain, mixed solid and liquid sleet, solid snow);
- infiltration (water movement down into soil porosity and cracks in rock);
- transpiration (liquid water converted to vapor and emitted from plants);
- liquid flow (water currents, floods);
- solid flow and fracture (long-term deformation of glacier ice under gravity pressure);
- sublimation (solid ice conversion to vapor phase); and
- ablation (solid ice conversion to liquid and vapor phases).

The energy sources that cause the phase transformations between solid, liquid, and gas are the radiation from the sun, the rotation of the Earth in space, the winds of the Earth, and the gravitational attraction of the water mass to the Earth. As these phase changes occur back and forth between the three types, some of the energy of the system is locked up inside the different phases and it can come back out as the latent heat of condensation/vaporization (water vapor condenses to liquid water, or water to gas, which equals about 2257 kJ/kg, or 600 cal of heat energy), and the latent heat of fusion/freezing (solid ice melts to liquid water, or water freezes to ice, which equals about 335 kJ/kg, or 80 cal). In general these energy conversions can be ignored unless calculations are needed to more precisely understand the absorption and release of these energies in understanding phase transformations as important aspects of climate change and other such relevant aspects. These heat energies can drive other parts of the system to affect the temperature-controlled winds, for example.

## 1.2 REGIONAL CLIMATE CONTROLS

Most of the water of the Earth is visibly in the oceans of the world, where some 96.5% is held (Fig. 1.3), and only 2.5% of the total amount of water on the planet is freshwater. Of that total amount of freshwater available for humans and other life to utilize, glaciers hold about 69% of the total, water underground holds about 30%, and that leaves only about 1.2% as surface water in rivers and lakes. If you then count up how that small 1.2% is distributed,

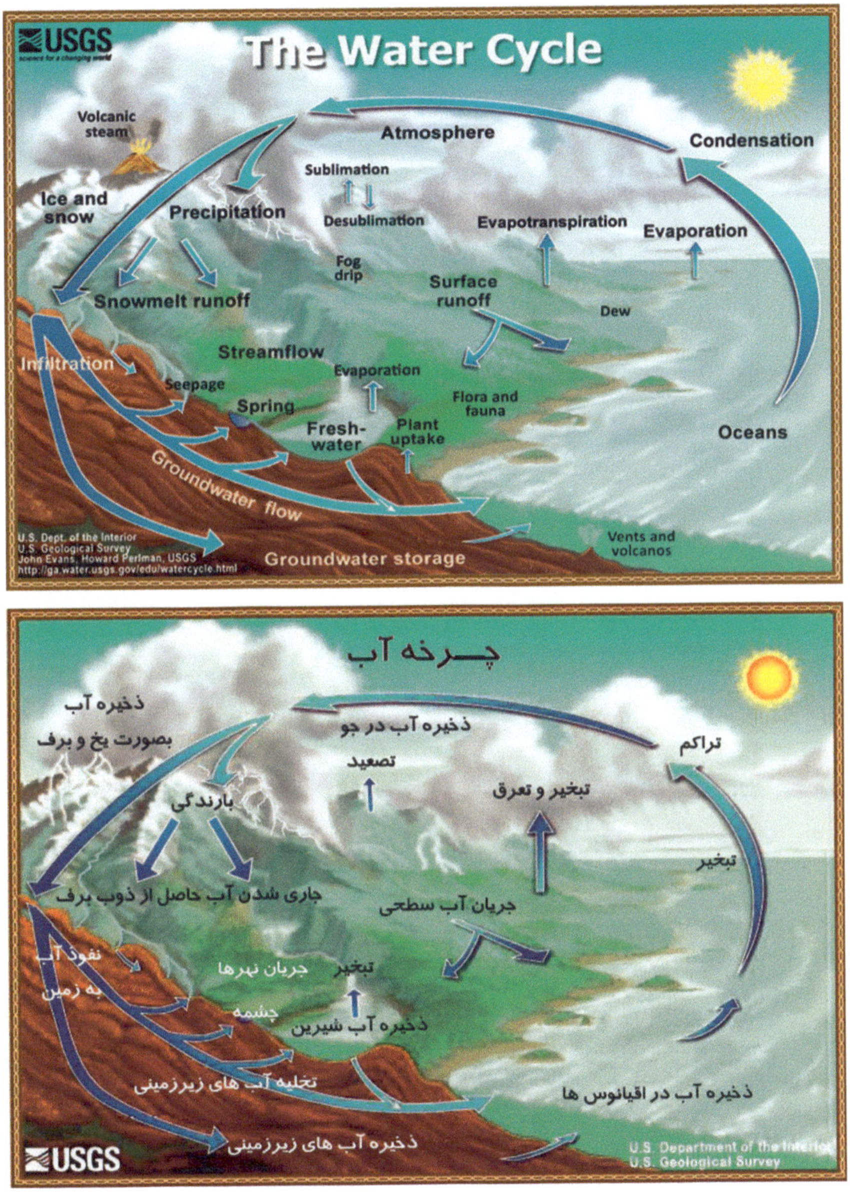


(A)

(B)

FIG. 1.1 The hydrologic (water) cycle of the world (A) in English (from US Geological Survey on-line source http://water.usgs.gov/edu/graphics/watercyclesummary.jpg (accessed 15.02.15)); (B) in Farsi;

*(Continued)*

(C)

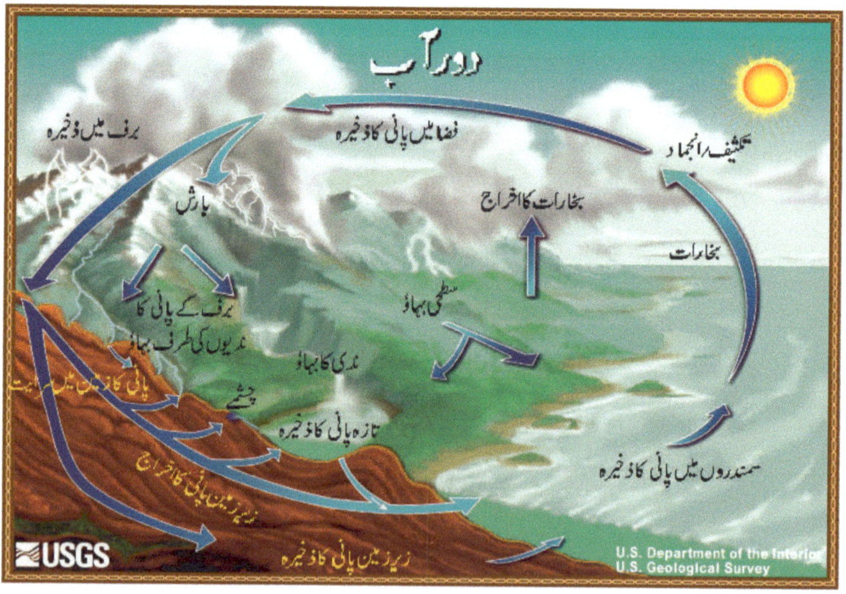


(D)

FIG. 1.1, CONT'D (C) in Pashto; (D) in Urdu; and

(E)

FIG. 1.1, CONT'D (E) in Russian.

you can see that again about 69% of it is in underground ice, lakes have about 30% of it, and rivers have only about 0.5% of the remainder, along with smaller amounts in the soils, the atmosphere, swamps, and living things. No wonder water is so precious!

The hydrological cycle in Central and Southwest Asia receives the greater part of its moisture dominantly from storms that originate in the Atlantic Ocean, the Mediterranean Sea, or from the Indian Ocean (Fig. 1.4). The wind sources that import these storms into the region are:

- westerly winds that blow mainly in winter from the Atlantic and Mediterranean;
- monsoon winds that blow from the south in summer from the Arabian Sea and Indian Ocean;
- the wind of 120 days (*badi sado bist roz*) blows out of Central Asia in summer from the north and west into Afghanistan and loops south and southeast into the low pressures of the monsoon that are dominantly along the border with Pakistan (Fig. 1.4A);
- katabatic winds—cold air drainage blows down from the Tibetan Plateau and other high-altitude mountain areas of the Hindu Kush mountains (Fig. 1.5); and
- anabatic wind—hot air rises up from lowland deserts that abound all around Afghanistan (Fig. 1.5).

The natural water supplies of any country are generally that which comes from the sky, either on an annual basis to produce river runoff, or the portion that is stored underground. The underground water is commonly derived from precipitation that has fallen, infiltrates through the soil and downward, where it is held underground, perhaps for a very long time. Thus in order to understand the watersheds and their importance in arid

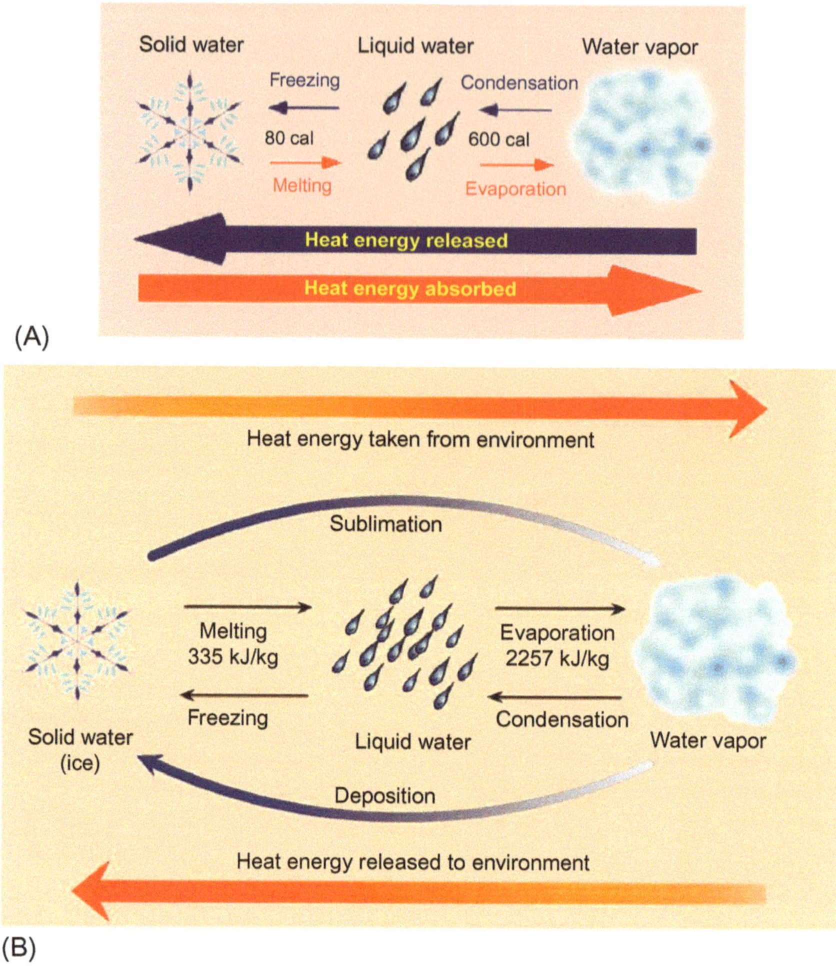


FIG. 1.2 Phase transformations of the $H_2O$ system and the heat energy taken on or released between the three phases of liquid water, gaseous water vapor, and solid water ice (A) in calories and (B) kilo-joules per kilogram.

countries such as Afghanistan and its neighboring countries of Pakistan, Iran, Turkmenistan, Uzbekistan, and Tajikistan, we must first consider the overall climate of the region, with its strong variations in precipitation, as we focus on the all-important hydrology of the Hindu Kush, Pamir, and Western Himalaya, mainly in terms of the causes and effects of glacier ice, the rivers, and the lakes.

The climates of the Earth are controlled largely by the amount of precipitation and temperature delivered at different times and places as a function of latitude, altitude, and position on a continent relative to ocean water masses. Mountainous and desertic countries such as Afghanistan, and all its neighboring countries, the interiors of most of which are far from oceanic moisture sources and subject to the vicissitudes of variable winds and erratic supplies

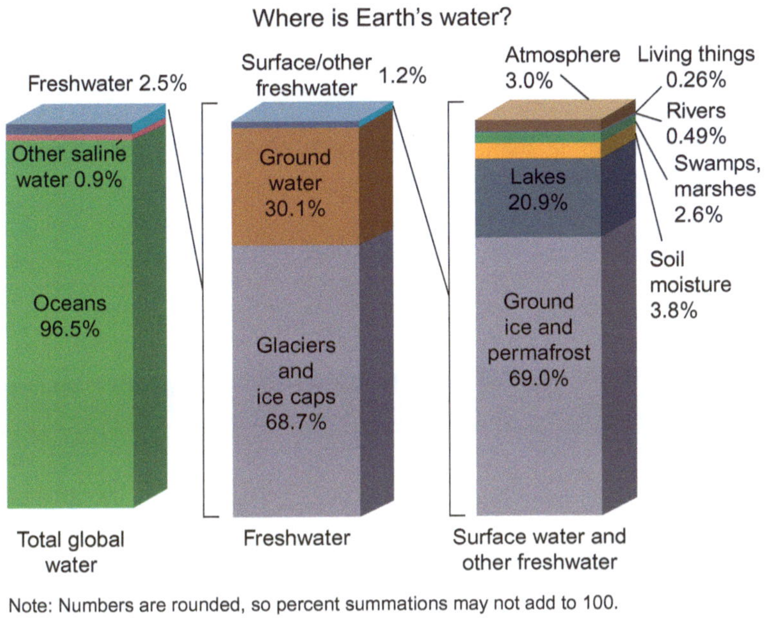


FIG. 1.3 The distributions of the world's water by its physical locations. *From Shiklomanov, I., 1993. World fresh water resources. In: Gleick, P.H. (Ed.), Water in Crisis: A Guide to the World's Fresh Water Resources. Oxford University Press, New York, http://www.envirothon.org/files/2014/The_Water_Cycle.pdf.*

of precipitation; all are characterized also by extremes of what is called *continentally* wherein winters can be quite cold and summers excessively hot and dry so that any water derived from snow melt and rainstorm precipitation is essential but evaporates quickly.

In general, Afghanistan and its neighboring countries are dominantly arid to semiarid regions, except in the frontal or foothill mountains where more *orographic precipitation* is caused when air masses are forced to rise and the attendant cooling causes condensation of moisture out of the air (Fig. 1.6). The amount of precipitation increases to the northeast in Afghanistan, and to the north in both Tajikistan and Pakistan, in response to the higher altitudes there (Fig. 1.7). Average annual precipitation is commonly <210 mm in many areas, declining to <110 mm in the southwestern deserts, and increasing to >1000 mm in the high mountains. In all, over 80% of the Afghanistan's water resources have their origin in the mountains >2000 m in altitude, which function as a natural storage of snow and ice that supports perennial flow in all the major rivers in summer. Such is also true of much of northern Pakistan and southwestern Tajikistan, although percentages and altitudes vary (Fig. 1.7).

In Afghanistan, the main period of precipitation in winter is borne on the Westerly winds and extends from November to May, but is shortened in the south to December through April. About half of the precipitation occurs in winter (January to March), mostly as snow. A further third of the moisture falls in spring (April to June). In the summer season some monsoonal precipitation makes it into extreme southeastern Afghanistan along the border with Pakistan, although occasionally monsoon moisture sources go well into the Hindu Kush in Central and Northern Afghanistan as well.

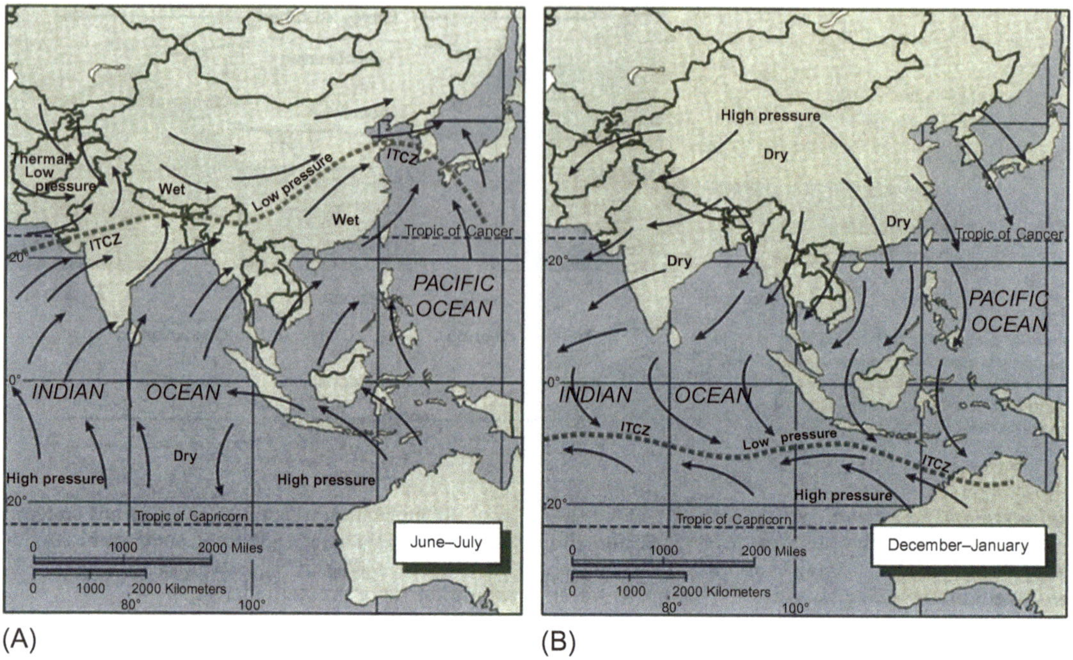


FIG. 1.4 Seasonal maps of Asia in (A) summer and (B) winter showing dominant wind directions and positions of the intertropical convergence zone (ITCZ) of the stormy, moisture-laden air masses that rise because of their warmth and cause monsoonal precipitation. The thermal low pressure in the summer season over Pakistan and part of Afghanistan shows the patterns of winds in the region at that time. *Map by permission from the Center for Afghanistan Studies.*

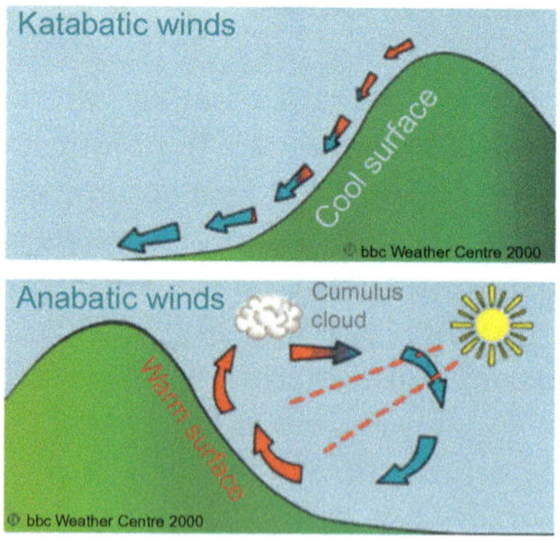


FIG. 1.5 Cold descending katabatic winds (upper) and warm rising anabatic winds that are typical in mountainous desert areas. *From the BBC Weather Centre on line.*

Precipitation in the summer in Pakistan is monsoonal of course, and it can be torrential, with major flooding being common, especially because Pakistan has a dearth of appropriate flood-control dams and other such necessary infrastructure. In the winter season, precipitation in Pakistan is dominantly from the Westerly winds, just as it is in most seasons elsewhere in the region in Iran, Turkmenistan, Uzbekistan, and Tajikistan.

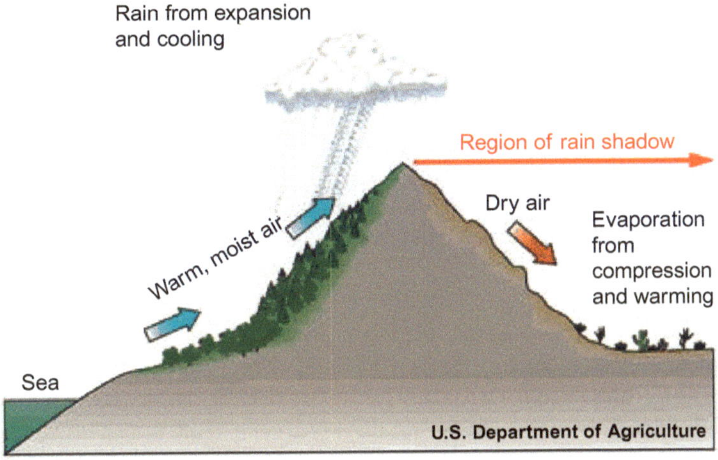


FIG. 1.6 Orographic rainfall that rains out on one side of a mountain and then the descending air on the other rain-shadow side warms and dries.

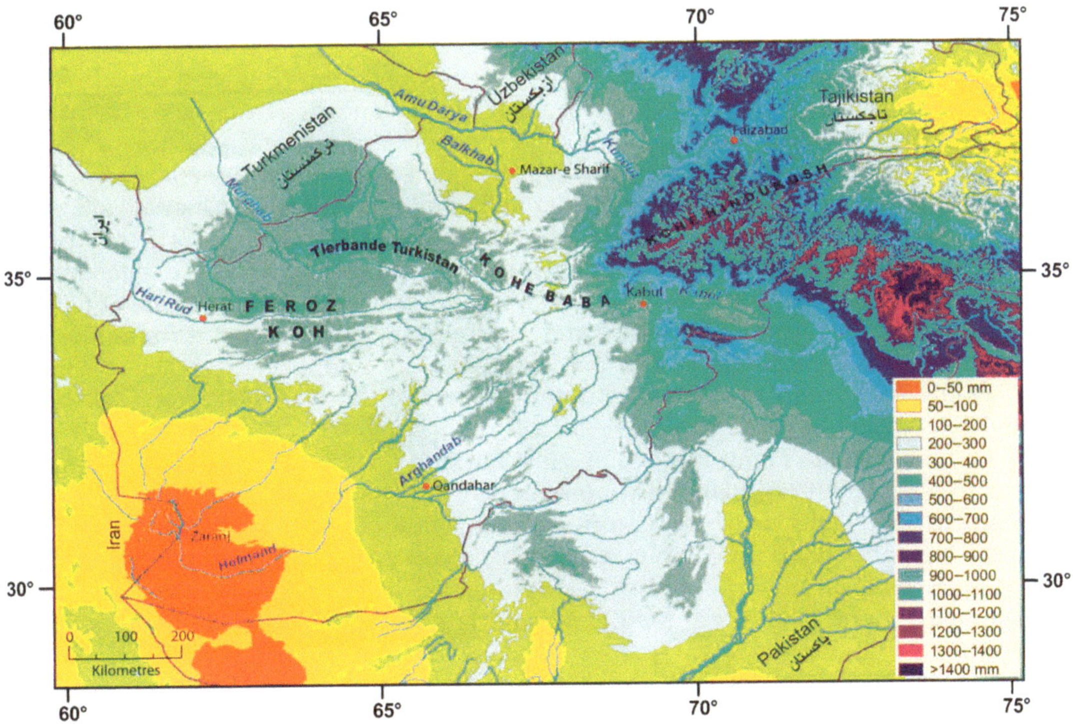


FIG. 1.7 Precipitation in Afghanistan and the nearest parts of the neighboring countries. *From Breckle, S.-W., Dittman, A., Rafifiqpoor, M.D., (Eds.), 2010. Field Guide Afghanistan: Flora and Vegetation, Scienta Bonnensis, Bonn, Germany; Shroder, J., 2014. Natural Resources in Afghanistan: Geographic and Geologic Perspectives on Centuries of Conflict. Elsevier.*

The water cycle is absolutely critical to all these countries, of course, just as it is in Afghanistan. Furthermore, the forces in the hydrological cycle have been nearly totally responsible for shaping most of the structural geomorphic landforms that abound in the region (Shroder, 2014), as well as the largest share of the natural hazards of the region. As overall semiarid to arid nations, a number of places in all of the countries of this region have no real surficial drainage; they are *arheic*. Elsewhere, however, the drainage net is well developed and provides the basic water source to the people, even as it is also responsible for the general character of most of the landscapes except in the driest of the broad lowland areas (Registan).

The watershed boundaries in Afghanistan begin high on ridge and mountain tops, generally, but not entirely, in the middle and northeast of the country; thereafter the rivers flow radially outward over the borders into neighboring countries. This position at the top of the watersheds places Afghanistan geomorphically as a source of snow and glacier ice storage and melt-water river delivery downstream that combine to produce most of the geomorphology or landforms of the country. Below the land surface of the nation the fractures in the bedrock, and the natural porosity in the sediment-filled basins throughout the nation (Kabul, Jalalabad, Seistan, northern basins), have considerable storage of underground water that is obtained through gently inclined karez tunnels, as well as boreholes and pumps.

Similarly, in Pakistan the high mountains receive plentiful precipitation, particularly from the winter Westerly winds, but also to a certain extent in the foothills to the Himalaya, whereas from there to the south is where most of monsoonal precipitation occurs. In general in Pakistan then, the precipitation received in the country can be divided into two main seasons: the summer monsoon that comes in from the east and southeast from July to September, and the winter Westerly disturbances that come into the country from across Iran and Afghanistan from December to March. In spite of severe monsoonal flooding in 2010 and 2011 in Pakistan, recent analyses show a significantly decreasing trend of precipitation all over the country, with periodic prolonged droughts that will pose severe risks to agriculture and the water-management sectors in Pakistan. The overall climate of Pakistan ranges from the high altitude glacier ice and tundra in the north, down through the more temperate and subtropical climates in the foothills to the semiarid steppes and arid deserts of Baluchistan south of Afghanistan and in Sind to the southeast along the border with India.

In Tajikistan, the high Pamir Mountains catch much of the Westerly precipitation that can come in all seasons but especially in winter and spring. The weather and climate are quite variable between conditions of strong continentality with its extremes of cold and hot temperatures, high-altitude, humid mountain environments, warm and cold steppe climates, and lowland, arid deserts. This presents a number of different climate zones, from subtropical, moderate temperate and Mediterranean-type climates, to high tundra and glacial environments.

None of the other countries in the region (Iran, Turkmenistan, and Uzbekistan) have significant glacier-water storage. For this reason they thus lack any significance to the discussion of transboundary water resources associated with Afghanistan rivers, and so will not be discussed further here.

## 1.3 MOUNTAIN TOPOGRAPHY OF THE REGION

The highland mountain terrains of Central and Southwest Asia that host the glaciers and snow that provide so much of the local water supplies begin with the so-called "Pamir Knot," which is located in Central Asia. The Pamir Mountains are formed by the junction or "knot" of the Tian Shan, Karakoram Himalaya, Kunlun, and Hindu Kush ranges. They are among the world's highest-altitude cluster of mountains. They are also known by the Chinese name of Congling 葱嶺 or "Onion Mountains." The region of the Pamir Knot is centered in the Tajikistan region of Gorno-Badakhshan. Parts of the Pamir knot also lie in the countries of Kyrgyzstan, Afghanistan, and Pakistan. South of Gorno-Badakhshan, the Wakhan Corridor of Afghanistan runs through the Pamir region, along with the Hindu Kush and Hindu Raj mountains, which also include the northern extremes of the Khyber Paktunkhwa (old North-West Frontier) Province and the northern extremes of the Northern Areas of Pakistan, part of which (Siachen Glacier) in the Karakoram Himalaya is contested with India. As a whole this region is characterized by being the highland source of numerous rivers that emanate from the melt of the profuse snow and glaciers of the region, coupled with plentiful precipitation from the Westerly and Monsoonal moisture sources.

### 1.3.1 Snow and Glaciers in the Mountains

Snow in mountains that does not melt away each summer will change to glacier ice in 5–7 years. Glacier ice moves downhill because of gravity. Such glacier ice is long-term storage of potential water for use in later years. Because of climate change in the past few decades, however, glacier ice in Afghanistan and the region in general has been melting away (Haritashya et al., 2009; Shroder and Bishop, 2010a,b). Once it is gone, it will not come back for a very long time—maybe never. Some glacier ice in Pakistan has been melting away, but in the highest Karakoram and Nanga Parbat, some glaciers have been growing because of global warming producing greater evaporation from the sea, which produces increased snow at high elevations. All glaciers depend on snow for accumulation to the eventual ice that they become. Then they can flow downhill to warmer locations at lower altitudes where they melt away in what is accounted for by measures of their *mass balance* (measures of *accumulation* against *wastage*) (Fig. 1.8). This melting wastage is controlled mainly by summer temperatures, which themselves are controlled by the amount of sunlight that falls on the ice to melt it, and the cloudiness that occurs to reduce the sunlight.

The accumulation of snow in the highlands in winter both melts and runs off into the rivers in spring and summer, or it remains throughout the year and becomes in a few years, a glacial river of ice. About 3000 separate glaciers exist in Afghanistan with an estimated area of ~2700 $km^2$, but they were once far more numerous than now and they were also much larger more than 11,700 years ago in the Pleistocene (Ice Age) time. At present, the glaciers of Afghanistan are concentrated in the higher northeastern drainages of the country (Amu Darya, Panjshir—Kunar—Kabul). High seismicity, frequent avalanches of snow and ice as well as rocks, coupled with widespread stagnation, retreat, and downwasting of the glacier ice have produced numerous debris-covered ice and rock glaciers, many of which have been

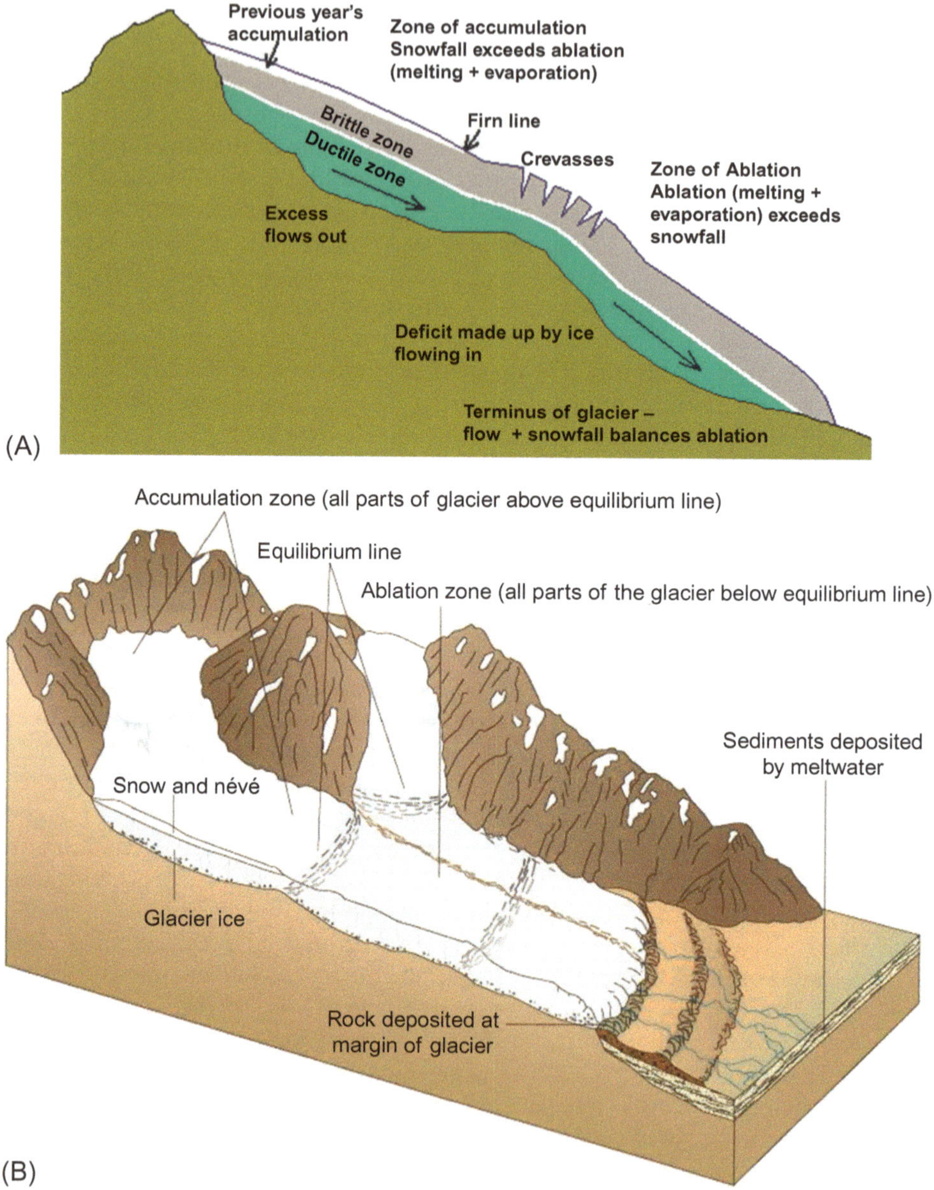


FIG. 1.8 (A) Cross section of mountain glacier showing uppermost zone of accumulation and lowermost zone of ablation. Névé or firn is recrystallized snowflakes; the firn line or equilibrium line is the place on the glacier every year in balance between the uppermost zone of accumulation of the new snow, and the lower zone of wastage where the ice melted away. (B) Oblique profile of configuration of a mountain glacier with a cross section at its edge. *Diagrams by permission, Center for Afghanistan Studies.*

essential sources of melt-water in past late summers (Bishop et al., 2014). Some of these important sources have probably melted out and dried up in recent years. The largest remaining glaciers in the Wakhan Pamir of Afghanistan have been mapped in detail recently (Haritashya et al., 2009), and assessed with comprehensive remote sensing with satellites in space, with the result that the glaciers can be seen through their pervasive retreat and downwasting to have mainly negative mass balances in recent decades, unlike some nearby higher-altitude glaciers in the Karakoram Himalaya of Pakistan that have been growing larger in some places (Copland et al., 2012).

Pakistan has over 5000 glaciers on its territory, amounting to perhaps 15,000 $km^2$ of thick ice cover. This large amount of glacier ice storage is a long-term valuable resource that provides tremendous summer and fall melt-water downstream on the Indus and many of its tributary drainages. Because of some lack of true scientific understandings about glaciers in the Himalaya in recent years, coupled with unprecedented media hysteria about all glaciers seemingly melting entirely away in a few decades, some startlingly inaccurate scare statements have been made. Nonetheless, the world's climates are changing quickly and the people of Afghanistan, Tajikistan, and Pakistan must develop good understandings of changes to come so that they are not caught off guard concerning their vital water resources.

Tajikistan is considered the main glacier center of Central Asia, with some 8000 glaciers occupying an area of about 8400 $km^2$, which is about 6% of the land area of the country (Kayumov, n.d. but post-2009). Seven glaciers in the high mountains of the Pamir are more than 20 km long, but most to all have been retreating and downwasting in recent years. The largest, the Fedchenko Glacier, has retreated by 44 $km^2$, which is 6% of its total area, and downwasted as much as 50 m in its lower regions. This is exemplary of the glaciers of the regional condition, and indicates the seriousness of the ongoing climate change, which so threatens regional water supplies.

Glaciers are markedly efficient erosion agents in Central and Southwest Asia, and high in the mountains of Afghanistan they have produced many spectacular landforms that have been mapped and discussed elsewhere (Shroder and Bishop, 2010a; Shroder, 2014). All of the stored ice is also a vital source of melt-water downstream, especially in late summer and early fall when second cropping can provide important additional food to hungry people. If this observed trend of glacier and ground-ice diminution in Afghanistan and throughout the region continues as it is projected to do, the implications to the arid countries are obviously most serious.

### 1.3.2 Devegetation

Throughout the region of Afghanistan and the surrounding countries, the removal of vegetation through deforestation, dearbification (Shroder, 2014), overgrazing, and widespread despoliation of landscapes is a scourge of environments that adds hugely to the diminution of water supplies. In spite of the attendant decrease in transpiration with plant loss, the fact that devegetation contributes so much to soil erosion because plant roots hold the soil in place against soil loss, the result is that devegetation contributes to a direct decrease in infiltration of soil water, surface runoff is accelerated and adds to ever greater soil erosion, and downstream floods increase as well.

### 1.3.3 Rain and Snowmelt on Hillsides

The splash of raindrops on the soil picks up little soil particles and moves them in the splash of the water (Fig. 1.9). At the top of the hill slope, the overland sheet flow produces *sheet erosion*. Lower down the slope, the rill flow produces *rill erosion*, and where the rills flow together, *gully erosion* occurs (Fig. 1.10). Some water soaks (infiltrates) into ground and adds to essential ground-water storage that can be accessed by karez and tube wells. Water that runs off the top of hill slopes first forms thin sheets of water that cover nearly all the ground in what is termed, *sheet flow* or *overland flow*. Some little way down the hill slope, however, the thin water sheets begin to collect more in slightly deeper depressions

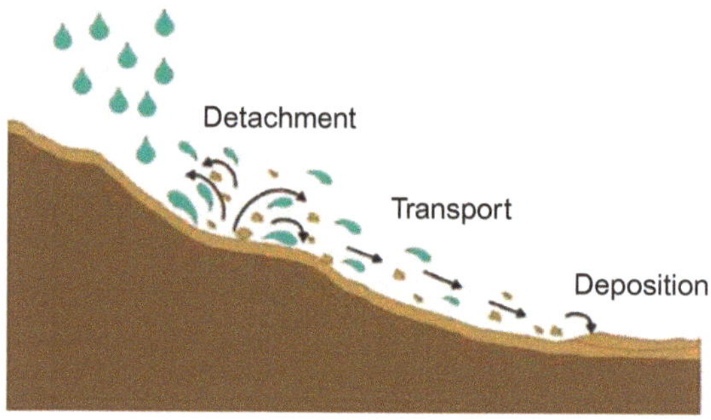


FIG. 1.9 Illustration of rainsplash erosion and soil-particle mobilization, transport, and deposition. *Diagrams by permission, Center for Afghanistan Studies.*

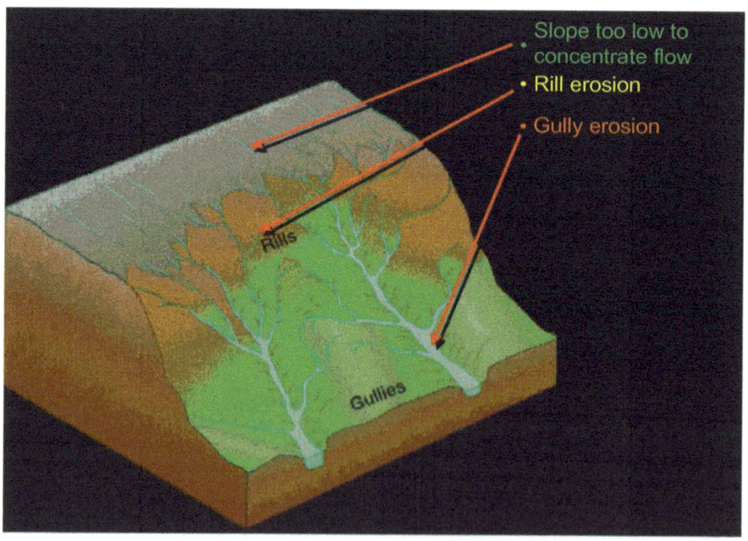


FIG. 1.10 Areas of sheet flow and sheet erosion on the upper slopes, passing downhill into zones of rills and rill erosion, to the lowermost gullies where the runoff water is concentrated. *Diagrams by permission, Center for Afghanistan Studies.*

FIG. 1.11 Photograph of gully in Afghanistan where a small check dam of rocks has been installed to help prevent further soil-erosion transport downhill. Many such check dams and slope terraces were built with money from foreign military agribusiness development teams (ADTs) but because the indigenous knowledge of reasons to do this is general lacking among the tribesmen, the prolongation or maintenance of such intelligent actions into the future is not likely to occur (by permission of Douglass Wissing).

of a few centimeters that are known as rills to produce *rill flow*. The rills join together further downslope and produce deeper *gully flow*, which themselves join together further downhill to produce *stream flow* or *river flow*. This water that runs off on the hillsides causes erosion of soil in which plants grow. All of this erosion increases, as noted above, where devegetation accelerates the processes of erosion. This soil erosion is a slow and terrible scourge that slowly destroys agricultural production and is insidious because people tend not to notice it, except over the very long term. It is a most serious hazard in the Asian regions, and should be protected against by construction of hillside benches and terraces, and with small rock check-dams in the low places where the water runs (Fig. 1.11). These check-dams trap the soil erosion and let the runoff water soak into the ground to save it for use later.

## 1.3.4 Snow and Ice Melt Becomes River Water

The melt waters of snow and ice flow downhill to become river water. Because glacier ice in the mountains also grinds up the rock that it moves over, a huge amount of sediment from glaciers gets into the river water where it is carried downstream. This sediment, as well as all that generated by the common soil erosion fills up behind dams and reduces water storage and water use behind the reservoir dam so the sediment must be diverted away before it gets to the dam site if this can be engineered, or excavated and removed later if possible. Several engineering techniques have been developed to stop or remove excess sediment in reservoirs but these techniques have not been much used, if at all in Central and Southwest Asia.

## 1.4 DRAINAGE SYSTEMS

Every stream, or segment of a river, is surrounded by its *drainage basin*, which is the total area of the land surface that contributes water into the stream. These drainage basins are also referred to as *watersheds*, or the area which sheds its water. The high-point line or region that separates one drainage basin from another is a *drainage divide* (Fig. 1.12). Drainage basins range in size from less than a square kilometer to vast areas of subcontinental dimension (Fig. 1.13). For example, the huge Indus River drainage basin starts near Mt. Kailas in the Himalaya of Tibet in China, before flowing across the Himalaya Mountains of northwestern India and into the Karakoram Himalaya of northern Pakistan, past the Nanga Parbat Himalaya, and out onto the lowland plains and plateaus at the Tarbella Dam site at the front of the mountains. In Afghanistan, the Kabul River and the Panjshir River flow together at the Sarobi *confluence*, before going on in the Jalalabad Basin to receive the Kunar River tributary and thereafter flowing across the border with Pakistan to receive the Swat River tributary before joining the Indus River at Attock. The drainage basin of the Kabul River thus includes most of eastern Afghanistan and part of Khyber Pakhtunkhwa Province in Pakistan (Table 1.1).

Similarly, the headwaters of the large Panj River tributary between Afghanistan and Tajikistan begin in the high-altitude Wakhan Corridor of Afghanistan before flowing generally to the northwest as a major tributary of the Amu Darya and ultimately draining into the Aral Sea. Several other major tributaries of the Amu Darya begin either in Afghanistan or in Tajikistan and contribute greatly to the Nile-sized river flow. The Helmand River, Hari Rud, Murghab, and a number of other river systems rise entirely within Afghanistan and flow out over the borders into the surrounding countries.

The relative impacts of the different moisture sources are not well understood because this remote region lacks any sort of dense network of weather stations or river gauges.

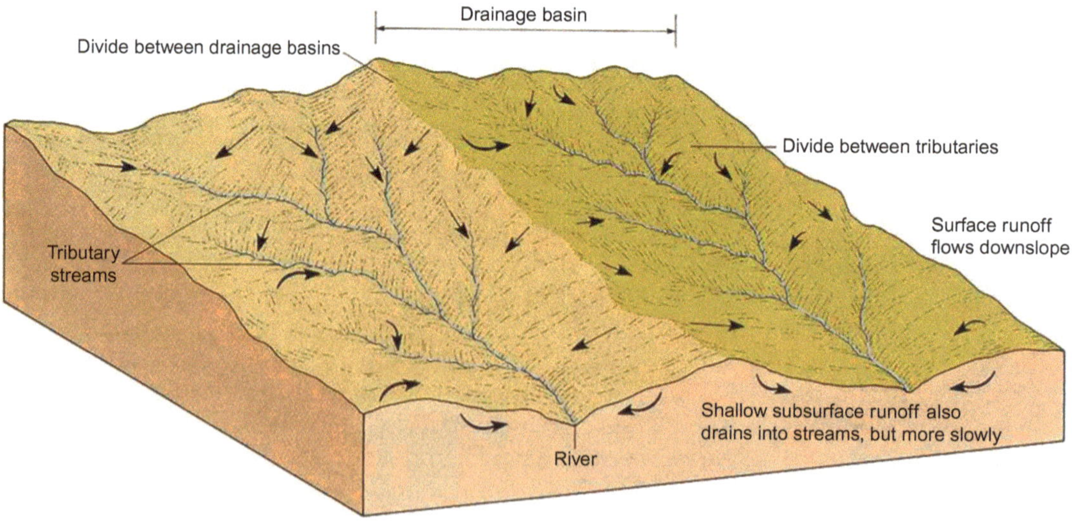


FIG. 1.12 Drawing of drainage basins separated by drainage divide between the basins. *Diagram by permission, Center for Afghanistan Studies.*

FIG. 1.13 (A) Map of main regional drainage basins of Central and Southwest Asia (Kabul to Indus, Helmand, Hari Rud—Murghab, Amu Darya to Aral Sea) that are associated with the largest rivers of Afghanistan. (From the map product, Transboundary Basins of Afghanistan http://www.transboundarywaters.orst.edu/research/Countries%20at%20Risk/Maps/Basemaps/Afghan%20Map%202.jpg. Product of the Transboundary Freshwater Dispute Database, College of Earth, Ocean, and Atmospheric Sciences, Oregon State University. Additional information about the TFDD can be found at: http://www.transboundarywaters.orst.edu.)

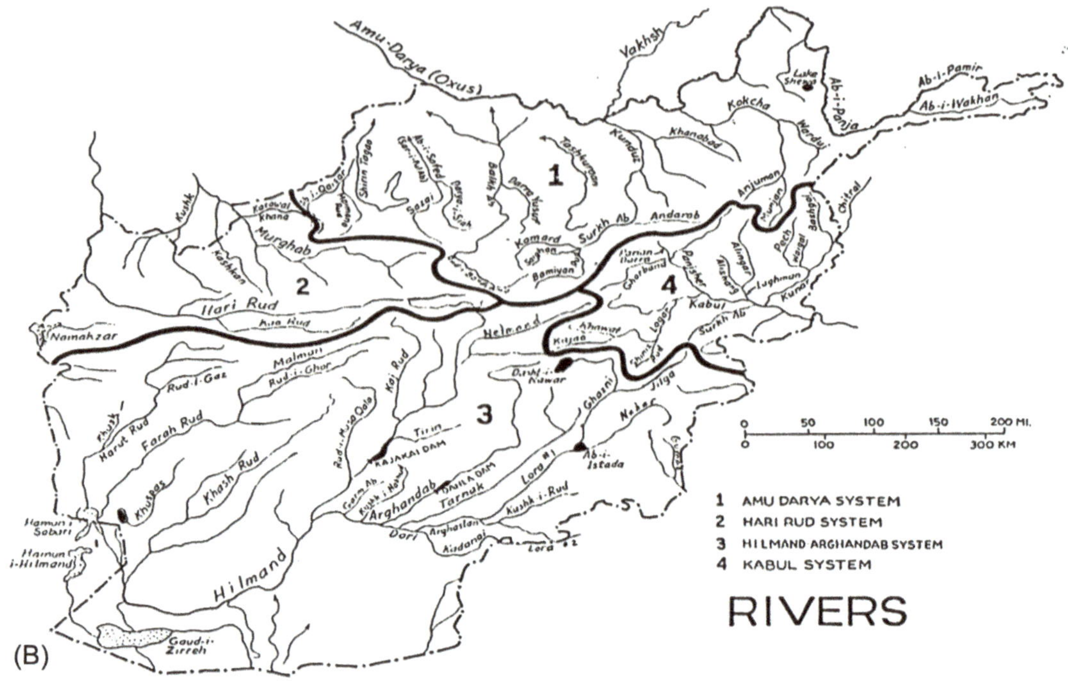


FIG. 1.13, CONT'D (B) Map of the main drainage basins of Afghanistan. *From Figure 5.2, Shroder, J., 2014. Natural Resources in Afghanistan: Geographic and Geologic Perspectives on Centuries of Conflict. Elsevier; Dupree, L., 1973. Afghanistan. Princeton University Press, Princeton, NJ.*

TABLE 1.1 Transboundary Regional River Basins of Afghanistan

| Basin | Afghanistan River | % in Afghanistan | km² |
|---|---|---|---|
| Aral Sea | Amu Darya | 14 | 166,392 |
| Indus | Kabul | 8 | 71,314 |
| Helmand | Helmand | 78 | 312,286 |
| Hari Rud | Hari Rud | 33 | 38,895 |
| Murghab | Murghab | 42 | 38,751 |

*Data from Transboundary Freshwater Dispute Database, Oregon State University; see Fig. 1.13A.*

Accordingly a combination of validated, remotely sensed climate parameters must be used to characterize the spatiotemporal distribution of rainfall, snowfall, and evapotranspiration in order to quantify their relative contributions to mean river discharges. Rainfall amounts can be calculated from calibrated, orbital, high-resolution Tropical Rainfall Measuring Mission data, and snow-water equivalents were computed from a snowmelt model, which was based on satellite-derived snow cover, surface temperature, and solar-radiation data

(Bookhagen and Burbank, 2010). Three key aspects of spatiotemporal precipitation patterns through the Himalaya region have been detected, which include: (1) strong decouplings of moisture occur between the Indus-Ganges foreland lowlands along the southern mountain fronts, versus that in the mountains; (2) the Indian summer monsoon is responsible for more than 80% of annual precipitation in the central Himalaya and Tibetan Plateau, but the geological syntaxial bend of the western Himalaya and Indus River between Skardu and Gilgit around Nanga Parbat in northern Pakistan receives only about 50% of its annual precipitation during the summer season; and (3) snowmelt contributions to river discharge differ markedly along the Himalaya range, and presumably also to the western Himalaya and the Hindu Kush—Pamir, although this latter range grouping has not yet been studied in detail by Western scientists. Thus as a fraction of the total river discharge, snowmelt constitutes up to 50% of the discharge in the far western (Indus area) catchments, with the main mountainous part of the Indus catchment itself having an annual snowmelt contribution of some 66%, and the lowland Indus through Pakistan has a May–October rainfall contribution of 25–65%.

Similar variations in the Pamir Knot may be expected as the interplay varies between southern monsoonal flows interacting with the dominant Westerly storm tracks. As a result, changes in the timing or amounts of snowmelt due to increasing temperatures or decreasing winter precipitation could have far-reaching societal implications (Bookhagen and Burbank, 2010). The controls on snow accumulation in these regions is a function of: (1) catchment hypsometry [the percentages of land areas in a catchment determined by its different amounts of altitude ("bins")]; (2) precipitation magnitude determined by distances toward the hinterland; and (3) seasonality of the precipitation determined by percentages of winter versus summer precipitation, and southern monsoonal versus Westerly storm tracks.

## 1.5 CONCLUSION

The hydrologic cycle of Central and Southwest Asia is thus seen as a complex interplay between highly variable storm tracks that are subject to several spatial and temporal perturbations at many scales, all of which can cause severe problems. The changes in snow and glacier cover, the long-term and increasing devegetation and associated soil erosion and decreasing water infiltration, the rapid and destructive flood runoff; all result in problems for the future of water usage in the region in general and Afghanistan in particular.

The result of this analysis of the general workings and actions of the hydrological cycle in this region enables elucidation and delineation of the chief problems with aqueous resources in the area. This, however, is not a panacea or a solution to the problems, but it does begin to help define the nature of the situation and provide greater understanding of the issues for further detailed analysis. Given the need to produce greater clarity of the hydrologic cycle for the government bureaucracies to help their people, it does help raise the hydro-cognizance of the region. Inasmuch as promotion of greater water awareness is very much in our purview, we see this as a most desirable attribute for any of the societies existing in semiarid and arid climates that depend so acutely on water.

## References

Bishop, M.P., Shroder, J.F., Ali, G., Bush, A.B.G., Haritashya, U., Roohi, R., Sarikaya, M.A., Weihs, B.J., 2014. Southwest Asia Regional Center: Afghanistan and Pakistan. Global Land Ice Measurements from Space (GLIMS). Springer Verlag, pp. 509–548.

Bookhagen, B., Burbank, D., 2010. Toward a complete Himalayan hydrological budget: spatiotemporal distribution of snowmelt and rainfall and their impact on river discharge. J. Geophys. Res. Earth Surf. 115 (F3), http://dx.doi.org/10.1029/2009JF001426.

Copland, L., Sylvestre, T., Bishop, M.P., Shroder, J.F., Seong, Y.B., Owen, L.A., Bush, A., Kamp, U., 2012. Expanded and recently increased glacier surging in the Karakoram. Arct. Antarct. Alp. Res. 43 (4), 503–516. http://dx.doi.org/10.1657/1938-4246-43.4.503.

Haritashya, U.K., Bishop, M.P., Shroder, J.F., Bush, A.B.G., Bulley, H.N.N., 2009. Space-based assessment of glacier fluctuations in the Wakhan Pamir, Afghanistan. Clim. Change 94, 5–18.

Kayumov, A., n.d. but post-2009. Glaciers resources of Tajikistan in conditions of climate change. World Meteorological Organization, https://www.wmo.int/pages/prog/www/OSY/Meetings/GCW-IM1/glaciers.pdf, 31 pp. (accessed 03.02.15).

Shroder, J., 2014. Natural Resources in Afghanistan: Geographic and Geologic Perspectives on Centuries of Conflict. Elsevier, Amsterdam.

Shroder Jr., J.F., Bishop, M.P., 2010a. Glaciers of Pakistan. In: Williams, R.S. Jr., Ferrigno, J.G. (Eds.), Satellite Image Atlas of Glaciers of the World—Asia, pp. 201–257. U.S. Geological Survey Professional Paper 1386-F4.

Shroder Jr., J.F., Bishop, M.P., 2010b. Glaciers of Afghanistan. In: Williams, R.S. Jr., Ferrigno, J.G. (Eds.), Satellite Image Atlas of Glaciers of the World—Asia, pp. 167–199. U.S. Geological Survey Professional Paper 1386-F3.

CHAPTER

# 2

# Hydrogeography (Drainage Basins and Rivers) of Afghanistan and Neighboring Countries

*J. Shroder*

**University of Nebraska at Omaha, Omaha, NE, United States**

OUTLINE



## Abstract

The river basins and rivers of Afghanistan rise in the high Hindu Kush, mainly in the central highlands and northeast high mountains. Climate controls of river flow are determined particularly by winter snows in much of the middle and higher altitude area, but also by the summer monsoon in the east. The Amu Darya, Kabul, Helmand, Hari Rud-Murghab, and northern area rivers, together with a few smaller rivers and seasonal streams flow outward from their Afghanistan headwater sources into the surrounding countries of Pakistan, Iran, Turkmenistan, Uzbekistan, and Tajikistan. Three decades of war from the late 1970s to the present have stopped hydrographical data collection, until recent reestablishment with the result that some figures are suspect, but general knowledge of annual water volumes is becoming more precise. The concomitant loss of water wisdom from the older knowledgeable people does continue to produce problems, as do errors in the collection and transmission of the historical hydrographic data.

**Keywords:** River systems, Panj-Amu Darya, Kabul River, Helmand River, Hari Rud-Murghab River System, Water wisdom, Climate controls



http://dx.doi.org/10.1016/B978-0-12-801886-6.00002-1

## 2.1 INTRODUCTION

The focus of this chapter is on the rivers that have their headwaters, or major parts of their channels, inside or bordering Afghanistan (Figs. 1.13 and 2.1). Because a number of these river systems ultimately debouch into oceans and seas far from the borders of Afghanistan, and are exceptionally important to the neighboring countries into which these waters flow, it is necessary to here expound on the details that will help people to understand their aqueous resources. The basic physical courses and characters of these rivers are essential to understand for numerous reasons; for example, some rivers are nearly overwhelmed with sediment derived from melt of glaciers and snow avalanches, whereas other sediment-laden rivers are more the product of desert flash floods. Both can have copious sediment at different seasons that can cause numerous problems with sedimentation in lakes and reservoirs. Another essential character of rivers in this region is the timing of water delivery downstream, which can be a function of the annual melt season of snow and ice, or of the time of the actual strongest precipitation in spring, summer, or fall seasons. The quantity of water or discharge in each river is essential to consider, as is where the previously collected data might be available, and when that original data collection occurred.

## 2.2 WATER WISDOM AND WAR

The three or more decades of war in Afghanistan that started in the late 1970s caused suspension of the water-resources data collection in the country, as well as destroying many of the older records of hydrological data (Campbell, 2015). In addition, the long upheavals of wars brought in from outside, or invasions, civil war, government change, and general chaos from the disastrous excursions into communism and the know-nothing Taliban years have greatly eroded water wisdom, in addition to destroying the water infrastructure of dams, canals, hydroelectric capabilities, and so much of the agricultural capability of the country. Almost all of the water-measuring equipment, communication infrastructure, and even the intellectual qualities of the country were destroyed as well. Water wisdom is a fragile resource, capable of rapid destruction or removal by the vagaries of war, as the people who hold it age away from capability, or are killed or incapacitated. It is not so quickly recovered, so considerable retraining is required to bring formerly wise water people back up to modern standard, while also training the new generations.

In fact, the prewar hydrological data collection was not at all that robust anyway, having only been started in about the 1950s as a result primarily of the needs of the irrigation schemes of the Helmand Valley Project in southwest Afghanistan (Heckmiller, 1961). Later this was expanded to 169 stream gauges nation-wide, with a great deal of Cold-War-motivated funding and assistance from the US International Cooperation Administration, which was the predecessor organization to the US Agency for International Development (USAID), as well as from the US Geological Survey (USGS). Being largely driven by foreigners then, and after 30 years of disruptions caused by hostilities in the country, as a result by the beginnings of the 21st century, Afghanistan's hydrologic expertise and capabilities were severely limited. Any Afghan hydrologists, or even general scientists who had not already fled the country for work elsewhere, were entirely isolated from both the international scientific communities, as

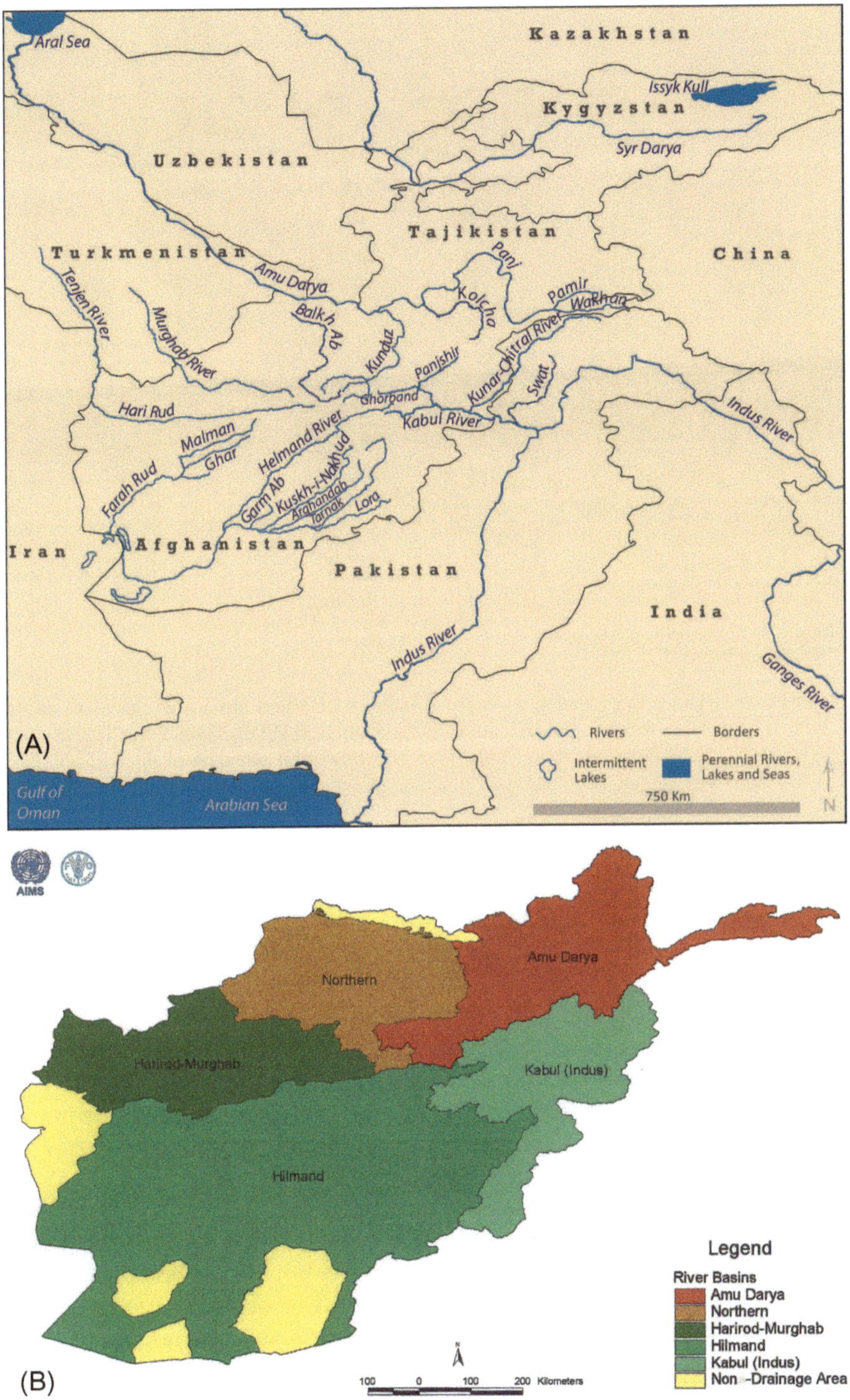


FIG. 2.1 River basin map of Afghanistan: (A) Map of rivers of the greater drainage area beginning in Afghanistan and (B) Main river basins of Afghanistan. *From Favre and Kamal (2004).*

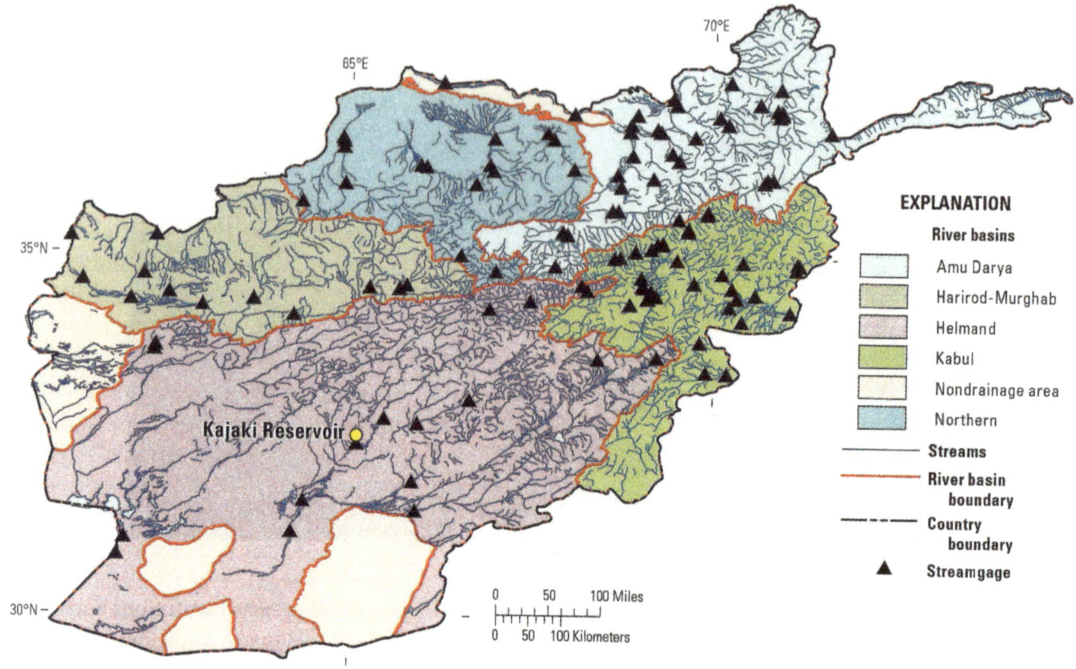


FIG. 2.2 Locations of Afghanistan's major drainage basins, reinstalled stream-gauge stations and the Kajaki Reservoir, a major source of hydropower and irrigation in the Helmand River Basin. *From Thomas, J. Mack of the US Geological Survey; Campbell, 2015. A dry and ravaged land: investigating water resources in Afghanistan. Earth 60(1 and 2), 49–55.*

well as any sorts of technological advancement made elsewhere, with the result that knowledge of water was rather weak and fairly nonfunctional. Accordingly, following the invasion of Afghanistan by the Western Coalition of troops in late 2001, considerable effort and money were expended to improve hydrological institutional capacity in the government and the universities, and to restore capabilities in water science. Among many of these efforts has been the rebuilding of the hydrological data collection network, upon which personnel from the USGS have been financing, and expending considerable time and effort to get up and running again (Fig. 2.2). At the present time about 127 stream gauges have been re-established in the country and many Afghan graduate students are now obtaining higher degrees on water in India, the United Kingdom, Germany, and Japan.

It is widely known that reliable streamflow information is required for wise hydrologic science and management of water resources (Campbell, 2015). Obviously it is essential to keep track of surface-water availability and quantities for direct consumption and irrigation. Because streamflow is generally both a major source of discharge from, and recharge into, ground-water systems, streamflow records are thus a key component to understanding the sustainability of use of underground water as well. Furthermore, in the growing world of declining water availability and the increasingly necessary accountability of where the water in any given place is going, quantifying streamflow is a vital component for maintaining old water treaties with neighboring countries, as well as for negotiating new understandings. Inasmuch as all of the water treaties of Afghanistan are hopelessly out-of-date, and moreover, signed on both sides with governments that are long-gone, it is imperative that new hydrologic data are collected to enable wise council and decisions at the international treaty tables with at least five countries in the region (Pakistan, Iran, Turkmenistan, Uzbekistan, and Tajikistan) that share transboundary water resources with Afghanistan (Favre and Kamal, 2004).

In any assessment of the hydrology of a region, in addition to the actual stream gauging, controlling conditions to measure the amount of moisture available is an obvious requirement. Such begins first with information about precipitation, which of course, is established with detailed assessments of short-term meteorology and long-term climate.

## 2.3 DETAILS OF CLIMATE CONTROLS IN AFGHANISTAN

The physical geography of Afghanistan is characteristically fairly spectacular, from the rather huge ~7 km of relief from lowest altitudes in the northwest Amu Darya (258 m) region to the highest at Naushaq (7492 m) mountain in the northeast. This >7 km of maximum relief in Afghanistan is exemplary of the potential energy in the environment of the ground surface, as well as the high atmospheric energy associated with the hydrologic cycle that drives so many of the natural processes at work in the country. Coupled with the exceptionally strong geologic forces of plate tectonics in the region; the result is combined exo- and endo-genetically driven, hydrological sources of considerable significance to the area.

Afghanistan is situated in the subtropical, dry zones of Southwest Asia that have continental types of climate characterized by desert, steppe, and highland temperature and precipitation regimes (Humlum, 1959; Shroder, 2014). The new Köppen-Geiger climate map of Afghanistan (de Bie et al., 2007; Shroder, 2014) shows Bsh subtropical steppes; Bsk midlatitude steppes; Bwh subtropical deserts; Bwk midlatitude deserts; Csa humid subtropics; Csb Mediterranean mild with dry summer; Dsa humid continental with severe dry winter; Dsb humid continental with severe warm dry summers; Dsc humid continental with severe cool summers; ET tundra with no true summer (also includes some ice) (Fig. 2.3).

Afghanistan is thus a mostly rather arid country with extreme minimum of perhaps <50 mm of precipitation in the Seistan Basin of the southwest, and around 1100–1400 mm in the mountains of the central Hindu Kush to the Wakhan Corridor to the northeast (Figs. 1.7 and 2.4). Precipitation measured about 300 mm along much of the northern border. Some of the tropical rainfall measuring mission (TRMM) data show much of those same trends, but especially the general monsoonal variation, whereas the snow-water equivalent measured by satellite radar shows the same general trends in a different remote-sensing medium (Fig. 2.4). In general, precipitation increases with altitude, especially in the high Hindu Kush and Pamir mountains to the northeast, but also to the northwest toward the Turkmenistan regions (Breckle et al., 2010; Shroder, 2014). The overall amount of moisture that falls over the country from which to estimate river flows has a number of rather sketchy guesstimates that were based upon somewhat simplistic methods. For example, in 1992 Nathan Berger

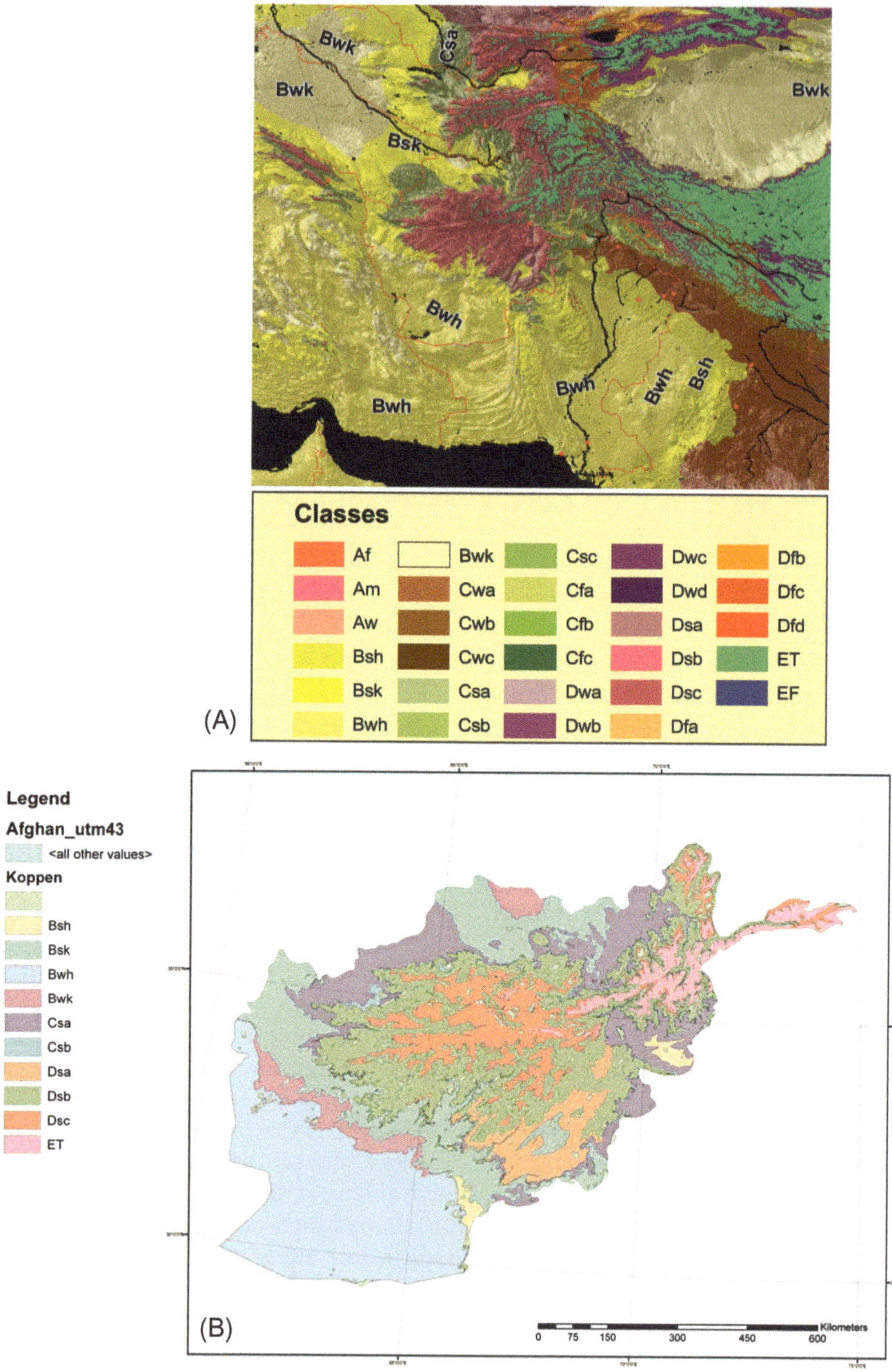


FIG. 2.3 (A) Revised Köppen-Geiger climate map of the Afghanistan region showing Tajikistan and Pakistan highland areas from where much of the moisture of the region emanates. (B) Detailed climate map of Afghanistan alone. *From de Bie, C.A.J.M., Skidmore, A.K., Toxopeus, B., Venus, V., 2010. An Updated Köppen-Geiger Climate Classification of the World Using Very High Resolution Interpolated Climate Surfaces of Monthly P and T Data From 1950 to 2000. ITC, Enschede. http://www.itc.nl/personal/debie/An%20updated%20K%F6ppen.htm (accessed 18.02.15); Shroder, J., 2014. Natural Resources in Afghanistan: Geographic and Geologic Perspectives on Centuries of Conflict. Elsevier, Netherlands.*

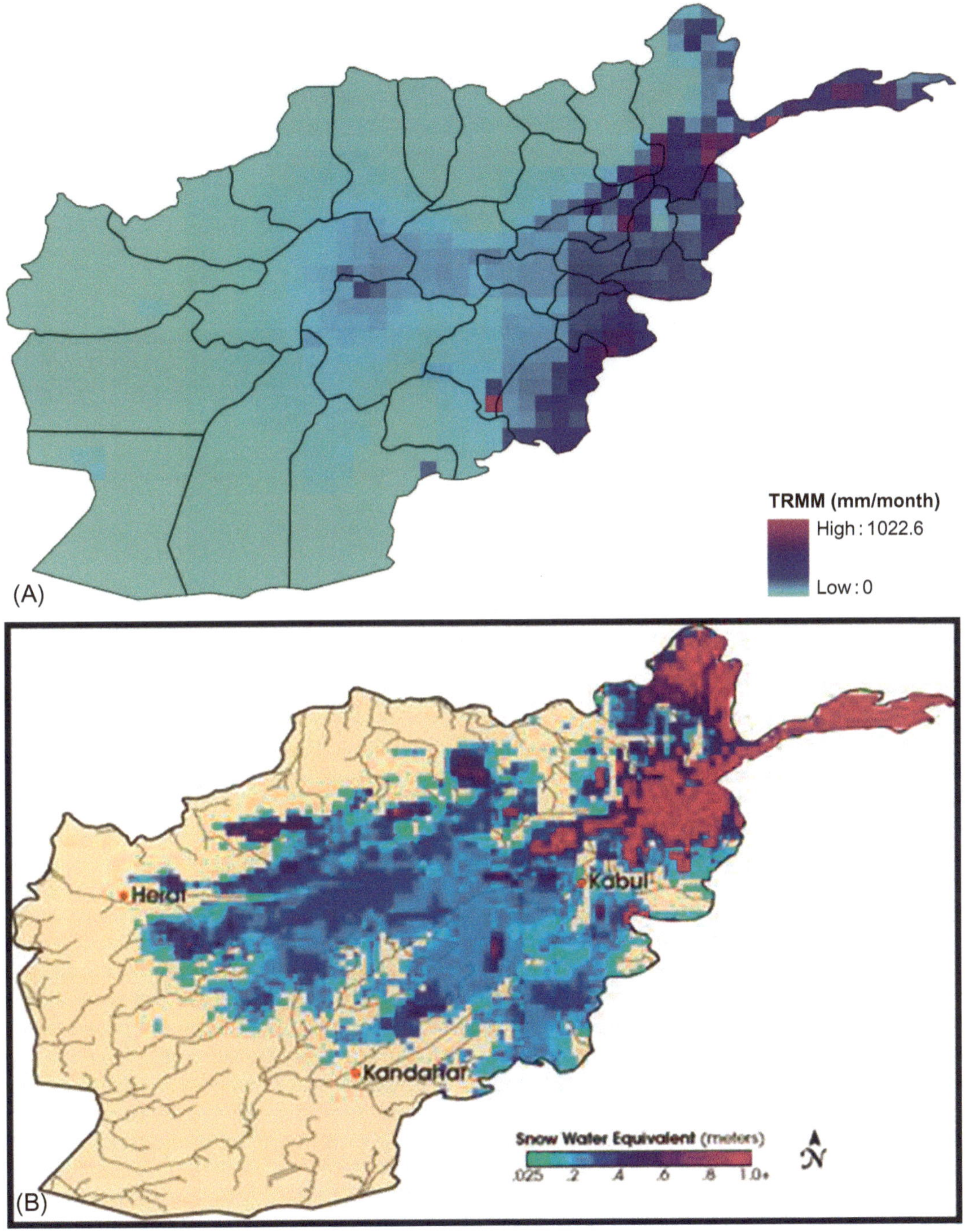


FIG. 2.4 Maps of precipitation in Afghanistan obtained by (A) the TRMM at ~4.3 km pixel resolution and (B) snow-water equivalent over Afghanistan on Mar. 13, 2005 based on MODIS (moderate resolution imaging spectroradiometer) combined with FEWS NET (famine early warning network) data on relative humidity, atmospheric, temperature at the surface, wind speed, and other factors (Riebeck, 2006).

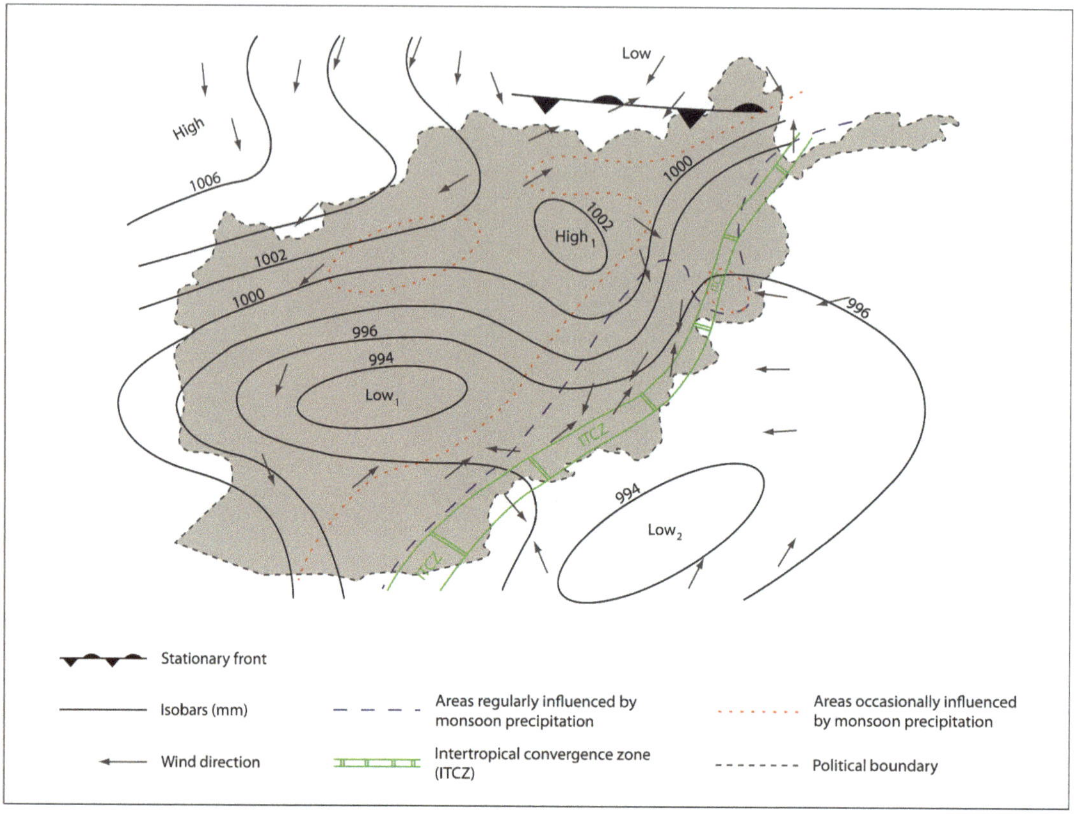


FIG. 2.5 Map of Afghanistan showing average positions of summer air pressure with the intertropical convergence zone (ITCZ) of low monsoonal pressure, wind directions, and areas of penetration of the monsoon rains into the Hindu Kush. *Based in part on Sival (1977) and Shroder (1989).*

estimated that some 236 km$^3$ fell over the whole country but FAO (1996) decided with its own apparently different thinking that the number would better be ~180 km$^3$. In any case, some measure of moisture arrives in both winter and summer.

Although winter Westerly storms with precipitation are fairly common, and most the summer monsoon penetration of the tropical convergence zone average from the southeast in Pakistan is the most active moisture source in most years (Fig. 2.5). Associated precipitation is concentrated in southeast Afghanistan along the Durand Line border with Pakistan, but periodic incursions of moisture-laden air masses over the Hindu Kush to the northwest and northeast can initiate torrential rains at altitude that generate catastrophic rapid, wet debris flows (*sailab*, *duab*) sweeping down dry channels from the mountains with massive slurries of mud, rocks, and debris (Shroder, 2014) (Fig. 2.5).

In the winter season, of course, much of the precipitation comes down as snow, but in the summer season only the high altitudes generally above 5000 m will ensure that any moisture arrives in frozen form. In either case with snow and ice precipitation, of course, after the

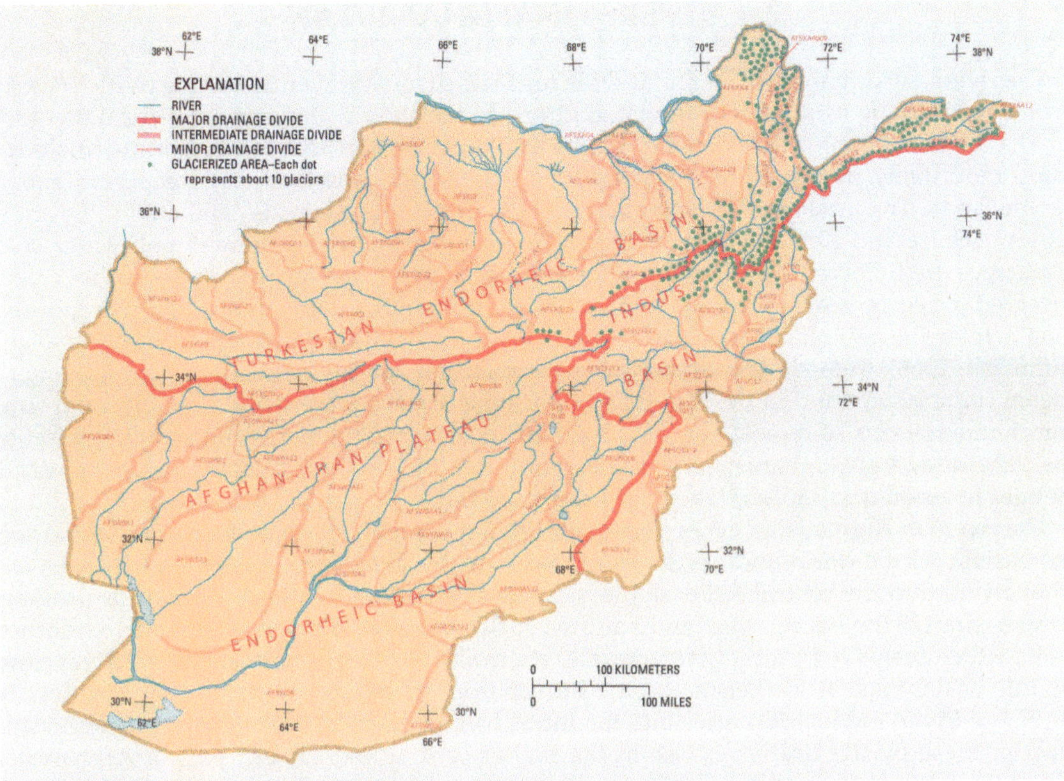


FIG. 2.6 Map of the major drainage divides of Afghanistan showing the glacierized regions (each dot represents about 10 glaciers). *From Shroder Jr., J.F., Bishop, M.P., 2010. Glaciers of Afghanistan. In: Williams Jr., R.S., Ferrigno, J.G. (Eds.), Satellite Image Atlas of Glaciers of the World—Asia. United States Government Printing Office, Washington, DC, pp. 167–199, U.S. Geological Survey Professional Paper 1386-F3.*

passage of 3–5 years without melting away, thick snow and ice accumulations will recrystallize enough to begin flowing downhill as a glacier. Glacier-ice masses in the high Hindu Kush and Pamir of Afghanistan constitute a long-term system of ice storage that serves as a vital late summer—early fall form of delayed melt-water release for irrigation in second cropping (Fig. 2.5). The problem with this traditional outlook is that climate change has been melting away altogether too much of the lower-altitude ice of Afghanistan so that some of smaller glaciers have disappeared entirely (Shroder, 1980, 2014; Shroder and Bishop, 2010; Bishop et al., 2014) (Fig. 2.6). Many of these smaller glaciers had existed for centuries high in their deep cirques as diminutive remnants of once larger ice masses (Porter, 2004), but still able to survive because they were shadowed for most of the time by the high mountains that surrounded them. In the more modern times with climatic warming underway, their continued existence at relatively lower altitudes compared to the higher altitude examples in the nearby western Himalaya of Pakistan, has become no longer possible in many cases. The results are less melt-water downstream later in each year.

## 2.4 RIVERS OF AFGHANISTAN

The rivers of Afghanistan are remarkable for their diversity, even as their activity is such a strong agent in the formation of fluvial landforms, and their waters are the critical lifeblood for the Afghan people. The larger rivers of arid Afghanistan rise in the Hindu Kush and Pamir mountains, mainly from snow and glacier melt, and flow on fairly steep gradients to the lowlands. The total amount of water that falls in the mountains is estimated to be about 150 $km^3$ (Favre and Kamal, 2004). Maximal flows in spring and summer, melt-water river discharges are followed by fall and winter minima, when some large rivers even have been observed in recent years to dry up completely. Most runoff is delivered from melting of seasonal winter snow or older glacial ice and high sediment loads are characteristic (Westfall and Latkovich, 1966). In the lowlands the total precipitation that falls is about 30 $km^3$. Ephemeral stream channels abound in the lower elevations and are commonly filled by torrential summer thunderstorms so typical of many desert areas, especially where monsoon precipitation has penetrated into Afghanistan from the south. In total, the precipitation in the country is thought to amount to 180 $km^3$ (Favre and Kamal, 2004).

The rivers in Afghanistan generally are controlled spatially by three location of the drainage basins: (1) the north and northwestern flow of the Nile-sized Amu Darya and several other rivers into the central Asian depressions of the Turkestan *endorheic* (interior drainage flow) basin; (2) the strong west and southwest flow of the Afghan-Iran Plateau endorheic basin, which seems to be largely structurally controlled, with the main Helmand River flowing into basins such as the Seistan on the Iranian border; and (3) the southeastward flow by the Kabul River and its tributaries into the Indus basin system in Pakistan (Figs. 2.1 and 2.5). Most of these master drainages rise in the higher central and eastern part of Afghanistan where the altitude is great enough to trap moisture arriving in the westerly winds, as well as the Indo-Pakistani monsoonal moisture coming from the south or southeast. Among all of the major rivers that rise in the highlands of Afghanistan, the waters of the *exorheic* Kabul River are the only ones to reach the ocean as they join the Indus River at Attock in Pakistan. The other transboundary rivers that exit the country's borders include the Amu Darya and its tributaries that form the northern border of the country and drain into the Turkestan endorheic basin with the drastically declining Aral Sea as the final sink, the Murghab River and the Hari Rud in the northwest that flow also into the Turkestan endorheic basin, and the Helmand River in the southwest, which drains into closed depressions in the Seistan Basin along the Iran-Afghanistan border (Shroder, 2014).

All these rivers of Afghanistan are generated in the highlands from melt waters derived from snow, or glacier ice and the permafrost ice inside the profuse rock glaciers of the region (Bishop et al., 2014), before passing out through mountain gorges and into wider valleys downstream where so many flights of fluvial terraces testify to strong downcutting (Shroder, 2014). In these regions a number of flat-floored mountain valleys also have been produced from the remnants of infilled prior lake basins that were formerly impounded behind a variety of natural dams produced by moraines, landslides, and *biohermal* (algae) travertine deposition, some of which have since been breached and drained, or filled up to overflowing with sediment. Profuse smaller drainages also occur that are generally ephemeral stream valleys. These commonly drain out of the mountains carrying plentiful sediment discharges of varying types (stream flow, rapid, and wet debris flows), which at the mountain fronts result in the common alluvial fans and of dramatically different gradients and sizes. Most of the alluvial-fan regions have networks of channels across them that are partly natural, and partly diversions made by human inhabitants seeking to divert seasonal water for irrigation. These alluvial fans and river floodplains grade out further into the lowlands as widespread, gently sloping alluvial plains, before ending in large ephemeral lake basins or swamps, many of which are at least partly covered with younger dune field regs (corruption of Arabic *erg* to *reg*, as in Registan). Such extensive dunes, as well as the common blankets of loess dust were mobilized out of the clastic alluvial sediments that had been brought into the basins originally by river flow into the lakes before they dried out. In prior centuries, historical records indicate that much of the lower-altitude uplands in antiquity and later were well mantled with soil and extensively forested or grass and shrub-covered, which would have allowed beneficial infiltration of precipitation waters and slowed runoff. In addition to the dramatically increased *deforestation* in Afghanistan in the past three decades or so, and other forms of *desertification*, flooding of several kinds also seems to have increased in Afghanistan (Shroder, 2014).

Afghanistan has hydrologic regimes that vary because of different runoff controls throughout the year. Some of the rivers, for example, have a pluvial regime, in which rain and snowmelt produce a late winter-spring maximum, and a late summer—early fall minimum when many dry up completely. Most of these small water courses carry no water at all from April-May to November-December, and in recent years some of the larger rivers such as the lower Helmand and the middle Kabul have dried up nearly completely as well. Torrential pluvial regimes occur in the many dry water courses (*darrah, juie, nullah, wadi*) in Afghanistan, especially at times of rapid snowmelt or summer rainstorms. Rivers with a *nival* (snow) regime in the high mountains of northeast Afghanistan tend to have a minimum discharge in winter when they freeze completely, and a summer maximum when the snow and ice melt to produce bank-full conditions. Those with a monsoon regime in the southeast along the border with Pakistan also have a summer maximum, but from increasing precipitation at that time derived from oceanic sources.

The estimates of the actual availability of water and its use in Afghanistan are laden with crude or incorrect information and a high degree of variable estimates by as much as one third in the numbers. This variance is so large that some of the estimates were apparently based upon what appear to be rather unreliable assumptions, or weak estimation methods, or both. In part this is because lack of reliable data collecting over the past three decades of war, and perhaps also, in part because of changes in some hydrological forecasting methods. In any case, in reference to the mean annual surface-water flow of Afghanistan, or total river discharge in the country, the United Nations FAO (1996) estimates were values of 55–57 $km^3$ for the whole country, whereas the values estimated later by Favre and Kamal (2004) were ~84 $km^3$ (Table 2.1). This apparent increase of 28 $km^3$ is apparently either derived from unknown citational sources, is the result of including waters that exit into Pakistan, or it might be an outright mistake, because Afghanistan has been subject to a recent severe drought cycle so that one would expect an overall decline of water resources, not an increase. In the absence of any better data, however, that of Favre and Kamal (2004) will be utilized except for some minor changes imposed on the data where clear changes were needed because some data are from Pakistan.

In spite of the problems with the numbers the estimates of total river discharge was thought to amount to about 47% of the total precipitation that is shared with the neighboring countries as transboundary water (Favre and Kamal, 2004), perhaps amounting to as much as

TABLE 2.1 Mean Annual Volume in Cubic Kilometer of River Water Discharge in Afghanistan by Drainage Basin

| River Basin | River Name | Discharge ($km^3$) | |
|---|---|---|---|
| Amu Darya | Abi Panj | 36.4 | |
| | Kokcha | 5.7–6.2 | |
| | Kunduz | 3.3–6.0 | |
| Amu Darya total | | | 45.4–48.1 |
| Kabul | Kabul | 2.52–0.4 | |
| | Panjshir | 3.13–2.9 | |
| | Kunar | 15.25–14.9 | |
| Kabul total | | | 20.9–18.2 |
| Northern Basin | Tashkurghan (Khulm) | 0.06–0.07 | |
| | Balkhab | 1.65–1.5 | |
| | Shirin Tagab | 0.1–0.1 | |
| Northern Basin total | | | 1.81–1.67 |
| Helmand Basin | Farah Rod | 1.25–1.3 | |
| | Harut Rod (Adraskan Rod) | 0.21–0.2 | |
| | Khash Rod | 0.17–0.49 | |
| | Ghazni Rod | 0.35–0.08 | |
| | Helmand (at Kajaki Dam) | 6.0–6.0 | |
| | Musa Kala | 0.22–0.29 | |
| | Arghandab | 0.82–8.9[a] | |
| | Lower Helmand | 0.11–0.4 | |
| Helmand total | | | 9.13–17.66 |
| Hari Rud-Murghab | Murghab | 1.35–1.6 | |
| | Hari Rud | 1.6–0.7 | |
| Hari Rud-Murghab total | | | 3.06–2.3 |
| Grand total | | | 80.3–87.93 |

The first discharge numbers are from Favre and Kamal (2004) from imprecisely known locations, and the second numbers are from on-line US Geological Survey Data collected largely in the 1960–70s and earlier (Williams-Sether, 2008; Olson and Williams-Sether, 2010). Variation in discharge numbers are caused by imperfections in the data collection, missing data from certain tributaries, some difference in locations of data collection sites, and probably climate variations over time. The main river basin totals are minimum figures because some tributaries that flow only into Pakistan are not included.

[a]*The data on the Arghandab River are possibly incorrect.*

TABLE 2.2 Mean Annual Surface-Water Flow in Afghanistan (Favre and Kamal, 2004; Shroder, 2014)

| River Basin | Surface Area (%) | Mean Annual Water Flow ($km^3$) | Total Annual Flow (%) |
|---|---|---|---|
| Amu Darya | 14 | 48 | 57 |
| Kabul | 12 | 22 | 26 |
| Helmand | 43 | 9.3 | 11 |
| Hari Rud-Murghab | 18 | 3 | 4 |
| Northern Basin | 13 | 1.8 | 2 |

Some variance is apparent in mean annual water flow between this table and Table 2.1, which is explained in the legend of Table 2.1.

over 80 $km^3$. But because 65% of the water is thought to be used inside the country, and those two percentages are incompatible (47% and 67% add to an impossible 112%), perhaps some ~50–55 $km^3$ has been thought to be available for use inside Afghanistan, of which perhaps 25–30 $km^3$ could be impounded for use inside Afghanistan, rather than let it flow out over the borders to be used by other countries. This fact has produced much consternation in Pakistan, for example (Anon., 2011).

The proportional distribution of total river flow by drainage basin (Table 2.2) is 57% from the Amu Darya that covers only 14% of the national territory, 26% from the Kabul River that covers some 12% of the country, and 11% from the Helmand River system that measures almost half (43%) of the total area of the country. Four percent of the total river flow comes from the northern river basins, and 2% from the Hari Rud-Murghab rivers. A number of exceptionally dry areas in extreme southwest and west Afghanistan, and as noted directly above, the sand-dune areas along the Amu Darya in the north, have no significant drainage whatsoever (Favre and Kamal, 2004).

The total annual volume of water used for direct human and animal consumption is estimated to be not more than about 0.2 $km^3$. Favre and Kamal (2004) used FAO data to use a rate of 10,000 $m^3$/ha for a total irrigated area of about 2.4 million ha, which amounts to about 24 $km^3$ of annual irrigation water that is utilized. They also noted that the government of Afghanistan has divided their land into four classes of irrigation water: (1) rivers and streams—86%; (2) springs—7.9%; (3) karez—7.0%; and (4) shallow and deep wells—0.5%.

### 2.4.1 Kabul River System

The ~700 km-long Kabul River rises in the mountains of central Afghanistan west of the city, whereas one of its main tributaries, the Panjshir comes from the Hindu Kush to the northeast and a tributary to it, the Ghorband rises in the Hindu Kush to the west. The Kabul River's largest tributary, the Kunar River enters Afghanistan from northwest Pakistan directly south of the Wakhan Corridor where it rises from large glaciers in the Karakoram Himalaya, as well as other glaciers in the eastern Hindu Kush and Hindu Raj of Pakistan. The Kabul River derives from only about 12% of the land area of Afghanistan, but has about 26% of the annual flow of the country. The river flows through or along the borders of 11 provinces with a population of over 7 million, and supports over 300,000 ha of intensively irrigated areas and high-value crops in Afghanistan. The Kabul River goes on to flow into Pakistan just north of the Khyber Pass, where it supports some 50,000 ha of agricultural crops before its eventual confluence with the Indus River at Attock. The river basin also includes four major hydropower dams. The river basin itself covers 53,000 $km^2$ inside Afghanistan and 14,000 $km^2$ in Pakistan before its confluence with the Indus.

The Kabul River has had highly variable discharges of 33–460 $m^3/s$, but in recent drought years has run nearly completely dry several times in the city of Kabul with its population of ~3 million people, which is a most serious shortfall, especially as water demand in Kabul City and the river basin as a whole is expected to increase in the near future. Because of mountain snow and glacier melt waters, the Panjshir and Kunar tributaries provide the bulk of flow in the lower Kabul where it can flow with a mean annual discharge of as much as 700 $m^3/s$ (22 $km^3$/year) before it enters Pakistan. Meanwhile, Pakistan is benefiting from water flows from Afghanistan but provides no financial support for flow-control structure or management of the river within Afghanistan (Fipps, 2007). Pakistan has also significantly increased its water use for power production, municipal use, and agriculture in the past 30 years and has a higher water demand than can currently be met, with the result that they do not want additional dams on the Kabul River in Afghanistan, as has been planned by the governments of Afghanistan and India (Anon., 2011).

## 2.4.2 Helmand-Arghandab River System

This important and longest (~1150 km) river system in Afghanistan drains a wide region extending from the Paghman Mountains in central part of the country where part of the Kabul River also rises, southwest into the Seistan depression and the Iranian border to encompass about 43% of the total land area of the nation. Its largest tributary, the Arghandab River that flows through Kandahar, joins the Helmand near Lashkargah. Two major dams in this system were built in the 1950s; the Kajaki Dam on the Helmand about 70 km upstream from Girishk, and the Dahla Dam on the Arghandab 50 km northeast of Kandahar. Total flow of the Helmand in nondrought years was estimated at about 7 $km^3$/year (Fipps, 2007), with mean discharges of 90–4000 $m^3/s$ and maxima of 18,000–20,000 $m^3/s$. The main branch of the Helmand in its delta in the Seistan depression forms the Iranian border where the waters drain off into *hamun* (*hamoon*) wetlands and lakes that are discussed in detail in Shroder (2014). On a number of occasions in the recent past (1990s and 2000s) Iran has entered Afghanistan to divert Helmand waters to its advantage. In addition, Iran has never compensated Afghanistan for Helmand River water as provided for in a 1973 treaty.

In this region the dominantly north-south running western border of Afghanistan with Iran jogs about 90 km east to allow the Iranians greater control of the Helmand tail-waters of the region. This jog, which was detrimental to the Afghans, who had to be bribed to even nominally accept and sign off on it, was a deliberate creation of the British-dominated Goldsmid Commission in 1872–73, who were reacting negatively to a country where they had fought and lost the First Anglo-Afghan War of 1838–42, and were about to fight the Second in 1878–80.

### 2.4.3 Hari Rud-Murghab River System

This 1100-km-long fluvial system represents about 12% of the water resources of Afghanistan and is dominant in the heavily irrigated regions around Herat. The waters of the Hari Rud mainly raise in central Afghanistan in the Kohi Baba Range above Bamiyan and flow in a nearly straight line west down a dominantly strike-slip, fault-guided valley trace toward the Iranian border, before looping north to form Afghanistan's far northwest border with Iran where it is renamed the Tejen. This river flows between Iran and Turkmenistan, before its last remnants are diverted off into canals or pass out into the sands of the Karakum desert. The Murghab River in this basin also flows out of Afghanistan into the Karakum desert as well, at first in Afghanistan through a strong canyon whose narrow width does not allow much room for irrigated agriculture along its banks. The average annual discharge of the Hari Rud is about $55\,m^3/s$, but during a spring flood in 1939 the discharge went up to $1090\,m^3/s$. The mean annual discharge of the Murghab was about $41\,m^3/s$ but during a flood in 1886 a discharge rate of as much as $367\,m^3/s$ was measured.

### 2.4.4 Amu Darya River System

The Amu Darya (Oxus of antiquity) originates in the Wakhan Corridor of Afghanistan as three separate drainages, the Aq Su, the Abi Pamir, and Abi Wakhan that are confluent downstream. For example, the northern Pamir tributary joins the southern Wakhan tributary in the Wakhan Corridor to form the Panj River of Badakhshan, which constitutes most of the national border of Afghanistan with Tajikistan. The Aq Su loops out of Afghanistan, north though Tajikistan to join the Panj downstream in its lower reaches. Where this Panj River is confluent with the Kokcha River further downstream coming in from southern Badakhshan, marks where the Amu Darya begins, which there goes on the form the rest of the entire southern border of Tajikistan and Uzbekistan with Afghanistan, as well as the eastern portion of the border of Turkestan with Afghanistan.

Estimates vary because of the lack of reliable measurements or preservation of past flow records, but most of the Nile-sized flow of the Amu Darya originates in Tajikistan at about 61% of the flow, and in Afghanistan at ~30% (Table 2.3). The downstream republics of Uzbekistan

TABLE 2.3 Average Water Generation and Use in the Amu Darya River Basin

| Country | Discharge to Amu Darya ($km^3$/year) | % of Total | Irrigated Land (Mha) | Water Allocation ($km^3$) | % of Total Used |
|---|---|---|---|---|---|
| Afghanistan | 24 | 30 | 1.2 | – | – |
| Tajikistan | 49.0 | 61 | 0.5 | 9.5 | 15.2 |
| Uzbekistan | 4.8 | 6 | 2.3 | 29.6 | 48.2 |
| Turkmenistan | 0.82 | 1 | 1.7 | 22 | 35.8 |
| Kyrgyzstan | 1.6 | 2 | 0.1 | 0.4 | 0.6 |

Estimates for Afghanistan vary because records are incomplete and the discharge of border rivers can be apportioned in different ways.
*From Fipps, G., 2007. Appendix to the transboundary water policy of Afghanistan: transboundary water issues. http://gfipps.tamu.edu/Afghanistan/Transboundary%20Water%20Issues26-04-07.pdf (accessed 18.02.15)*

and Turkmenistan together use about 84% of the total flow of the Amu Darya but contribute only ~7% of the total volume, which fact is producing considerable consternation in the governments of the upstream nations of Tajikistan and Afghanistan. In comparison Afghanistan currently uses <10% of its total contribution (1.5–2 $km^3$) and only about 2% of the total discharge. Because of this limited use of water by Afghanistan, the acute shortages of drinking and irrigation waters in northern Afghanistan, large areas of Kunduz, Balkh, Jowzjan, and Faryab regions, indicate the possibilities of building a number of gravity-fed, water diversion structures from the Amu Darya to service these regions better (Fipps, 2007).

The average discharge of the Amu Darya appears to be around 2000 $m^3$/s/year. The Kokcha itself, one of the major rivers of northern Afghanistan, has measured discharge rates of ~5.7–6 $km^3$/year, whereas the Kunduz River tributary to the Amu Darya also measures similar amounts with a mean discharge of 3.3–6 $km^3$/year. All other rivers in this northernmost drainage basin of Afghanistan have waters in insufficient quantities to travel from their mountain sources to be confluent with the Amu Darya; instead the waters are diverted off for irrigation and what is left behind soaks into the desert sands that are peripheral to the Amu Darya.

## 2.5 CONCLUSION

As Afghanistan reestablishes its hydrograph measuring stations for river-discharge assessment in order to recover from 30 years of no data as a result of the incessant war, the desire has been to bring the nation back up to some level of intellectual parity with surrounding nations. The result of this long war disruption, along with climate change, is that discharge estimates for the rivers of Afghanistan, based as they are on only incomplete historical records, can be somewhat variable and ambiguous, or even quite incorrect. Thus the total renewable water resources in Afghanistan have been quite variously estimated with as much as a 65% variability: 55 $km^3$/year (FAO, 1997); 57 $km^3$/year (Azimi and McCauley, 2002); 65 $km^3$/year (Nathan Berger, 1992); 75 $km^3$/year (Qureshi, 2002); and 84 $km^3$/year (Favre and Kamal, 2004). This high variability exists because of probable collection errors, inclusion of underground water recharge in a few cases, and probably outright errors in transcription. Another part of the reason for these variable and ambiguous numbers is also the fact that in the 1970s, the Government of Afghanistan released water-resource monitoring information according to a somewhat quirky scheme of river-basin divisions that were not based on actual hydrologic divides, but more on direction of surface flow inside Afghanistan only, rather than where the water ended up (Uhl and Tahir, 2003). Finally, many different sets of measurement unit types have been utilized by a number of different entities (million meters cubed, $Mm^3$; billion meters cubed, $Bm^3$; billion cubic meters, BCM; and cubic kilometers, $km^3$; for example, 1 billion cubic meters = 1 cubic kilometer), all of which units require careful conversion to maintain accuracy. The result has been a plethora of variable water accounting figures, some of which are incompatible, incorrect, or just rather difficult to interpret.

Finally, the hydrology of Afghanistan is an essential attribute that has received increasing scientific attention because intensive study of this critical resource is a necessity for dealing with that which must be preserved to ensure an adequate future. For a dominantly semiarid to arid country, with a treasure house of natural water renewal systems in its highlands, establishment of a robust hydro-knowledge base is a critical part of maintaining the peace in a volatile region. The beginnings of such valuable resource information are at hand and increasing; this chapter provides some baseline information that seems to be reasonably accurate. Within the coming decade, it is likely that this hydro-knowledge will increase, especially if neighboring countries with more advanced hydro-knowledge and hydro-engineering capabilities can see the value of helping Afghanistan to increase its own hydro-information capabilities. The good will so obtained would go a very long way to solving many problems related to access to adequate water supplies in this difficult region.

## References

Anon., 2011. Sharing water resources with Afghanistan. Dawn. http://www.dawn.com/news/673055/sharing-water-resources-with-afghanistan (accessed 17.02.15).

Azimi, A., McCauley, D., 2002. Afghanistan's Environment in Transition. Asian Development Bank, Manila. http://www.adb.org/publications/afghanistans-environment-transition (accessed 18.02.15).

Breckle, S.-W., Dittman, A., Rafiqpoor, M.D. (Eds.), 2010. Field Guide Afghanistan: Flora and Vegetation. Scientia Bonnensis, Bonn, Germany, 861 pages.

Bishop, M.P., Shroder Jr., J.F., Ali, G., Bush, A.B.G., Haritashya, U., Roohi, R., Sarikaya, M.A., Weihs, B.J., 2014. Remote sensing of glaciers in Afghanistan and Pakistan. In: Kargel, J.S., Leonard, G.J., Bishop, M.P., Kääb, A., Raup, B. (Eds.), Global Land Ice Measurements From Space. Springer-Praxis, New York, NY. (Chapter 23), ISBN-10: 354079817X, ISBN-13: 978-3540798170.

Campbell, J., 2015. A dry and ravaged land: investigating water resources in Afghanistan. Earth 60 (1 and 2), 49–55.

de Bie, C.A.J.M., Skidmore, A.K., Toxopeus, B., Venus, V., 2007. An Updated Köppen—Geiger Climate Classification of the World Using Very High Resolution Interpolated Climate Surfaces of Monthly P and T Data from 1950 to 2000. ITC, Enschede. http://www.itc.nl/personal/debie/An%20updated%20K%F6ppen.htm (accessed 18.02.15).

FAO, 1996 (also Klemm, W., 1997). Water resources and irrigation. Promotion of Agricultural Rehabilitation and Development Programmes for Afghanistan. Food and Agriculture Organization of the United Nations. FAO Project TCP/AFG 4552.

FAO, 1997. Afghanistan (Climate and Water Resources). Food and Agriculture Organizations of the United Nations, Rome. 5 pp. http://www.fao.org/nr/water/aquastat/countries_regions/afghanistan/index.stm (accessed 18.02.15).

Favre, R., Kamal, G.M., 2004. Watershed Atlas of Afghanistan. 1st edition - Working Document for Planners. FAO and Afghanistan Information Management Agency (AIMS).

Fipps, G., 2007. Appendix to the transboundary water policy of Afghanistan: transboundary water issues. http://gfipps.tamu.edu/Afghanistan/Transboundary%20Water%20Issues26-04-07.pdf (accessed 18.02.15).

Heckmiller, I.A., 1961. Stream-gaging problems in Afghanistan. Int. Assoc. Sci. Hydrol. Bull. 6 (3), 35–37.

Humlum, J., 1959. La geographie de l'Afghanistan. Etude d'un pays Aride Scandanavian University Books, Copenhagen.

Nathan Berger, Inc., 1992. Afghanistan water constraints overview analysis. Final Report for U.S. Agency for International Development; Contract No. 306-0205-C-009385-00. 119 pp. http://pdf.usaid.gov/pdf_docs/PNABL957.pdf (accessed 18.02.15).

Olson, S.A., Williams-Sether, T., 2010. Streamflow characteristics of streamgages in Northern Afghanistan and selected locations. US Geological Survey Project Product Number 179; Data Series 529. 523 pp. http://pubs.usgs.gov/ds/529/pdf/ds529.pdf (accessed 09.02.15).

Porter, S.C., 2004. Late Pleistocene glaciation of the Hindu Kush, Afghanistan. Quaternary glaciations extent and chronology—part III: South America, Asia, Africa, Australia, Antarctica. Dev. Quaternary Sci. 2 (C), 1–2.

Qureshi, A.S., 2002. Water resources management in Afghanistan: the issues options. Working Paper 49; Pakistan Country Series No. 14. 30 pp. http://www.afghaneic.net/library/hydrological%20surveys/wor49.pdf (accessed 18.02.15).

Riebeck, H., 2006. Aiding Afghanistan: MODIS snow maps. http://earthobservatory.nasa.gov/Features/Afghanistan/afghanistan4.php (accessed 17.02.15).

Shroder Jr., J.F., 1980. Special problems of glacial inventory in Afghanistan. In: World Glacier Inventory Proceedings Riederalp Workshop, September 1978, pp. 142–147 (IAHS-AISH) Publication No. 126, Hydrological Sciences Bulletin.
Shroder, J., 2014. Natural Resources in Afghanistan: Geographic and Geologic Perspectives on Centuries of Conflict. Elsevier, Netherlands.
Shroder Jr., J.F., Bishop, M.P., 2010. Glaciers of Afghanistan. In: Satellite Image Atlas of Glaciers of the World—Asia, Williams, R.S. Jr., Ferrigno, J.G. (Eds.), United States Government Printing Office, Washington, DC, pp. 167–199 U.S. Geological Survey Professional Paper 1386-F3.
Shroder, J.F., 1989. Slope failure: extent and economic significance in Afghanistan and Pakistan. In: Brabb, E.E., Harrod, B.L., Balkema, A.A. (Eds.), Landslides: Extent and Economic Significance in the World. Rotterdam, Netherlands, pp. 325–341.
Sival, T.R., 1977. Synoptic-climatologic study of the Asian summer monsoon in Afghanistan. Geogr. Ann. 59A, 67–87.
Uhl, V.W., Tahir, M.Q., 2003. Afghanistan: An Overview of Groundwater Resources and Challenges. Uhl, Baron, Rana & Associates, Inc., Washington Crossing, PA. 55 pp. http://www.vuawater.com/Case-Study-Files/Afghanistan/Afghanistan_Overview_of_GW_Resources_Study-2003.pdf (accessed 15.02.15).
Westfall, A.O., Latkovich, V.J., 1966. Surface waters investigations plan for Afghanistan: administrative report. United States Geological Survey and the Water and Soil Survey Department, Ministry of Agriculture and Irrigation, Royal Government of Afghanistan. 62 pp.
Williams-Sether, T., 2008. Streamflow Characteristics of Streams in the Helmand Basin, Afghanistan. U.S. Geological Survey, Reston, VA. US Geological Survey Project Product Number 167; Data Series 333, 349 pp. http://pubs.usgs.gov/ds/333/pdf/ds333_v1.1.pdf (accessed 09.02.15).

CHAPTER

3

# Ground-Water Geology of Afghanistan

*L. Sinfield*, J. Shroder*[†]

*US Navy, San Diego, CA, United States [†]University of Nebraska at Omaha, Omaha, NE, United States

OUTLINE





*Transboundary Water Resources in Afghanistan*
http://dx.doi.org/10.1016/B978-0-12-801886-6.00003-3

*A little water is a sea to an ant* ***Afghan proverb***

## Abstract

Ongoing tectonic continental collision between the Indian subcontinent and Eurasia is the main driving force molding and creating snow- and rain-fed mountain watersheds and ground-water basins in Afghanistan. All large cities, as well as rural, agricultural-based populations of Afghanistan draw water from aquifers within tectonically controlled, alluvium-filled structural basins. The underlying geologic fabric controls ground-water occurrence and changing climate will dramatically change watershed natural conditions Also human-caused contamination and salinization of aquifers, recharge reduction along alluvial aquifers and canal systems, and overuse of fresh-water resources, negatively affect many downstream and cross-border rivers and alluvial aquifers. Conflict increases impacts by reducing government ability to control and manage aquifer over-pumping, especially from high capacity, engine-powered well pumps that supply water for illicit opium poppies. To overcome obstacles, first steps are to understand natural underlying geologic and hydrologic conditions and limitations controlling water availability. Sustainable approaches to managing water resources in Afghanistan must be directed by underlying natural conditions. Cities, populations, and economics will always be focused on these structurally controlled valleys and their alluvial aquifers. Relatively inexpensive, potable, and bacteriologically safe drinking water occurs in deep aquifers that cannot be ignored and developing this valuable resource correctly will significantly improve health and economics in Afghanistan. Technology to access water sources must be applied correctly because the new mud-lubricated, rotary drill-rigs can drill deeper, and with larger well pumps, can increase detrimental water withdrawals. Irrigation and watershed management programs for Afghanistan should use: (1) new passive-seismic technology to conduct inexpensive yet effective ground-water exploration, (2) mud-lubricated, rotary drilling techniques for Afghan cities, and (3) improved karez repair and rehabilitation. Robust governance, focused security, and capacity-building goals, based on underlying natural conditions and limitations, help reduce threats and improve downstream and transboundary water quality and quantity issues.

**Keywords:** Ground-water basins, Alluvial aquifers, Karez, Passive-seismic exploration, Percussion drilling, Rotary drilling, Boreholes, Over-pumping, Geothermal energy, Geochemistry, Arsenic, Water management

## 3.1 INTRODUCTION

Underground water is a critical resource for the continued development of the Afghanistan economy and for improvements to the health of the Afghan people, both within the growing urban population and the agricultural areas. The economic lifeblood of Afghanistan is agrarian-based, with more than 80% of the Afghan people working in agriculture with arable land comprising only 12% of the total land area. Agriculture and population are concentrated along all the main river valleys of the country (Mahmoodi, 2008) with a smaller percentage of rain-fed lands outside of the river valleys. All major aquifers of Afghanistan occur in and beneath these river basins (Figs. 3.1 and 3.2) with water availability fueling the economic engine of Afghanistan, such as it is.

Presently, although estimates vary, more than 70% of the urban population and 80% of the rural population have no access to potable drinking water and continue to drink unsafe

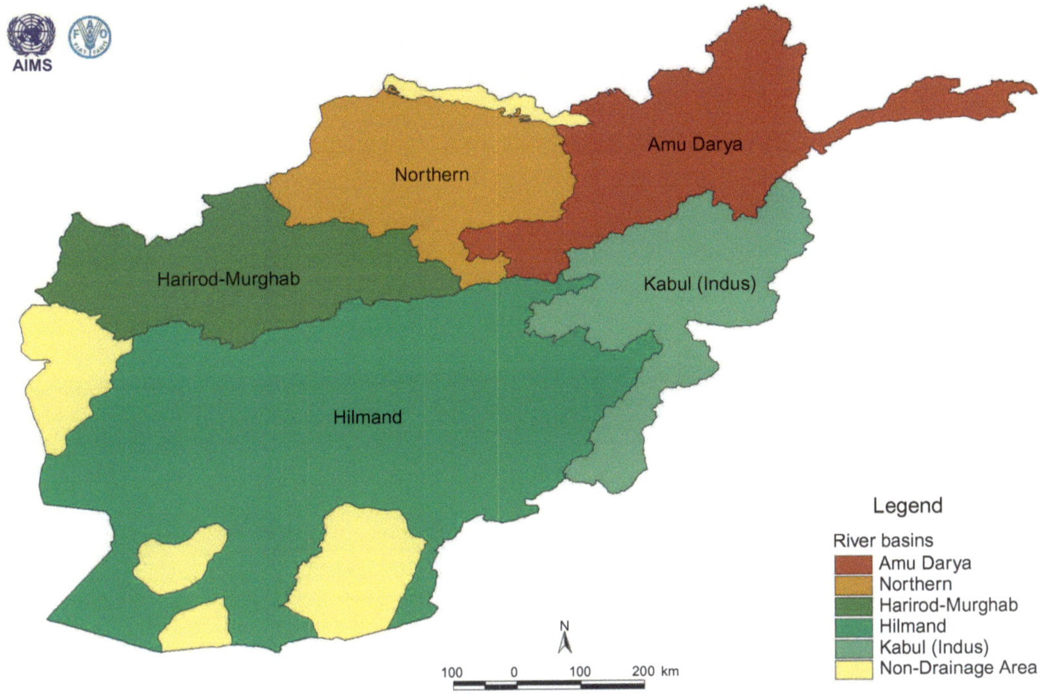


FIG. 3.1 Watersheds of Afghanistan. *From Favre, R., 2004. Watershed Atlas of Afghanistan: Working Document for Planners, 5 Volumes, first ed. Watershed Consultant Golam Monowar Kamal, AIMS Focal Point, Kabul.*

surface water or shallow ground water affected by human sewage. In cities, piped drinking-water distribution systems are less than 50% efficient due to a lack of maintenance and shortage of power to operate pumps and disinfection systems (Mahmoodi, 2008). When power is only available a few hours a day, pumping stations cannot maintain water pressures within city water distribution systems, increasing probability of backflow and siphoning of bacteriologically contaminated water and sewage into water systems. Chlorine disinfectants, which can alleviate some of the bacteriological contamination issues, cannot be maintained at effective concentrations in the distribution system under these conditions.

Using river water and providing surface-water treatment to produce potable and safe drinking water for growing cities in Afghanistan is expensive and technologically complex. Treatment at this level requires a reliable electricity supply, complex transportation, and intricate logistical networks to supply required chemicals, filters, and parts, and highly trained water treatment operators and engineers. Surface-water treatment at the municipal level is currently not viable in Afghanistan.

Ground-water sources have another set of issues. Unfortunately, many of the shallow alluvial aquifers are considered to be "ground water under the influence of surface water" and are greatly influenced by bacteriological contamination from dumping of sewage and other waste on the ground. Bacteriological contamination can be from direct surface-water recharge, from ephemeral and characteristically dry channels (Fig. 3.3), or directly beneath many villages from open dumping on the ground or in sewer gutters (Figs. 3.4 and 3.5).

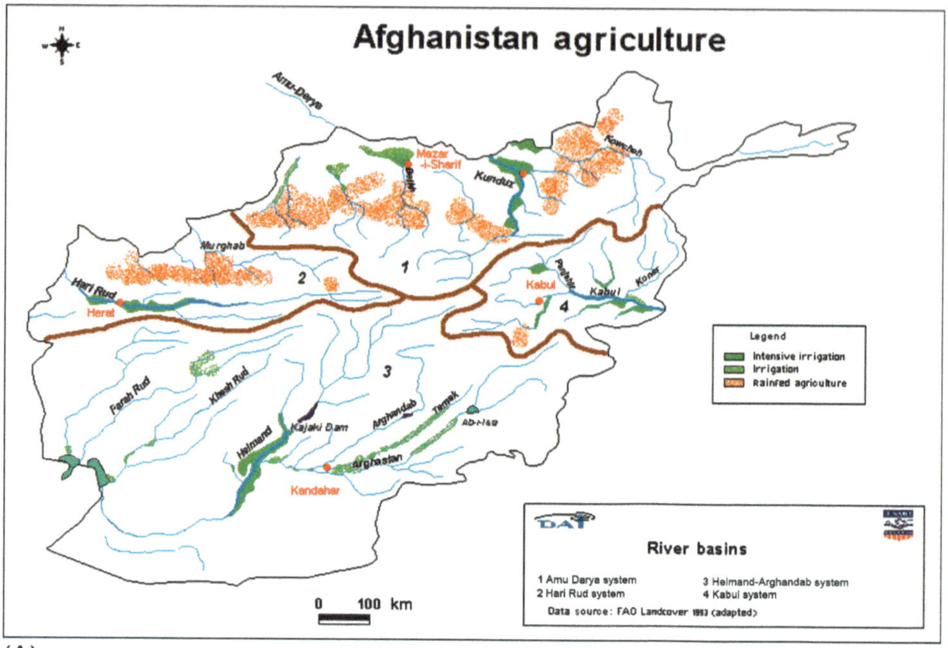


(A)

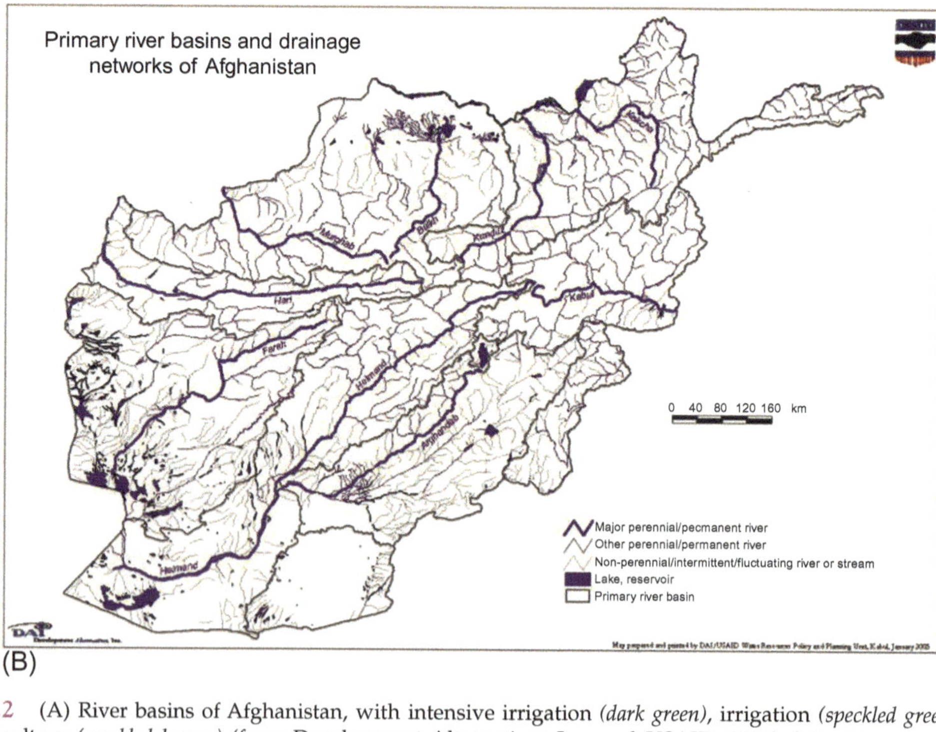


(B)

FIG. 3.2 (A) River basins of Afghanistan, with intensive irrigation *(dark green)*, irrigation *(speckled green)*, rainfed agriculture *(speckled brown)* (from Development Alternatives Inc. and USAID, 2003). http://www.vuawater.com/images-international-case-studies/AflanduseLg.gif, (accessed 25.12.15). (B) Primary river basins and drainage networks of Afghanistan (from Development Alternatives Inc. and USAID, 2003). http://www.vuawater.com/International-Case-Studies/Case-Studies-International-Asia.html (accessed 25.12.15).

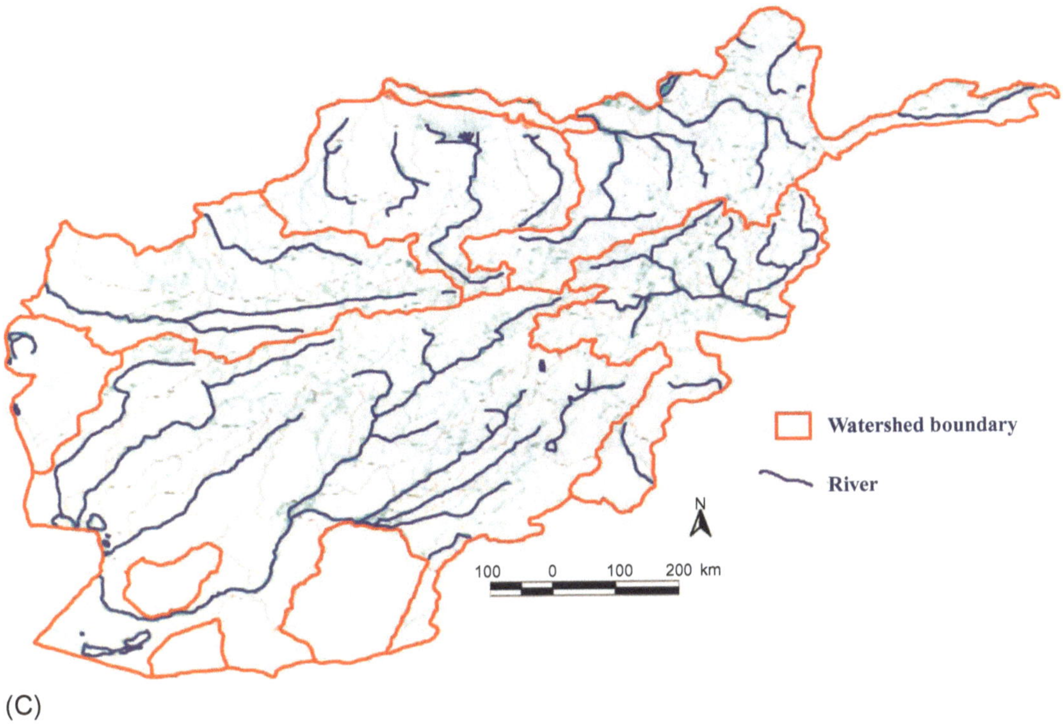


FIG. 3.2, CONT'D (C) Watershed boundaries for the major rivers of Afghanistan. *Modified from Favre, R., 2004. Watershed Atlas of Afghanistan: Working Document for Planners, 5 Volumes, first ed. Watershed Consultant Golam Monowar Kamal, AIMS Focal Point, Kabul.*

FIG. 3.3 Photo of the dumping of raw sewage into a dry stream channel. *Photo by L. Sinfield, 2012.*

FIG. 3.4 Photos of concrete sewage gutters in the city of Kandahar where the first photo (A) shows children washing clothes in the gutter (*juie*). No other sewage system occurs in the city. Most of the gutters drain to the main canals within the city, which are also used for drinking water. Also, note in the lower left (B) the blue water tank on top of the building, which is filled when the water distribution system has pressure. *Photos by L. Sinfield, 2012.*

If sewer infrastructures does exist, operation and maintenance of these systems are low priorities for many Afghan municipalities and governmental institutions, which also can cause releases of sewage (Fig. 3.6).

To provide safe drinking water, the Danish Committee for Aid to Afghan Refugees (DACAAR) has some success in Afghanistan with using sand-filled biofilters (Fig. 3.7) to filter and treat bacteriologically impacted water, but the units are small, produce limited quantities of water, and are only suitable for a single household per unit. Providing drinking water

FIG. 3.5 (A) Photo of hand-operated well in Kandahar near sewer gutter. Hand-operated wells (India Mark III type) typically tap the shallowest aquifers beneath the city of Kandahar (*Photo by L. Sinfield, 2012*). (B) Children selling vegetables over open sewer ditch from where they can get contaminated water to sprinkle of the vegetables to retard desiccation (*Photo by Dennis Bruhn, UNEP, 2012*).

using deeper ground-water sources is simpler, requires less maintenance than a surface-water treatment plant, is more affordable, and in many cases, results in safe potable water for a large city populations. Sustainable ground-water development strategies using new exploration techniques, coupled with good management of surface water that recharges those aquifers, will ensure that Afghans can continue improving their health and wellbeing. Water quality and quantity in rivers will impact aquifers, the health of populations, and the economic development of not just Afghanistan, but downstream countries as well that rely on the upper watersheds of their major rivers that originate in Afghanistan.

FIG. 3.6 Photo of raw sewage seeping upwards from a septic system. Maintenance and repair of sewer systems are not a high priority for many Afghan institutions. *Photo by L. Sinfield, 2012.*

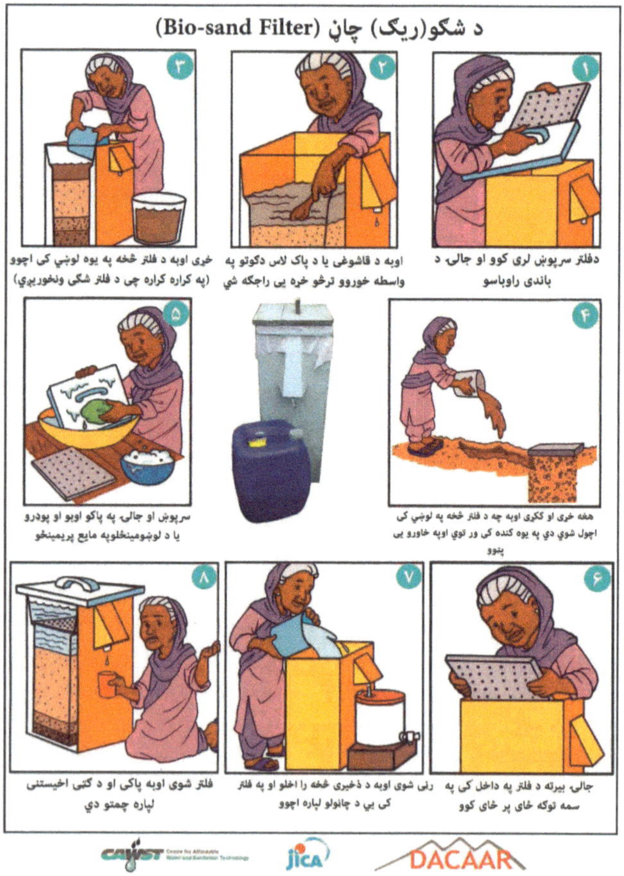


FIG. 3.7 DACAAR Biosand filter poster with cartoons on what it is and how to use it with instructions in Dari language. Used for bacteriological contaminant filtering and treatment of household water. *From www.DACAAR.org (accessed 17.12.15).*

## 3.2 GEOLOGIC OVERVIEW

With snow- and rain-fed mountain sources, the rivers of the Afghanistan region follow the dominant tectonic fabrics of the underlying Afghan geologic structure. This fabric ultimately controls snow and rainfall patterns, the flow of surface water, ground-water recharge, the occurrence of highly productive deep alluvial aquifers, and the quantity and quality of both surface water and underground water.

A major east-west geosuture along the Hari Rud River splits the waters of Afghanistan into either north and south, or southwest flowing watersheds, all emanating from the central mountain massif. Major aquifers, all within Quaternary and Neocene structural basins or valley-fill sediments, follow the same pattern because most all occur within the river valleys of these watersheds. The aquifers are recharged via infiltration into the coarse-grained valley sediments from rivers and streams descending from the high mountains.

Afghanistan has a long and complex tectonic history that has resulted in a complicated geologic structure with some of the most varied geology in the world (Abdullah et al., 1980; Ruleman et al., 2007; Shroder, 2014). According to Abdullah et al. (1980), most of the central mountain massif of Afghanistan is underlain by a sedimentary, igneous, and metamorphic rock complexes of varying ages. In general, most river valley floors consist of Holocene to middle Pleistocene conglomerate and sandstone and late Pleistocene loess with Holocene and late Pleistocene fan alluvium and colluvium along the sides of the valleys.

Overall, the geology, terrain, surface water, and ground water of the region are controlled and driven by one dominating process that has affected the pre-existing accreted terrains: The ongoing Himalayan Orogeny—the continental collision between the Indian subcontinent and Eurasia (Fig. 3.8).

### 3.2.1 Patchwork of Geologic Blocks

As noted in the second chapter of Shroder (2014), Afghanistan is composed of a patchwork of different crustal and oceanic blocks; a number of exotic terranes that were welded onto the southern margin of Eurasia with a series of successive accretionary events beginning in the Paleozoic and continuing to the present (Ruleman et al., 2007). The structure of Afghanistan is dominated by the Mesozoic (Cimmeride) and Tertiary (Himalayan) orogenic episodes that have created the mountainous core (massif) and the recent sedimentary deposition in the adjacent areas. According to Banks (2014):

> During the late Permian, a number of tectonic plate fragments (micro-plates) broke away from the southern "super-continent" of Gondwanaland. One of these, the Afghan micro-plate, collided with the Eurasian continental plate in the Mesozoic. This "Cimmeride" (also loosely referred to as "Hercynian") orogenic episode commenced around the late Triassic and was complete by the Jurassic/Early Cretaceous (200–150 million years ago). The Cimmeride orogeny created the Paropamisus-Band-e Turkestan mountains, to the south of Faryab (Whitney, 2006).
>
> Around the Early Cretaceous, the Indian plate disengaged from Gondwanaland and subsequently collided with the Eurasian plate in the Paleogene (late Paleocene, early Eocene) resulting in further orogenesis, crustal thickening and crustal displacement (broadly referred to as the so-called Himalayan orogenic episode). South of the Hari-Rud fault, the remnant of the Afghan micro-plate (the Afghan Block) has been (and is still being) squeezed south-westward at rates in excess of 1 cm/year by this crustal shortening (Whitney, 2006).

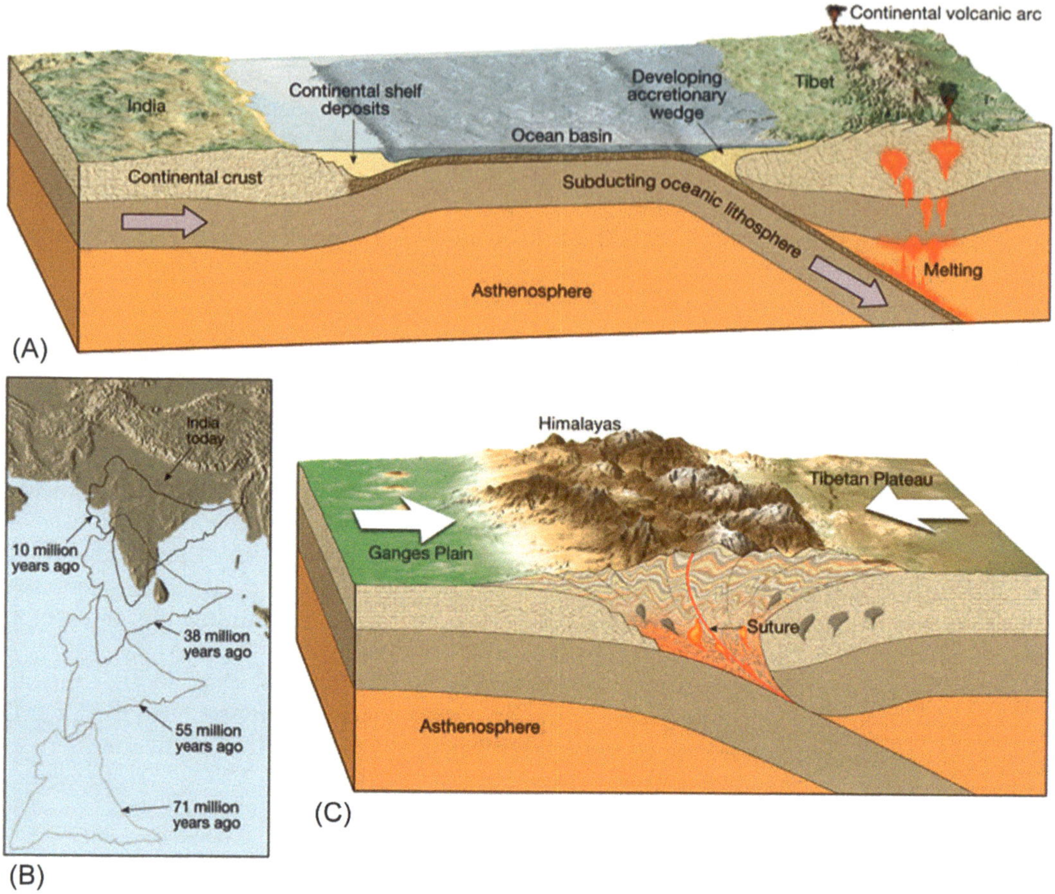


FIG. 3.8 (A) Ongoing continental collision between the India and Eurasia (Tibet) plates at some time prior to the ongoing plate-tectonic collision of the present time. (B) Progressive plate-tectonic migration of Indian subcontinent over the past 55 million years to impact with Eurasia, and force up the collisional mountains of the Hindu Kush and Himalaya. (C) Mountains of the Hindu Kush and Himalaya just southwest of the Tibetan Plateau suture between the two colliding plates. *These images were created by the Earthscope National Office, a National Science Foundation funded project. For more information, visit www.earthscope.org (accessed 11.07.15). (Per illustration publication rules on website).*

These blocks have been classified as four distinct seismotectonic provinces (Ruleman et al., 2007; Shareq, 1981), each with different geologic histories, structures, and ground-water basin types (Fig. 3.9). The provinces, from the oldest to youngest, are as follows:

- The "North Afghan Platform" is a pre-Himalayan Orogeny, continental margin of the Eurasian plate that occurs north of the Hari Rud fault zone and west of the Central Badakhshan fault systems, and includes the Tajik basin.
- The narrow "Middle Afghanistan geosuture zone" occurs immediately south of the North Afghan Platform and is part of the current right-lateral Hari Rud fault zone.
- South of the geosuture, the "Afghan Block" includes numerous exotic folded, faulted, partially metamorphosed, and deformed blocks that comprise the mountains and plains of the Farah, Helmand, and Arghandab watersheds.

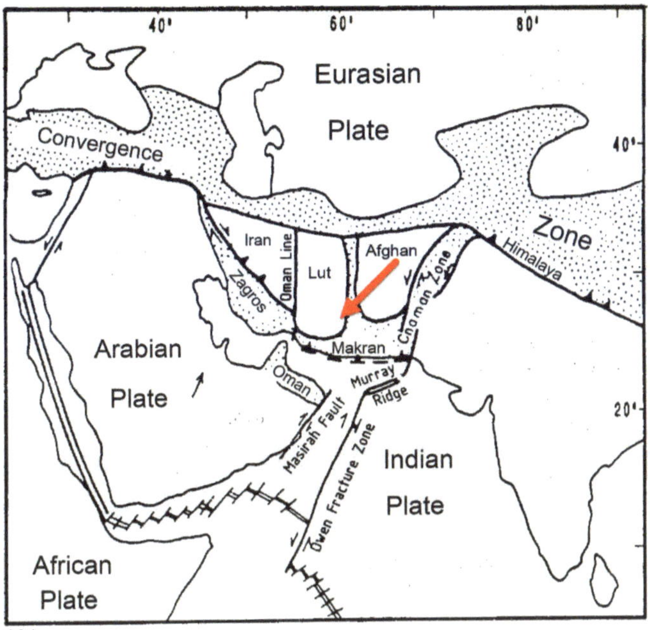


(A)

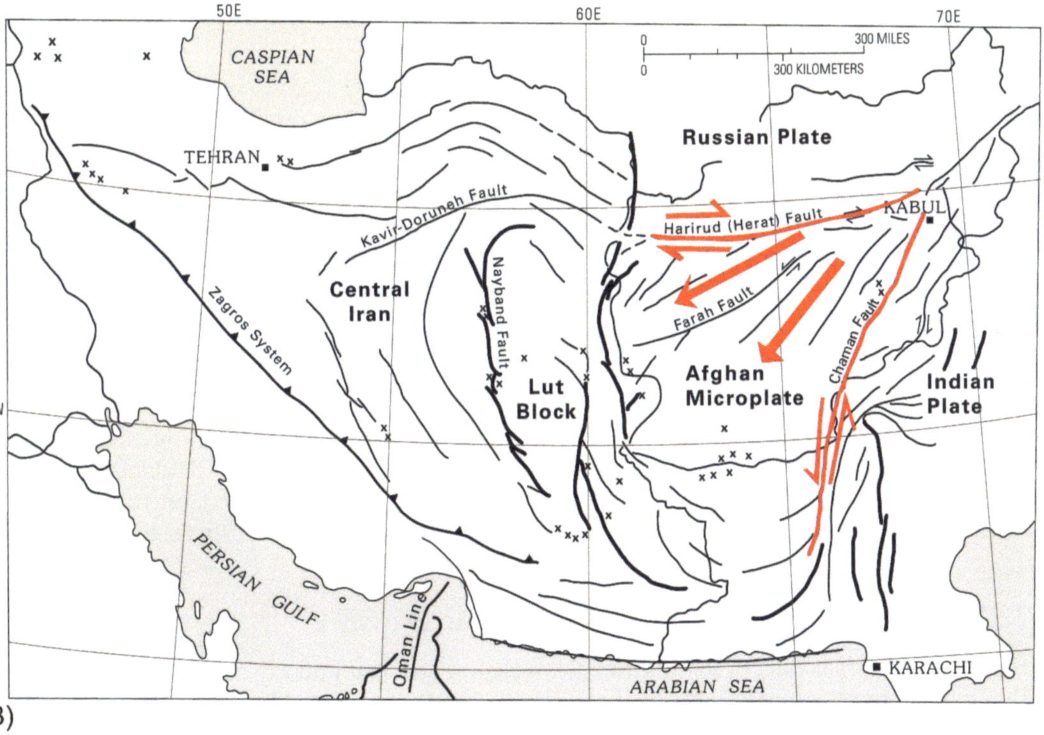


(B)

FIG. 3.9 (A) Associated plates and faults, subduction zone, sutures, and transform faults that occur around Afghanistan (*From Edgell, H.S., 1992. Basement tectonics of Saudi Arabia as related to oil field structures. In: M.J. Rickard et al. (Eds.), Basement Tectonics 9. Kluwer Academic Publishers, Dordrecht, pp. 169–93, 1 table, 13 figures.*). (B) The Afghan Block, between the Hari Rud and Chaman faults, is being pushed to the southwest into the Lut Depression in Iran (*Modified from Whitney, J.W., 2006. Geology, water, and wind in the lower Helmand Basin, southern Afghanistan: U.S. Geological Survey Scientific Investigations Report 2006-5182, 40 pp.*).

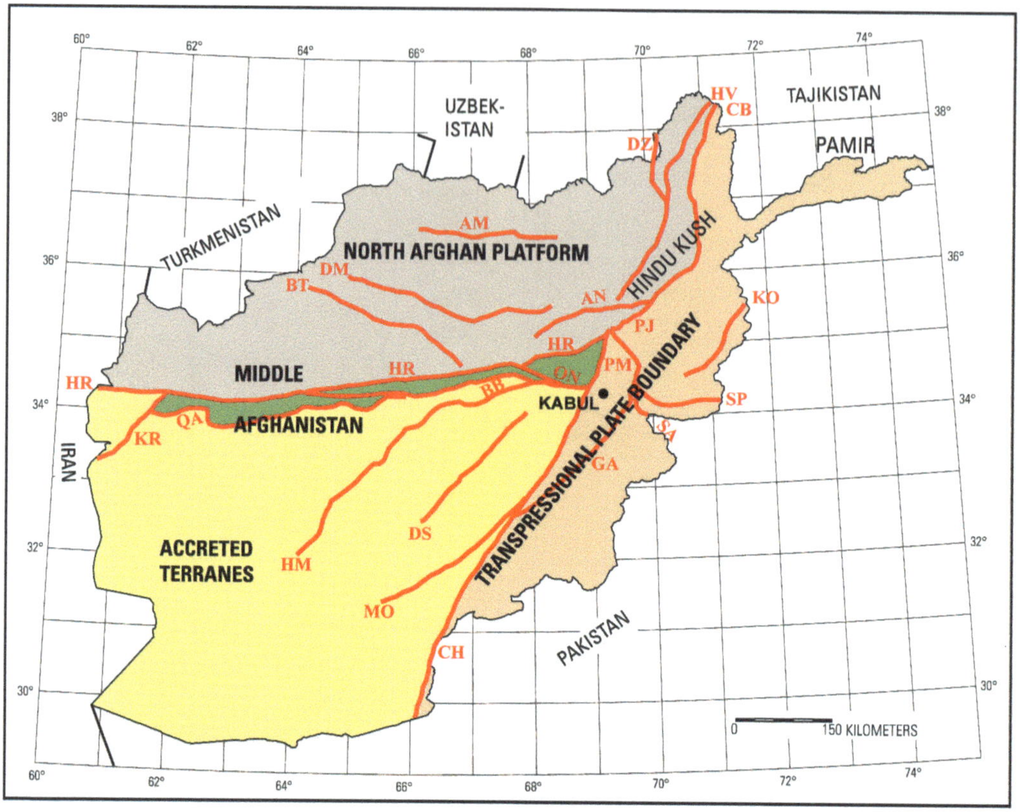


FIG. 3.10 Seismotectonic province map of Afghanistan showing the four main structural parts of the region (North Afghan Platform; Transpressional Plate Boundary along the border with Pakistan; accreted Terrains of southwest Afghanistan, and the middle Afghanistan shear zone). *From Ruleman, C.A., Crone, A.J., Machette, M.N., Haller, K.M., Rukstales, K.S., 2007. Map and Database of Probable and Possible Quaternary Faults in Afghanistan. USGS, Reston, VA. U.S. Geological Survey Open-File Report 2007-1103, 39 pp., 1 plate.*

- The "Transpressional Plate Boundary" occurs south and east of the Afghan Block and includes the eastern portion of the Hindu Kush and Pamir mountain ranges, and the Sulaiman fold and thrust belt of southern Afghanistan and Northern Pakistan (Figs. 3.9 and 3.10). The northern edge is defined by the left-lateral Chaman to Central Badakhshan fault systems (Ruleman et al., 2007) and includes the exotic Kabul crustal block.

## 3.2.2 North Afghan Platform Seismotectonic Province

Structurally, this is part of the old Eurasian craton and is relatively stable, mildly deformed with a basement of metamorphic and igneous rocks that formed the southern margin of the Eurasian continental plate adjacent to the Paleo-Tethys Ocean (Shareq, 1981; Tapponnier et al., 1986). This platform was accreted prior to the Permian-Triassic, during the Carboniferous-Permian Hercynian Orogeny. Deposited upon this margin is a relatively flat-lying Jurassic clastic sequence, which is in turn overlain by a Cretaceous carbonate platform and Tertiary alluvial and evaporite strata as thick as 6 km at the northern Afghan border (Brookfield and Hashmat, 2001). According the Abdullah et al. (1980), most of the watershed is underlain by relatively undeformed Neocene, Paleocene, and Late Cretaceous carbonate-terrigenous strata composed of irregularly interbedded clays, siltstone, and shale, with single beds and lenses of limestone, dolomite, gypsum, and anhydrite. Lower slopes are also covered by Quaternary to Holocene eolian sand and loess deposits forming gentle rolling hills.

According to Banks (2014):

The North Afghan Platform is typically divided into two distinct areas:

- the northern Murghab-Upper Amu Darya Basin, which is the main, subsiding sedimentary basin that has been accumulating Neogene and Quaternary deposits; and
- the area of the southern Paropamisus-Band-e Turkestan Uplift.

These two features appear to have acquired their tectonic character of respective subsidence and uplift both during and after the Himalayan-Alpine orogeny, although it appears probable that they evolved as down-warped ("intracratonic rift sag basin"—Klett et al., 2006) or uplifted structures during Jurassic-Paleogene times.

The area of the Paropamisus-Band-e Turkestan Uplift is subdivided into three distinct fault blocks, which step down towards the north into the Murghab-Upper Amu Darya Basin. These are: (1) Qala-e Naw Fault Block; (2) Maimana Fault Block; and (3) Shebergan Fault Block.

### 3.2.3 Middle Afghanistan Geosuture Seismotectonic Province

The Middle Afghanistan geosuture (Hari Rud Geosuture), is a major east-west-trending, continental-scale suture forming the boundary between the North Afghan Platform and the accreted terranes of the Afghan Block and other terranes to the south. The Hari Rud River flows along a major downfaulted graben within the geosuture south of the Paropamisus Mountains with intermontane alluvial-filled aquifers within the trough. Folds in the rocks of the geosuture have east-west strikes cut by numerous subparallel and oblique faults with local imbricate structures.

### 3.2.4 Afghan Block Seismotectonic Province

The Afghan Block, a series of extensively deformed microplates, was accreted south of the North Afghan Platform along the southern edge of the Eurasian continental plate during the Mesozoic Cimmerian Orogeny, which involved closing of the Paleo-Tethys Ocean. The Middle Afghanistan Geosuture forms the northern edge and the Chaman fault forms the southeastern boundary of the Afghan Block. Many mountain ranges of this block have northeast-trending linear, steep-sided folds indicative of Cimmerian Orogeny thrust faulting. To accommodate the ongoing continental collision, the contraction between the Indian and Eurasian plates, the Afghan Block may be actively extruded to the southwest between the left-lateral Chaman and the right-lateral Hari Rud fault systems (Tapponnier et al., 1986). The Farah Rud, Helmand, and Darafshan fault systems may be suture zones within the Afghan Block, reactivated as left-lateral strike-slip faulting during the Himalayan orogeny with minor faults acting as transfer zones between these faults.

As a result of this southwestern movement and associated faulting, many river valleys and ground-water basins in this block are wedge-shaped and spearheaded north-eastwards and widening southwesterly into the Seistan depression indicating extension between the diverging strike-slip faults (Figs. 3.10–3.12). Alluvial-filled structural basins host large aquifers and occur along the Helmand (within Farah, Tarnak, Arghistan and Arghandab rivers) watersheds within in the Afghan Block.

### 3.2.5 Transpressional Plate Boundary Seismotectonic Province

Near the border with Pakistan, the NE-SW-oriented left-lateral Chaman fault forms the southeastern boundary of the accreted terrain of the Afghan Block and the northern edge of the wide Transpressional Plate Boundary between the Eurasian and the Indian plates to the south and east. This area is associated with the closing of the Neo-Tethys Ocean and the collision of India into Eurasia. Characterized by a complex fold and thrust belt of the western end of Himalayan Orogenic belt, the deformation within the Transpressional Plate Boundary is driven by the ongoing continental collision between the northward-moving

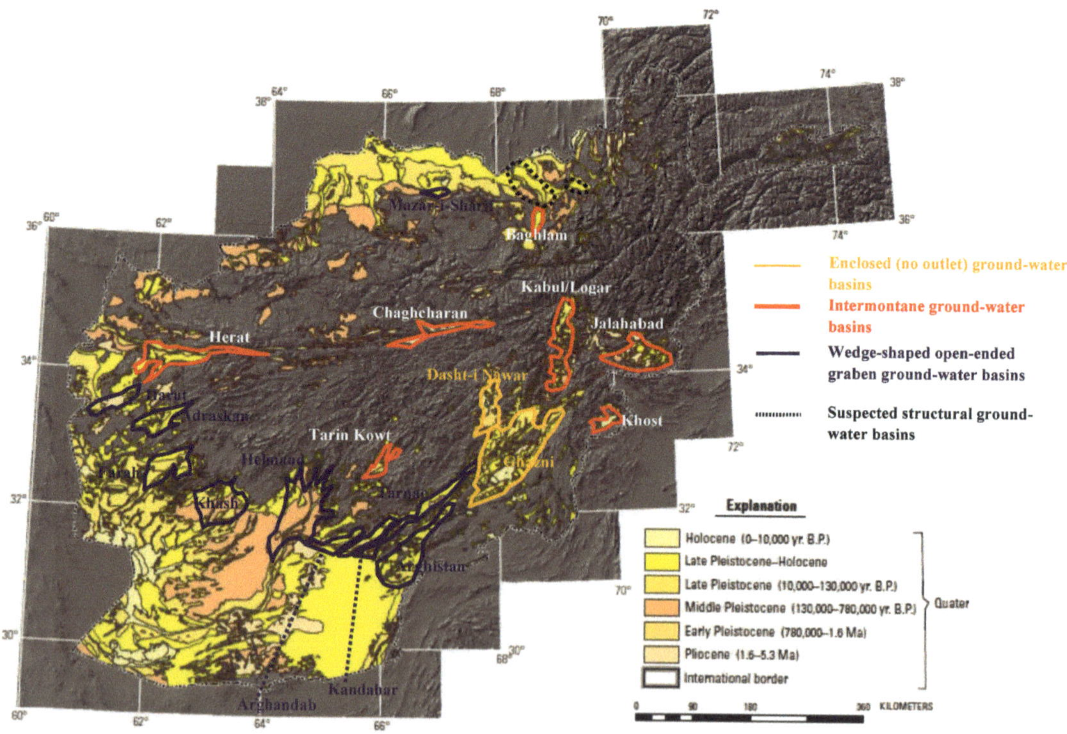


FIG. 3.11 Structural fault-bounded ground-water basins of Afghanistan. Many more of these ground-water basins will be discovered and mapped properly in future as further intensive ground-water geology is done for the many basins in the country. *Modified from the Quaternary fault map of Afghanistan, after Ruleman, C.A., Crone, A.J., Machette, M.N., Haller, K.M., Rukstales, K.S., 2007. Map and Database of Probable and Possible Quaternary Faults in Afghanistan. USGS, Reston, VA. U.S. Geological Survey Open-File Report 2007-1103, 39 p., 1 plate.*

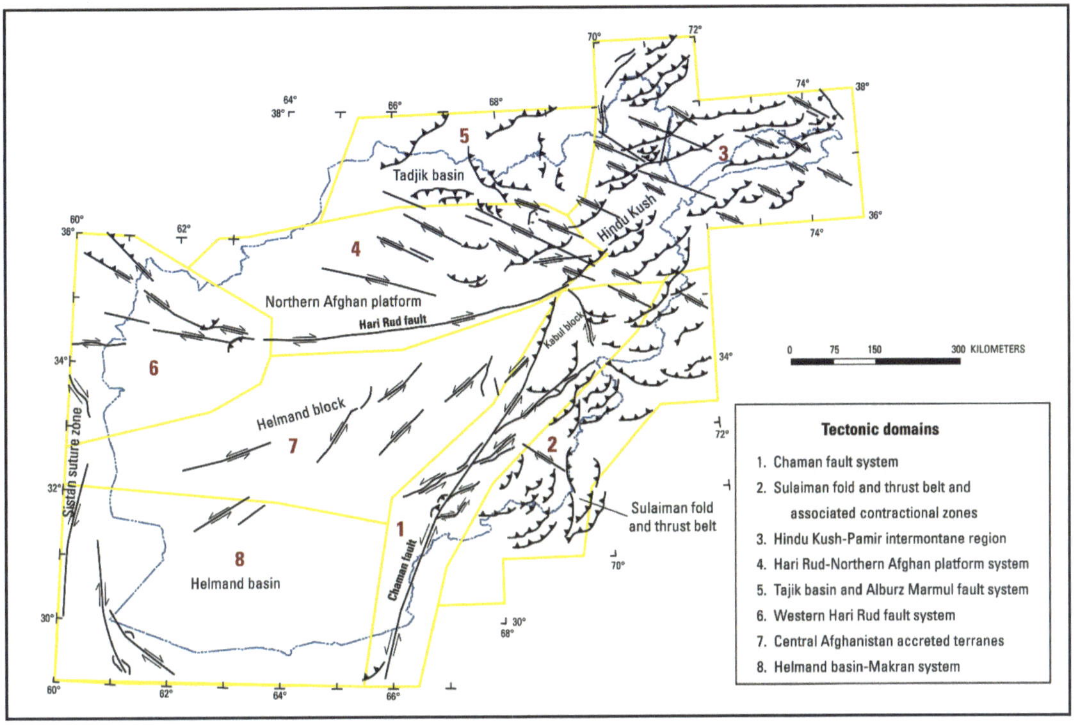


FIG. 3.12 Quaternary fault map of Afghanistan. *From Ruleman, C.A., Crone, A.J., Machette, M.N., Haller, K.M., Rukstales, K.S., 2007. Map and Database of Probable and Possible Quaternary Faults in Afghanistan. USGS, Reston, VA. U.S. Geological Survey Open-File Report 2007-1103, 39 pp., 1 plate.*

Indian subcontinent and Eurasia, spectacularly manifested in the east as the Himalayan Mountain range (Ruleman et al., 2007; Wheeler et al., 2005).

Associated alluvial-filled extensional basins also host large aquifers and are bounded by normal faults (via transpression) located at northwest-trending, strike-slip systems as major thrust-fault stepovers (Ruleman et al., 2007). The exotic crustal Kabul Block, which occurs within the Transpressional Plate Boundary Block, is bounded to the east and west by the right-lateral Sorobi fault and the left-lateral Chaman-Paghman fault system and to the north by the Hari Rud-Panjshir fault. These extensional basins, which are expressed as structural grabens, host the most prolific aquifers of Afghanistan.

## 3.2.6 Current Faulting

Active faulting still continues within Afghanistan. Ruleman et al. (2007) have noted that the most active system, the Chaman fault system along the Afghanistan/Pakistan border, has over 10 mm/year of movement and is associated with the Moqur fault system with less movement of 1–10 mm/year. This faulting is important for the Tarnak River basin and associated aquifers, because they occur in the structural trough formed by the Moqur fault system (Fig. 3.12).

## 3.3 GROUND-WATER OCCURRENCE

According to Uhl and Tahiri (2003), the most prolific aquifer systems in Afghanistan are within Quaternary and Neogene-Age alluvial deposits of major rivers and intermontane valleys (Fig. 3.12). Modifying Uhi and Tahiri's (2003) concept of these intermontane aquifers:

- The largest aquifers occur within alluvium-filled, fault-bounded structural graben basins formed by tectonic transpression, and provide water to the largest cities within Afghanistan, including Kabul, Kandahar, Mazar-i-Sharif, Jalalabad, Qalat, and Herat. These structural basins are either fully fault-bounded intermontane grabens (Kabul, Baghlan, and Jalalabad), or wedge-shaped half-grabens that open up to the south or north discharging into and across large plains (Kandahar, Tarnak, Mazar-i-Sharif, and Farah).
- Within the mountains, thin alluvial sediment aquifers occur in narrow valleys between steep valley walls of the middle and upper reaches of all the mountain-fed rivers of the region (Fig. 3.13).

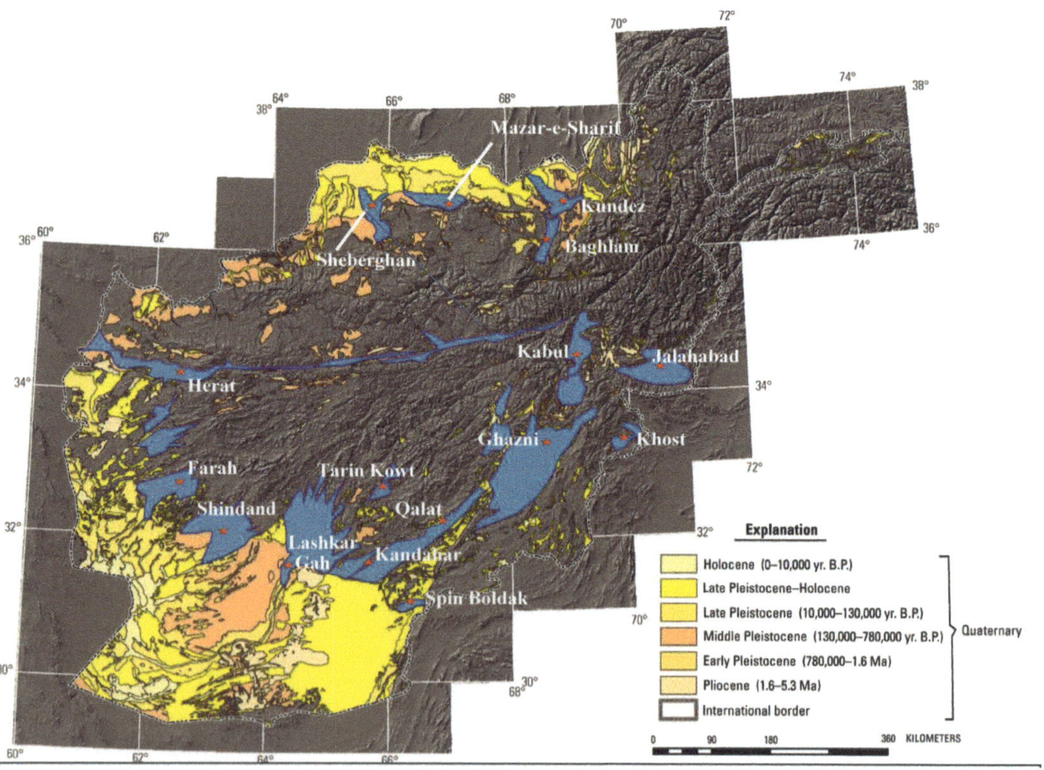


FIG. 3.13 Fresh-water aquifers *(blue)* and cities of Afghanistan (FAO, 1965a,b). *Modified from Quaternary fault map of Afghanistan, Ruleman, C.A., Crone, A.J., Machette, M.N., Haller, K.M., Rukstales, K.S., 2007. Map and Database of Probable and Possible Quaternary Faults in Afghanistan. USGS, Reston, VA. U.S. Geological Survey Open-File Report 2007-1103, 39 pp., 1 plate.*

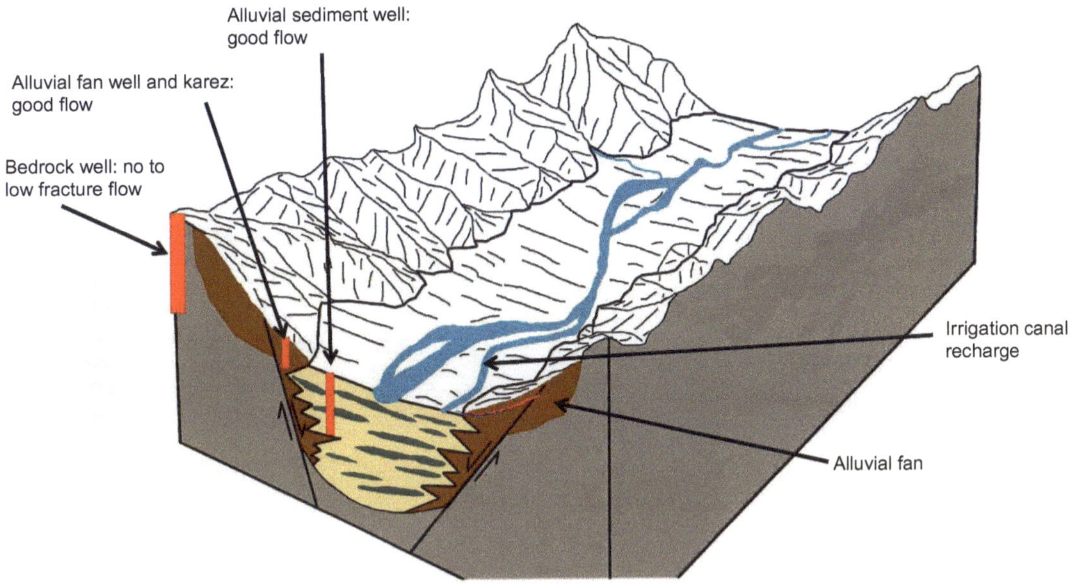


FIG. 3.14 Narrow alluvial aquifer in mountain valley underlain by a graben structure typical of Afghanistan. Note lenticular zones of water-bearing sands and gravels in basin center where ancient river channels sediments have been deposited.

- Within the southern and northern plains of Afghanistan, fresh water is generally restricted to river valley-fill aquifers of the lower Helmand, Northern, Amu Darya, and Murghab-Kushk wa Kaskan river basins (Fig. 3.14). Typically, only brackish ground water occurs outside of the river valleys in these areas.
- Carbonate rock aquifer systems occur on the northern flank of the Hindu Kush Mountain Range and along portions of the Helmand River in the Oruzgan Province.
- Limited fractured bedrock aquifers supply springs at higher elevations.

Figs. 3.13–3.15 show cross sections through different sections of the river valleys and associated underlying aquifers. For many rivers in Southern Afghanistan, thin alluvial aquifers in narrow valleys occur in the upper mountainous portion of the watershed, with transitions to fully fault-bounded, intermontane basins downstream, to wedge-shaped basins as the river flows out to the southern plains, and to river valley-fill aquifers in the most downstream portions of the watersheds.

Ground-water quality also follows this pattern. According to the published hydrogeological map (Anon., 1976), the majority of the territory of Afghanistan, especially around the mountainous central massif, is underlain by fresh ground water with total dissolved solids (TDS) concentrations of less than 1000 mg/L, with a hydrochemical composition typically of calcium-magnesium bicarbonate, sodium-calcium bicarbonate, or calcium-sulfate-bicarbonate. In low-lying desert areas away from the mountains, some tendency to salinization occurs, with TDS concentrations of greater than 1000 mg/L, with recorded

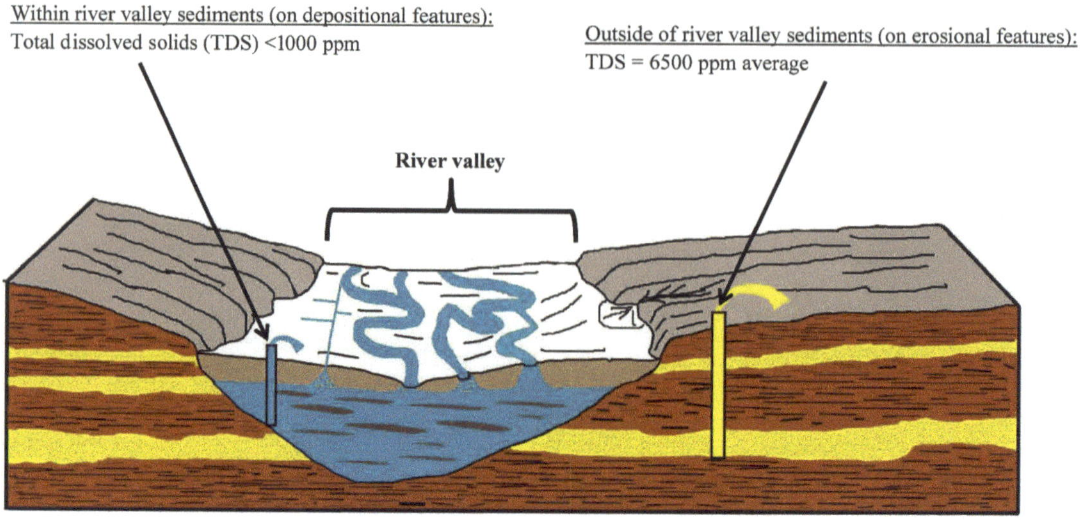


FIG. 3.15 Bank-to-bank only, alluvial aquifer in river valley.

concentrations up to approximately 3000 mg/L (Banks and Soldal, 2002). This is particularly the case that occurs in three areas:

- the deserts in the southwest of the country: the southern Kandahar, Helmand and Nimroz Provinces;
- the deserts in the extreme west of the country: the western Farah and Herat Provinces; and
- the lowlands of the Northern Platform.

Only scarce documentation exists regarding the ground-water development potential of the bedrock aquifer systems in Afghanistan. The yield potential of the crystalline rocks (granites, schist, gneiss, etc.) is expected to be significantly lower than the unconsolidated aquifer systems. Other sedimentary and igneous rock units, which underlie large parts of the country may have development potential, but as yet have not been explored in any detail.

## 3.3.1 North Afghan Platform Seismotectonic Province Ground Water

The Northern, Amu Darya, and Murghab-Kushk wa Kaskan[1] watersheds occur in this block and fresh-water aquifers are mostly confined to the area between the river valley banks and fresh water zones close to mountain recharge zones (UN, 1986). All the wells have been constructed along the river valley floors and directly adjacent areas. Wells attempted in the

[1] Considered to be also part of the Hari-Rud watershed since the rivers of the Murghab-Kushk wa Kaskan join the Hari-Rud downstream in Turkmenistan, but included here as part of the northern watersheds since the Murghab-Kushk wa Kaskan watershed shares common geologic, hydrologic, and geochemical characteristics with the Northern-Amu Darya watersheds.

Tertiary strata and hillsides above the valley floors have been dry or have low yields of highly brackish water. Alluvial aquifers beneath Baghlan and Mazar-i-Sharif may occur within structural basins (Fig. 3.12) as shown by Quaternary faulting along the basin margins (Ruleman et al., 2007).

Recharge in the area was described by Banks (2014) as:

> rainfall recharges aquifers in the hilly and mountainous zones in the south of the Province. These aquifers in turn provide spring discharge and ground-water baseflow to the main river systems draining north (Almar, Qaysar, Maimana, Shirin Tagab) and east (Chechaktu). The rivers also, of course, carry seasonal fluxes of snowmelt and rainfall. As these river systems pass from the pre-Neogene rocks in the south, onto the Neogene and Quaternary terrain towards the north, the ground-water levels are relatively deep and the rivers commence infiltration of recharge into the Neogene and Quaternary aquifers. Although ground-water levels become shallower as the terrain falls towards the north, ground-water levels are still typically 15 to 40m below ground level along much of the Maimana and Shirin Tagab valleys… Recharge by infiltration of river water persists along much of the Maimana and Shirin Tagab valleys, as far as Andkhoi. A zone of fresher ground-water thus exists along the river corridors of the main rivers.

Also, Banks (2014) noted that direct infiltration of precipitation is severely limited over much of the Faryab Province with estimates of potential direct recharge in excess of 100mm/a in mountainous areas, a few tens of mm/a in foothill areas, to <10mm in the Neogene/Quaternary sections.

Oil and gas exploration drilling in this area has indicated that both shallow and deeper aquifers contain brackish and saline ground water. United Nations (1986) noted that in most of the low-lying plains area between the Amu Darya River to the north and the line of the towns/cities of Dawlatabad-Sheberghan-Balkh-Khulm in the south, conditions are not favorable for ground-water development. The Balkh River valley in the northern basin is filled with fluvial deposits saturated with fresh ground water and may be a structural basin. Two aquifers are of significant productivity: Sholgara in the south and Mazor-Balkh as where the river exits from the mountains. The water supply system of Mazar-i-Sharif is based on deep wells drilled into the Balkh River aquifer. The water is of good quality only where fed by infiltration of the Balkh River water. The Saripul River valley is filled with alluvial-colluvial deposits (sand, gravel, cobbles and boulders) saturated with fresh water from Saripul to Sheberghan from where the Sheberghan water supply originates. The Shirin Tagab River valley and its Qaysar River tributary have fluvial sand and gravel saturated with fresh ground water. The water supply systems for Maimana, Shirin Tagab, and Dawlatabad are based on wells tapping this aquifer.

United Nations (1986) reports that carbonate rocks of Upper Jurassic-Cretaceous and Upper Cretaceous-Paleocene (Northern Basin) and Miocene age occur within the Hindu Kush and along its northern flanks. The carbonate massif to the north (Northern watershed and the western part of the Amu Darya watershed) is comprised of limestone and dolomite locally interbedded with sandstone and conglomerate. Sinkholes, caves and caverns also occur. Significant springs from ground-water base flow issue from the carbonate massif on the northern flank of the Hindu Kush mountain range and a number of these springs form the headwaters of the rivers in north central and northwest Afghanistan (Uhl and Tahiri, 2003).

### 3.3.2 Hari Rud Geosuture Seismotectonic Province Ground Water

Two major aquifer types occur within the Hari Rud River valley graben within the Geosuture Block:

- thin alluvial sediment aquifers in narrow valleys that occur between steep valley walls of the mountain massif within the upper reaches of the Hari Rud River, including the area of Chaghcharan; and
- enclosed intermontane basin and valley aquifers of the city of Herat.

### 3.3.3 Afghan Block Seismotectonic Province Ground Water

Four major aquifer types occur within the Afghan Block:

- thin alluvial sediment aquifers in narrow valleys that occur between steep valley walls of the mountain massif and include middle and upper reaches of mountain-fed rivers of the region;
- enclosed intermontane basins including the Tarin Kowt basin that drains to the southwest to the Helmand River and the Dasht-i-Nawar basin that is a fully enclosed structural basin (extinct volcanic caldera) with no outlet;
- these same rivers flow southward into wedge-shaped structural half-graben basins that open to southern plains beneath the city of Kandahar and the middle Helmand, Tarnak (Qalat), Khash, Adraskan, Arghandab, Harut, and Farah Rivers; and
- within the southern plains, fresh water is commonly restricted to river-valley-fill aquifers of the lower Helmand River and lower reaches of other river basins.

The Registan Desert of southern Afghanistan is a vast sand desert, comprised primarily of both stabilized and active eolian sand dunes and is an unsettled area with very little ground-water exploration work conducted (Fig. 3.16). Given the sparse rainfall and high evapotranspiration rates, ground water in this desert region is expected to be primarily brackish and saline. Numerous small interdune depressions or *nawars* occur throughout the region where finer-grained sediments have been deposited as a result of ephemeral fluvial reworking of sand-dune sediment or deposition and reworking of wind-blown loessic dust. Known as dry-lake playas in the western USA, these similar Registan nawars, because of their impermeability, may briefly hold shallow surface waters following rare rainstorms that quickly evaporate before much infiltration is possible downward to recharge the ground water. The extreme eastern area is characterized by igneous rocks and some Paleocene sandstones, silts, and conglomerates along the Pishin Lora River valley. The far eastern Lora Valley, in contrast to the desert, reportedly contains limited quantities of fresh ground water (Uhl and Tahiri, 2003) (Fig. 3.17).

The Dasht-i-Margo Desert to the west of the Registan Desert is also largely unexplored. The ground-water development potential is probably expected to be good along the Helmand River Valley, and possibly in the limestone formations in the very extreme south of the Afghanistan along the border with Pakistan (Chaghay Area).

Uhl and Tahiri (2003) noted that the Tarnak River valley is a tectonic trough stretching northeast to southwest and is filled with Quaternary and Neocene deposits with multi-layered aquifers. Detailed geophysical and hydrogeological investigations were carried out in the 1970s under a UNDP program in the upper Tarnak River valley. A substantial amount of ground water is held in storage in two river valley aquifer systems, the Shinkay

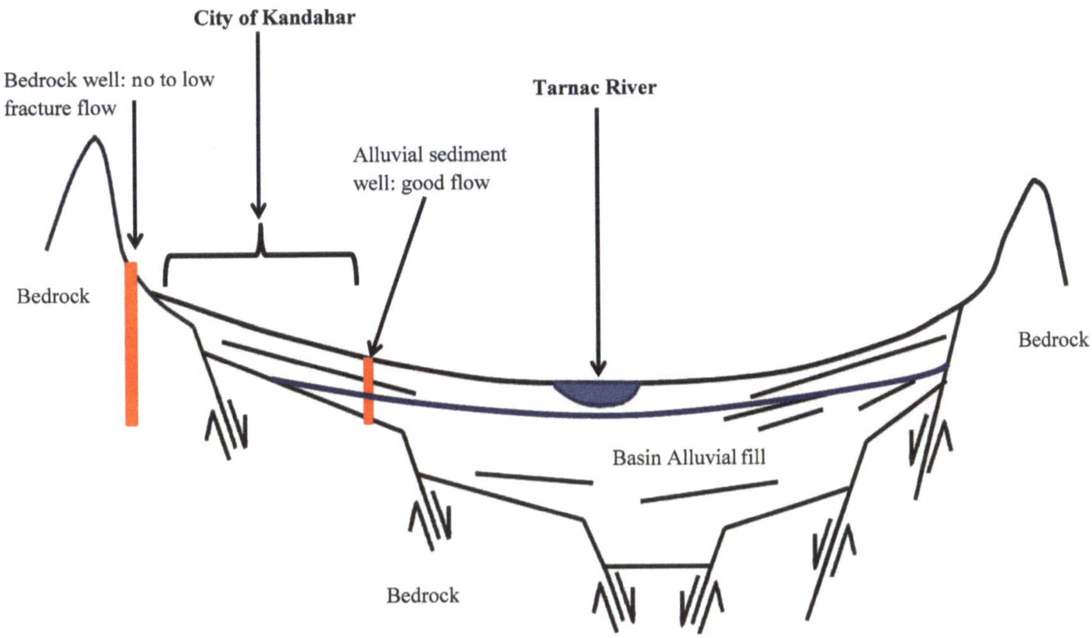


FIG. 3.16 Generalized cross-section of alluvial ground-water basin in multiple-step graben structure.

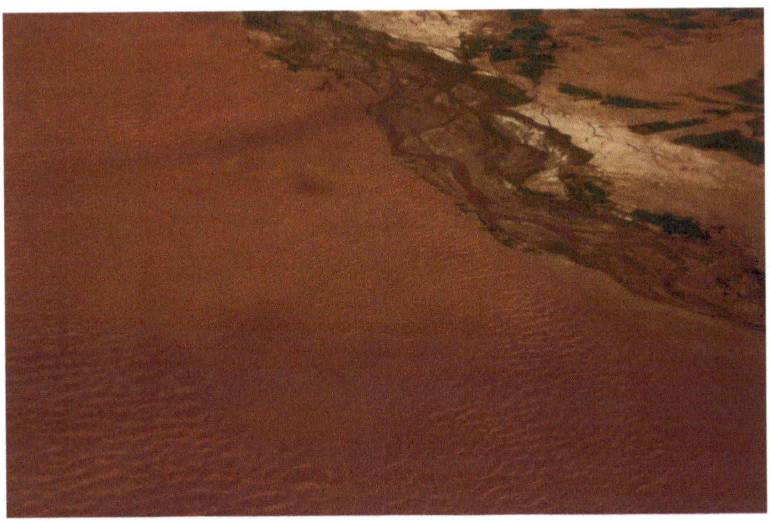

FIG. 3.17 Photo of the Registan Desert with the braided Dori River forming the boundary between the red dune fields of the Registan Desert and the agricultural fields of Kandahar Province to the north. *Photo by L. Sinfield, 2012.*

and the Sengasi Moqur, with the Shinkay Aquifer containing a volume of over 1400 million $m^3$ of water stored in the first 50 m of saturation (see section below on flowing Qalat artesian well).

A Quaternary aquifer system underlies the Dasht-i-Nawar volcanic caldera basin. Fractured volcanics invariably overlying the older deposits are water bearing. In valleys and the undrained depression of Dasht-i-Nawar, the volcanics are overlain by younger alluvium and alluvial-fan deposits. The discharges of a few springs vary from 0.2–3 to 5–18 L/s. The maximum discharge of 400 L/s has been recorded from Saraw Spring, with ground water discharging in isolated streams with yields of up to 40 L/s over a distance of 300 m. The yield of karezes in the southern and eastern parts of the superimposed basin ranges from 3 to 15 L/s. The water is of a bicarbonate type with a mixed cation composition, the dissolved-solids content is 0.2–0.5 g/L (Uhl and Tahiri, 2003).

### *3.3.3.1 Kandahar*

As with the aquifers in the Kabul area, the alluvial aquifers beneath the city of Kandahar also occur within an extensional and translational, fault-bounded basin. Recent drilling and installation of municipal water wells in the city indicate the presence of fault-bounded, deep alluvial aquifers beneath. Camp Dresser and McKee (CDM) (2003) defined three unconfined aquifers beneath Kandahar, referred to as the Shallow, Upper, and City Aquifers. Both the United Nations Habitat (at the Kandahar Public Works yard) and Afghan Urban Water Supply and Sewage Corporation (AUWSSC) found a fourth deeper aquifer (Deep Aquifer) indicated on cross sections A-A′ through D-D′ (Figs. 3.18–3.22). A subsurface

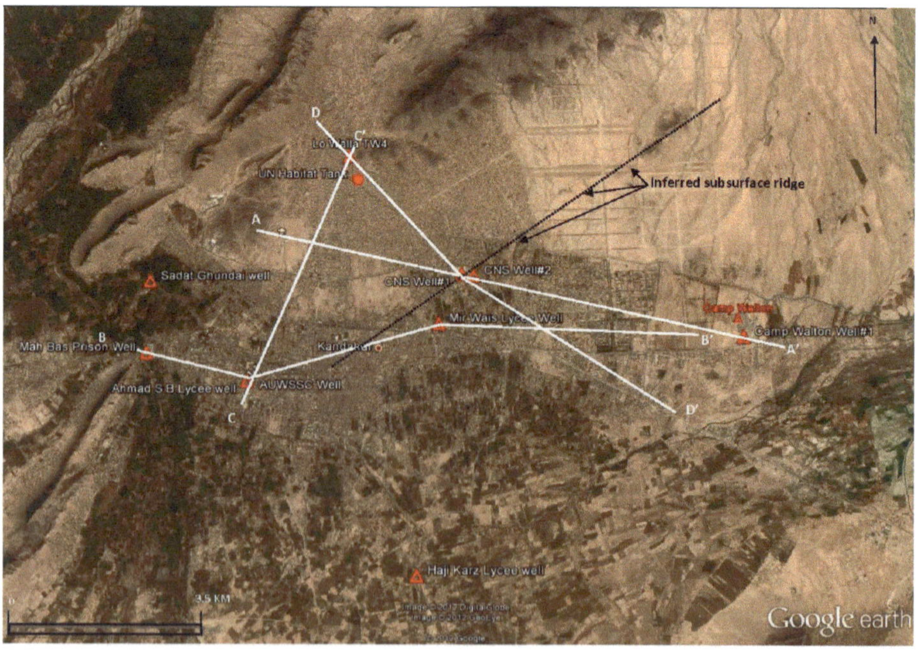


FIG. 3.18 Cross-section location map of Kandahar city plotted on Google earth image.

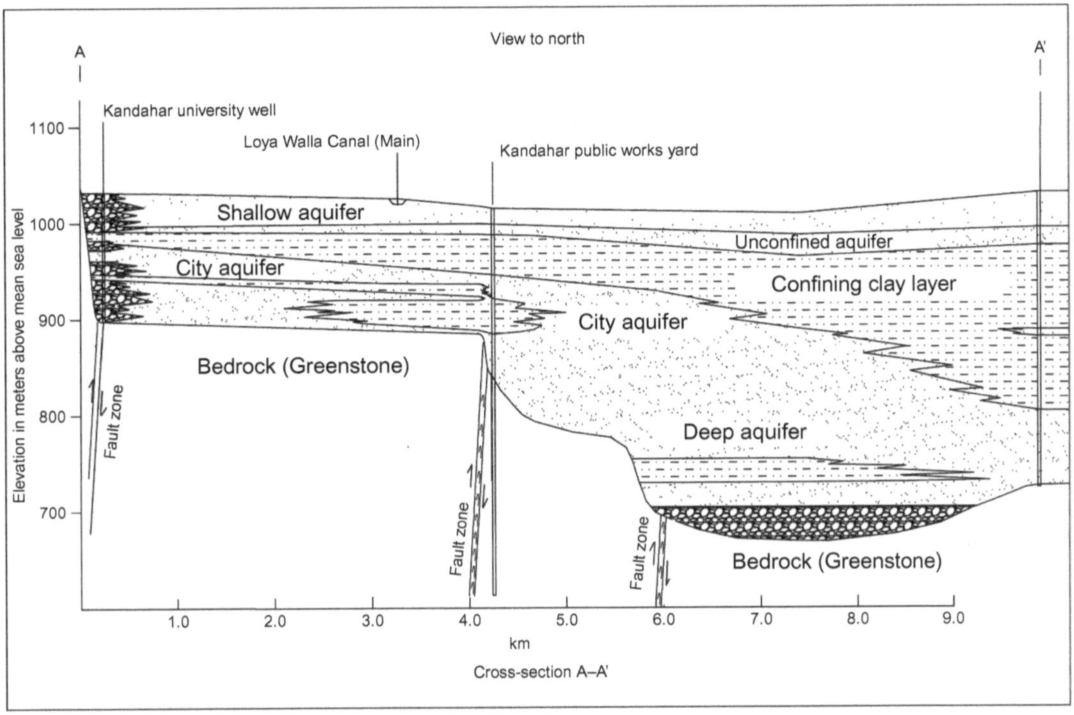


FIG. 3.19 Kandahar city cross section A-A′.

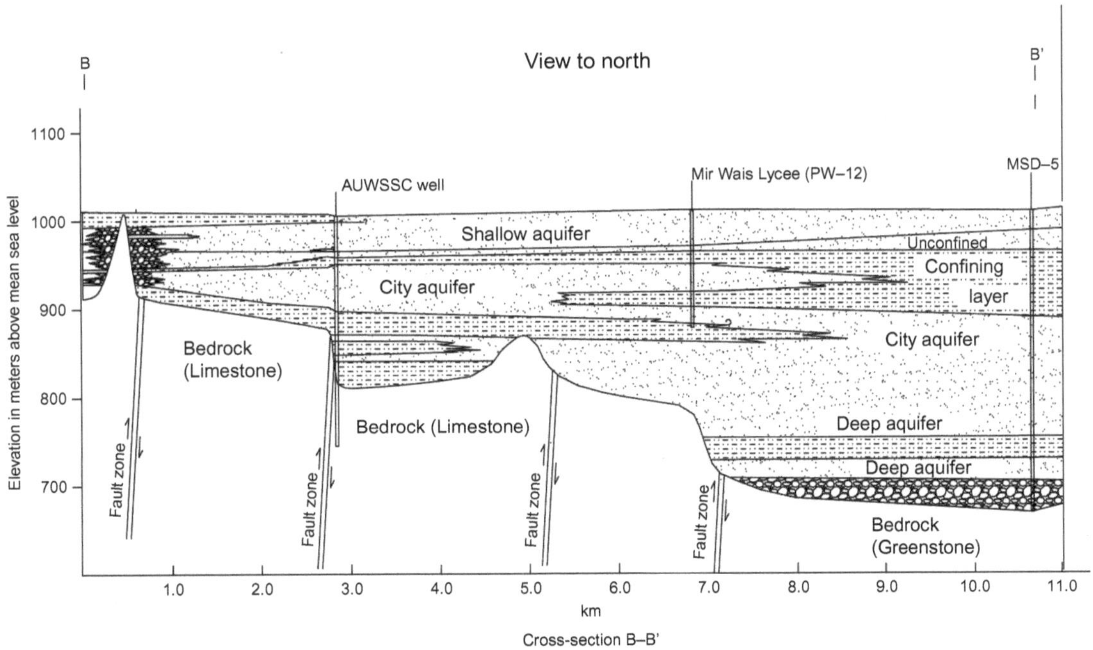


FIG. 3.20 Kandahar city cross section B-B′.

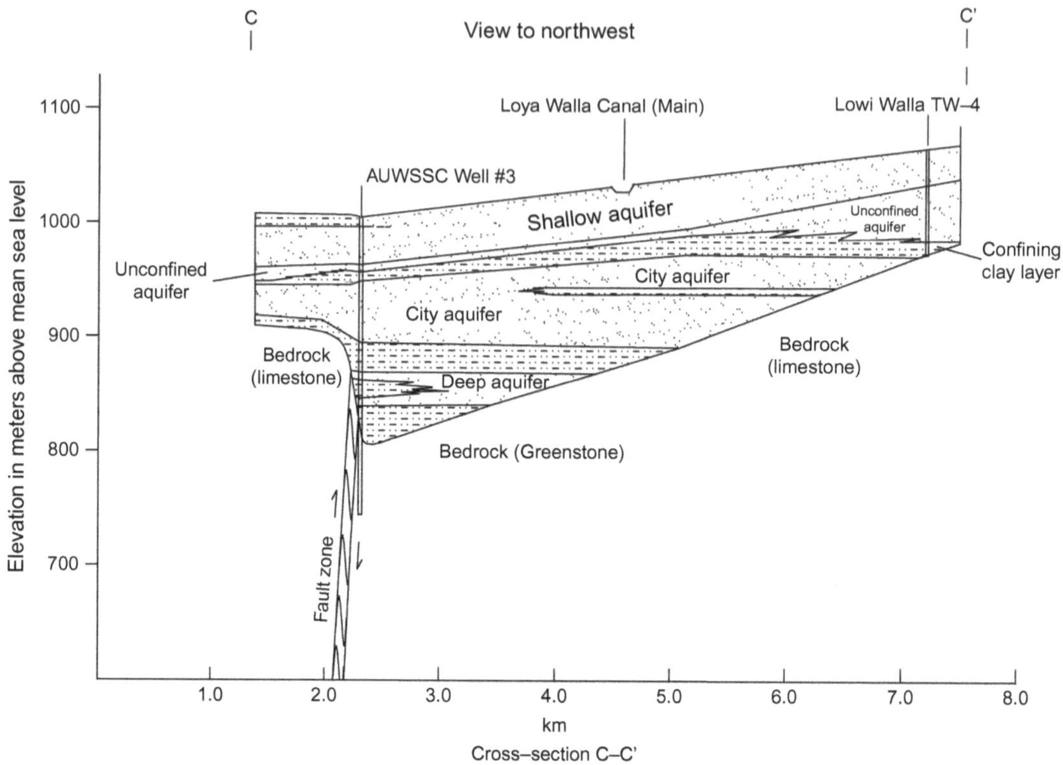


FIG. 3.21 Kandahar city cross section C-C′.

northeast-trending bedrock "step" marks the edge of the deeper alluvial aquifer to the south. According to CDM (2003), leaking local canals built by the United States in the 1950s within the city, are the main source of recharge for the Shallow and Upper aquifers (Fig. 3.22). The Deep Aquifer may be recharged by the Tarnak River to the northeast.

As shown in Fig. 3.18, the canal system crosses near locally steep-sided folded hills surrounding Kandahar with coarse boulder-type colluvium along the base of each hill. As shown by isoconcentration contours of TDS in the Shallow Aquifer (Fig. 3.23), these canals may recharge the shallow aquifer (Fig. 3.24). The TDS contours may also indicate that salinization of shallow aquifers by agricultural drainage may also be occurring.

## 3.3.4 Transpressional Plate Boundary Block

The Indus watershed occurs within the Transpressional Plate Boundary Geologic Block, which is part of the upper watershed of the Indus River of Pakistan, and includes nine watersheds: Kabul, Chak wa Logar Rod, Ghorband wa Panjshir, Alingar, Kunar, Sardihwa Ghazni Rod, Shamal (Khost), Gomal, and Pishin Lora (Favre, 2004). The headwaters of the Arghistan are also in the block, but it drains to the Helmand River downstream.

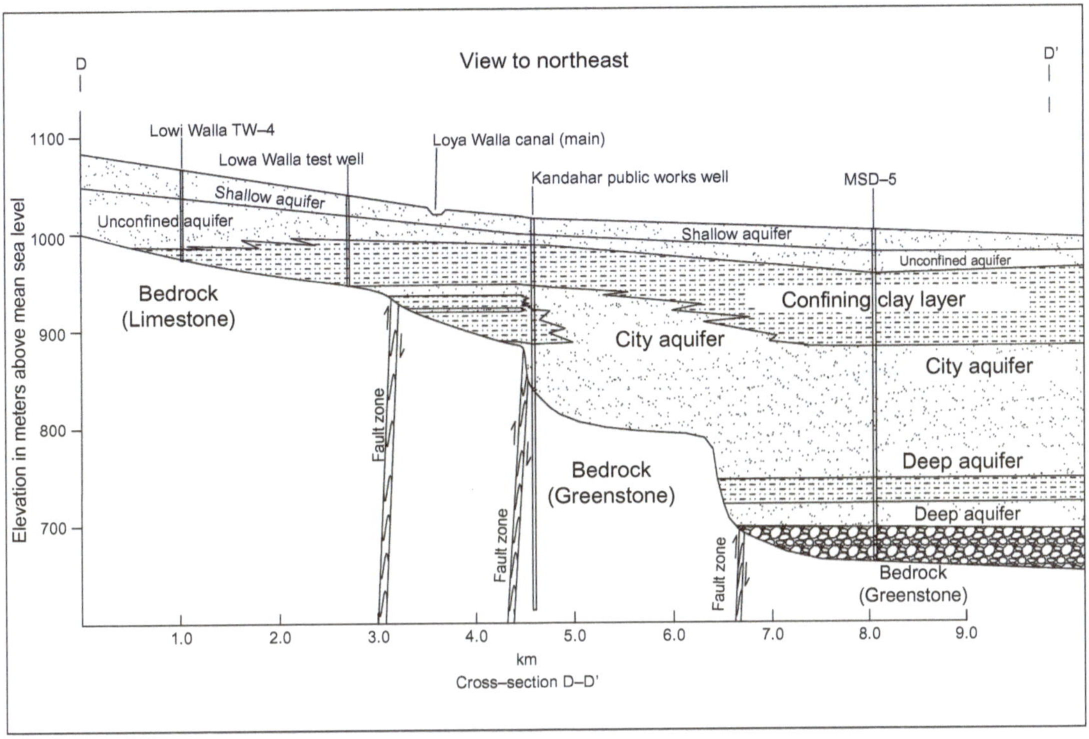


FIG. 3.22 Kandahar city cross section D-D′.

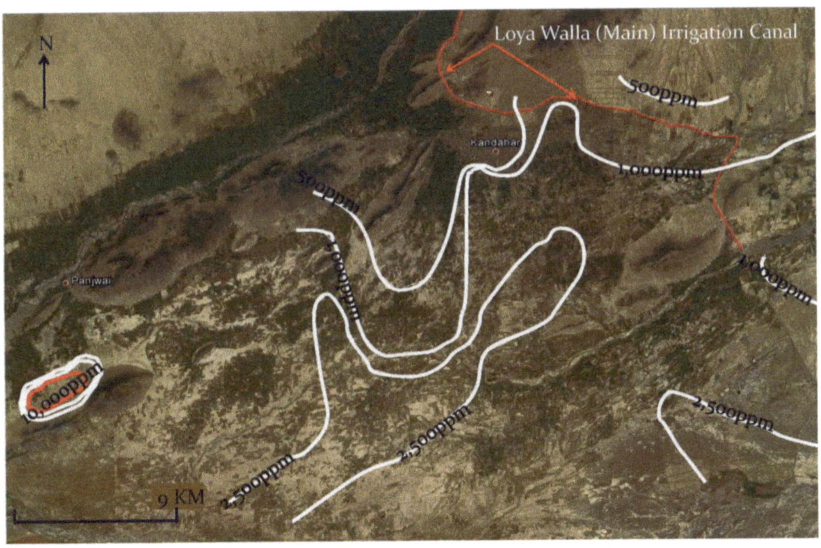


FIG. 3.23 Total dissolved solids isoconcentrations in shallow ground water near Kandahar city. Plotted on basemap image from Google Earth.

FIG. 3.24 Main Loya Walla canal that was built by the United States in Kandahar city in the 1950s. This canal and other canals in the city may be the main source of recharge for the Shallow aquifer. *Photo by L. Sinfield, 2012.*

Three types of aquifers occur in this block:

- intermontane structural graben basins filled with Quaternary and Neogene alluvial sediments, including basins beneath Kabul, Panjshir, Logar, Jalahabad, and Khost watersheds. One intermontane basin is enclosed with no outlet, the Ghazni Rod basin;
- thin alluvial sediment aquifers in narrow valleys that occur between steep valley walls consisting of folded Proterozoic to Tertiary igneous, metamorphic and sedimentary rocks of the upper reaches of the Logar, Kunar, Panjshir, Kabul, and other rivers of the block; and
- limited fractured bedrock aquifers.

The four Quaternary and Neocene sediment aquifer systems within the Kabul Basin (Upper Kabul River Aquifer, Paghman River Aquifer, Lower Kabul River Aquifer, and the Logar River Aquifer) are described elsewhere in this book (see Chapter 4).

### 3.3.4.1 *Jalahabad Basin*

The Jalalabad Basin is situated in the southern part of the Nangrahar Province. The UN (1986) noted that the aquifer extends from Daronta in the west to Lowyah Dakfah in the east and can be subdivided into two parts. One part extends along the course of the Kabul River and is composed of fluvial sand and gravel. The second part along the margins of the basin consists of gravel, cobbles, boulders, and conglomerate in a loam-sand matrix, which varies greatly in thickness and yield and is difficult to drill through.

### 3.3.4.2 *Darya-i-Logar Basins*

The US Army Corps of Engineers (USACE, 2002) noted at the mouth of the Darya-i-Logar Valley, the alluvium constitutes a belt 3000–4000 m wide and 100–150 m thick. In the middle reaches of the river, near Safed Sang, the deposits range in thickness from first few meters to more than 40 m. The aquifer is unconfined and water levels range from 0.8 to 17.25 m. The unit ground-water discharge at Kandwana is 230 L/s with specific yields of wells varying between 0.3 and 12 L/s. Coefficients of permeability range from 60 to 120 m/day, in some places up to 540 m/day. The water is fresh, of the bicarbonate and bicarbonate-magnesium types, with the dissolved-solids content of 0.7–2.0 g/L and temperature ranging from 13 to 17°C.

#### 3.3.4.3 *Shamal (Khost) and Gomal Basins*

The Shamal (Khost) and Gomal basins occur in the southern portion of the transpressional boundary area and their rivers connect with the Indus downstream of Afghanistan in Pakistan. No ground-water data are available for these areas.

#### 3.3.4.4 *Sardihwa Ghazni Rod (Nawar Depression) Basin*

According to Uhl and Tahiri (2003), the Sardihwa Ghazni Rod (Nawar Depression) is an enclosed fresh water basin with no outlet covering an area of about 1600 $km^2$ and much of the low-land area of this basin is underlain by Quaternary deposits. This closed basin represents an important ecological resource as it is located along a major flyway for migratory birds. It is our understanding that very little ground water is used for irrigation purposes in this closed basin. Any development of ground water within this basin should be planned carefully and ecological water needs should be a major consideration.

## 3.4 GROUND-WATER GEOCHEMISTRY

Ground-water geochemistry data are provided by Saffi (2007) of DACAAR. These data indicate ground-water geochemistry for most of the alluvial ground-water basins are dominated by the sodium bicarbonate with some magnesium bicarbonate ground water. Piper plots (Figs. 3.25–3.28) indicate that fresh magnesium bicarbonate ground

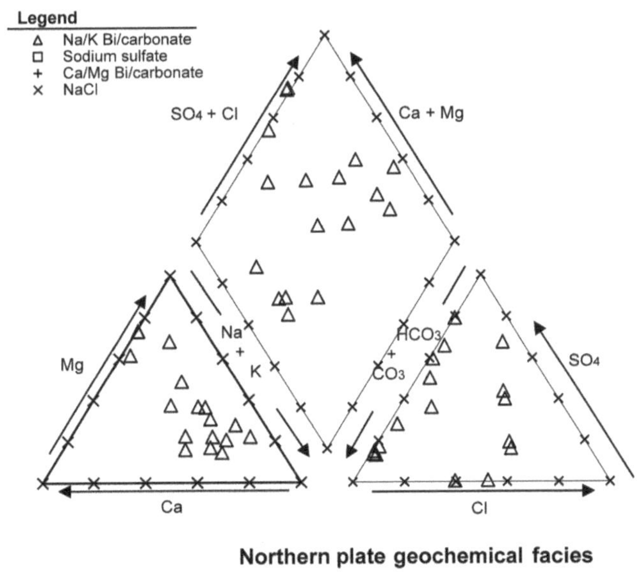


FIG. 3.25 Piper diagram of ground-water geochemistry of North Afghan Block.

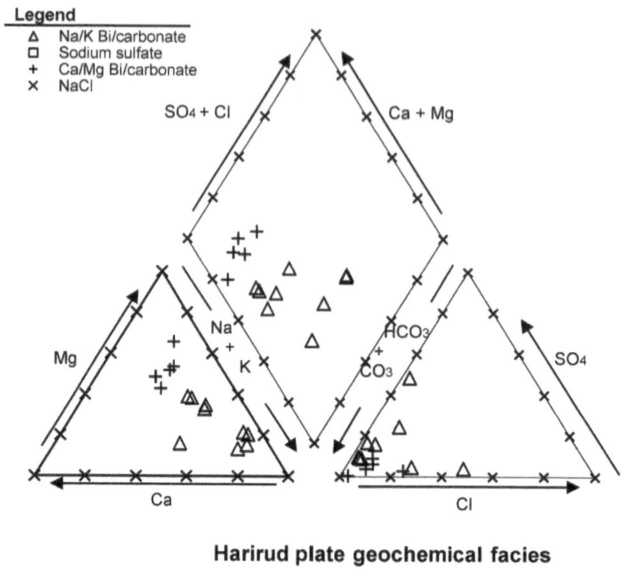


FIG. 3.26 Piper diagram of ground-water geochemistry of Hari Rud Block.

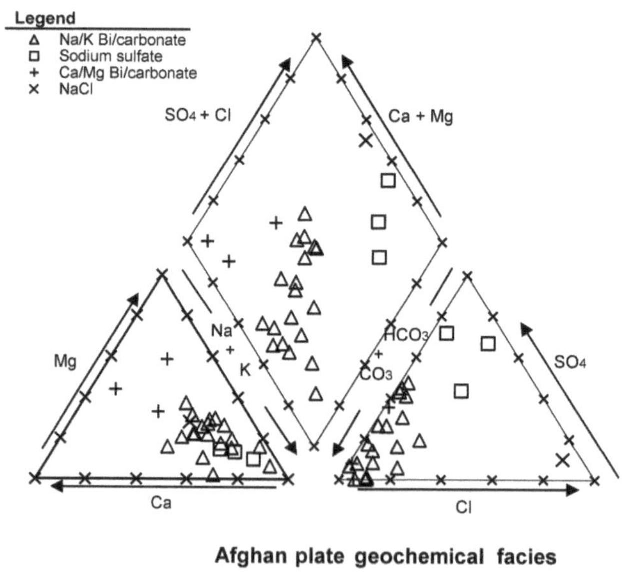


FIG. 3.27 Piper diagram of ground-water geochemistry of Afghan Block.

water trends toward the sodium apex (sodium bicarbonate), indicating either ion exchange with montmorillonite clays or evaporative processes, adding and/or concentrating sodium and the removing calcium. Magnesium probably originates from the dolomite and andesitic rocks in the region.

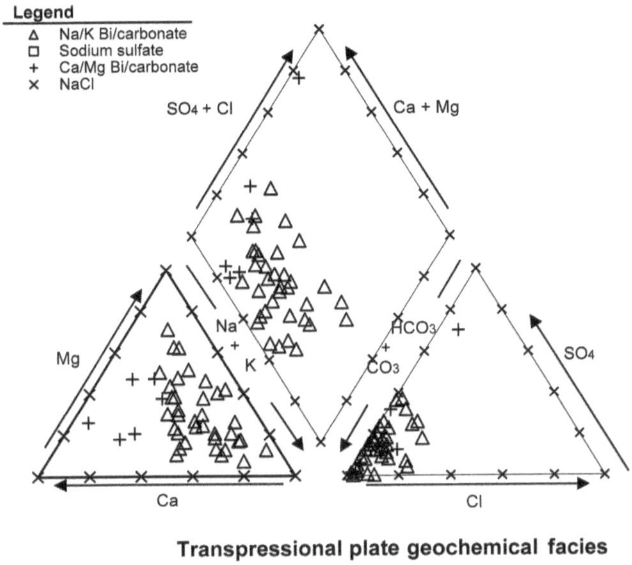


FIG. 3.28 Piper diagram of ground-water geochemistry of Transpressional Block.

## 3.4.1 North Afghan Platform Seismotectonic Province Ground-Water Geochemistry

Ground water in the North Afghan Platform seismotectonic province is predominately brackish with a TDS median of 1756 ppm, and ranging from 150 to 4599 ppm (Table 3.1; Fig. 3.25). Banks (2014) found that "the main mechanism for salt accumulation in both rivers and ground water in the central and northern parts of Faryab is by evaporative processes—including the evapotranspiration of water used for irrigation and the periodic flushing of evapoconcentrated, rainfall-borne and wind-blown salts from the soil zone. We fully acknowledge that other processes may be important, including dissolution of paleoevaporite salts, such as halite or gypsum, from Neogene and other deposits."

In Qala-e-Naw, a few ground-water measurements collected during the rainy season found wells with fresh water, but by June, these same wells had brackish water, indicating a seasonal fresh-water lens may form during the period of peak rivers flows in adjacent dry washes and rivers. Bacteriological contamination is also high, with 43% of the ground-water samples testing positive for total coliform, which is indicative of ground water under the influence of surface water. Linked with seasonal recharge of the shallow aquifer, ground waters in these river valleys are highly subject to bacteriological contamination from the typical open defecation to sewage dumping along the rivers.

TDS, sulfate, boron, and fluoride (as $F^-$) exceeded recommended limits of the World Health Organization (WHO) in over 50% of the ground-water samples, which may be the result of the high evaporate content of the geologic material of the province. During the peak period of the rainy season (February to April), rivers have their highest flows with fresh water.

TABLE 3.1 North Afghan Platform Ground-Water Geochemistry Summary

| Parameter | Range | Median | WHO-Recommended Limit | Percentage Exceeding WHO Limit |
|---|---|---|---|---|
| pH | 6.0–8.6 | 7.6 | 6.5–8.0 | 26% above 8.0 |
| Total hardness (as Ca) | 235–1328 ppm | 287 ppm | 300 ppm | 28% |
| Total alkalinity (as $CaCO_3$, Alkaphot) | 145–750 ppm | 250 ppm | – | – |
| Total dissolved solids | 150–4599 ppm | 1756 ppm | 1000 ppm | 58% |
| Sulfate | 4–5412 ppm | 370 ppm | 250 ppm | 59% |
| Sodium | 1–788 ppm | 104 ppm | 200 ppm | 26% |
| Chloride | 0–9318 ppm | 177 ppm | 250 ppm | 43% |
| Nitrate | 1–58 ppm | 6.3 ppm | 50 ppm | 4% |
| Nitrite | 0–0.2 ppm | 0 ppm | 0.2/0.3 ppm | 4% |
| Boron (as $BO_2^-$) | 0.09–1.7 ppm | 0.65 ppm | 0.5 ppm | 56% |
| Bromide (as $Br^-$) | 0.02–0.94 ppm | 0.13 ppm | – | – |
| Fluoride ($F^-$) | 0.28–7.1 ppm | 2.3 ppm | 1.5 ppm | 54% |
| Iron (total) | 0–1.4 ppm | 0.21 ppm | 0.3 ppm | 18% |
| Manganese | 0–11.5 ppm | 0.01 ppm | 0.4 ppm | 20% |
| Total coliform | Absent to present | Absent | Any detection is over limit | 43% (present) |
| Fecal coliform | Absent to present | Absent | Any detection is over limit | 39% (present) |

## 3.4.2 Hari Rud Geosuture Seismotectonic Province Ground-Water Geochemistry

Analytical data from well samples collected in the Hari Rud geosuture province (Table 3.2) indicate that water is predominately fresh with TDS median of 639 ppm and ranging from 186 to 2368 ppm. TDS, sulfate, sodium, chloride, boron (as $BO_2^-$), and fluoride (as $F^-$) only exceeded the WHO-recommended limits in 13–29% of the ground-water samples and as with the North Afghan Platform, may indicate some input from evaporites or mixing with geothermal waters.

## 3.4.3 Afghan Block Seismotectonic Province Ground-Water Geochemistry

Analytical data from well samples collected in the Afghan Block seismotectonic province (Table 3.3) indicate that ground water within this watershed is predominately fresh with a TDS median of 791 ppm, and ranging from 196 to 9727 ppm. The highest TDS concentrations occur in the southern part of Helmand Province outside of the river valley floor. TDS, sulfate, sodium, chloride, boron (as $BO_2^-$), and fluoride (as $F^-$) exceeded the WHO-recommended limits in only 16–41% of the ground-water samples, with sodium having the highest number of exceedances (41%).

TABLE 3.2 Hari Rud Geosuture Ground-Water Geochemistry Summary

| Parameter | Range | Median | WHO-Recommended Limit | Percentage Exceeding WHO Limit |
|---|---|---|---|---|
| pH | 7.0–8.7 | 7.80 | 6.5–8.0 | 32% above 8.0 |
| Total hardness (as Ca) | 83–200 ppm | 93 ppm | 300 ppm | 0% |
| Total alkalinity (as $CaCO_3$, Alkaphot) | 65–270 ppm | 190 ppm | – | – |
| Total dissolved solids | 186–2368 ppm | 639 ppm | 1000 ppm | 12% |
| Sulfate | 11–530 ppm | 180 ppm | 250 ppm | 27% |
| Sodium | 54–374 ppm | 103 ppm | 200 ppm | 29% |
| Chloride | 3–512 ppm | 72 ppm | 250 ppm | 22% |
| Nitrate | 1–168 ppm | 10.6 ppm | 50 ppm | 13% |
| Nitrite | 0–0.1 ppm | 0.00 ppm | 0.2/0.3 ppm | 0% |
| Boron (as $BO_2^-$) | 0–0.6 ppm | 0.45 ppm | 0.5 ppm | 21% |
| Bromide (as $Br^-$) | 0–0.6 ppm | 0.05 ppm | – | – |
| Fluoride ($F^-$) | 0–2.8 ppm | 0.95 ppm | 1.5 ppm | 14% |
| Iron (Total) | 0–0.25 ppm | 0.01 ppm | 0.3 ppm | 0% |
| Manganese | 0–0.10 ppm | 0.002 ppm | 0.4 ppm | 0% |
| Total coliform | Absent to present | Absent | Any detection is over limit | 19% (present) |
| Fecal coliform | Absent to present | Absent | Any detection is over limit | 33% (present)[a] |

[a]*One of three samples collected.*

TABLE 3.3 Afghan Block Ground-Water Geochemistry Summary

| Parameter | Range | Median | WHO-Recommended Limit | Percentage Exceeding WHO Limit |
|---|---|---|---|---|
| pH | 6.8–8.6 | 7.80 | 6.5–8.0 | 24% above 8.0 |
| Total hardness (as Ca) | 12–390 ppm | 139 ppm | 300 ppm | 7% |
| Total alkalinity (as $CaCO_3$, Alkaphot) | 8–1500 ppm | 230 ppm | – | – |
| Total dissolved solids | 196–9727 ppm | 791 ppm | 1000 ppm | 37% |
| Sulfate | 1–2020 ppm | 178 ppm | 250 ppm | 29% |
| Sodium | 7–2080 ppm | 159 ppm | 200 ppm | 41% |
| Chloride | 1–1320 ppm | 43 ppm | 250 ppm | 27% |
| Nitrate | 0–81 ppm | 4.0 ppm | 50 ppm | 2.2% (1 sample) |
| Nitrite | 0–4.3 ppm | 0.01 ppm | 0.2/0.3 ppm | 13% (above 0.3 ppm) |

*(Continued)*

TABLE 3.3 Afghan Block Ground-Water Geochemistry Summary—Cont'd

| Parameter | Range | Median | WHO-Recommended Limit | Percentage Exceeding WHO Limit |
|---|---|---|---|---|
| Boron (as $BO_2^-$) | 0–1.4 ppm | 0.30 ppm | 0.5 ppm | 20% |
| Bromide (as $Br^-$) | 0–2.9 ppm | 0.10 ppm | – | – |
| Fluoride ($F^-$) | 0–11 ppm | 0.75 ppm | 1.5 ppm | 16% |
| Iron (Total) | 0–2.0 ppm | 0.02 ppm | 0.3 ppm | 2.5% (1 sample) |
| Manganese | 0–1.0 ppm | 0.00 ppm | 0.4 ppm | 2.3% (1 sample) |
| Total coliform | Absent to present | Absent | Any detection is over limit | 20% (present) |
| Fecal coliform | Absent to present | Absent | Any detection is over limit | 11% (present) |

TABLE 3.4 Transpressional Plate Boundary Ground-Water Geochemistry Summary

| Parameter | Range | Median | WHO-Recommended Limit | Percentage Exceeding WHO Limit |
|---|---|---|---|---|
| pH | 6.5–9.0 | 7.80 | 6.5–8.0 | 31% above 8.0 |
| Total alkalinity (as $CaCO_3$, Alkaphot) | 26–580 ppm | 163 ppm | – | – |
| Total dissolved solids | 173–1445 ppm | 459 ppm | 1000 ppm | 5% |
| Sulfate | 1–1240 ppm | 86 ppm | 250 ppm | 2% |
| Sodium | 23–336 ppm | 100 ppm | 200 ppm | 9% |
| Chloride | 2–888 ppm | 22 ppm | 250 ppm | 6% |
| Nitrate | 0–19 ppm | 2.3 ppm | 50 ppm | 0% |
| Nitrite | 0–0.2 ppm | 0.00 ppm | 0.2/0.3 ppm | 0% |
| Boron (as $BO_2^-$) | 0.1–3.9 ppm | 0.30 ppm | 0.5 ppm | 21% |
| Bromide (as $Br^-$) | 0–0.78 ppm | 0.09 ppm | – | – |
| Fluoride ($F^-$) | 0–2.1 ppm | 0.50 ppm | 1.5 ppm | 2% |
| Iron (Total) | 0–0.9 ppm | 0.00 ppm | 0.3 ppm | 2% |
| Manganese | 0–1.3 ppm | 0.00 ppm | 0.4 ppm | 6% |
| Total coliform | Absent to present | Absent | Any detection is over limit | 8% (present) |
| Fecal coliform | Absent to present | Absent | Any detection is over limit | 25% (present) |

### 3.4.4 Transpressional Plate Boundary Seismotectonic Province Ground-Water Geochemistry

Transpressional Plate Boundary province ground water is predominately fresh (Table 3.4) with a TDS median of 459 ppm, and ranging from 173 to 1445 ppm. Of these parameters, boron (as $BO_2^-$) had the highest number of exceedances (21% of the samples), with the rest remainder of the parameters having exhibiting very low exceedance rates. Boron detected in the ground-water samples may be have originated from the erosion of saline-derived sources rocks, or may be associated with geothermal activity in the area.

## 3.5 NATURAL AND HUMAN-CAUSED CONTAMINANTS IN GROUND WATER IN AFGHANISTAN

Numerous visitors over the years in Afghanistan have noted the many problems of obtaining pure water free from natural or human-caused contamination. Now that Kabul has a population estimated to be ~6 million (Rasmussen, 2014) because of centripetal migration into a seemingly safer city environment, the need to look after the shallow ground water upon which so many people depend especially is growing progressively more acute. A number of studies of ground and surface water have been done by different foreign-aid agencies trying to provide high quality technical data on the many contamination and overdraft problems (Nembrini, 2002; Broshears et al., 2005; Houben and Tunnermeier, 2005; JICA, 2008; Mack and Capelli, 2010: detailed in the following Chapter 4 herein). Similar detailed and high-level studies of ground water have been done in a few other areas of Afghanistan (Banks, 2014), but in general the problematic security situation has kept many other good studies from being done as they should have been already because of the obvious nation-wide acute needs (Tables 3.5 and 3.6). Wide varieties of contaminants in the ground water have been

TABLE 3.5 Major Deficiencies Categories in Ground and Surface Water in Afghanistan

1. Weak water knowledge and limited capacity building in water issues (Table 3.6).
2. Poor integrated water-resources management (Table 3.7).
3. Limited initial technical or conservation measures for water (Table 3.8).
4. Low supporting measures with regards to water (Table 3.9).
5. Few instruments exist for integrated water-resources management (Table 3.10).

Each of these deficiency categories has numerous specific deficiency elements that are detailed in the subsequent associated tables.
*From Saffi, M.H., Kohistani, A.J., 2013. Water Resources Potential, Quality Problems, Challenges and Solutions in Afghanistan: An Evaluation of the Past and Present Situation of Water Resources and Consequences on the Basis of Previous Water Resource Investigation and Present National Groundwater Monitoring Wells Network Data Management, Mapping, Visualization and Assessment. DACAAR, Kabul. http://www.dacaar.org/Publications/6%20Research%20and%20Studies/WASH/Report%20on%20Water%20Resource%20Porential_opt.pdf (accessed 09.11.15).*

TABLE 3.6 Deficient Elements Concerning Weak Water Knowledge and Limited Capacity Building About Water in Afghanistan

1. Only limited investment in the building of hydrological and hydrogeological expertise.
2. Lack of professionals in the water sector.
3. Poor integration, coordination, data collection, information sharing, or dissemination of water resource information or data.
4. Poor alternative research, assessment, or consultancy on water resources.
5. Lack of monitoring, management, development, protection, or sustainability of water resources.

*From Saffi, M.H., Kohistani, A.J., 2013. Water Resources Potential, Quality Problems, Challenges and Solutions in Afghanistan: An Evaluation of the Past and Present Situation of Water Resources and Consequences on the Basis of Previous Water Resource Investigation and Present National Groundwater Monitoring Wells Network Data Management, Mapping, Visualization and Assessment. DACAAR, Kabul. http://www.dacaar.org/Publications/6%20Research%20and%20Studies/WASH/Report%20on%20Water%20Resource%20Porential_opt.pdf (accessed 09.11.15).*

assessed, with elevated levels of nitrate ($NO_3$), boron, dissolved solids, and fecal contamination (*Escherichia coli*) being the most noticeable at first (Broshears et al., 2005), but later efforts by DACAAR (Saffi and Kohistani, 2013) added in other serious health hazards such as nitrite ($NO_2$), fluorine, sodium, uranium, chromium, and especially, arsenic.

## 3.6 ARSENIC IN GROUND WATER IN AFGHANISTAN

The naturally occurring native element arsenic (As) can be harmful to human health in drinking water as it causes various tumors, cancers, skin diseases, and other medical problems. The hydrogeochemistry of arsenic in ground water is complex, as it can be released from sulfide mineralization, or in the common geothermal waters of Afghanistan. It is commonly sorbed onto iron (III) oxyhydroxide minerals and can be released under moderately reducing conditions by reductive dissolution of these minerals. In neutral-alkaline conditions it can be rather soluble and can be concentrated by evaporative processes, which of course, prevail in much of Afghanistan (Banks, 2014).

Although the dangers of As in drinking water have been known for many decades, ever since high levels of poisonous As were first recognized about 1995 in Bangladesh and the Ganges Delta of India (~1995) (Nickson et al., 1998; Acharyya et al., 1999; Mukherjee et al., 2006), the UN and other agencies have been concerned about possible similar negative situations in Afghanistan (UNICEF, 2012). Accordingly DACAAR did much of the sampling from its 261 ground-water monitoring wells in 22 provinces, and UNICEF added data from 4 additional provinces, with only 8 provinces left unstudied because of security problems and remoteness.

A total of only 699 samples were collected and processed, with some 80 samples that exceeded 10 μg/L of As marked for special attention (Fig. 3.29; Table 3.5). This is in marked contrast to the higher 50 μg/L of As in Afghanistan that is a now the proposed new standard for the country, perhaps because so many exceedances were discovered in the initial studies. For example, Kabul had the highest levels of As in the ground-water wells, with a high point of 900 μg/L of As. This is twice as high as nearby Logar at 450 μg/L in its highest of 43 negative wells, with 282 wells tested. Logar's high values were attributed to ancient copper mining, which is possible, but it is more likely that the large quantity of copper sulfides outcropping in the province, mostly unmined, are the geogenic source. Similarly Ghazni had 171 wells tested, with 56 wells being too high, and its highest well was at 300 μg/L of As; Farah had 44 samples and one with a high of 300 μg/L of As as well. Neither of these places, nor Kabul had any old mining that could have produced abnormally high As readings, so miscellaneous (unknown) geogenic processes were attributed to be the causes for these excessive values (Ravenscroft et al., 2011; Saltori, 2004a,b).

## 3.7 GROUND WATER, AFGHANS, AND GEOLOGY

Traditional Afghan ground-water systems, such as karezes (qanats), springs, and shallow wells, supply more than 15% of Afghanistan's irrigation water and much of the drinking water supply. Irrigation water is also supplied by lifting ground water from shallow wells using a Persian wheel or "arhat" (Qureshi, 2002).

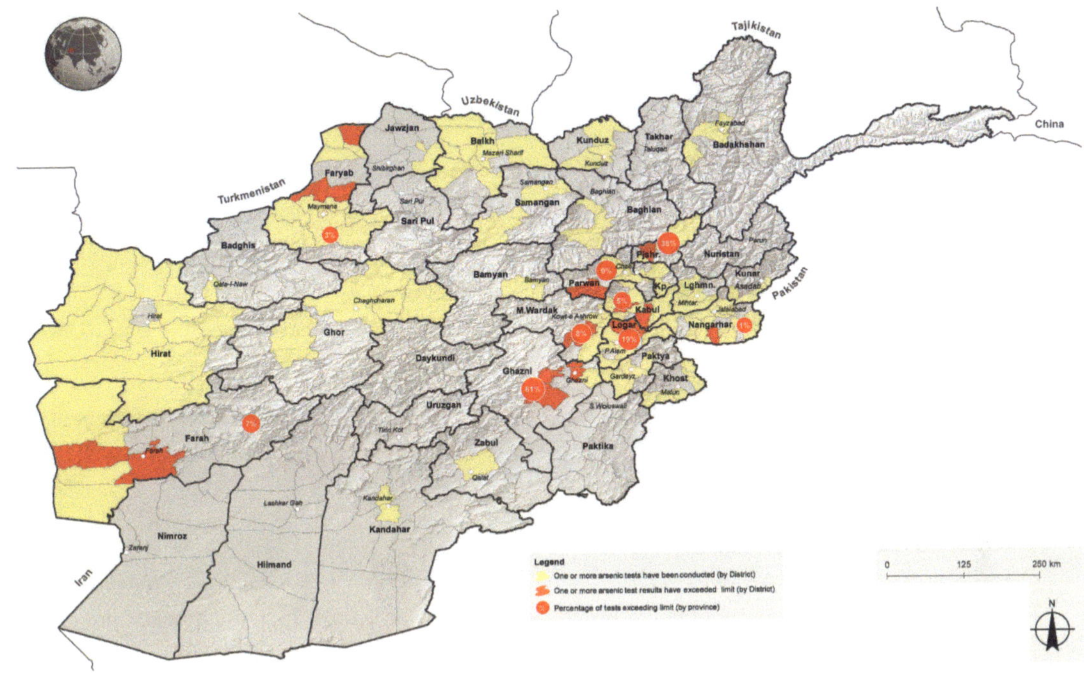


FIG. 3.29 Map of wells tested for arsenic in Afghanistan over the past decade by DACAAR and UNICEF (2012). Areas in yellow indicate that one or more tests have been conducted by the district; areas in pink indicate one or more arsenic test results have exceeded the arsenic limit by district; and the red circles indicate the percentage of tests exceeding the limit by province The highest are Ghazni Province at 61%, Pansheer at 38%, Logar at 19%. Parwan Province is 9%, Wardak Province is 8%, and Kabul is 5%; Farah is 7%, Faryab is 3%, and Nangarhar is 1%. *Notes*: (a) The WHO guideline is 10 µg/L and the Afghanistan proposed standard is 50 µg/L. (b) Arsenic determination was carried out by the Wagtech Arsenator field kit. (c) Water samples were collected from DACAAR ground-water monitoring wells in their target provinces and districts during 2005–11, and reflect the status of contamination thereof. (d) Arsenic contamination of ground water in Afghanistan is of natural, geological origin (geogenic). (e) Short or acute health effects can occur within days of exposure. Long or chronic effects occur over many years. Long-term exposure to arsenic has been linked to cancer of the bladder, lungs, skin, kidneys, nasal passages, liver and prostate.

## 3.7.1 Karezes

Karezes, known as qanats or foggara in other areas, are underground tunnel systems that can be considered as an underground aqueduct (Fig. 3.30). They tap shallow ground-water lenses and aquifers and channel the water by gravity to downstream fields and villages. Karezes have been used utilized for thousands of years in Afghanistan, Iran, China, and North Africa, and are common in the south and southwest of Afghanistan and less so in the north of the country. Approximately 6741 karezes occur in Afghanistan and have irrigated about 163,000 ha of land (Qureshi, 2002; Denning et al., 2013).

Vertical maintenance shafts (*chah*) connect the karez to the surface, allowing periodic cleanout of the tunnels (Fig. 3.31A). These vertical shafts are commonly surrounded by circular piles of cleanout dirt that make them look like rows of bomb craters. In places, especially where the substrate lacks internal strength, collapses may occur that require abandonment of parts of the karez

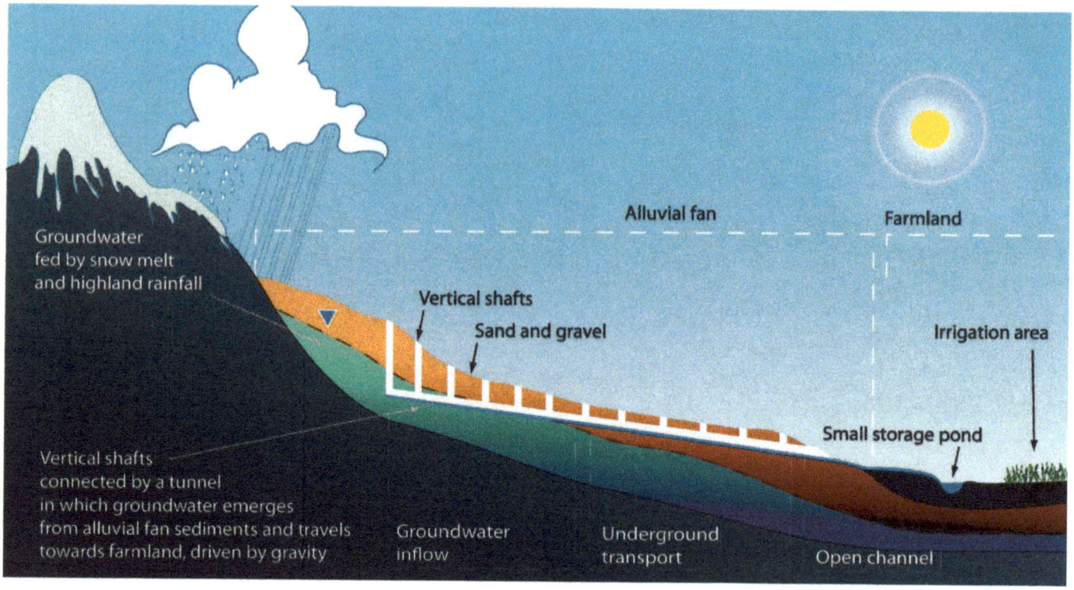


FIG. 3.30 Diagram of typical and traditional karez. *From Fig. 5.33 in Shroder, J.F., 2014. Natural Resources in Afghanistan: Geographic and Geologic Perspectives on Centuries of Conflict. Elsevier, Amsterdam; by permission from Earth magazine.*

tunnel and offset laterally to avoid the collapsed portion. In such places older and newer karez segments may exist side-by-side, with some of the older vertical shafts being filled in (Fig. 3.31B). Many karezes can have several branches that feed water via gravity to one main tunnel and may be several miles long. Karez water is used for both irrigation and drinking water purposes.

Two different locations of karez seem to occur in the Kandahar Province:

- karezes on valley floors within agricultural areas (Fig. 3.32); and
- karezes constructed on valley alluvial fans (Fig. 3.33).

Valley floor-karez commonly parallel irrigation canals and ditches and may be used to capture water leaking from the canals. It appears that the karez on the alluvial fans also parallel ephemeral or active stream channels in some places (Fig. 3.33) and indicate that these karez may also be capturing shallow or perched ground water as well. This water may be migrating from along the stream channel into the alluvial fan and then into the karez tunnel. In both cases, the cleanout-soil spoil piles adjacent to each karez shaft are approximately the same size, and indicate a reasonably consistent depth of the karez. Therefore, some karez appear to be tunnels that follow the topography, perhaps with a slightly varying gradient underground, as opposed to those upslope-deepening tunnels with only a uniform gradient underground (Fig. 3.34). In addition, in comparing karez to the locations of ephemeral channels, the majority of karez occur along the main dry channel and along the distal portion of alluvial fans (Fig. 3.32). Based upon these observations, a revised conceptual model of a typical Kandahar karez is presented in Fig. 3.34. In future ground-water research in Afghanistan, perhaps Afghan water specialists can prove or disprove this hypothesis with careful below-ground surveys, as well as interviews with people who dig karez.

FIG. 3.31 (A) Photograph of karez cleanout shafts along valley floor. Karez along irrigation canal or ephemeral or perennial rivers can take advantage of leaky canals and infiltration from nearby losing or infiltrating streams to gain shallow ground-water recharge (*Photograph by L. Sinfield, 2012*). (B) Photograph of multiple generations of karez cleanout shafts showing filled and abandoned portions, together with more recent open shafts (*Photograph by U.S. military*).

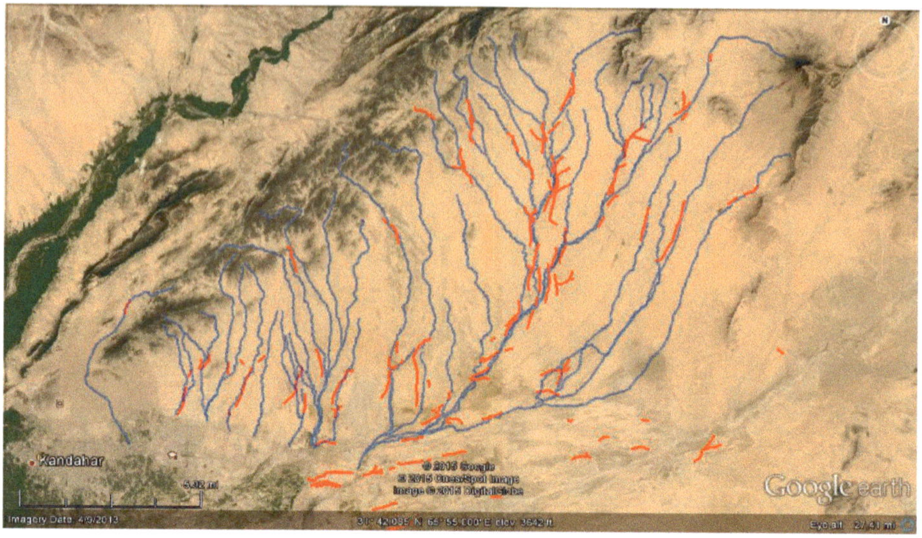


FIG. 3.32 Comparison of karez *(red)* and dry *(ephemeral)* channel locations *(blue)* on alluvial fans plotted on basemap image from Google Earth. Note that the largest number of karez occur along the main ephemeral channels and along the distal portion of the alluvial fans.

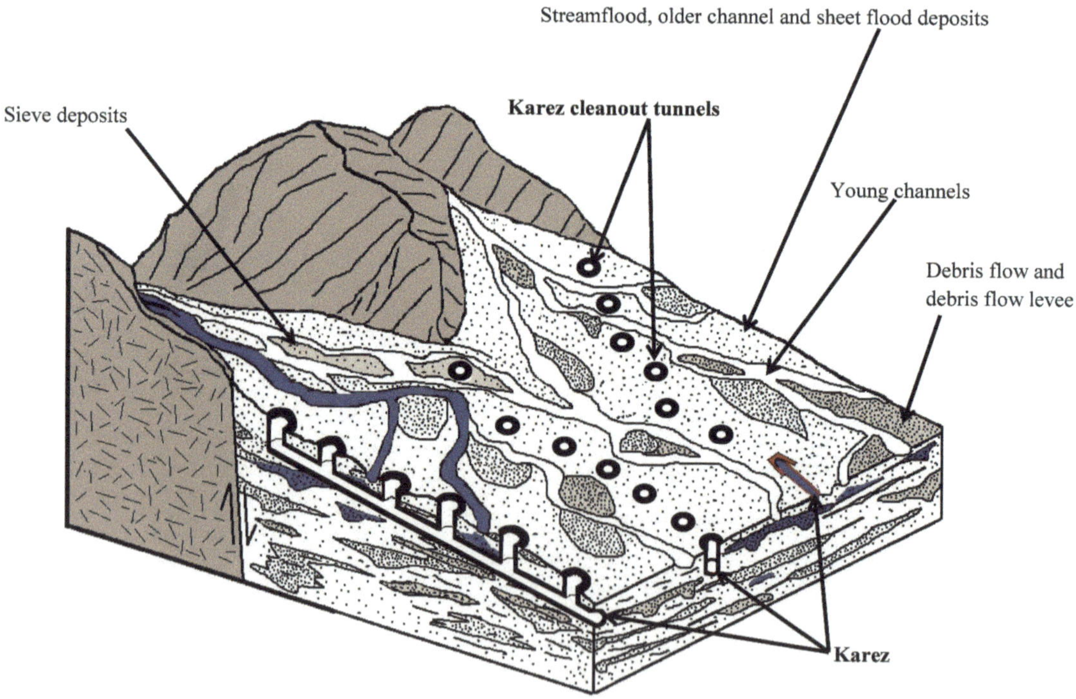


FIG. 3.33 Conceptual model of karez in Kandahar area. The karez tunnels can parallel the sloping topography at a consistent depth and not deepen up the alluvial fan so that they collect water from a variety of perched aquifers and other lenticular water sources provided by losing or infiltrating ephemeral streams as well.

FIG. 3.34 Comparison between the old and new karez construction models, with the old concept of a near horizontal tunnel is shown in black, and the new concept above it with an exaggerated steeper gradient in white and red. The new concept, however, must intersect various perched aquifers or infiltration zones from surficial ephemeral streams to be effective.

Arid alluvial fans are composed of interbedded sheetflood and fluvial sediments with steep, channelized flow, braided distributaries, and underfit channels. Deposits are stratified, and composed of angular grains with poor to moderate sorting and rounding, and clean of fines. A shallow karez that follows fan topography will have a high probability of intersecting highly permeable buried channel deposits that are hydraulically connected to active fan stream channels.

This new conceptual model helps in optimizing karez repair and rehabilitation. Karez occurring along main ephemeral channels that have multiple feeder channels, distal portions of alluvial fans (with large upstream watersheds), and along river valley bottoms, should be given priority for repair as they likely have a higher probability of success. In addition, karez within river valleys should be given top priority, because they have much lower evaporation rates than open canals, and capture water leaking from both canals and overwatered agricultural fields, as well as stream channels.

### 3.7.2 Small Wells

Historically and up to current times, karez, springs and shallow hand-dug open wells have been used to access ground-water resources for both irrigation and domestic uses in Afghanistan. DACAAR maintains a database of wells within Afghanistan and has installed many shallow wells throughout Afghanistan. Review of the DACAAR database indicates that most of the wells in Northern Afghanistan have been installed in Kabul, Murghab-Kushk wa Kaskan, Amu Darya, and Northern river valleys. Most of the wells in Southern Afghanistan have been installed in the Hari Rud, Farah, Arghandab, and Helmand river valleys as well as the low-lying valleys plains between the Tarnak and Arghandab rivers near the city of Kandahar. Very few wells have been installed outside of these wide-river valley floors.

One reason for the predominance of shallow alluvial valley wells is due to the local Afghan percussion drilling method, which uses a tripod and a diesel-powered winch to pound well borings through unconsolidated sediments (Fig. 3.34). This type of drilling method cannot penetrate consolidated sediments or rock. Most boreholes within Afghanistan are drilled using this rather antiquated tripod method. Only recently has modern mud-rotary drilling equipment been used, but due to relatively high costs, these rotary rigs (Fig. 3.35) have been exclusively in large municipal areas or by opium poppy growers in the Helmand Province.

FIG. 3.35 (A) Photograph of percussion-cable-tool rig in Afghanistan whereby the drill bit is raised and dropped successively by the engine and cable on the drum wheel to pulverize the rock or sediment beneath to force a hole into the ground. (B) View of the tripod used to hoist the drill bit and cable. (C) View of the engine and cable on the drum that is periodically wound up and released to drop down and push a hole into the ground as sediment is forced aside or brought out in an attached bailer. *All photos by L. Sinfield, 2012.*

In most of urban areas of Afghanistan, shallow-drilled wells are used to obtain water for drinking and other household activities and many occur right along city streets and canals (Fig. 3.5A). As noted above, shallow ground water can be influenced by surface water and is commonly subject to bacteriological contamination from sewage dumping. As the water levels continue to fall, around 0.5–3 m each month depending on location, poorer families are unable to extend the depth of their wells, and are forced to acquire water from communal wells. Many of these wells have already dried up and residents (generally women and children) are forced to long distances for water (Qureshi, 2002), or purchase trucked water at a premium (six times distribution water prices, Eng. Tawab. Balakarzai, 2010, personal communication) to meet their daily water demands (Fig. 3.36).

It is clear that Afghan society needs to learn a great deal about interconnections between surface and ground water (Tables 3.5–3.6), as well as much more about how to study and understand their own water resources, both above and below the ground surface, so that they can alleviate problems caused by major and minor deficiencies. Such things as technical and conservation measures, public awareness, integrated ground and surface water management, or solutions and policy-related options are all deficiencies in need of remediation everywhere (Tables 3.8–3.11).

FIG. 3.36 Rotary drilling rig that uses a mud lubricant to allow the bit to spin against the rock and sediment to grind a hole and flush out the well cuttings. *Photo by L. Sinfield, 2012.*

**TABLE 3.7** Deficient Elements of the Poorly Integrated Water-Resources Management in Afghanistan

1. Quite weak to nonexistent water-resources monitoring systems (from both qualitative and quantitative viewpoints).
2. Limited to nonexistent data information systems or databases.
3. Poor national water-quality standards (primary and secondary), and weak national water resources regulation act (both surface and ground water).
4. Weak water safety plan (quality control and quality assurance).
5. Limited to nonexistent policy-relevant research or alternative solutions to water-resources management.
6. Poor water-resources management, development, protection, and sustainability.
7. Very weak environmental security.

*From Saffi, M.H., Kohistani, A.J., 2013. Water Resources Potential, Quality Problems, Challenges and Solutions in Afghanistan: An Evaluation of the Past and Present Situation of Water Resources and Consequences on the Basis of Previous Water Resource Investigation and Present National Groundwater Monitoring Wells Network Data Management, Mapping, Visualization and Assessment. DACAAR, Kabul. http://www.dacaar.org/Publications/6%20Research%20and%20Studies/WASH/Report%20on%20Water%20Resource%20Porential_opt.pdf (accessed 09.11.15).*

**TABLE 3.8** Deficient Elements Concerning the Limited Initial Technical or Conservation Measures for Water in Afghanistan

1. Limited to no ground-water recharge by surface water during peak flow for aquifer recharge and recovery.
2. Almost no rain-water harvesting.
3. Only limited construction of water-storage infrastructure (dams, reservoirs, trenches, or injection wells).
4. No installations of desalination plants where the areas have saline water.

*From Saffi, M.H., Kohistani, A.J., 2013. Water Resources Potential, Quality Problems, Challenges and Solutions in Afghanistan: An Evaluation of the Past and Present Situation of Water Resources and Consequences on the Basis of Previous Water Resource Investigation and Present National Groundwater Monitoring Wells Network Data Management, Mapping, Visualization and Assessment. DACAAR, Kabul. http://www.dacaar.org/Publications/6%20Research%20and%20Studies/WASH/Report%20on%20Water%20Resource%20Porential_opt.pdf (accessed 09.11.15).*

**TABLE 3.9** Deficient Elements Concerning the Poor Supporting Measures in Afghanistan With Regard to Public Awareness and Participation About Ground and Surface Water

1. Low awareness of the inhabitants on the importance of water (qualitative and quantitative).
2. No encouragement of public participation in water issues.
3. No stress to develop campaigns to increase public awareness about water issues.

*From Saffi, M.H., Kohistani, A.J., 2013. Water Resources Potential, Quality Problems, Challenges and Solutions in Afghanistan: An Evaluation of the Past and Present Situation of Water Resources and Consequences on the Basis of Previous Water Resource Investigation and Present National Groundwater Monitoring Wells Network Data Management, Mapping, Visualization and Assessment. DACAAR, Kabul. http://www.dacaar.org/Publications/6%20Research%20and%20Studies/WASH/Report%20on%20Water%20Resource%20Porential_opt.pdf (accessed 09.11.15).*

**TABLE 3.10** Major Deficiencies Exist in Afghanistan Concerning Integrated Ground and Surface Water Management in That Few Instruments Exist for the Integration of Such Activities

1. Fragmented institutional arrangements.
2. Poor creation of enabling environment for effective policies and strategies for water-resources monitoring.
3. Poor formulation of enforcement regulation.

*From Saffi, M.H., Kohistani, A.J., 2013. Water Resources Potential, Quality Problems, Challenges and Solutions in Afghanistan: An Evaluation of the Past and Present Situation of Water Resources and Consequences on the Basis of Previous Water Resource Investigation and Present National Groundwater Monitoring Wells Network Data Management, Mapping, Visualization and Assessment. DACAAR, Kabul. http://www.dacaar.org/Publications/6%20Research%20and%20Studies/WASH/Report%20on%20Water%20Resource%20Porential_opt.pdf (accessed 09.11.15).*

TABLE 3.11 Major Categories of Solutions and Policy-Relevant Options to Major Deficiencies Observed Concerning Water Resources in Afghanistan

1. Management issues and improvements.
2. Enabling environments.
3. Initiation of technical and conservation measures.
4. Capacity building.
5. Knowledge building.

The detailed lists pertinent to each category are listed in the associated tables tied to each of these categories.
*From Saffi, M.H., Kohistani, A.J., 2013. Water Resources Potential, Quality Problems, Challenges and Solutions in Afghanistan: An Evaluation of the Past and Present Situation of Water Resources and Consequences on the Basis of Previous Water Resource Investigation and Present National Groundwater Monitoring Wells Network Data Management, Mapping, Visualization and Assessment. DACAAR, Kabul. http://www.dacaar.org/Publications/6%20Research%20and%20Studies/WASH/Report%20on%20Water%20Resource%20Porential_opt.pdf (accessed 09.11.15).*

## 3.8 QALAT FLOWING ARTESIAN WELL

A water well was drilled by the U.S. Navy Seabee well drilling team in 2011 in the Tarnak valley basin (TVB) near Qalat, Afghanistan, was drilled through successive alluvial (river), and possibly colluvial (landslide and earthquake) and eolian (wind-blown) sediment layers of mixed clays, silts, sands, and gravels that characterize this broad, arid valley. The well was completed and cased to a depth of 363 m from the well top at 1580 m. Water in the well is from a confined, artesian aquifer, which means that the water rises under pressure from the aquifer >305 m to flow from the well without a pump, in excess of 12.6 L/s (Fig. 3.37). This so-called "gusher" in this arid land constitutes an unforeseen cornucopia to local Afghan people where such water is seen as lifeblood itself. The fact that tapping such a deep artesian horizon for water is well beneath the unconfined, water-table water above it, and into which local karez wells have been tunneled is an added benefit because it will not affect them. Furthermore casing the well additionally separates the deep, confined water from the shallow, unconfined,

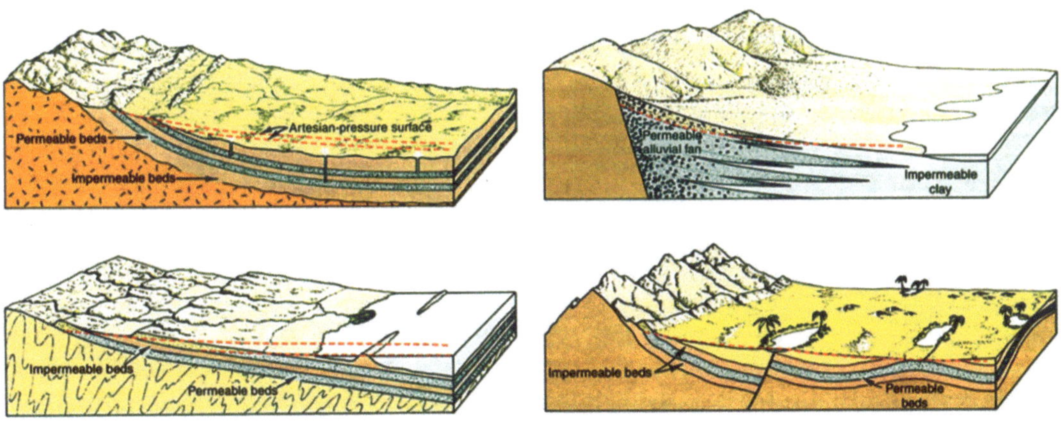


FIG. 3.37 Conceptual drawings of four types of artesian water that can occur in variable permeable water-bearing layers underground typical of various parts of Afghanistan. The artesian well at Qalat is most like the example in the upper right where the artesian water recharge zone is higher in the mountains to the north and the water flows down underground in permeable gravels buried beneath clays.

water-table water in which the karez function, which is essential to protect the shallow aquifers used by the local people.

The Tarnak River itself flows from NE to SW and is trained in the valley along the traces of one of the strands of great Chaman-Quetta strike-slip fault system (CQF) where all the land to the north of the fault is moving sideways and SW relative to the opposite south side that is moving to the NE. Depending upon exactly where the well is located relative to where the underground sediments are shattered by the fault, is likely to determine some of the future of this well to continue to deliver water. In any case, the presence or absence of this major fault is probably only incidental to the water production of the well and will not be considered further here because it is much easier to understand the ground water in general geohydrological terms having to do with the typical artesian aquifers characteristic of this region that are explained further below. That doesn't mean that the fault might not be important, but at this point we do not have enough information to go any further in our thinking on how faulting could affect this artesian water delivery.

So in a simple hydrogeological assessment of the Tarnak artesian basin, for example, the rocks in the mountains on the north side of the fault are dominantly Paleozoic and Mesozoic (~200–150 million year) carbonates, whereas the mountains to the south and east are largely either ancient volcanic rocks, perhaps 100 million years old, and very much older metamorphic rocks. Carbonate rocks can yield good water supplies, as can many volcanic rocks but the metamorphic rocks will not ever yield much water. These rock types, broken and fractured by the CQF could be a source of some of the water flow in the Qalat basin, but actually with the topographic and most likely subsurface nature of the TVB, any water from these marginal higher areas would not be necessary to provide the necessary hydraulic head (pressure) to produce the artesian water. Instead these older and more compact rocks around the periphery of the TVB can be thought of as providing most of the outside rim of the bathtub that contains the gravel and sand artesian aquifers beneath Qalat.

The regional geology of the Qalat area was mapped by Soviet geologists in the 1970s, which was then compiled and re-presented by US Geological Survey geologists, Sawyer and Stoesser (2005). The overall surficial geology of the region of the well head is comprised of various outcroppings of conglomerate and sandstone (clastics) of Middle and Late Pleistocene, and Holocene in age, which makes these deposits range in age from no older than about a half million years, right up to those of more modern times. These coarse clastic sands and gravels are porous and permeable enough, as well as being quite extensive in the TVB, so that water infiltrating into them from the surface recharge zones many tens of kilometers further away toward the bathtub rim, and over 300 m higher up in the basin to the northeast, providing the necessary head. Such an altitude difference between recharge zone and production of water at Qalat would be more than sufficient to produce the flowing artesian nature of the well.

One thing to keep in mind in dealing with poorly educated Afghan villagers, who will try to use water from this well probably indiscriminately into the future, is that most artesian aquifers are quite finite and will lose their pressure over time, as well as perhaps, their total water quantity delivery. Thus this is not a wonderful resource that will last forever unchanged, unless the resource is protected from over-production and/or later over-pumping. If possible, good records of water production quantity should be kept into the future to enable wise water use and better protection of the resource, as well as to help provide useful data for the development of other such nearby wells.

Another point to be made with this artesian water is that such water is generally "old," having soaked into the recharge area perhaps thousands of years ago. Such "old" water moving around underground over a long period of time can encounter and dissolve solids (common calcium carbonate or $CaCO_3$; common rock salt or NaCl; iron; sulfur; etc.) that may be quite undesirable. No mention was made of any dissolved solids in the well driller's report, so this water may be younger and less filled with such contaminants, which is a good thing. It also likely means that other artesian water "pay zones" occur in the Tarnak Valley, but care must be taken to not drill other wells too close nearby and into the same artesian water depth underground for fear of negative consequences to the original well (loss of pressure, cross contamination, water-discharge reduction, etc.). Used properly and sparingly this water and its pressure could last for generations; used to excess might dry it up within a few years. Records should be kept to enable greater knowledge of the region for future generations.

## 3.9 GEOTHERMALLY IMPACTED GROUND WATER

According to Saba et al. (2004), "Geothermal systems of Afghanistan are mainly associated with the fault and fracture networks, seismic activity, and young magmatism encountered at this boundary and its associated branching fault systems." Thus major geothermal manifestations are located along the Hari Rud to Panjsher geosuture and the Chaman-Quetta-Moqor fault systems in the active terrain of central Afghanistan. Geothermal manifestations in these areas are mostly marked in the fracture systems of active faults, within graben or half-graben basins and linear faulted valleys or wide valleys of the southern structural components of Afghanistan. The fairly common occurrence of geothermal waters in the much faulted and geologically active terrains of Afghanistan indicates that in future these energy resources are likely to receive greater exploitation. Based upon the geothermal activity and the deep-seated structure of geosutures and associated faults, ground water will more than likely have high TDS concentrations within:

1. the Paropamisus Mountains along the Hari Rud geosuture, extending to active faults along the southern edge of the mountains adjacent to the Hari Rud River valley;
2. mountain-front aquifers of the low-lying Shinkay Hills along the Pakistan/Afghan border along the Chaman-Quetta-Moqor fault system;
3. deep aquifers of the Tarnak river valley within the active Moqor fault system and associated splays; and
4. aquifers potentially associated with the accreted terrain of the Helmand fault system within the upper slopes of Helmand River valley, Dorafshan fault system near Tarin Kowt, and with the Farah Rud River and fault.

## 3.10 PASSIVE-SEISMIC TECHNOLOGY FOR AFGHANISTAN

A relatively new geophysical method, the horizontal-to-vertical, ambient-noise, seismic method also known as "Passive Seismic," and abbreviated as HVSR seismic method, or H/V method, has a great potential in mapping the fault-bounded aquifers typical of Afghanistan.

According to Lane et al. (2008), "Estimating sediment thickness and the geometry of the bedrock surface is a key component of many hydrogeologic studies. The horizontal-to-vertical (H/V) ambient-noise seismic method is a novel, non-invasive technique that can be used to rapidly estimate the depth to bedrock. The H/V method uses a single, broad-band three-component seismometer to record ambient seismic noise. The ratio of the averaged horizontal-to-vertical frequency spectrum is used to determine the fundamental site resonance frequency, which can be interpreted using regression equations to estimate sediment thickness and depth to bedrock." The method takes advantage of the seismic velocity differences (based upon rock density) between bedrock and the overlying alluvium. The method is generally able to differentiate between the two materials and provide an estimated depth to bedrock. The product from the field investigation would be depth-to-bedrock data points across a river valley, which could be contoured to produce a map of depth to bedrock and/or elevation of the top of bedrock. This map can be used to target locations for ground-water accumulation and further exploration (Fig. 3.38).

The passive-seismic method requires only a small seismometer in the field and a laptop for interpretation in the office (Fig. 3.39). The seismometer selected has an integrated GPS and operates on 2 AA-size batteries which may last several weeks of use. No other

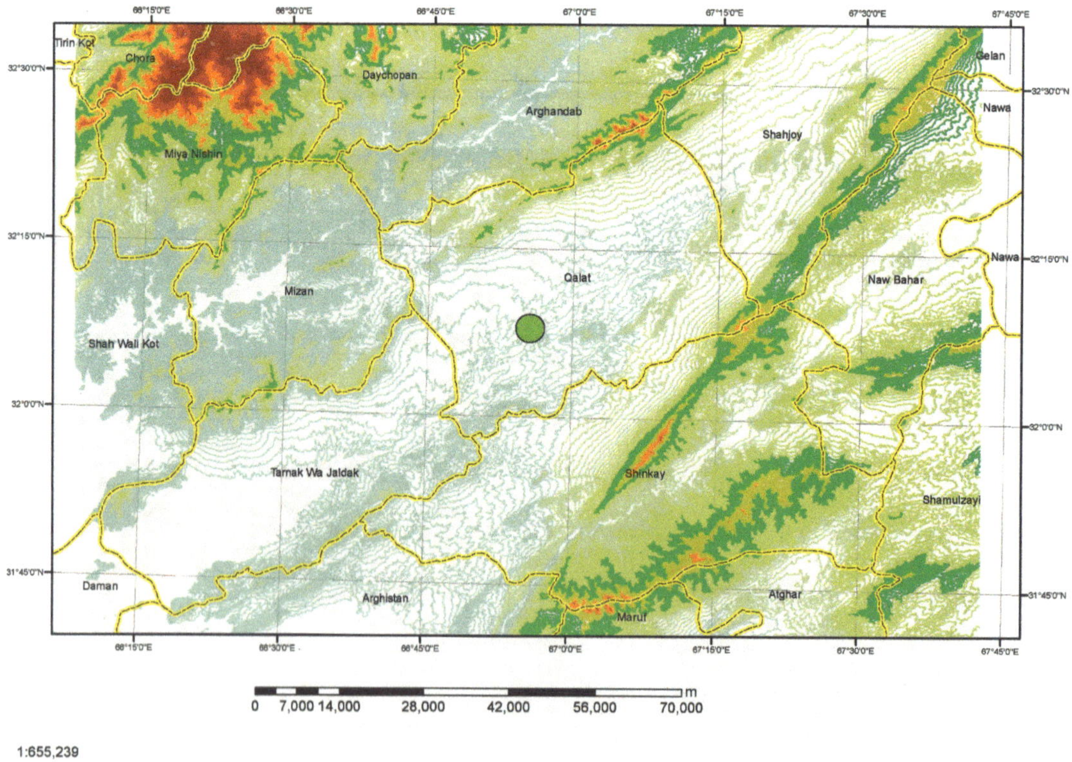


(A)

FIG. 3.38 (A) Location of the Qalat artesian well on lower regions of alluvial fan and alluvial plain in the Tarnak Valley where the slope to the south of the alluvial fan show the likely subsurface configuration of the permeable artesian aquifer also sloping down to the south in the substrate.

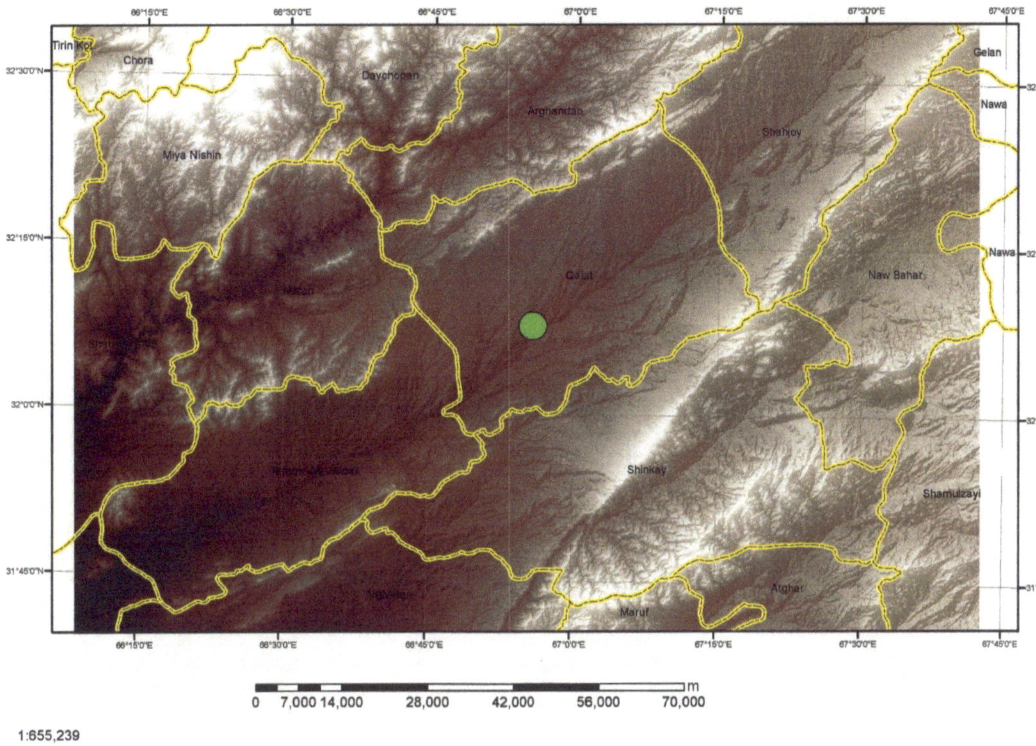


(B)

FIG. 3.38, CONT'D (B) Shaded relief map of the location of the Qalat artesian well showing the surface fluvial configuration of the region with the most likely water recharge zones existing to the north and northeast.

FIG. 3.39 Passive-seismic equipment that includes a three-component seismometer connected to field laptop computer for H/V seismic surveys. *From Lane Jr., J.W., White, E.A., Steele, G.V., Cannia, J.C., 2008. Estimation of bedrock depth using the horizontal-to-vertical (H/V) ambient-noise seismic method. In: Symposium on the Application of Geophysics to Engineering and Environmental Problems, April 6–10, Philadelphia, PA. Environmental and Engineering Geophysical Society, Denver, CO, pp. 13. http://water.usgs.gov/ogw/bgas/hvseismic/ (accessed 23.12.15).*

equipment is required in the field and everything can be operated by one student. Readings take approximately half an hour. This is important from a security perspective in Afghanistan and would require only a small security footprint. In addition, the method will work in a noisy urban environment. Total cost for the equipment is approximately $30,000 USD. This passive-seismic technique could be a "game changer" in water resources development for many third-world countries such as Afghanistan, Pakistan, and Tajikistan. For the fault-bounded aquifers of Afghanistan, this tool will be essential for effective ground-water exploration and ground-water development of the country.

## 3.11 CONCLUSIONS

Natural geological forces within the region have created the conditions for the formation of ground water and ground-water basins. These same forces also drive the sustainability of the aquifers beneath Afghanistan. Ongoing continental collision has created alluvial-filled tectonic depressions and valleys throughout Afghanistan, which host good quality aquifers beneath all the major cities in the region. These structural ground-water basins are fully enclosed intermontane basins, narrow mountain river valleys, or wedge-shaped open grabens. Upper watersheds near mountain recharge areas, deep alluvial aquifers in river valleys, and structural aquifer basins are all dominated by fresh good-quality sodium/magnesium bicarbonate waters, with colluvium, alluvial fans, and plains by brackish low-quality sodium bicarbonate to sodium sulfate waters. In the northern and southern plains of Afghanistan, fresh ground water is commonly restricted to alluvial fill in river valleys, and is generally under the influence of surface water and highly subject to bacteriological contamination from the dumping of raw sewage. Typically, poor-quality ground water occurs within large plains outside the river valleys within shallow brackish aquifers, areas subject to current or past geothermal activity along active fault systems, and near the northern and southwestern boundary of Afghanistan including the Northern, Amu Darya, and Helmand watersheds. Human-caused salinization of aquifers, reduction of recharge along river alluvial aquifers and canal systems, and overuse of fresh water resources will negatively affect many downstream and cross-border alluvial aquifers and rivers.

Using awareness of the natural conditions and limitations as a guide, application of new technology can help map, develop, and correctly manage ground-water resources of Afghanistan and downstream users of the water. New and relatively inexpensive, passive-seismic instruments can be used to map aquifer boundaries, subsurface faults or bedrock "steps," and aquifer depths. Based upon these new exploration techniques, deep water wells should be installed for major municipalities and district centers to improve water quality and health for many Afghans. Using a new conceptual model of karez, repair and rehabilitation of the underground water tunnels should be optimized based upon the location of main active portions of channels within large alluvial fans and karez along river bottoms. Karez will have less evaporative losses than open irrigation canals and should be encouraged in river valleys. These techniques can help deliver better water to more Afghans, but may also increase ground-water withdrawals. Implementation of an Irrigation and Watershed Management Program with an emphasis on optimization of deep ground-water well drilling in cities, karez repair, and passive-seismic technology for ground-water exploration, along with a robust governance and capacity-building goals, will help reduce these threats and improve downstream and transboundary water quality and quantity issues.

## References


Abdullah, S.H., Chmyriov, V.M., Dronov, V.I. (Eds.), 1980. Geology and Mineral Resources of Afghanistan, 2 Volumes. British Geological Survey, Nottingham. Geological Survey of Afghanistan and British Geological Survey Occasional Publication No. 15, reprinted 2008.

Acharyya, S.K., Chakraborty, P., Lahiri, S., Raymahashay, B.C., Guha, S., Bhowmik, A., 1999. Arsenic poisoning in the Ganges delta. Nature 401, 545. http://dx.doi.org/10.1038/44052.

Anon., 1976. Hydrogeological Map of Afghanistan, 1:2,000,000. Government of Afghanistan.

Balakarzai, A.T., 2010. Assessment of Domestic Water Consumption From a Socioeconomic Perspective in Kandahar, Afghanistan. Unpublished Master's Thesis, Asian Institute of Technology, Bangkok, Thailand.

Banks, D., 2014. Hydrogeological Atlas of Faryab Province. NORPLAN for the Afghan Ministry of Rural Rehabilitation & Development (MRRD).

Banks, D., Soldal, O., 2002. Towards a policy for sustainable use of groundwater by nongovernmental organizations in Afghanistan. Hydrogeol. J. 10, 377–392. http://dx.doi.org/10.1007/s10040-002-0203-y.

Brookfield, M.E., Hashmat, A., 2001. The geology and petroleum potential of the North Afghan Platform and adjacent areas (northern Afghanistan, with parts of southern Turkmenistan, Uzbekistan and Tajikistan). Earth Sci. Rev. 55 (1–2), 41–71.

Broshears, R.E., Akbari, M.A., Chornack, M.P., Mueller, D.K., Ruddy, B.C., 2005. Inventory of Ground-Water Resources in the Kabul Basin. USGS, Reston, VA. U.S. Geological Survey Scientific Investigations Report 2005-5090, http://pubs.usgs.gov/sir/2005/5090/sir20055090.pdf (accessed 08.11.15).

Camp Dresser and McKee (CDM), 2003. Kandahar Groundwater Resource Assessment. Study Report for U.S. Agency for International Development. USAID, Washington, DC.

Denning, G.M., Steele, W.B., Johnson, W.C., Chauvin, K.W., Omelicheva, M.Y., Dobson, J.E., Herlihy, P.H., Mandel, R.D., Hoopes, J.W., Macpherson, G.L., Haufler, M., 2013. Final Report of the University of Kansas Research and Educational Support for U.S. Army Programs and Initiatives at Fort Leavenworth, Kansas [Karez]: 1–48. http://www.google.com/url?sa=t&rct=j&q=&esrc=s&source=web&cd=8&ved=0ahUKEwiij_2mnYTKAhVC4CYKHaBCDSUQFghOMAc&url=http%3A%2F%2Fwww.dtic.mil%2Fcgi-bin%2FGetTRDoc%3FAD%3DADA602982&usg=AFQjCNFOApMeFCrHFE3emxR6ic9RyDKMNg&sig2=lgJg2bPtDlNDXQZFMjmXUA (accessed 01.01.16).

Favre, R., 2004. Watershed Atlas of Afghanistan: Working Document for Planners, 5 Volumes, first ed. Watershed Consultant Golam Monowar Kamal, AIMS Focal Point, Kabul.

Food and Agriculture Organization of the United Nations, 1965a. Survey of Land and Water Resources: Afghanistan. Geology. vol. II Food and Agriculture Organization of the United Nations, Rome.

Food and Agriculture Organization of the United Nations, 1965b. Survey of Land and Water Resources: Afghanistan. Hydrology. vol. III Food and Agriculture Organization of the United Nations, Rome.

Houben, G., Tunnermeier, T., 2005. Hydrogeology of the Kabul Basin Part II: Groundwater Geochemistry and Microbiology. Foreign Office of the Federal Republic of Germany, Hannover. http://www.susana.org/_resources/documents/default/2-1437-hydrogeologykabulbasin2en.pdf (accessed 08.11.15).

Japanese International Cooperation Agency (JICA), 2008. The study on groundwater resources potential in Kabul Basin in the Islamic Republic of Afghanistan. In: 6th Joint Technical Committee (JTC), March. Sanyu Consultants, Nagoya.

Klett, T.R., Ulmishek, G.F., Wandrey, C.J., Agena, W.F., 2006. Assessment of Undiscovered Technically Recoverable Conventional Petroleum Resources of Northern Afghanistan. USGS, Reston, VA. U.S. Geological Survey Open-File Report 2006-1253.

Lane Jr., J.W., White, E.A., Steele, G.V., Cannia, J.C., 2008. Estimation of bedrock depth using the horizontal-to-vertical (H/V) ambient-noise seismic method. In: Symposium on the Application of Geophysics to Engineering and Environmental Problems, April 6–10, Philadelphia, PA. Environmental and Engineering Geophysical Society, Denver, CO, pp. 13. http://water.usgs.gov/ogw/bgas/hvseismic/ (accessed 23.12.15).

Mack, T., Capelli, K., 2010. Afghanistan's Kabul Basin Faces Major Water Challenges. USGS, Reston, VA. U.S. Geological Survey Newsroom Article Release, http://www.usgs.gov/newsroom/article.asp?ID=2521#.Vj_pdE3lvIU (accessed 08.11.15).

Mahmoodi, S.M., 2008. Integrated water resources management for rural development and environmental protection in Afghanistan. J. Dev. Sustain. Agric. 3, 9–19.

Mukherjee, A., Sengupta, M.K., Hossain, M.A., Ahamed, S., Das, B., Nayak, B., Lodh, D., Rahman, M.M., Chakraborti, D., 2006. Arsenic contamination in groundwater: a global perspective with emphasis on the Asian scenario. J. Health Popul. Nutr. 24 (2), 142–163. http://www.bioline.org.br/pdf?hn06020 (accessed 06.11.15).

Nembrini, P.G., 2002. Cities in War: Thirsty Cities: Kabul Water Supply: Evolution Since the 1992-94 Civil War. Geneva Foundation, Tacoma Occasional paper No. 7.

Nickson, R., McArthur, J., Burgess, W., Ahmed, K.M., Ravenscroft, P., Rahman, M., 1998. Arsenic poisoning of Bangladesh groundwater. Nature 395, 338.

Qureshi, A.S., 2002. Water Resources Management in Afghanistan: The Issues and Options. International Water Management Institute, Colombo. Working Paper 49, Pakistan Country Series No. 14.

Rasmussen, S.E., 2014. Kabul—the fifth fastest growing city in the world—is bursting at the seams. The Guardian. http://www.theguardian.com/cities/2014/dec/11/kabul-afghanistan-fifth-fastest-growing-city-world-rapid-urbanisation (accessed 08.11.15).

Ravenscroft, P., Brammer, H., Richards, K., 2011. Arsenic Pollution: A Global Synthesis. Wiley, Chichester.

Ruleman, C.A., Crone, A.J., Machette, M.N., Haller, K.M., Rukstales, K.S., 2007. Map and database of probable and possible quaternary faults in Afghanistan. USGS, Reston, VA. U.S. Geological Survey Open-File Report 2007-1103, 39 pp., 1 plate.

Saba, D.S., Najaf, M.E., Musazai, A.M., Taraki, S., 2004. Geothermal Energy in Afghanistan. Center on International Cooperation, New York University; Center for Policy and Development Studies, New York; Kabul. http://www.mindfully.org/Energy/2004/Afghanistan-Geothermal-Energy1feb04.htm (accessed 07.11.15).

Saffi, M.H., 2007. Groundwater Resources at Risk in Afghanistan. Scientific Investigation Report in Afghanistan. Danish Committee for Aid to Afghan Refugees, Kabul. http://www.dacaar.org/resources (accessed 07.11.15).

Saffi, M.H., Kohistani, A.J., 2013. Water Resources Potential, Quality Problems, Challenges and Solutions in Afghanistan: An Evaluation of the Past and Present Situation of Water Resources and Consequences on the Basis of Previous Water Resource Investigation and Present National Groundwater Monitoring Wells Network Data Management, Mapping, Visualization and Assessment. DACAAR, Kabul. http://www.dacaar.org/Publications/6%20Research%20and%20Studies/WASH/Report%20on%20Water%20Resource%20Porential_opt.pdf (accessed 09.11.15).

Saltori, R., 2004a. Arsenic Contamination in Afghanistan: Preliminary Findings. DACCAR, Kabul. pp. 1–15.

Saltori, R., 2004b. Arsenic contamination in Afghanistan: preliminary findings. In: Conference on Water Quality—Arsenic Mitigation, Taiyan, 23–26 November. UNICEF and Water Sanitation Group, Kabul.

Sawyer, D.A., Stoesser, D.B., 2005. Geologic map of quadrangle 3266, Ouruzgan (519) and Moqur (520) quadrangles, Afghanistan. USGS, Reston, VA. U.S. Geological Survey, 1:250,000.

Shareq, A., 1981. Geological and geophysical investigations carried out in Afghanistan over the period 1972–1979. In: Gupta, H.K., Delany, F.M. (Eds.), Zagros, Hindu Kush, Himalaya-Geodynamic Evolution. American Geophysical Union Geodynamics Series, 3. AGU, Washington, DC, p. 7586. 1 pl., scale 1:2,500,000.

Shroder, J.F., 2014. Natural Resources in Afghanistan: Geographic and Geologic Perspectives on Centuries of Conflict. Elsevier, Amsterdam.

Tapponnier, P., Peltzer, G., Armijo, R., 1986. On the mechanics of the collision between India and Asia. In: Coward, M.P., Ries, A.C. (Eds.), Collision Tectonics. Geological Society of London, London, pp. 115–157. Geological Society Special Publications No. 19.

Uhl, V.W., Tahiri, M.Q., 2003. Afghanistan: An Overview of Groundwater Resources and Challenges. Uhl, Baron, Rana Associates; Basic Afghanistan Services, Washington Crossing, PA; Kabul. http://www.vuawater.com/Case-Study-Files/Afghanistan/Afghanistan_Overview_of_GW_Resources_Study-2003.pdf (accessed 07.11.15).

UNICEF, 2012. Arsenic Contamination of Drinking Water in Afghanistan (Map), WASH Section. https://www.humanitarianresponse.info/sites/www.humanitarianresponse.info/files/WASH_WQ_ARSENIC_NATIONAL.pdf (accessed 06.11.15).

United Nations, 1986. Ground Water in Continental Asia (Central, Eastern, Southern, South-Eastern Asia). United Nations, New York. Natural Resources/Water Series No. 15.

United States Army Corps of Engineers Topographic Engineering Center (USACE-TEC), 2002. Water Resources Areal Appraisal of Afghanistan. USACE-TEC, Alexandria, VA.

Wheeler, R.L., Bufe, C.G., Johnson, M.L., Dart, R.L., 2005. Seismotectonic Map of Afghanistan, With Annotated Bibliography. USGS, Reston, VA. U.S. Geological Survey Open-File Report (OFR) 2005-1264.

Whitney, J.W., 2006. Geology, Water, and Wind in the Lower Helmand Basin, Southern Afghanistan. USGS, Reston, VA. U.S. Geological Survey Scientific Investigations Report 2006-5182, 40 pp.

CHAPTER

# 4

# Development of Water Resources in the Kabul River Basin

N. Eqrar*, J. Shroder†

*University of Kabul, Kabul, Afghanistan †University of Nebraska at Omaha, Omaha, NE, United States

OUTLINE



## Abstract

The greater Kabul River Basin (KRB), as the seat of government in Afghanistan and a key region of population growth, needs a high level of water management in order to maintain livability. The obvious growing water needs for a burgeoning population, and the concomitant increasing sewage contamination of surface and ground-water supplies in a city with a quite limited to no sewage system means that serious and modern water-management efforts must be undertaken to avoid a catastrophic drought condition, contagion or other water-diminishment problem. The World Bank's multisectoral decision-support system for the KRB, if implemented in a timely fashion by the new Government of Afghanistan has a good likelihood of succeeding in helping stabilize a difficult and an all too commonly, violence-prone situation.

**Keywords:** Logar-Upper Kabul subbasin, Panjshir subbasin, Lower Kabul subbasin, Hydrogeology, Moisture seasonality, Ground-water aquifer, Reservoir planning, Hydropower, Sewage infiltration, Aquifer contamination, Lake Hashmat Khan, Decision-support system



http://dx.doi.org/10.1016/B978-0-12-801886-6.00004-5

## 4.1 INTRODUCTION

The basin of the Kabul River is the seat of the Government of Afghanistan (GoA) and home to millions of people who must have reasonable access to plentiful water, providing such can be obtained. In recent years the population expansion, growing industrialization and revived agriculture after decades of war in the country have put great pressure on water resources, which have been studied by a variety of foreign and domestic agencies to bring relief from problems with a variety of surface and subsurface water sources. Water is fairly widely recognized as the key to Afghanistan's future (Fipps, 2006). As Azimi and McCauley (2002) and Shroder (2014) have noted, the annual ground-water recharge in Afghanistan is about 18 $km^3$, whereas the total annual extraction of this water has only been about 3.7 $km^3$; this means that a great deal more can be extracted but perhaps it is not exactly where it is most needed (Watershed Atlas of Afghanistan, 2004). As of about 2006, the whole country only had sanitation systems for about 5% of rural residents and 16% of urban dwellers (Fipps, 2006). At the same time improved drinking water was available only to about 11% of rural residents and 19% of the urban folks. In addition, the Ministry of Energy ad Water has noted that 85% of the population is involved in irrigation-dependent agriculture, 98% of all the water diverted from the rivers is used by agriculture, and 60% or more of the diverted water is then lost to seepage because of the very poor on-farm water management. These same irrigation canal systems also provide the drinking water available to most of the population (Fipps, 2006). Campbell (2015) has detailed all the efforts of the U.S. Geological Survey to gain information on updating ground-water-level monitoring networks, as well surface water resources so that long-term sustainability of the vital resources will be better understood.

The Kabul Basin, however, is already well known as being in the unenviable position of diminution of water quantity and quality progressively with urban development, land use, climate change, socio-economic development and frequent drought events (Saffi, 2011). Mack et al. (2009) and Mack and Capelli (2010) have emphasized that in the next 50 years drinking water needs in the Kabul Basin are expected to increase sixfold, mainly due to population increase from returning refugees. Furthermore the increasing air temperatures associated with global climate change are likely to reduce water resources in the Kabul Basin and estimates are that at least 60% of the shallow groundwater wells in the city will be ineffective and are likely to become inoperative. The situation is growing so serious that numerous pioneering studies have been undertaken to seek solutions through detailed studies that are summarized in this chapter (Banks, 2002; Gellasch, 2014a; Himmelsbach et al., 2006; Houben and Tünnermeier, 2005; Mack et al., 2013; Niard, 2006; Tünnermeier and Houben, 2005; Williams, 2009; World Bank, 2010) (Table 4.1). Studies have assessed climate trends, water use, surface and ground-water availability and water quality by integrating numerous forms of data, which include surviving historical climate and water records, surface and ground-water analyses, satellite imagery, large-scale maps, geologic investigations and estimates of public-supply and agricultural water uses, in order to provide a comprehensive overview of the water resources in the Kabul Basin.

Surface water resources and underground water are commonly treated as separate entities by many people, but in fact they are part and parcel of the same thing and must be considered

TABLE 4.1 Background Factors Used in Scoping Strategic Options for the Development of the Greater Kabul River Basin That is in so Much Trouble

1. Development is required of an integrated basin-planning framework for analyzing and prioritizing water-resources development options in Afghanistan, as well as applying them to the Kabul River Basin
2. Long periods of conflict have destroyed much water capability, crippled the economy and exacerbated poverty
3. Afghanistan currently has a strikingly low level of development of water resources and correspondingly low levels of water-related services, including urban and rural water supplies, irrigation, hydropower and other uses
4. Water-resources development and management is central to Afghanistan's economic growth and poverty-reduction efforts
5. Since the new government came to power in 2002, basin perspective and integrated water management have been cornerstones of its approach to water-resources development
6. Central issues to make real progress include: (a) increase investment in water infrastructure to remove constraints of low and highly variable streamflows; reduce impact of droughts and unpredictable precipitation; and provide a base for integrating with growing and broader national economy; (b) analyze and prioritize projects in multisectoral basin framework; (c) strengthen capacity at weak Ministry of Energy and Water; (c) help Afghanistan to gain sufficient technical capacity to participate effectively and with a knowledge-based approach in transboundary-waters discussions with other riparian countries
7. Specific planning objective is to analyze medium- and long-term options for development of water resources in Kabul River Basin
8. Kabul River seems to have adequate water resources overall, but distribution of water in time and space are inadequate to meet demand
9. All development options are potentially viable projects and include: (a) 13 dam and reservoir projects, 8 with capacities of 1171 MW, 3 of which are run-of-river; (b) five ground-water aquifers; (c) 14 irrigation areas; and (d) one major transbasin conveyance link connecting Panjshir to Kabul watershed. If all projects are undertaken by the different sectors without integrated planning, they will not constitute a rational, sustainable or efficient investment plan, and may result in serious water conflicts, foregone benefits and increased costs
10. The DSS for the Kabul River Basin has an optimization model that maximizes net economic benefits of water used in the different sectors

The approach used is a multisectoral decision-support system (DSS), which is used to analyze and assess various development options that are based on: (a) cost; (b) water demands; (c) water availability; (d) economic impacts; (e) long-term consistency with development goals and trends; (f) sustainable use of water resources; (g) knowledge base of water demand, development and conveyance options and hydrological system configuration; and (h) mathematical models.

*From World Bank, 2010. Scoping Strategic Options for Development of the Kabul River Basin: A Multisectoral Decision Support System Approach. Sustainable Development Department, South Asian Region, The World Bank. http://www-wds.worldbank.org/external/default/WDSContentServer/WDSP/IB/2010/04/12/000333037_20100412001029/Rendered/PDF/522110ESW0Whit1anistan0Final0Report.pdf (accessed 07.06.15).*

in concert by professional hydromanagers. This means that the interactions and feedbacks between the two subsystems above and below the surface have to be engineered together, which is the approach taken here.

## 4.2 GEOLOGICAL BACKGROUND

The Kabul River Basin (KRB) in the northeast quarter of Afghanistan is situated at the intersection of three major translational and extensional, fault systems that have enabled and shaped three interconnected, largely fluvial aquifers (Logar, Kabul, Paghman) that are 20–70 m thick and composed of coarse sand to gravelly colluvial detritus (Böckh, 1971; Himmelsbach et al., 2006; Tünnermeier and Houben, 2005). High permeabilities occur where the coarse clastics are uncemented, but deeper buried clastics may be slightly cemented to produce lower permeabilities. A common uppermost loess cover of lower permeability serves as an important protective layer that should be preserved against excavation for use as building materials (brick making) and such. The main ground-water recharge occurs after snowmelt in March to May and direct infiltration from the rivers in the basin.

The surface of the KRB is host to groups or regions of largely mountain runoff-fed tributaries that collectively form the river system (Fig. 4.1), which is subdivided into a number of subbasins of greater and lesser importance (Fig. 4.2). The northern, or largely left flank of the overall basin is quite mountainous, with peaks >6000 m that provide much of the flow of the river system, and which offer numerous rock-bound sites for impoundments to facilitate hydropower generation and irrigation of appropriate flatter agricultural basin. Such plains, foothills and lower mountain ranges are characteristic of the southern portions of the basin, where some 20% of the irrigated areas of Afghanistan exist (World Bank, 2010). The principal subbasins of the Kabul River system (Fig. 4.2) have been delineated on the basis of their climate, hydrology and physiographic characteristics into the Logar-Upper Kabul, the Panjshir and the Lower Kabul.

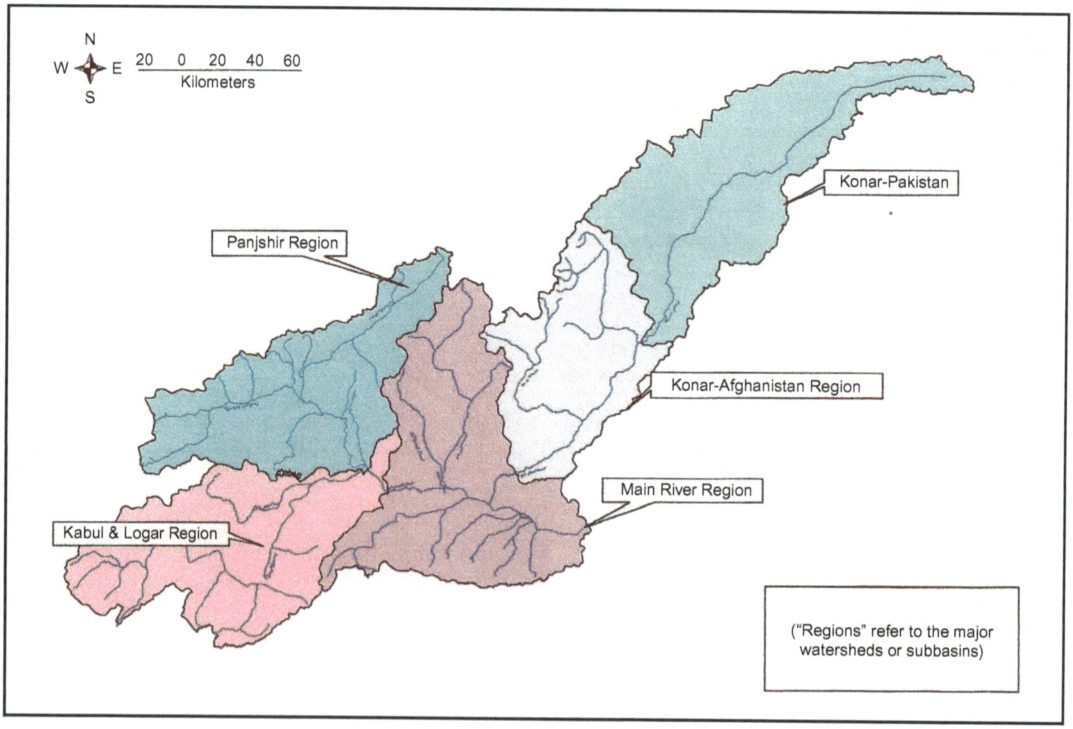


FIG. 4.1 Map of the chief tributaries and geographic configuration of the Kabul River system, with its five major regions: (1) Kabul and Logar Region; (2) Panjshir Region; (3) Main River Region; (4) Konar (Chitral)—Pakistan Region; (5) Konar (Kunar)—Afghanistan Region. *From World Bank, 2010. Scoping Strategic Options for Development of the Kabul River Basin: A Multisectoral Decision Support System Approach. Sustainable Development Department, South Asian Region, The World Bank. http://www-wds.worldbank.org/external/default/WDSContentServer/WDSP/IB/2010/04/12/000333037_20100412001029/Rendered/PDF/522110ESW0Whit1anistan0Final0Report.pdf (accessed 07.06.15). Source: Toosab and RCUWM (Toosab Consulting Engineers and Regional Center for Urban Water Management), 2006. Integrated Water Resources Management of the Kabul River Basin.*

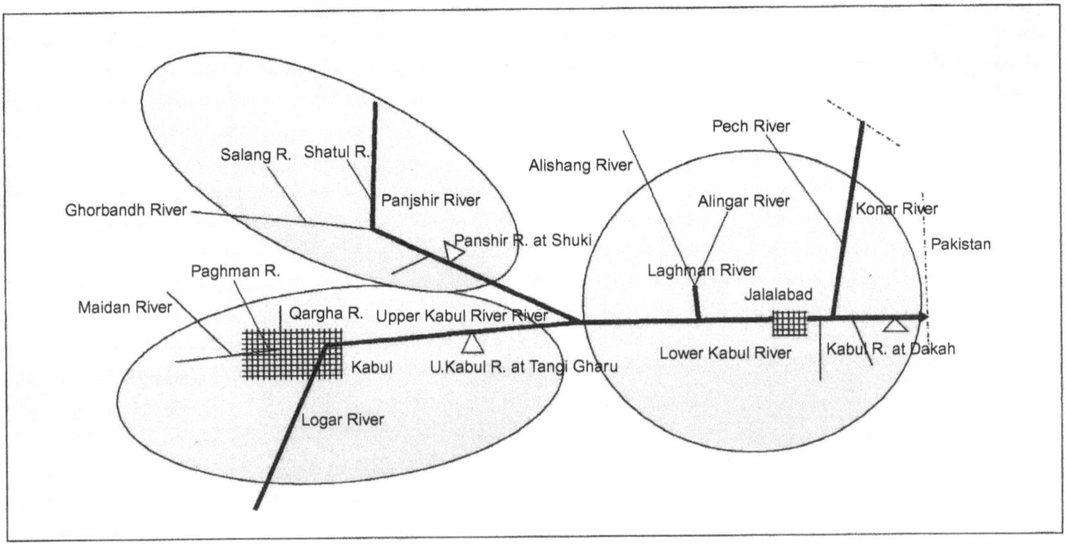


FIG. 4.2 Map of the principal subbasins of the Kabul River system: (1) Logar-Upper Kabul subbasin; (2) Panjshir subbasin; (3) Lower Kabul subbasin. The three *open triangles* are existing hydrologic measuring stations (World Bank, 2010).

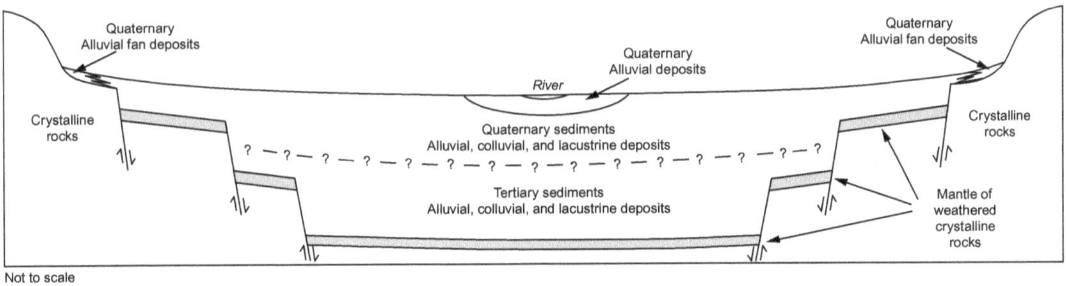


FIG. 4.3 Generalized geological cross section of the Kabul Basin showing the surficial and subsurface configuration of the rocks, sediment and underground water storages in the pores and fissures of the materials in the Basin. In general, the less-cemented, porous and permeable sediments of the sediments deposited in the basin host the greatest amounts of water, with the older and deeper, cemented sediments, as well as the relatively impermeable crystalline rocks of the surrounding mountains holding the least (BGR, 2005).

The geology and hydrogeology of the Kabul Basin controls the surficial and underground water resources (Figs. 4.3–4.8). This downfaulted sedimentary basin is a fairly typical situation of an irregular basin of largely crystalline (igneous and metamorphic) rocks without much associated ground water that has trapped and held various colluvial (mass movement), alluvial (fluvial), lacustrine (lake) and eolian (wind-blown) sediments of differing clastic grain sizes. The coarser-grained sizes (sands and gravels) of the basin can hold significant water as aquifers of essential capability. The seasonality of water delivery to the environment and to human use in the Kabul Basin is shown in Fig. 4.9.

The city of Kabul was originally founded in antiquity as Kabura, which was an Achaemenid outpost some 2500 years ago, and then became dominated successively by the Bactrian

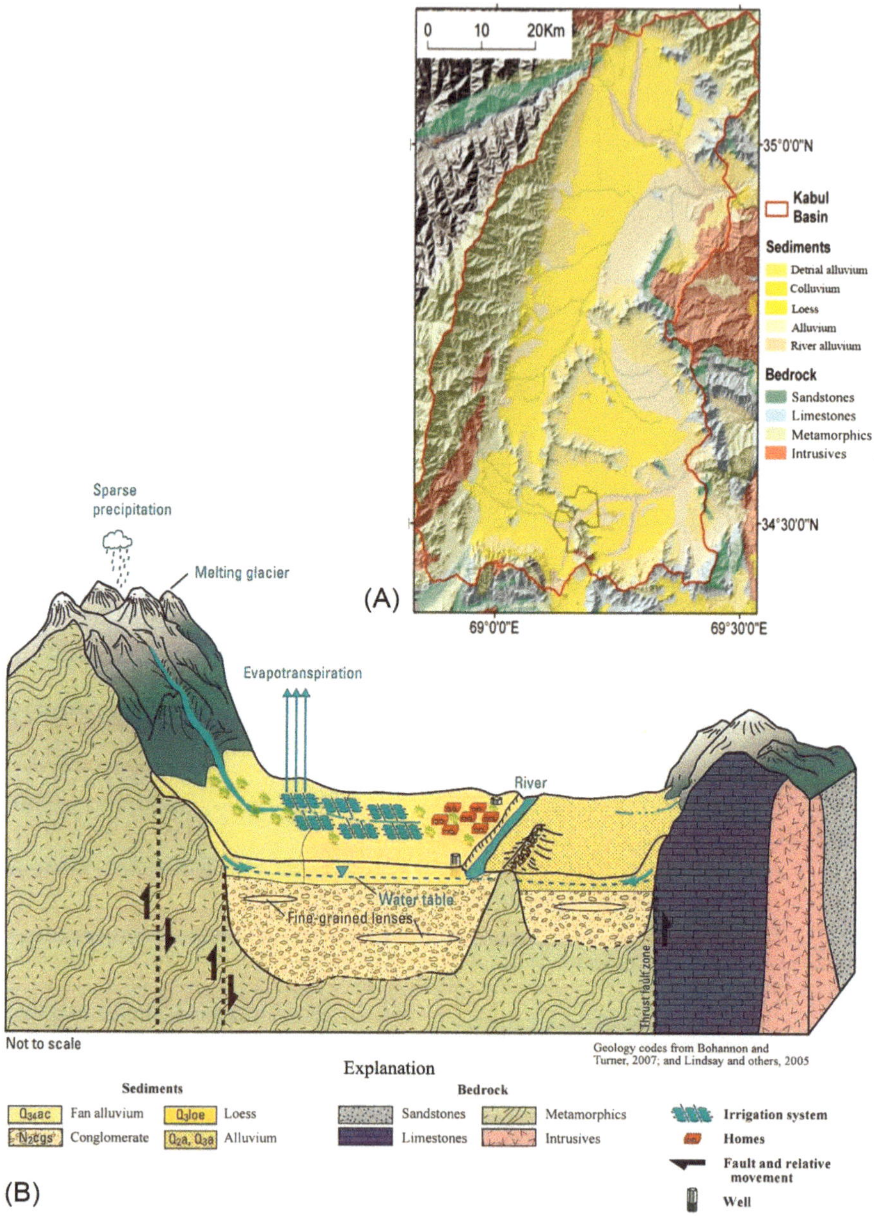


FIG. 4.4 (A) The main rock types in the Kabul Basin (Bohannon, 2007). (B) The different aquifer systems in Kabul City, pore aquifer (sediment) and fracture aquifer (bed rocks) (USGS, 2009). Generalized hydrologic cycle operative in the Kabul Basin, with its sparse precipitation at higher altitudes at basin margins, infiltration of recharge waters at basin edges and adjacent to rivers, and water storage in the varied sands and gravels of the sedimentary basin occurring in the basin of largely downfaulted crystalline (igneous and metamorphic) rocks of the region (Mack et al., 2009).

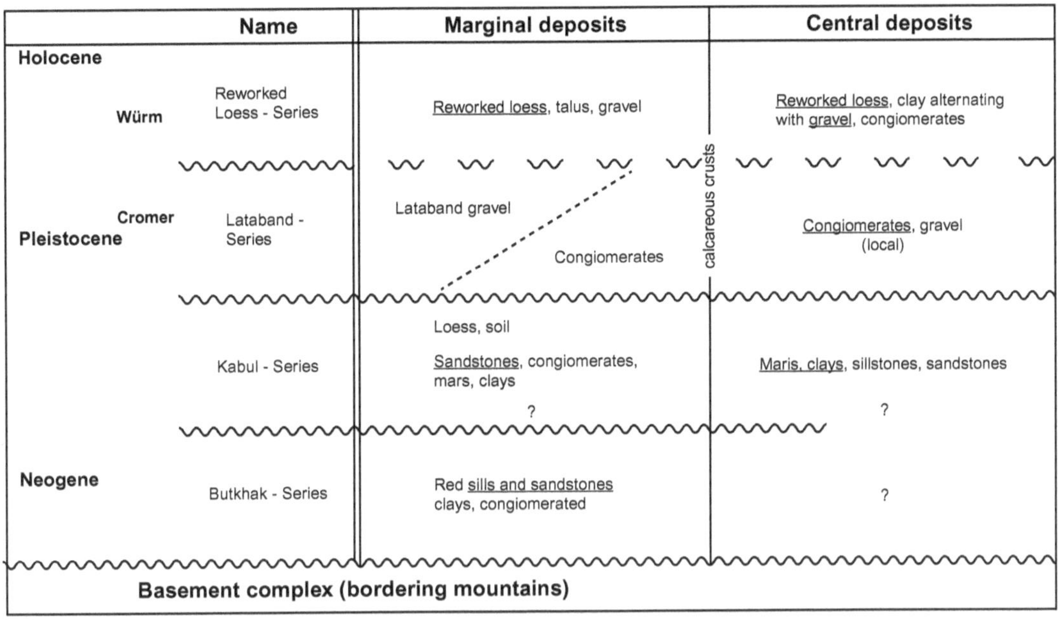


FIG. 4.5 Schematic stratigraphy of the Kabul Basin sedimentary rocks (from Hommilius, 1973), where the various sediment units in the basin can be formally recognized and more easily characterized in terms of their aquifer conditions and water yields (BGR, 2005).

Greeks, Kushan Buddhists, Indian Hindus and finally the Arabs after the Muslim conquests. The first Afghan rulers were the Ghorids in the 12th century (Lonely Planet, 2015). During these thousands of years we may presume that people obtained their water by the standard classical means of sending family representatives to river access sites and a few hand-dug wells to obtain daily supplies. Not until the latter half of the 20th century would Kabul have the beginnings of rudimentary piped water supplies (Nembrini et al., 2002).

The availability of water in Kabul was never much of a problem because the resources above grade and underground were relatively rich, with several productive aquifers ready to harvest after the expenditure of the necessary effort to dig or drill a well to them (Nembrini et al., 2002). Shallow ground water from wells <30 m deep is the main source for water supply in Kabul, Afghanistan (Houben et al., 2009; Tünnermeier and Houben, 2005; Mack et al., 2013). Thus in the late 1970s, almost 70% of the population in the city had decent access to water and a great many households were connected to the water distribution networks. A number of new water-development networks were planned then and in various stages of construction, with assistance of several foreign donors. This all began to terminate with the onset of the communist coup in 1978 and the subsequent Soviet invasion in 1979 as the country fell into waxing and waning episodes of extreme violence and the later civil war in the 1990s.

Instability and war in Kabul for much of the past four decades had a tremendously negative effect on the infrastructure as development of water supplies largely ceased, with much of the pre-war water-delivery capabilities destroyed by various types of explosions or simple breakdowns that were left unrepaired. Nembrini et al. (2002) noted that such violent conflicts and frequent political change had disastrous consequences on the public infrastructure. This

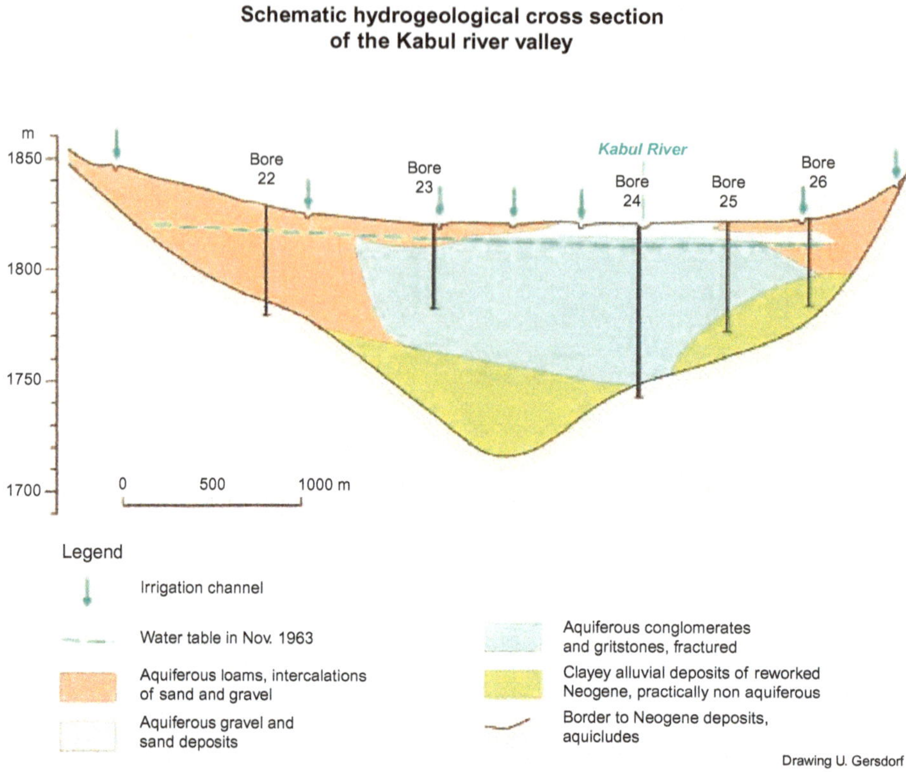


FIG. 4.6 Kabul Basin geological cross section showing borehole derived stratigraphy and ground-water condition mapped a half century ago (BGR, 2005).

was exemplified in Kabul particularly by the disruption of the power supply and the increasingly difficult access to water, which had immediate negative consequences on the public health of the entire town.

Under the Taliban rule of the late 1990s in Kabul, their near complete lack of technical competence in infrastructure maintenance, coupled with their no-nothing approach to governance, led to widespread breakdown of the urban infrastructure. Following the invasion of the Western Coalition troops in 2001 to restore a measure of normal modern rule in the country, in short order foreign donors and relief agencies arrived to improve the situation, particularly in urban water delivery because that was a glaringly obvious need everywhere in the country (Tables 4.2 and 4.3).

Challenges to the water supply that were soon discovered included:

- the tense security situation limited the mobility of water-cognizant personnel;
- dilapidated infrastructure occurred as a result of decades of war and neglect;
- high levels of non-revenue-bearing water existed, estimated at about 40%, including water from illegal connections;
- inappropriate water-pipe materials such as the asbestos concrete had been used in the older pipes, especially that provided by Russian contractors;

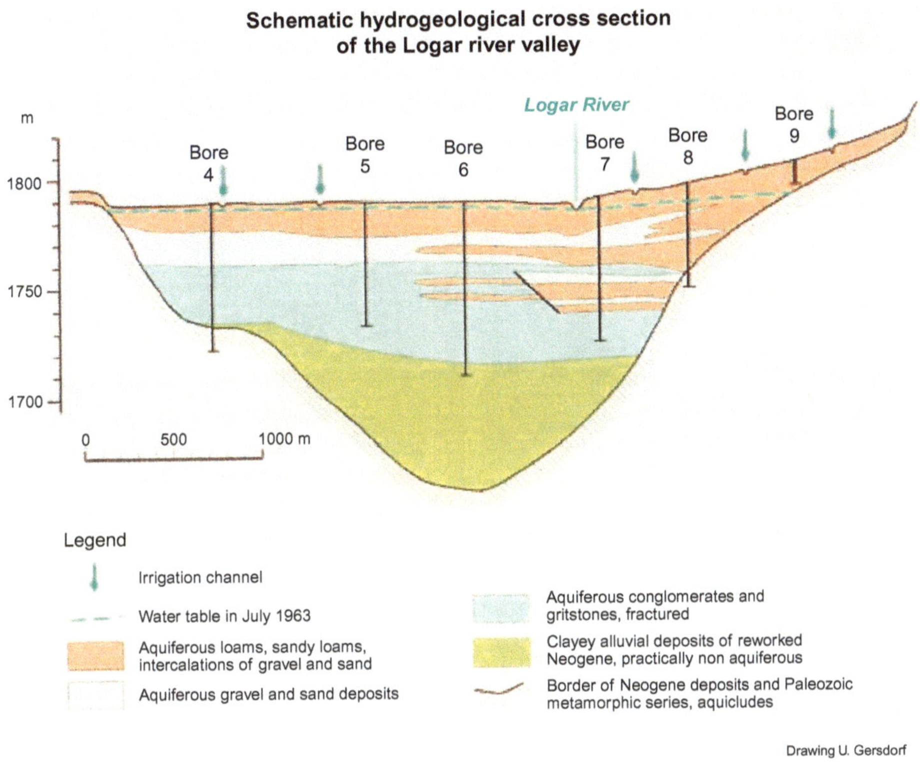


FIG. 4.7 Logar Basin stratigraphic configuration showing interpretation of configuration of sediment layers and associated underground water (BGR, 2005).

- a lack of qualified, water-cognizant personnel existed;
- extreme poverty was widespread;
- traditional social norms were a problem, especially concerning women that precluded their essential participation in necessary water-gathering activities;
- pollution of much of the shallow ground water occurred because of lack of sanitation and rampant open surficial defecation;
- poor service quality of the remaining piped water supplies, particularly because of the interruption of service caused by an unreliable and intermittent electric-power supply; and
- a delay in the establishment of large water works due to absence of competent or knowledgeable contractors capable of such installation.

By 2005, about 15% of the population of Kabul or a half million people had been connected to piped water supplies through an estimated 30,000 connections to households, whereas some 120,000 houses in Kabul had a water well in their courtyard, at least in those located on the extensive sedimentary plains of the city, but not those on the bedrock hills rising above the plain (Nembrini et al., 2002). In any case, these shallow wells are commonly polluted and subject to drying up in droughts as the water table declines below the well bottom. About a quarter of these wells have fallen dry in recent years, and roughly 40% of the remainder could

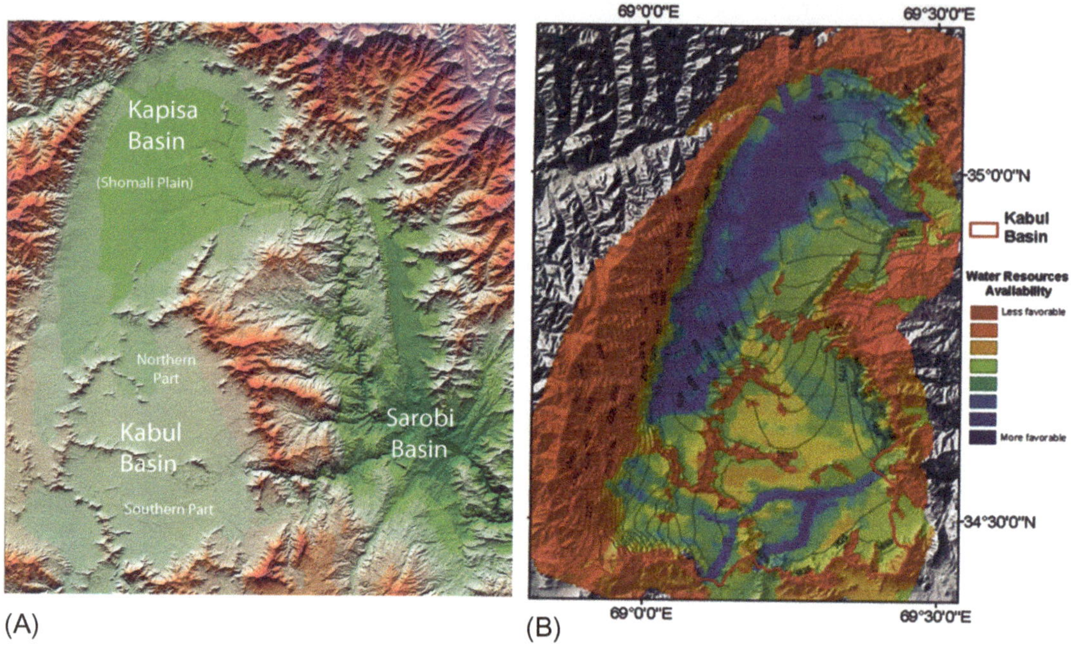


FIG. 4.8 (A) Geographic layout of the various topographic-geohydrologic subbasins of the Greater Kabul Basin is labeled on the digital elevation model. The main or central Kabul Basin, with its Northern and Southern parts are shown in the lower central part of the greater basin, the Paghman-Darulaman Basin is on the lower southwest side, and the Kapisa Basin with the Shomali Plain on its surface is in the north (after Fig. 3.19 from Shroder, 2014). (B) Water-resources availability in the Kabul-Kapisa basins of the Greater Kabul Basin, based upon an integration of estimated hydraulic properties, stream leakage, irrigation leakage, inflows from the mountain fronts, and aquifer thickness. Simulated 25m head contours are shown. The areas most favorable to production of ground water are clearly the northern parts of the Greater Kabul Basin. The Kabul-Logar river system occurs in the lower half of the map, where the course of the main river and its tributaries are delineated by the *deep-blue* colors. (From USGS abstract at Second Hindu Kush Geology Conference, 2007.)

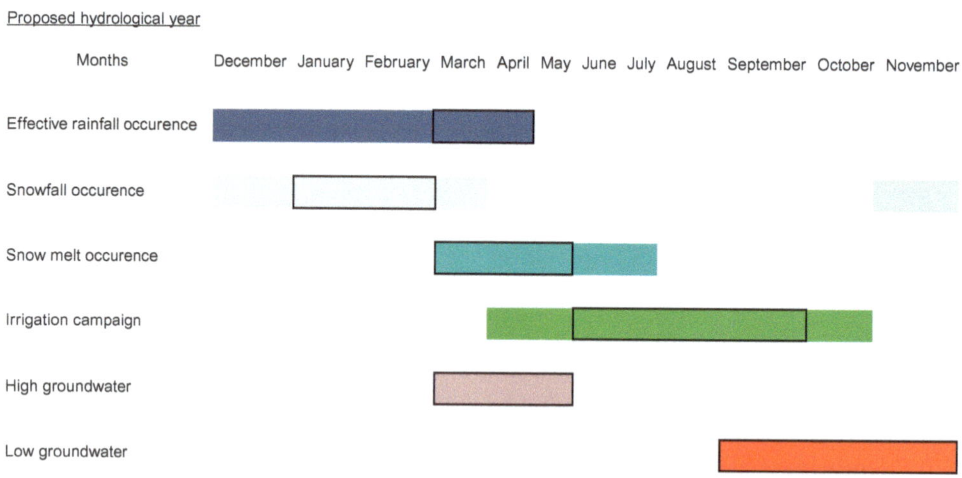


FIG. 4.9 The seasonality of moisture in the Greater Kabul Basin is shown in this breakdown of when snow and rain can be expected, as well as when irrigation is underway, and the condition of the underground water at different times.

TABLE 4.2 Strategic Key Findings for Necessary New Development of Water Resources in the Kabul Basin, Afghanistan

1. *Critical conveyance needs*—Water conveyance link (canal, tunnel) needed to bring water from Panjshir subbasin to Kabul population of 4 million that could serve 8 million, plus 43 $Mm^3$/year for Aynak copper mine
2. *Critical storage projects*—Multipurpose water storage facilities needed in both Panjshir and Logar-Upper Kabul subbasins
3. *Irrigation*—Maximum irrigation development is 184,000 ha, under assumption of 1-year drought, whose recurrence probability is ~10% (once in 10 years)
4. *Tradeoff between agriculture and urban water supply*—Cost of Kabul urban water supply increased by 25% when water from Logar River used exclusively to increase Logar subbasin irrigation by 73%
5. *Hydropower production*—Mixed hydro-thermal electricity system in Kabul, with new dam cascade on Panshir River (Gulbahar Power and Irrigation Dam) and Kabul River (Naglu Hydropower Plant, Sarobi II Power Plant) can meet maximum projected energy demand. Yet if hydropower is only source, additional storage for power production in Kunar also required
6. *Kunar (Konar) storage*—Such water storage required to meet higher winter energy demand, and for compensating generation shortfalls elsewhere in system
7. *Baghdara versus Panjshir projects*—Updated cost estimates needed to assess resettlement and rehabilitation costs between Baghdara and Panjshir hydro and irrigation projects for prioritizing investment options in Kabul Basin

See Table 4.5 and Fig. 4.16 for locations of named dams.
*From World Bank, 2010. Scoping Strategic Options for Development of the Kabul River Basin: A Multisectoral Decision Support System Approach. Sustainable Development Department, South Asian Region, The World Bank. http://www-wds.worldbank.org/external/default/WDSContentServer/WDSP/IB/2010/04/12/000333037_20100412001029/Rendered/PDF/522110ESW0Whit1anistan0Final0Report.pdf (accessed 07.06.15).*

TABLE 4.3 Strategies for Accelerating Development of Water Resources in Afghanistan

1. Key data gaps and new investment plans need to be developed in Kabul River Basin
2. Multisectoral analysis and planning is needed for all other major river basins in Afghanistan
   - Address lack of clarity on priority options that produce delays and poor resource use dedicated to project preparation;
   - Understand national water-resource situation, interbasin linkages and dependencies in energy and food-grain production;
   - Assess priority options for financing national water sector;
   - Understand levels of current and future water usages fundamental to developing appropriate GoA position in each shared international river basin.
3. Build capacity for planning and investment preparation as a cross-cutting imperative
   - Close data gaps in hydrological record;
   - Improve quality of preparation studies to investment-grade standards;
   - Update project costs planned in 1970s to modern standards.
   - Develop appropriate set of planning tools for each basin;
   - Address constraints of limited staff and skills base;
   - Constituting critically needed, multidisciplinary, Afghan transboundary water team so that various efforts of related capacity building can be focused and coordinated.
4. Strengthening institutions for decision making for multisectoral water resources issues
   - Supreme Council for Water Affairs Management and Ministry of Energy and Water must closely coordinate;
   - Finalization of Kabul River Basin development plan must be managed effectively and new plans must be developed for the other river basins;
   - Coordinate decision making within government on multipurpose water infrastructure investments;
   - Build consensus on and insuring acceptance of investment plans by different sectors.
5. Make progress on first steps

In order to offset the Taliban and warlord violence, it is critical to accelerate delivery of water-resource projects to show the competence of the GoA in such procedures, and in moving directly toward implementation with coordinated international assistance. Four parallel tracks are proposed.
*From World Bank, 2010. Scoping Strategic Options for Development of the Kabul River Basin: A Multisectoral Decision Support System Approach. Sustainable Development Department, South Asian Region, The World Bank. http://www-wds.worldbank.org/external/default/WDSContentServer/WDSP/IB/2010/04/12/000333037_20100412001029/Rendered/PDF/522110ESW0Whit1anistan0Final0Report.pdf (accessed 07.06.15).*

fall seasonally or permanently dry because of increased withdrawals and climate change if they are not deepened to overcome these problems (Mack et al., 2009).

Also in 2005 the public water-supply system provided about 60,000 $m^3$ of water through piped distribution networks from three main aquifers:

- Logar River aquifer well field southeast of the main city center;
- Afshar well field that draws from the Paghman River aquifer in the west of the city; and
- Alauddin well field that pumps from the Upper Kabul River aquifer in the south.

In 2013 the US Agency for International Development (USAID, 2013) noted that the then current average supply of water in Kabul was extremely limited at only 15 L/day, and the piped-water system of the city that was operated by the Afghanistan Urban Water Supply and Sewerage Company could only cover <20% of the population. Furthermore the antiquated and ailing distribution system over the years had become highly unreliable and inefficient, with water typically delivered every other day for only 12 h. This archaic system forced households and businesses to store water for use during the interim days. Consequently, at the behest of the GoA the German Development Bank (KfW) early on spearheaded the Kabul Urban Water Supply program to meet the goals established in the Afghan National Development Strategy, which planned to increase piped water to at least 50% of the population in the city.

The KfW set up a four-part scheme involving assistance and funding from USAID, the French Development Agency and the governments of Germany and Greece to expand the Logar II well field and pumping station, construct a new water-storage reservoir, as well as installation of complete hydraulic equipment for the reservoir at 5000 $m^3$ capacity (USAID, 2015). The original USAID (2013) objective specifications included:

- increased piped water coverage to ~1.4 M people (~29% of population in 2010 of ~4.5 M);
- double production from 60,000 to 120,000 $m^3$/day, or 44 M $m^3$/year;
- increase reservoir capacity from 32,700 to 73,000 $m^3$;
- increase distribution network from ~500 to 1300 km; and
- triple household connections from 30,000 to 90,000. D. Bank 2002 http://www.holymoor.co.uk/PartF.PDF (accessed 07.07.15).

By 2015 USAID had also planned to increase the new household connections up an additional 45,000 people to a total of 135,000 people, as well as achieving water quality that meets the standards of the World Health Organization by using continuous water disinfection. This would reduce health risks and improve livelihoods by providing an improved drinking water supply, while reducing interruptions of water supply in the new service areas and keeping the water on 24 h/day. This would be achieved in part by designating water-protection zones and adopting protection measures for the drainage area of the Logar well fields.

The steadily increasing population of ~4 million people at the present time consumes ~30–40 M $m^3$ of ground water per year (Houben et al., 2009), which compares exactly at the 27 L/day needed to the estimates of local water usage of 20–40 L of water per day, or the larger, 50–120 L/day used by the Western Coalition soldiers (Gellasch, 2014b). In Kabul itself the small number of people who have piped water usage utilize about 50–75 L/day/person, whereas the people who must use hand pumps and carry the water home use only 20–30 L/day/person (Eqrar, n.d.). The recharge of the area is estimated at 20–45 M $m^3$/ha in wet years, although the drought of 2000–05 prevented significant recharge, with the result that levels of the water table beneath Kabul have begun precipitous declines.

Historical levels of ground water quantity and quality show considerable decline in recent years because: (1) most springs and karez have dried up; (2) annual precipitation has decreased; (3) water quality has decreased seriously; (4) water-logging and salinization have increased in some areas; (5) water-table levels have declined in others; (6) evaporation has increased with rising global temperatures; and (7) marshes in Kabul Basin have dried up to leave surficial salt crusts (Saffi, 2011; Mack et al., 2013). Rates of decline of water-table level beneath Kabul were 0–0.7m/year from 2004 to 2008, but have doubled to 1.5m/year from 2008 to 2012 (Mack et al., 2013).

Only 20% of the population have access to tap water and most (86%) households have only simple cesspits for sewerage. Shallow ground water thus has tremendous amounts of pollutants due to improper waste and wastewater disposal. Elevated concentrations of nutrients such as boron, chloride, nitrate, sulfate, fecal bacteria, total dissolved solids indicate contamination by wastewater disposal into the aquifers (Himmelsbach et al., 2006; Gellasch, 2014b). High infant mortalities are attributed to high levels of ground-water contamination. The Water, Sanitation, Hygiene assessment by UNICEF (2013) notes that more than 41,000 people (mainly children) were dying from diarrhea in Afghanistan at that time; whereas 103/1000 was the infant mortality rate, the under-5 mortality was 149/1000; the diarrhea prevalence was 31%; under nutrition was 39%; stunted children were 64%; and underweight people were 67%.

Solutions to the situation of declining water levels in the Kabul Basin are manifold and include a variety of solutions of greater and lesser cost. Eqrar (n.d.) has listed four different solutions for the declining water-table levels: (1) Reinfiltrate river water during the wet season in the Kabul-Basin aquifers by artificial recharge methods (ponds, injection wells); (2) Build reservoir lakes for longer-term infiltration; (3) Tap deeper Neogene aquifers; and (4) Get water from elsewhere (Panjshir) by more expensive pipeline or more uncertain water-tanker trucks. In addition, the refurbishment of a major ephemeral wetland in the south part of the city, the Kole (Lake) Hashmat Khan, has been studied in some detail to help increase recharge.

### 4.2.1 Kole Hasmat Khan Recharge

The ephemeral lake and wetlands of Kole Hashmat Khan are situated in the south of Kabul. The area extends over an approximate surface area of 190ha, and is surrounded in the uplands by the Sherdarwaza-Welayatee series of metamorphic rocks, namely: gneiss, amphibolites, schist and phyllite of Precambrian age (Figs. 4.10–4.13). The surface of the wetland area zone material and the greater part of flat plain of this area are covered with fine clastic sediments of silty clay and fine sand, very dark brown or black colored, or sometimes gray, interspersed with sporadic angular particles of colluvial sediments from the rocky foothills.

The Logar River is the main water source, flowing in early winter and spring through a distributary or sub-stream, including seasonal snow melting and rain water for the lake. It has been important to various bird populations (Fig. 4.11). In terms of environmental threats in the city itself, the Kole Hashmat Khan wetland is the only remaining water body of an extensive wetland that once stretched across the city's southern plains. Seeking formal protection status for the area has become increasingly difficult in recent times, and finally, this lake, which is surrounded by dense reed beds, was one of the most important and well-protected wetlands until the onset of the war in 1979 (Fig. 4.13).

FIG. 4.10 Full lake of Kole Hashmat Khan from the south in Apr. 2004. *Photograph by N. Eqrar.*

FIG. 4.11 Birds at Kole Hashmat Khan. *Photograph by Saeed Naqibullah Mostafawi.*

Recently, the Kole Hashmat Khan wetland has begun to become completely drained of water by late summer nowadays, and climate change appears to be having a major impact as the water table is falling rapidly in this area (Fig. 4.14). The lake dried out entirely in 1999 due to the severe drought in the region. Prior to the war, it received water from a sub-stream or distributary of the Logar River, but this was informally dammed in various places and became silted up. In addition, water from the Logar River has been diverted for irrigation. The local community has encroached upon the lake, and the settlements have extended all around it. The surrounding wetland area has been occupied illegally throughout by warlord-types of people who are difficult to dislodge, and the large former wetland has been converted in part into an unplanned residential area, which has reduced the wetland to a minimum size (Table 4.4).

FIG. 4.12 View to the southwest over Kole Hashmat Khan during the winter wet season. *Photograph by Saeed Naqibullah Mostafawi.*

FIG. 4.13 View to the southwest in late spring of the receding water and thick reed beds. *Photograph by Saeed Naqibullah Mostafawi.*

#### *4.2.1.1 Surface Water*

Koli Hashmat Khan is a small, shallow (c.1.5m max. depth) L-shaped lake, c.2.5km long and 300–1000m wide. As noted, it has in the past been connected to the Logar River by a small sub-stream during the wet season (Wildlife Conservation Society/Afghanistan, Aug. 2010). To rehabilitate the Koli Hashmat Khan wetland, the local sub-stream should be redeveloped and opened up again to bring water from the main water source of the Logar River, whenever water flows in November until the end of April or the wet season of the year (Figs. 4.10–4.13).

The wetland area extends over an approximate surface that was 190ha in 1980. Based on a rough calculation of the wetland area, using ArcGIS and ArcScene tools, the area of the wetland has been calculated to be around 141.7ha in 2014, which means 49ha of the

FIG. 4.14 Same view as Fig. 4.13 in midsummer after all the water has evaporated and the reeds have been harvested. *Photograph by Saeed Naqibullah Mostafawi.*

TABLE 4.4 Key Points Concerning Kole Hashmat Khan to be Carried Forward and Implemented if Possible

*Existing problem at Kole Hashmat Khan wetland*

- Decreasing of wetland size due to illegal occupation.
- At present, no protection staff exist at Koli Hashmat khan wetlands.
- At present, grazing, and the collection of fuel reeds in the vicinity of Koli Hashmat Khan are open to anyone.
- A shortage of water resources exists.

*Recommended actions*

- Installation of meteorological and hydrological gauges stationed around the wetland area.
- Continuous collection of meteorological and hydrological data around the wetland area.
- A ground-water monitoring program of wetland area.
- Collection of water sampling of surface and ground water for analyzing and interpreting chemical, microbiological and physical characteristics of Koli Hashmat Khan wetland.
- Construction of small scale reservoir dams on the Logar River.
- Construction of subsurface dam with an isolating retaining wall and impermeable basement, or any harvesting methods to restore seasonal water flow during the wet season and use it during dry season.
- Rehabilitation of the local sub-stream to bring water during the wet season and distributed by gravity from the main source of the Logar River.
- Managed ground-water recharge or implemented pilot project as an artificially aquifer recharge activity around the wetland.
- Carefully delineate and map the boundaries of the protected area, dispose sign posts around the lake to inform the visitors on the status of the area.
- Establishment of a protection zone around wetland, to stop expansion of illegal residential buildings and transportation of liquid and solid waste.
- Strengthening institutional structures for restoration, rehabilitation and protection of the wetland.
- Encouraging of international association for financial supporting of this wetland.
- Encouraging, facilitating, and supporting national research on various aspects of the wetlands and rural life in the surrounding areas.
- Enhance public awareness, wetland importance, biodiversity and sustainable use.
- Finally remove physical stresses such as, grazing, logging, poor water quality, urban illegal development, solid waste dumping and residential waste liquid, restore water management, replant vegetation, restore vegetation management and different types of biodiversity.

wetland area have been lost due to occupation by the local people. The volume of water-storage capacity in Koli Hashmat Khan, will be around (8,763,126 $m^3$) or 8,763,126,000 L. In the meantime, various pollutants, including domestic sewage, agricultural effluents and the spreading of some industrial solid waste and liquid waste, are known to have entered the wetland area, which is expected by some to become one of the catastrophic local environmental phenomena of the future.

Because of negative balance between recharge and discharge due to low annual precipitation in the Kabul basin, all surface waters have now largely disappeared from the wetland. Summer evaporation also contributes to the drying up of the lake by late summer, and sometimes as early as July when winter precipitation has been low, or possibly as rain. The growth of the population, and the considerable extent of human utilization of the water, are the main important factors that have caused lowering of the water table. In fact, many people have drilled deep wells on the upper and lower sides of the Koli Hashmat Khan wetland for drinking and irrigation purposes. The over-utilization of the ground water from this area are some of the negative impacts on the dryness of this wetland (Figs. 4.14 and 4.15).

After 1999 the wetland has periodically become dry, in the dry season of the year, because of climate-change impact, low annual precipitation, change in precipitation type (mostly rain), limited snow coverage on the elevated areas of Sherdarwaza mountain, the depth of the water column has decreased about (50–80 cm; local farmers report in 2014), growth of the huge population caused illegal occupation of the wetland area, and finally over-irrigation drawn from the Logar River, the wetland and traditional local sub-stream became fully dry and most parts of the wetland were used as a playing ground and were also utilized as an overgrazing site for local people.

#### 4.2.1.2 *Snow Melting and Rain Water*

Seasonal snow melting and rainwater are minor sources of water for this lake, based on the meteorological data, the precipitation type has changed from heavy snows or earlier times, to rain showers due to global-warming phenomena and climate-change impacts, which are occurring all over the world. Due to rain showers during the wet season of the year, the siltation is also increasing and causing decreasing of water-storage capacity in the wetland. In addition, still the absence of any rain gauges, snow gauges, water level gauges or flow-water measurement gauges continues around this wetland, so that accurate and continuously collected data are not obtained.

FIG. 4.15 View similar to that of Fig. 4.10 in Oct. 2006 after the lake has been completely dry for some months. *Photograph by N. Eqrar.*

#### 4.2.1.3 *Ground Water*

The second alternative to help with the problems of Lake Hasmat Khan will be using the ground-water resources in the vicinity of this area, which need to be investigated outside of the wetland, even from southeast of Chaman Hozoori or the national playing ground, or around the areas of the Benihesar area (Bagh Afzal). Another option will be usable during the dry season of the year or whenever the wetland becomes dry, when the ground water flows from outside the wetland area, and must be circulated through a trench recharge or flow by ring-road canal around the wetland circumstance, to infiltrate slowly and raise up the water level of the wetland.

Additionally, one of the practical ways in a semi-arid country or water-scarce country is the buildup of a subsurface dam [a subsurface dam is a system to store ground water by a "cut-off wall" (dam body) set up across a ground-water channel] on valley area with a good isolated retaining wall and the establishing of any water harvesting or water diversion methods, to restore water during the wet season of the year, to be used during the dry season.

With proper water management, conservation and rehabilitation of this wetland, we will be able to develop the area as a natural garden for intelligent water management to help different water bird species, agriculture sources, tourism, recreation and hunting, rangeland and pastureland, all naturally recharging ground water into the other research areas downstream (Table 4.4).

## 4.3 NEW DEVELOPMENT OF THE KABUL RIVER BASIN

The World Bank (2010) has taken the lead in pushing the GoA to take better control of its water resources in order to formulate and undertake a new integrated development plan for the KRB using the approach of a multisectoral, decision-support system (DSS) (Tables 4.1–4.3). Accordingly, the work focuses on the multiple tasks necessary to:

**(1)** Analyze the medium- and long-term options for developing the water resources in the KRB for multiple important usages, including:
- domestic water supply
- industrial water supply
- hydropower
- mining
- irrigated agriculture
- environment.

**(2)** Collate information on the basin, including:
- existing water-resources development options
- potential water-resources development options
- water uses
- water demands.

The whole information that becomes available should be put together in a simple and user friendly DSS, in order to enable multisectoral analysis to optimize development options in the KRB by any of the concerned and relevant government ministries, as well as their development partners (World Bank, 2010). Such work, conducted in collaboration with the GoA, is quite expected to help strengthen the adoption of integrated approaches to planning of the KRB, and associated water-resource management.

## 4.3.1 Kabul River Basin Water Storage Options

In a mountainous country such as Afghanistan the greater KRB has a range of rather advantageous topography for the development of water storage projects (World Bank, 2010). The upper Panjshir River rises in the Central Hinu Kush and flows through the steep Panjshir Gorge to where it issues out onto the Shomali Plain, where it is confluent with the smaller Ghorband River (Figs. 4.16 and 4.17). The proposed Golbahar Dam and reservoir (Panshir I or R10; Table 4.5) would be established near the mouth of gorge. Downstream from the Shomali Plain, a second major gorge has been eroded in the bedrock uplift across the course of the lower Panjshir, where the second dam site of the Baghdara reservoir (R11; Table 4.5; Fig. 4.16) would be constructed. Several other dams could be constructed farther up the upper Panjshir or the Ghorband (Barak, R9; Totumdara, R8; Table 4.5; Fig. 4.16). A number of reservoirs are planned for the Logar and upper Kabul rivers (R1, R2, R4, R7); as well as another hydroelectric power dam and reservoir on the lower Kabul River, Sarobi II (R168) (Table 4.5; Fig. 4.16). Other barrages (R20, R21) and reservoirs without hydroelectric power (R17, R19) are planned for the Konar (Kunar) and Alingar tributaries to the Lower Kabul River. This is all planning for water demands as they are most likely to occur in the next decade or so (Table 4.6).

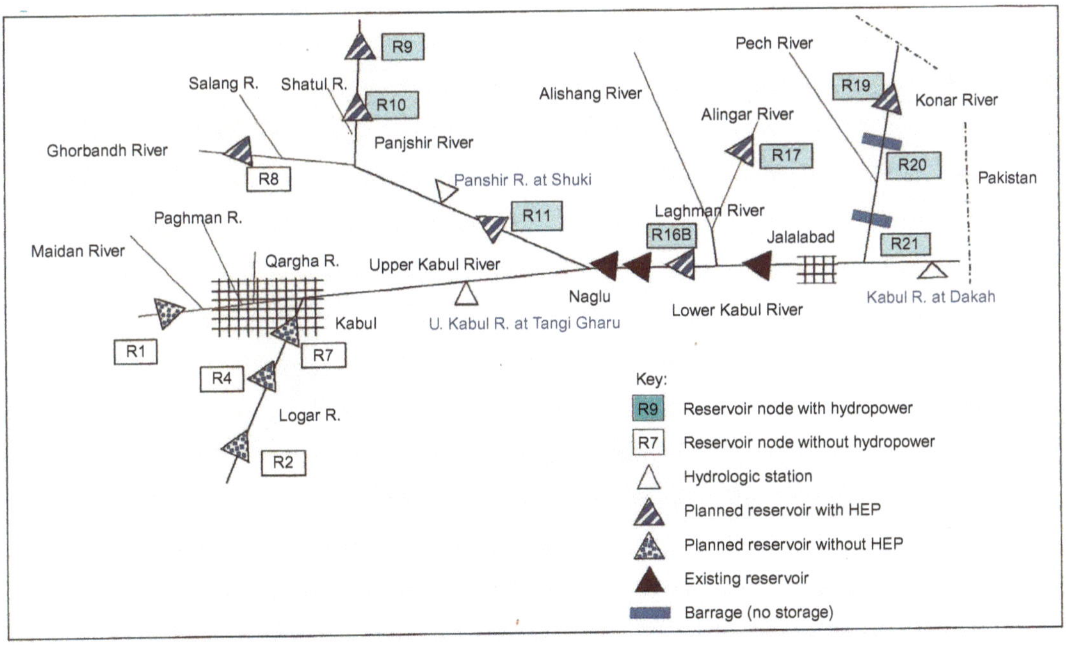


FIG. 4.16 Map of dam and reservoir water storage options on the greater Kabul Basin, which are listed in Table 4.5. *From World Bank, 2010. Scoping Strategic Options for Development of the Kabul River Basin: A Multisectoral Decision Support System Approach. Sustainable Development Department, South Asian Region, The World Bank. http://www-wds.worldbank.org/external/default/WDSContentServer/WDSP/IB/2010/04/12/000333037_20100412001029/Rendered/PDF/522110ESW0Whit1anistan0Final0Report.pdf (accessed 07.06.15).*

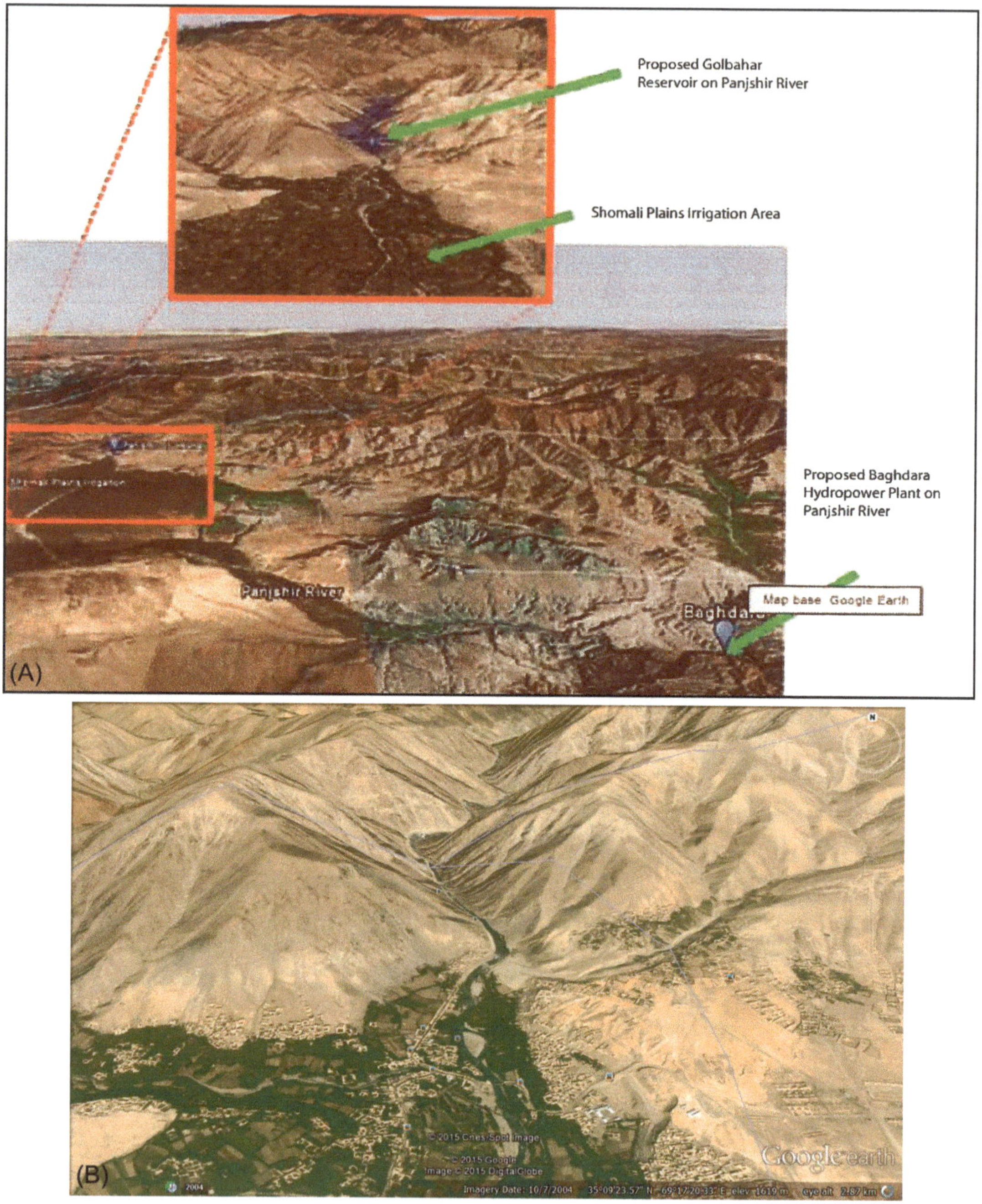


FIG. 4.17 Perspective map of the sites where one or more dams and water reservoirs could be constructed in the (A) upper and (B) Shomali plains irrigation area. *From Google Earth.*

TABLE 4.5 Proposed Water Resources Development Projects and General Dams Characteristic in Kabul Basin or Indus Basin

| | | Dam Height (m) | Gas Storage (Mm³) | Live Storage (Mm³) | Installed Capacity (MW) | Annual Cost (MUS$) | Capital Cost (MUS$) | Annual Unit Cost of Live Storage ($/m³) |
|---|---|---|---|---|---|---|---|---|
| *Panjshir subbasin* | | | | | | | | |
| Totumdara | R8 | 135 | 410 | 340 | na | 33.2 | 332 | 0.098 |
| Barak | R9 | 155 | 530 | 390 | 100 | 117.4 | 1174 | 0.301 |
| Panjshir I | R10 | 180 | 1300 | 1130 | 100 | 107.8 | 1078 | 0.095 |
| Baghdara | R11 | 40 | 400 | 330 | 210 | 60.7 | 607 | 0.184 |
| *Logar-Upper Kabul subbasin* | | | | | | | | |
| Haijan[a] | R12 | 50 | 220 | 200 | na | 7.2 | 72 | 0.036 |
| Kajab | R2 | 85 | 400 | 365 | na | 20.7 | 207 | 0.057 |
| Tangi Wardag | R4 | 65 | 350 | 300 | na | 35.6 | 356 | 0.119 |
| Gat | R7 | 20 | 500 | 440 | na | 5.1 | 51 | 0.012 |
| *Lower Kabul subbasin* | | | | | | | | |
| Sarobi II (ROR) | R16B | 200 | na | na | 210 | 44.2 | 442 | na |
| Laghman A | R17 | nd | 405 | 288 | 44 | 1251 | 125.1 | 0.434 |
| Konar A | R19 | nd | 1212 | 1010 | 366 | 94.8 | 948 | 0.094 |
| Konar B (ROR) | R20 | nd | na | na | 81 | 23.2 | 232 | na |
| Kama (ROR) | R21 | nd | na | na | 60 | 11.5 | 115 | na |

[a] *Near the Shahtoot site that is currently under study.*
*Notes:* Costs are for full development of the site; ROR, run-of-river; na, not applicable; nd, no data.
*Sources: Montreal Engineering Company, 1978. Kabul River Valley Development Project, Master Plan Report. Two volumes. Government of Afghanistan, Ministry of Water and Power, Kabul, vol. 1, chapter 4; Toosab and RCUWM (Toosab Consulting Engineers and Regional Center for Urban Water Management), 2006. Integrated Water Resources Management of the Kabul River Basin; and mission estimates.*

## 4.3.2 Irrigation Development in Greater Kabul Basin

The existing and potential areas of irrigation development that can take place in the greater Kabul Basin (Fig. 4.18) include (large green ovals) the Logar-Upper Kabul subbasin, the Panshir subbasin (Shomali Plain and Kapisa), and the Lower Kabul subbasin. These three main areas have been further subdivided into 14 individual areas that have been identified as water-demand nodes for agriculture and used as calculation devices for water planning (World Bank, 2010).

## 4.3.3 Environmental Protection in the Greater Kabul Basin

At the same time that megaplans are being constructed for the greater development of the Kabul Basin water scenarios, considerable effort is also being expended on considerations of the water quality in the basin, especially as it is being so adversely impacted by anthropogenic immisions (Houben and Himmelsbach, 2005; Himmelsbach et al., 2006; Mack et al., 2013).

TABLE 4.6 Water Supply and Demand Balance for the Greater Kabul River Basin, as Measured at Dakah, Close to the Pakistan Border in M m$^3$/Year

| | With Kunar | Without Kunar |
|---|---|---|
| | **Subbasin Flow** | **Subbasin Flow** |
| Average annual water supply | 19,206 | 5224 |
| Basin bulk water demands | 3874 | 2901 |
| Urban and rural drinking | 401 | 401 |
| Environmental flow requirement | 41 | 41 |
| Mining | 43 | 43 |
| Irrigation | 3389 | 2216 |

This is a planning document that can be used to calculate possible future water demands as they will occur in the next decade or so.
*From World Bank, 2010. Scoping Strategic Options for Development of the Kabul River Basin: A Multisectoral Decision Support System Approach. Sustainable Development Department, South Asian Region, The World Bank. http://www-wds.worldbank.org/external/default/WDSContentServer/WDSP/IB/2010/04/12/000333037_20100412001029/Rendered/PDF/522110ESW0Whit1anistan0Final0Report.pdf (accessed 07.06.15).*

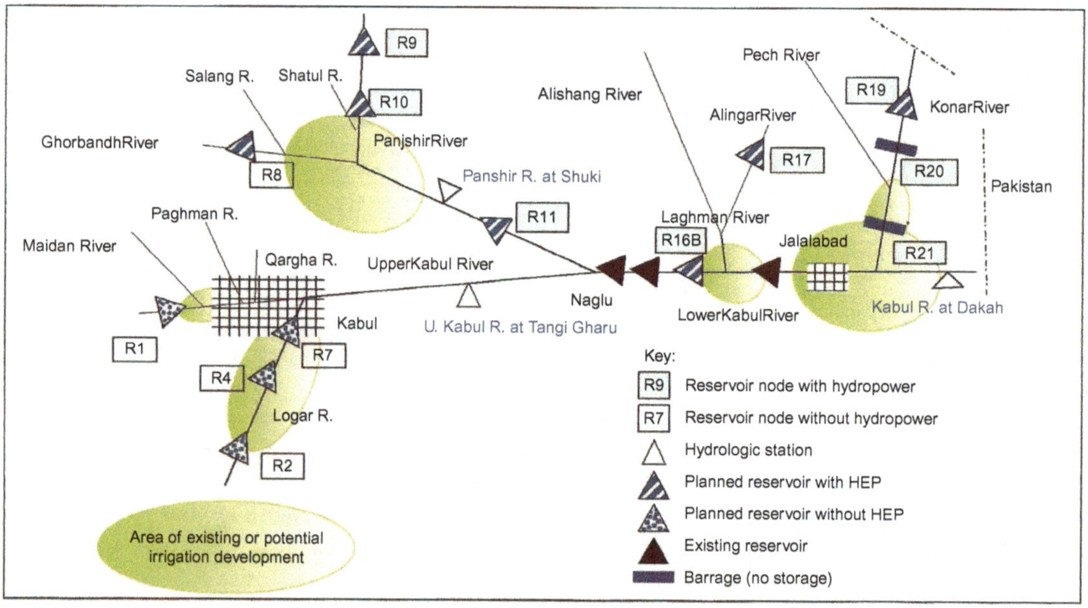


FIG. 4.18 Schematic map of existing and potential areas of irrigation development. *From World Bank, 2010. Scoping Strategic Options for Development of the Kabul River Basin: A Multisectoral Decision Support System Approach. Sustainable Development Department, South Asian Region, The World Bank. http://www-wds.world-bank.org/external/default/WDSContentServer/WDSP/IB/2010/04/12/000333037_20100412001029/Rendered/PDF/522110ESW0Whit1anistan0Final0Report.pdf (accessed 07.06.15).*

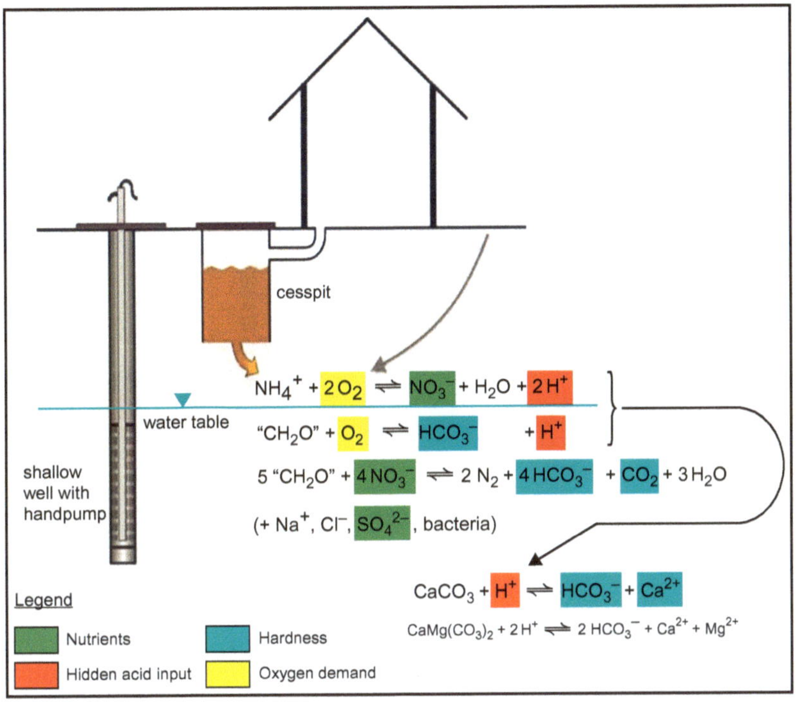


FIG. 4.19 Simplified diagram showing the anthropogenic impact on the quality of ground water that is associated with chemical processes arising from sewage contamination in Kabul City. *From Houben and Himmelsbach (2005).*

The chemical processes arising from sewage contamination are especially problematic in that they feed organic nutrients into the ground-water system that has chemical demands upon the oxygen content of the water, while also producing a hidden acid input, as well as adversely contributing to greater hardness of the water (Fig. 4.19).

The adverse anthropogenic impacts on the ground-water quality in the greater KRB aquifers have been a matter of considerable discussion among many people, educated Afghans and most foreigners for many decades. The all-too-common, water-borne diseases of Kabul (amoebiasis, cholera, giardia) have been well known to the domestic and foreign medical communities for somewhat longer. The pathenogenic and chemical pathways, in what has been informally termed by some biologists as a major "fecal environment" have been diagrammed and described in detail by several authors (Houben and Himmelsbach, 2005; Mack et al., 2013).

The adverse chemical reactions from sewage in the soil and shallow ground water that are inimical to human health also add additional salt (sodium—Na, chloride—Cl, sulfate—$SO_4^{2-}$) and bacteria into the ground water, which promotes the growth of the additional bacteria by the large supply of nutrients (Houben and Himmelsbach, 2005). A qualitative view of the material flows into the ground water of the Kabul Basin are instructive as to the number of pathways of water and pollution loads that enter and leave the basin (Fig. 4.19). The river inputs into this relatively closed mountain basin transport only a minor amount of dissolved solids or mineralization with them as they enter the basin. Slightly more mineralization of the waters occurs as a result of water-rock interactions within the aquifer, but the major reactions

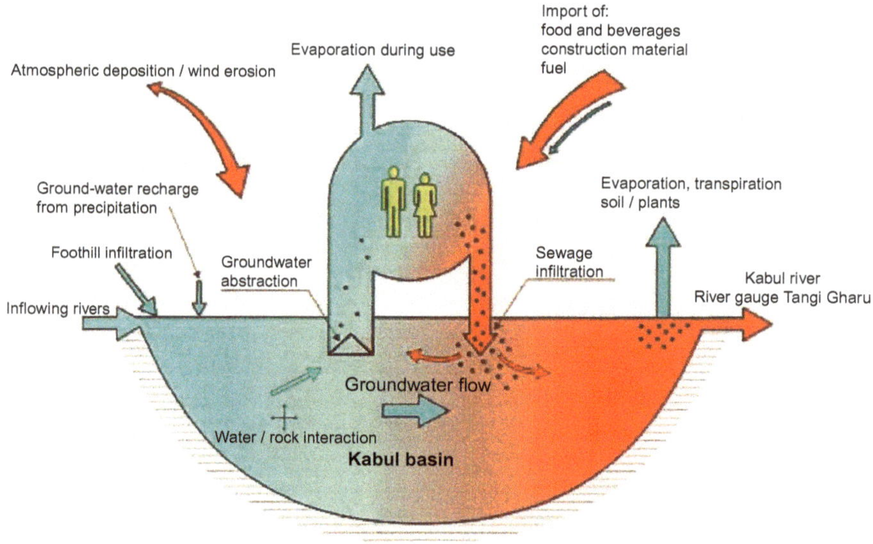


FIG. 4.20 Schematic diagram of material flows in ground water of the Kabul Basin. *From Houben and Himmelsbach (2005).*

in the shallow ground water are those from sewage (Fig. 4.20). The enrichment in the salt load contributed in this fashion is exacerbated further by the strong evaporation in the greater KRB, as well as by the import of food, drinks, construction materials, fuel and minerals, all of which contribute some part to the strong pollution sources in the region.

The major contamination of the waters of the Kabul Basin could be reduced by a climatologically wet period that lasted for several years to flush out minerals from the aquifers. Given the drought cycles that have been increasing with increasing frequency in recent years (Shroder, 2014), this would be an unlikely occurrence, however. Instead, a long-term solution would appear to be import a long-distance steady water supply from wetter areas to the north such as the Panjshir Valley.

In addition, the problem could at least be partly ameliorated by implementing a number of technical measures, along with stronger urban-planning controls (Houben and Himmelsbach, 2005), such as:

- regulated urban planning;
- regulated garbage/refuse collection in the urban centers and proper disposal;
- centralized and/or decentralized water supply (with treatment);
- centralized and/or decentralized sewage treatment; and
- sale of water from tanker trucks.

The ideal situation is for a centralized water supply and sewage disposal at a large and pervasive scale but this is unlikely to happen because of the frequent uncontrolled construction and widespread poverty. Rather, decentralized sources of supply to several city blocks by deep cased wells, coupled with sewage disposal in small treatment plants downstream are feasible in some urban districts, at least. In addition, some of the most strongly contaminated wells should be shut down and attention given to construction of drainage pits far away from good wells and never dug deeply enough to penetrate the ground-water-protective, loess-cover sediments.

## 4.3.4 Future of Greater Kabul River Basin

Future prediction is notoriously fraught with the arrival of reality that so damages the hopes of dreamers, but at least the process of futures formulation does help to envision different possibilities. In addition, the planning processes for water projects in Afghanistan generally best involves the trinity of integrated water-resource management, river-basin management and participation by multistakeholder platforms (MSP). These three planning issues are part of a push toward greater devolution of decision-making processes, both from the central government down to subbasin platforms, as well as from the government to the water users (Thomas, 2013). Setting up of the planning objectives and criteria generally begin with direct contact with the stakeholders who have the most legitimate interest in the outcome. Consultations with the MSP are the normal mechanisms to ascertain their viewpoints in order to translate these into specific development objectives and criteria. Such information then enables assessment of the various options and scenarios. In these cases in Afghanistan, however, consultation with all the MSP was not possible because of the prevailing situation of instability and violence in the country, with the result that a potential surrogate could be obtained by reference to the Water Law of 2009, as well as by assessment of it by those who have made it their business to make such law accessible to independent outside analysis (Santabanez and Bell, 2012; Thomas, 2013). With these and other government statements about the Water Law, a reasonable effort can be made to ascertain what the most realistic goals should be (Table 4.7), and what the consequences would be, as well as the best criteria and metrics that could be used to assess the options of the different scenarios (Table 4.8). The Kabul Basin megaplan designed by the World Bank (2010) thus is a measure of the thoroughness and technical sophistication applied to the seemingly intractable problems of water supply and sewage control in eastern Afghanistan (Fig. 4.21).

TABLE 4.7 Envisioning the Future of the Kabul River Basin: An Ideal Kabul River Basin in 20 Years Would Have These Attributes

| | |
|---|---|
| 1. | Reliability of agricultural water supply is increased and drought vulnerability is decreased in existing and new areas of irrigated agricultural development |
| 2. | Sufficient hydropower generating capacity is developed to ensure that demand in the basin is fully met in all seasons |
| 3. | Domestic users and industry in and around Kabul have secure access to safe, reliable and adequate water supply for future needs |
| 4. | All rural households have improved access to a secure and safe drinking water supply |
| 5. | Increased agroprocessing investment, coupled with an improved all-weather road system, gives farmers improved access to markets and higher-value options |
| 6. | No water quality problems emerge as a result of urban and industrial growth in water use and adequate environmental flows are maintained |
| 7. | Well-managed watershed areas are productive and environmentally protected, and the quality of livestock grazing is improved, ground-water recharge is improved, and soil erosion is minimized |
| 8. | Development activities are socially, environmentally and economically sound |
| 9. | Food security is achieved |
| 10. | Decrease in poverty measured as per capita annual average income that rises from current levels to \$ X per day |

*From World Bank, 2010. Scoping Strategic Options for Development of the Kabul River Basin: A Multisectoral Decision Support System Approach. Sustainable Development Department, South Asian Region, The World Bank. http://www-wds.worldbank.org/external/default/WDSContentServer/WDSP/IB/2010/04/12/000333037_20100412001029/Rendered/PDF/522110ESW0Whit1anistan0Final0Report.pdf (accessed 07.06.15).*

TABLE 4.8 Consequence Table—Criteria and Metrics to Assess Scenarios Options

| Type | Criteria | Indicator | Preference (H = Higher is Better, L = Lower is Better) | Units |
|---|---|---|---|---|
| Economic | Total net economic benefits | Economic benefits less costs | H | M$/year |
| | Agriculture | Agricultural benefits | H | M$/year |
| | Power | Electricity generation benefits | H | M$/year |
| | Agroprocessing | Value-added of agroprocessing in basin | H | M$/year |
| | Mining | Annual water supply matches production requirements | H | $Mm^3$ |
| | Employment | Total new full-time equivalent jobs | H | M# jobs |
| Social | Poverty | Change in no. people > $1($5?)/day | H | M# people |
| | Resettlement | People relocated due to development projects | L | Thousands# people |
| | Access to water | Percent rural population w/ access to safe drinking water | H | % |
| | Kabul water supply | Annual water production per capita | H | $Mm^3$/person |
| | Food security | Number of food secure districts | L | Number |
| Environmental | Environmental flows | Flow in Upper Kabul River through Kabul City | H | $Mm^3$ during tourist season |
| | Wildlife sanctuary | Restoration and maintenance of waterfowl habit | H | Species diversity (index, counts) |
| Institutional | Information | Reliable knowledge base (eg, hydro-met) and tools | H | Scale (1–5) |
| | Financing risk | Financing Risk Scale | L | Scale |
| | Technical risk | Technical Complexity Scale | L | Scale |

*From World Bank, 2010. Scoping Strategic Options for Development of the Kabul River Basin: A Multisectoral Decision Support System Approach. Sustainable Development Department, South Asian Region, The World Bank. http://www-wds.worldbank.org/external/default/WDSContentServer/WDSP/IB/2010/04/12/000333037_20100412001029/Rendered/PDF/522110ESW0Whit1anistan0Final0Report.pdf (accessed 07.06.15).*

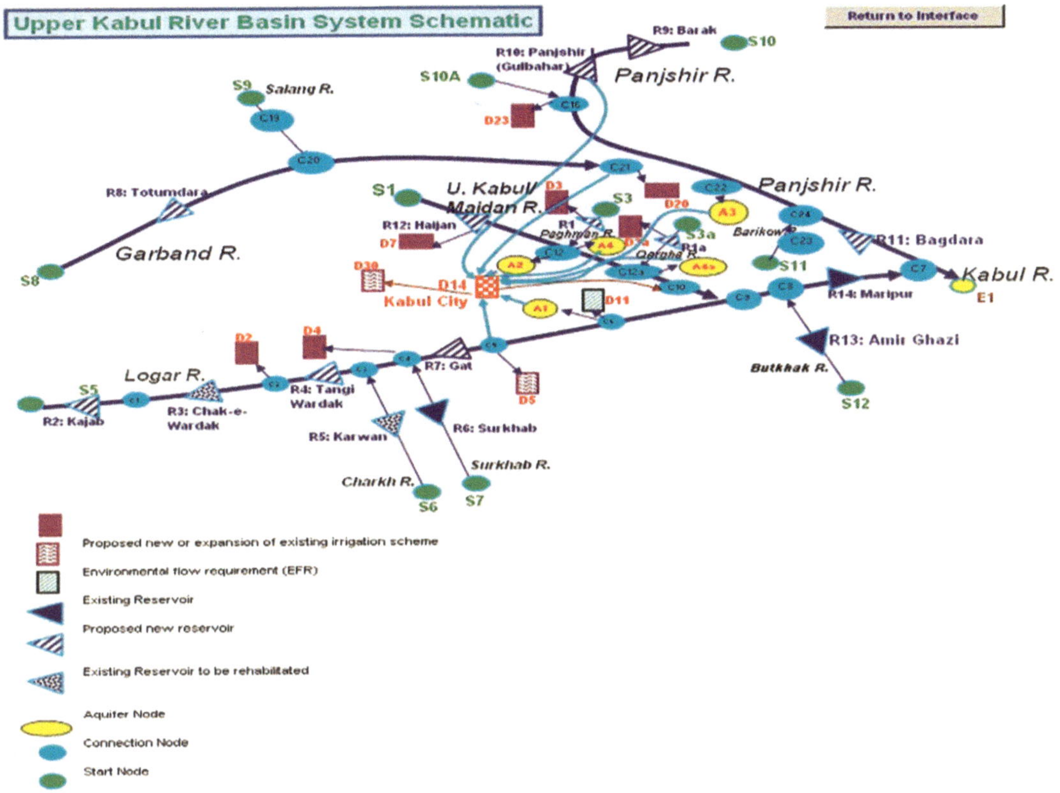


FIG. 4.21 Schematic diagram of the greater Kabul Basin with the river network conveyance options and development options as established by the World Bank (2010). This complex diagram exhibits a level of advanced hydro-analytics and technical skill that if implemented in larger measure, could dramatically alter the rather unsophisticated condition of the present day (World Bank, 2010).

## References

Azimi, A., McCauley, D., 2002. Afghanistan Environment in Transition. Asian Development Bank, South Asia Department. http://www.amazon.com/Afghanistans-Environment-Transition-Ali-Azimi/dp/9715614787 (accessed 07.08.15).

Banks, D., 2002. Water Assessment Mission to Afghanistan. Part F. Urban Water Supply Systems in Afghanistan. Kirkens Nødhjelp, Oslo, Norway. 20 pp. http://www.holymoor.co.uk/PartF.PDF (accessed 07.07.15).

Böckh, E.G., 1971. Report on the groundwater resources of the city of Kabul. BGR archive number 0021016.

Bohannon, R.G., Turner, K.J., 2007. Geologic map of quadrangle 3468, Chak Wardak-Syahgerd (509) and Kabul (510) quadrangles, Afghanistan: U.S. Geological Survey Open-File Report 2005-1107-A, 1 sheet, scale 1:250,000.

Campbell, J., 2015. A dry and ravaged land: investigating water resources in Afghanistan. Earth 60 (1&2), 48–55. http://www.earthmagazine.org/article/dry-and-ravaged-land-investigating-water-resources-afghanistan (accessed 07.14.15).

Danish, N., 2011. Water pollution silent killer in north Afghan district. Global Voices Asia. Afghan Investigative Journalism Fund project; IWPR; 6 pp. https://iwpr.net/global-voices/water-pollution-silent-killer-north-afghan (accessed 07.07.15).

Eqrar, N., n.d. (2006). Groundwater resources of the Kabul Basin (Afghanistan). Joint Afghan—BGR Project; Federal Foreign Office of Germany. Power Point; 21 slides. https://www.bgr.bund.de/EN/Themen/Wasser/Projekte/

abgeschlossen/TZ/Afghanistan/kabul_water_conference_hydrogeology_part1_pdf.pdf?__blob=publicationFile&v=2 (accessed 07.07.15).

Fipps, G., 2006. Advancing water and sanitation in Afghanistan. Water, Politics and Reconstruction—Realities and Lessons Learned in Afghanistan. US Embassy Documents; Afghan Reconstruction group, US Embassy—Kabul. http://gfipps.tamu.edu/Afghanistan/Advancing%20US%20Efforts%20on%20Water%20-%20Afghanistan.pdf (accessed 07.14.15). http://www.eli.org/sites/default/files/docs/pcnrm/Abstracts/Fipps.Afghanistan.pdf?q=pdf/pcnrm/Abstracts/Fipps.Afghanistan.pdf (accessed 07.14.15).

Frischmann, P., 2013. Afghanistan Resource Corridor Development: Water Strategy Final Kabul River Basin Report Version 4.0. Australian Aid; The World Bank. 70 pp. http://www.artf.af/images/uploads/RC_WaterStrategy.pdf (accessed 03.13.15).

Gellasch, C.A., 2014a. Hydrogeology of Afghanistan and its impact on military operations. In: Harmon, R.S., Baker, S.E., McDonald, E.V. (Eds.), Military Geosciences in the Twenty-First Century, vol. XXII. Geological Society of America Reviews in Engineering Geology, pp. 69–81. http://dx.doi.org/10.1130/2014.4122(08). http://reg.gsapubs.org/content/22/69.abstract (accessed 07.07.15).

Gellasch, C.A., 2014b. The role hydrogeology will play in stabilizing Afghanistan. In: Maryland Groundwater Symposium, 24 September 2014. Uniformed Services University of the Health Sciences. 24 pp. http://www.mde.state.md.us/programs/Water/Water_Supply/Source_Water_Assessment_Program/Documents/GWS-2014/m3.Gellasch.pdf (accessed 07.06.15).

Himmelsbach, T., Houben, G., Niard, N., Tünnermeier, T., 2006. Urban impacts on groundwater resources and quality in the Kabul Basin, Afghanistan. Geophys. Res. Abstr. 8 (01399). SRef-ID: 1607-7962/gra/EGU06-A-01399; European Geosciences Union.

Hommilius, J., 1973. Hydrogeology and Geology of Kabul basin, Germany.

Houben, G., Himmelsbach, T., 2005. Hydrogeology of the Kabul Basin Part I: Geology, aquifer characteristics, climate and hydrogeology. BGR (Federal Institute for Geosciences and Natural Resources). https://www.bgr.bund.de/EN/Themen/Wasser/Projekte/abgeschlossen/TZ/Afghanistan/hydrogeology_kabul_basin_1.pdf?__blob=publicationFile&v=3 (Accessed 3/5/2016).

Houben, G., Tünnermeier, T., 2005. Hydrogeology of the Kabul Basin—Part II: Groundwater Geochemistry and Microbiology. Foreign Office of the Federal Republic of Germany; BGR. http://www.susana.org/_resources/documents/default/2-1437-hydrogeologykabulbasin2en.pdf (accessed 07.06.15).

Houben, G., Niard, N., Tunnermeier, T., Himmelsbach, T., 2009. Hydrogeology of the Kabul Basin (Afghanistan), Part I: Aquifer and hydrology. Hydrogeo. J. 17, 665–677.

Lonely Planet, 2015. Kabul. http://www.lonelyplanet.com/afghanistan/kabul/history (accessed 07.08.15).

Mack, T., Capelli, K., 2010. Afghanistan's Kabul Basin Faces Major Water Challenges. U.S. Geological Survey, Reston, VA. http://www.usgs.gov/newsroom/article.asp?ID=2521#.VZtzbE3JDIV (accessed 07.07.15).

Mack, T.J., Akbari, M.A., Ashoor, M.H., Chornack, M.P., Coplen, T.B., Emerson, D.G., Hubbard, B.E., Litke, D.W., Michel, R.L., Plummer, L.N., Rezai, M.T., Senay, G.B., Verdin, J.P., Verstraeten, I.M., 2009. Conceptual model of water resources in the Kabul Basin, Afghanistan. Scientific Investigations Report 2009-5262; 82 pp., 10 appendices. http://pubs.usgs.gov/sir/2009/5262 (accessed 07.07.15).

Mack, T.J., Chornack, M.P., Taher, M.R., 2013. Groundwater-level trends and implications for sustainable water use in the Kabul Basin, Afghanistan. Environ. Syst. Decis. 33 (3), 457–467. http://link.springer.com/article/10.1007%2Fs10669-013-9455-4 (accessed 07.07.15).

Montreal Engineering Company, 1978. Kabul River Valley Development Project, Master Plan Report. Two volumes. Government of Afghanistan, Ministry of Water and Power, Kabul, vol. 1.

Mostafawi, S.N., Ostrowski, S., 2010. Wildlife Conservation Society, Afghanistan.

Nembrini, P.G., Jansen, P., Pinera, J.F., Luff, R., Bernard, O., Weber, M., Elliot, M.J., 2002. Cities in War: Thirsty Cities. Geneva Foundation, Tacoma, WA. 29 pp. http://www.thirstycitiesinwar.com/wp-content/uploads/2013/03/kabuloccasionalpaper7.pdf (accessed 07.08.15).

Niard, N., 2006. Hydrogeology of the Kabul Basin—Part III: Modelling Approach; Conceptual and Numerical Groundwater Models. Foreign Office of the Federal Republic of Germany; BGR. http://www.susana.org/_resources/documents/default/2-1437-hydrogeologykabulbasin3en.pdf (accessed 07.06.15).

Reddy, K.R., 2008. Enhanced aquifer recharge. In: Darnault, C.J.G. (Ed.), Overexploitation and Contamination of Shared Groundwater Resources. Springer Science + Business Media B.V., pp. 275–288.

Saffi, M.H., 2011. Groundwater natural resources and quality concern in Kabul Basin, Afghanistan: Scientific Investigation Report in Afghanistan. DACAAR (Danish Committee for Aid to Afghan Refugees), 122 pp.

http://www.dacaar.org/Publications/6%20Research%20and%20Studies/WASH/Groundwater%20natural%20resources%20and%20quality%20concern%20in%20Kabul%20Basin,%20Nov%202011.pdf (accessed 07.06.15).
Santabanez, M.P., Bell, M.A., 2012. Water Law in Afghanistan. Unofficial Translation of 2009 Afghanistan Water Law. College of Agricultural & Environmental Sciences, University of California, Davis, CA. 2 pp.
Shroder, J., 2014. Natural Resources in Afghanistan: Geographic and Geologic Perspectives on Centuries of Conflict. Elsevier.
Thomas, V., 2013. "Good" water governance models in Afghanistan: Gaps and opportunities. Afghanistan Research and Evaluation Unit (AREU); Policy Note Series. 2 pp. http://www.areu.org.af/EditionDetails.aspx?EditionId=626&ParentId=7&ContentId=7&Lang=en-US (accessed 07.11.15).
Tünnermeier, T., Houben, G., 2005. Hydrogeology of the Kabul Basin—Part I: Geology, Aquifer Characteristics, Climate and Hydrography. Foreign Office of the Federal Republic of Germany; Hannover, Germany, Federal Institute for Geosciences and Natural Resources, BGR.
UNICEF, 2013. Water quality & HWTS in Afghanistan. WASH (Water, Sanitation, Hygiene) Section, UNICEF Afghanistan Country Office; Power Point, 12 pp. http://hwts.web.unc.edu/files/2014/07/2013Nagpur_Day2_07_Taj.pdf (accessed 07.07.15).
USAID, 2013. Kabul urban water supply program. http://www.usaid.gov/news-information/fact-sheets/kabul-urban-water-supply-program (accessed 07.08.15).
USAID, 2015. Kabul urban water supply (AUWSSc). http://www.usaid.gov/news-information/fact-sheets/kabul-urban-water-supply-kuws (accessed 07.08.15).
Water Shed Atlas of Afghanistan, 2004. http://www.krbp.net/eng_reports/Watershed%20Atlas_Part%20I_II.pdf.
Williams, D.B., 2009. Finding water in the heart of darkness; Afghanistan's ongoing water challenges. In: Earth. vol. 54. American Geological Institute, Alexandria, VA, pp. 48–57.
World Bank, 2010. Scoping Strategic Options for Development of the Kabul River Basin: A Multisectoral Decision Support System Approach. Sustainable Development Department, South Asian Region, The World Bank. http://www-wds.worldbank.org/external/default/WDSContentServer/WDSP/IB/2010/04/12/000333037_20100412001029/Rendered/PDF/522110ESW0Whit1anistan0Final0Report.pdf (accessed 07.06.15).

CHAPTER

# 5

# $H_2O$ Hazards, Risks, and Disasters in Afghanistan and Surrounding Countries

*J. Shroder*

**University of Nebraska at Omaha, Omaha, NE, United States**

OUTLINE



## Abstract

Afghanistan is commonly considered by many experts as one of the places in the world that is most afflicted by natural hazards, especially by those directly or indirectly caused or strongly affected by the $H_2O$ system. Several forms of landslides (loess flows, debris flows, etc.), water floods, snow avalanches, and four kinds of droughts are among the most serious hazards that the citizens of Afghanistan must endure, and which kill or injure hundreds annually, and destroy extensive infrastructure nationwide. The Government of Afghanistan has traditionally been unable to cope with most of such natural disasters, beyond some fairly rudimentary rescue and remedial actions, but some of the slack has been picked up by the United Nations and U.S. Agency for International Development.

**Keywords:** Water-induced slope failure, Landslides, Loess flow, Debris flow, River flooding, Snow avalanche, Meteorological drought, Agricultural drought, Hydrological drought, Socioeconomic drought


http://dx.doi.org/10.1016/B978-0-12-801886-6.00005-7

## 5.1 INTRODUCTION

Hazards, risks, and disasters associated with water in Afghanistan include all sorts of problems associated with the solid and liquid phase varieties of the $H_2O$ system, in addition to problems associated with too much or too little of the essential resource. Where that water resource might be located in any given area, and what its condition or state might be in terms of solutes, sediment loads, pollution, and any other additives at any given time are essential to know. Whatever might be the situation will determine whether or not the water will constitute a welcome boon or a destroying agent. Learning how to deal with all the varieties of snow, ice, meltwater, snow avalanches, varied flood types, droughts, different sediment/water concentrations that constitute continua of landslide/debris-flow hazards, ground-water recharge possibilities, and normal seasonal variations are very much capabilities in need of considerable development in Afghanistan. The problem in Afghanistan more than in most places is that the landscape is configured in such a fashion that the solid and liquid phases of the water materials bring both life and death in almost equal measure in such a complex way that people do not very well understand the situations that lead to hazard. This technical ignorance about the water systems in Afghanistan is mainly because of a dearth of education, as well as an unfortunate deliberate ignorance bred out of an unquestioned obedience to manipulative, quasi-religious authorities who control much misguided public opinion.

In terms of general water-related hazards in Afghanistan, it must be noted that at the national level, the country is not water scarce, even though water can be classified as scarce in one basin and many subbasins (Beekma and Fiddes, 2011). On the other hand Afghanistan definitely can be considered water scarce if the needs for flows to maintain the environment are taken into consideration. Also it should be noted that Afghanistan is not particularly distinct or much different from other countries in the world's dry regions in terms of outcomes measured through most drought indicators. Afghanistan does have a high sensitivity to drought if socioeconomic conditions are considered, including the overwhelming dependence on agriculture for generation of employment and income. From another point of view, Afghanistan is also vulnerable to drought because of poor access to improved water resources and generally poor rural access.

On the other hand at the present time, Afghanistan appears to have a greater vulnerability to floods than to drought (Beekma and Fiddes, 2011). This combined vulnerability to both floods and drought, and the nature of the climate in the country is typical for semiarid and arid zones that have high inter-annual variability. The result is that such conditions warrant development of effective strategies for drought and flood-preparedness, and early warning of such coming problems. Such strategies could also help the country cope with gradual shifts resulting from climate change that are less sudden than the inter-annual changes, and are easier to prepare for or to adapt to. A system is needed for the timely detection of flood and drought risk that would provide scientifically accurate prediction, as well as focusing on supplying public information and devising participatory preparedness strategies. The revisions of the new Water Law in 2011 form an excellent enabling environment for the development of such systems.

Remedial catchment conservation and practices of water harvesting can be used to address both problems of flood vulnerability and drought sensitivity, but not much attention has been paid to these procedures in recent years. Instead however, remote sensing with satellite imagery will enable forecasting floods and droughts, as has been shown with data gathered through the Panj—Amu Darya River Basin Program that show a good correspondence of remote-sensing data with observed discharges (Beekma and Fiddes, 2011). Such work with remote sensing can compensate for the low density of ground-measurement locations. Still however, area-wide coverage of precipitation and automated discharge at various locations are needed, along with: (1) systematic surveys of glacial resources that incorporate cutting-edge techniques to account for the significant amount of debris-covered ice (Bishop et al., 2014a,b; Shroder and Bishop, 2010); (2) closer links should be established between the Afghanistan Ministry of Energy and Water with international agencies and foreign universities who can provide technical support; (3) a snow-survey program should be established to measure snow coverage, snow-depth measurements, and to make modeling studies of snow-water equivalent (SWE); (4) establishment of a low-cost network of meteorological stations to gather essential climatic data in complex topographies and improve the accuracy of seasonal forecasts; (5) long-term (20 years) time series should be established of normalized difference vegetation indices (NDVI) to allow for annual anomaly calculations to improve drought assessments; and (6) ground-validation data should be improved in order to better calibrate NDVI values to pattern types of different crops (Beekma and Fiddes, 2011).

Water can kill you in many ways, especially if you fall into a substantial amount of it and cannot swim, as is true of most Afghans who generally do not learn to swim at a young age because of a lack of either sufficient water or motivation. But drowning in Afghanistan is relatively rare anyway, unless one happens to live close to a major river or channel way where floods can occur, as they do with apparent increasing frequency in the country, probably because of watershed degradation and increased rapid runoff, or because of climate change, or both. Another little-known factor by which water kills is in its weight, especially when added to unstable hillslopes. For example, one cubic meter of water ($1\,m^3$) weighs a metric ton (1000 kg), which is quite heavy and well known to anyone who must carry any of it home from its source, which is the natural condition for most rural Afghans. This great weight is not the only factor by which water causes slope failure, however, and these other factors are discussed below in that section of this chapter. Water as hazard thus needs further explanation as it leads to so many risks and disasters. Afghanistan has an overwhelming abundance of water-induced, hazardous situations because of natural and human-caused environmental problems that need to be better understood in the context of transboundary water resources.

## 5.2 WATER-INDUCED SLOPE FAILURES

The failure of mountain slopes in Afghanistan takes on many forms, depending upon whether the rocks in any particular place are sedimentary or crystalline metamorphic and igneous types, and whether or not they are at higher or lower altitudes so that liquid water or solid ice are involved (Shroder, 1989; Shroder et al., 2011a,b). The causes of landslides in these regions, or indeed of any region, is some mix of passive, underlying combination of lithologies, structures, and geologic relationships, combined with some activating mechanisms that induce the slopes to overcome the shear resistance to failure and to thereby initiate movement downslope.

A common initiating cause of slope failure in Afghanistan and many other locations is high seismicity in which the earthquake shock causes high acceleration of the ground; the subsequent motions may lead to massive ground rupture. In fact as Wheeler and Rukstales (2007) have noted, a region encompassing south and central Badakhshan in northeastern Afghanistan has the highest number (21) of magnitude ≥7 historical earthquake epicenters in the world in the past 1.75 centuries, which means that one would expect to find an inordinate amount of large landslides in the region, which is the startling observation that can be made (Shroder et al., 2011a,b; Shroder, 2014).

Another causative or modifying factor of slope failure can be the liquid water or solid ice that applies its variable forces to increase shear force or reduce shear resistance until the slopes are ruptured (Table 5.1). In fact in the generally higher altitudes of much of the Hindu Kush in central or northeast Afghanistan a great many high-altitude rockslide complexes (Shroder et al., 2011a; Shroder, 2014; Bishop et al., 2014a,b) have multiple generations of movement wherein much ground ice has accumulated as ice cement or ice cores to produce rock-glacier forms. Such landforms are exceptionally prevalent on many of these several square km slope failures, and the presence of so much large rock debris from the high peaks as a result of the high seismicity provides an ideal, open porosity for the later buildup of permafrost ground ice, which also constitutes meltwater source regions. In fact, the melting of ground ice in one of these rock-glacier forms high up in the Paghman Range above Kabul was discovered by the author (Shroder and Giardino, 1978) to be the main source of the Paghman River, which supplies vital freshwater to Kargah Reservoir and part of the water supply of Kabul City.

## 5.2.1 Landslides

In general in Afghanistan landslides have no doubt occurred for a multiplicity of causes, as noted above. But water into the many unstable shales and other weak rocks in central and northern Afghanistan has certainly caused numerous slope failures (Shroder, 1989), and a number of recent events, as well as new discoveries, show the obvious direct water causation, or its high probability. Two examples come to mind; the large wet flow of loess over the town of Ab Barak in Badakhshan Province in 2014, and an observation of large mass-movement lobes across the active Chaman Fault in Zabul Province by H. Maher (email to JS, Sep. 5, 2014).

After heavy rainfall in the period from 25 to 29 Apr., the *Ab Barik loess flow* occurred on May 2, 2014, which sent at least two waves of wet loess mud across part of the village of Ab Barik (Unitar, 2014; Witze, 2014; Zahori and Ahmed, 2014). The slope failure buried some 300 houses and killed as many as 250 to perhaps 2700 people, who were buried beneath 10–30 m

TABLE 5.1 Causes of Slope Failure Related to Water

| | |
|---|---|
| 1. | Water weight ($1\,m^3 = 1$ metric ton $= 1000\,kg$) added to slope can exceed shear strength. |
| 2. | Water dissolves soluble cement, reducing cohesion and shear strength. |
| 3. | Water increases ground-water pore pressure to reduce shear resistance. |
| 4. | Freezing produces 9% expansion of water to ice to disrupt rock. |
| 5. | Absorption of water by clay-mineral expansion disrupts rock. |
| 6. | Water saturation destroys intergranular capillary tension. |
| 7. | Water-seepage pressure reduces shear resistance. |

*Modified from Shroder, J.F., Jr., 1971. Landslides of Utah. Utah Geol. Mineral. Surv. Bull. 90, 51 pp.*

FIG. 5.1 Low altitude oblique aerial photograph of Ab Barak loess flow in southern Badakhshan Province that occurred on May 2, 2014; view to south to southwest. The main slope failure has moved to the east, upstream, as well as to the west downstream, in two successive but different directions of movement, with the east (upstream to the left) motion being first, and the west (downstream to the right) being second. A marked basin of "profound concavity" (Shroder, 1976) occurs on the upper right of the hilltop that marks a basin or earlier strong erosion that was probably also a loess flow. *Courtesy The Atlantic.*

of mud (Fig. 5.1). The source area of the landslide was about 400 m long and 300 m wide, covering an area of about 98,000 $m^2$. The depositional area spread both up and down the valley, covering an area of about 134, 350 $km^2$ (Unitar, 2014).

The village of Ab Barik itself, some 19 km southwest of provincial capital Fayzabad, was built on a south-facing slope of resistant crystalline bedrock of probable early Paleoproterozoic gneiss that is faced by a massive and steep, 325-m high hill of wind-blown, clay and silt, loess dust that faces north over the town in the valley. Such situations of unstable loess above resistant and impermeable crystalline bedrock are a ready-made situation in Afghanistan for many prior slope failures because of buildup of water pressure within and beneath the thick loess-cover sediment above bedrock (Shroder et al., 2011b) (Fig. 5.2).

A pre-failure satellite image of the region (Google Earth™) reveals that midway down on that slope existed an oval-shaped area of cracked and hummocky or wavy loess that was about 280 m wide, and 230 long in the downslope direction (Fig. 5.3). This oval-shaped region provided an indication of incipient failure that appears to have been moved slightly before the subsequent more massive slope movement. Furthermore, across and up the slope to the west occurs another mass-movement basin of profound concavity (Shroder, 1976) that measures some 420 m wide and 620 m long, which is an indication of another prior slope failure that must have occurred hundreds to thousands of years ago. It is apparent that this hill of loess overlying crystalline rock, and the oval area of incipient failure on it, absorbed much of the heavy rain upon it to cause the rapid mass failure. Had anyone had the forethought to notice the sure landform signs of the impending disaster, then lives might have been saved.

The landslide that resulted measured some 800 m of movement down the hillside, moving north from about 150 m above the incipient failure oval area, to across the narrow valley bottom and over part of the village. Once the first wave of mud reached the valley bottom, it turned and actually moved east, up the valley bottom before coming to rest. Later, a second

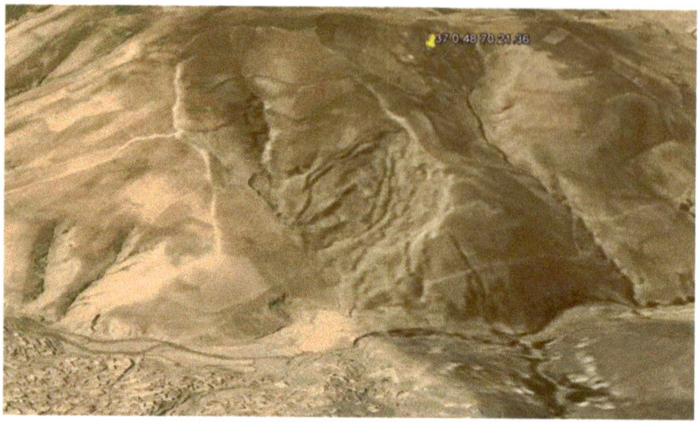

FIG. 5.2 Pre-failure scene from Google Earth™ showing approximately the same view direction as Fig. 13.1, but the scene was acquired prior to the loess-flow failure. A number of what appear to be incipient ground cracks, turf rolls, and eroded gullies occur in the middle of the slope and give an indication of coming slope-failure problems, as well as evidence that the hazard was looming in the future, and could have been predicted as a possible hazard zone, had anyone had appropriate knowledge and experience.

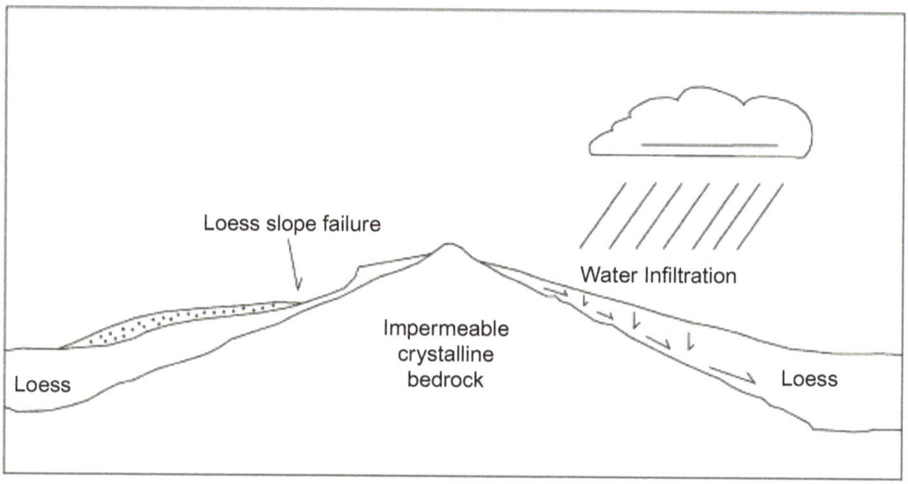


FIG. 5.3 Cross section of loess-sediment blanketed over crystalline bedrock, which is a close analog to the situation of the Ak Barak loess flow where the crystalline bedrock is exposed at the bottom of the loess hill across the valley above it. *From Shroder, J.F., Schettler, M.J., Weihs, B.J., 2011. Loess failure in northeast Afghanistan. J. Phys. Chem. Earth 36, 1287–1293, http://dx.doi.org/10.1016/j.pce.2011.03.001; Shroder, J.F., 2014. Natural Resources of Afghanistan: Geographic & Geologic Perspectives on Centuries of Conflict. Elsevier, Amsterdam.*

wave came down from the hill and turned to move west down valley, before coming to rest over more of the village.

Shroder et al. (2011b) had studied the situation of thick loess deposits occurring over crystalline bedrock and mapped the locations of 34 separate but similar such loess-failure events that had occurred in Takhar and Badakhshan Provinces in the past. An overwhelming proportion of such slope failures were observed to have occurred on north- to west-facing aspects, which would likely be the result of the wetter and colder microclimate on such slopes.

That fact, in addition to the incipient zone of pre-failure disruption shown in Fig. 5.2, as well as the basin of "profound concavity" (Shroder, 1976) adds further weight to the idea that this region is subject to repeated failures of this type, some of which could be predictable if appropriate indicators could be recognized by knowledgeable people. In fact as the Unitar (2014) people well noted, the new main scarp above the town is itself likely to continue to be quite unstable and further retrogressive slope failure can be expected.

The *Kuchnay Makhay rockslide—debris flow complex* crosses a prominent active zone of the Chaman Fault about 27 km southeast of Qalat in Zabul Province (Fig. 5.4). The 2.5 km-long landslide complex has the configuration of lobate slow debris-flow morphologies that can emanate from unstable sedimentary rocks, as well as the configuration of those features wherein plentiful precipitation can mobilize rock and debris to flow downslope, generally as either a rapidly moving slurry mixture of water and rocky debris, or as a slower moving mass such as an earthflow. A third possibility is for a high-speed emplacement initiated by a strong seismic shock that provides significant horizontal acceleration to produce a high-speed rock avalanche, the velocity of which requires no particular amount of water, in a manner envisioned by Shreve (1968), Melosh (1987), or a number of other models for long run-out zone landslides.

The landslide complex is characterized by three basal lobes that exhibit separate flow morphologies, together with upper hummocky topography, flow lineations, and two spatulate toes that are 25–30 m thick (Fig. 5.4). The third lobate toe on the north side of the complex is only about a km long, and is perhaps 20 m thick. The weak clays and shales of the Katawaz Basin flysch sediments of Eocene age are uplifted in the Sur Ghar Range at this location and appear to have failed as a result of reduction of shear strength because of excess water, probably from the summer monsoon some centuries in the past, and perhaps by seismic acceleration as well.

Evidence for an even more aqueous emplacement instead is the fact that the landslide complex is extensively trenched by water-eroded gullies, which attest to the prevalence of strong rainstorms in the region, even though it is otherwise largely desert to semiarid terrain. In addition, the landslide complex has three basal lobes (400, 600, and 335 m wide), the southern and wider two of which are spread out in a spoon or spatulate shape at the toe, which does give some impression of its possible once water-saturated nature. In addition, the lithologies exposed in the Sur Ghar mountains of the source region was mapped by Bohannon (2005) from Russian sources as clay and shale of Eocene age, which is certainly quite appropriate for either the slow or the rapid, wet debris-flow category, rather than the drier, rock-avalanche model of emplacement of the mass.

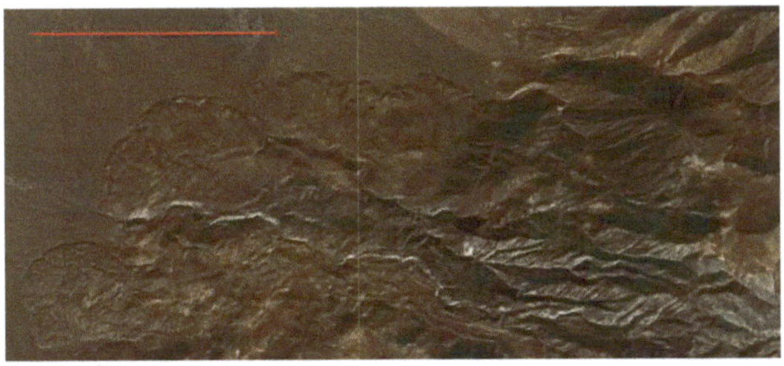

FIG. 5.4 Vertical-downward view of Kuchnay Makhay slope-failure complex that crosses the Chaman Fault zone. The *red bar* in the upper left is 1 km in length. North is up.

The only possible problem with the aqueous debris-flow model of emplacement is the fact that no water catchment basin exists in the source region, but this lack of evidence is mitigated somewhat by the fact that only a 150–200 m deep saddle-shaped depression exists in the head now, between moderately rounded peaks that occur to the north and south and extend up to 2550 and 2600 m altitude. The higher southern one of these rounded peaks actually has several mountaintop swales on it, which resemble the graben types of sackung failure that are probably reflections of intermittent movement on the Chaman Fault, which has produced these apparently undrained depressions at the top of the mountain region. In all probability the combination of similar mountaintop depressions was the source of water collection into the clays and shales, which allowed subsequent mass failure and flow once the unstable lithologies were water-saturated enough. The removal of the whole mountain mass would have left the saddle-shaped depression behind as it failed, perhaps also as a result of one or more accompanying seismic events.

### 5.2.2 Debris Flows

The viscous slurry-flow movement of mixtures of fine and coarse sediments and rocks downslope in channels at moderate to fairly rapid velocities has a long and rich history of observation and description (Costa and Wieczorek, 1987; Jakob and Hungr, 2005; and many others). First noticed and described by British explorers and geologists in the western Himalaya in the 19th century, the phenomena have come to be understood in such places as Afghanistan as a commonly recurring hazardous event that does much damage and deposits huge gravelly debris loads on the ubiquitous alluvial-colluvial fans at mountain fronts there (Shroder, 2014).

Basically debris flows are characterized by mixtures of rock and soil debris mixed with variable amounts of water (Figs. 5.5 and 5.6), so that the whole viscous slurry flows down gradient rather like wet concrete flowing out of a cement-mix truck. The typical debris-flow paths have an initiation or catchment zone where water and blocks of sediment can flow or fall in, a transport zone channel to flow down, and a final deposition fan zone. Debris flows commonly move in distinct surges or slugs of coarse-grained, boulder masses that are separated by more watery intersurge flow or hyperconcentrated streamflow (Fig. 5.6). This flow characteristic can be caused by the periodic stalling of the more sluggish flow in the channel that is followed by a break of the boulder dam and remobilization, and/or by aperiodic infall of blocks of sediment into the channel filed with a flow in progress.

Debris flows of this kind are by far the most common in the region, and are fairly typical of regions where the environment has been changed sufficiently over the long term by deforestation, overgrazing, soil erosion, and general catchment degradation so that precipitation water does not infiltrate slowly down into the soil, but instead runs off quickly to produce debris flows or flash floods. Although scientific measurements of magnitude and frequency of debris flows in Afghanistan have not ever been made in any systematic way, the likelihood is of an increase in debris flows through time as a result of environmental degradation (Shroder, 2012). This observation is compounded by the further likelihood of such problems as a result of climate change in the region (Shroder, 2014).

Santi et al. (2010) have investigated debris-flow impact, vulnerability, and response throughout the world, but also notably in the upper Indus River basin of the Hindu Kush

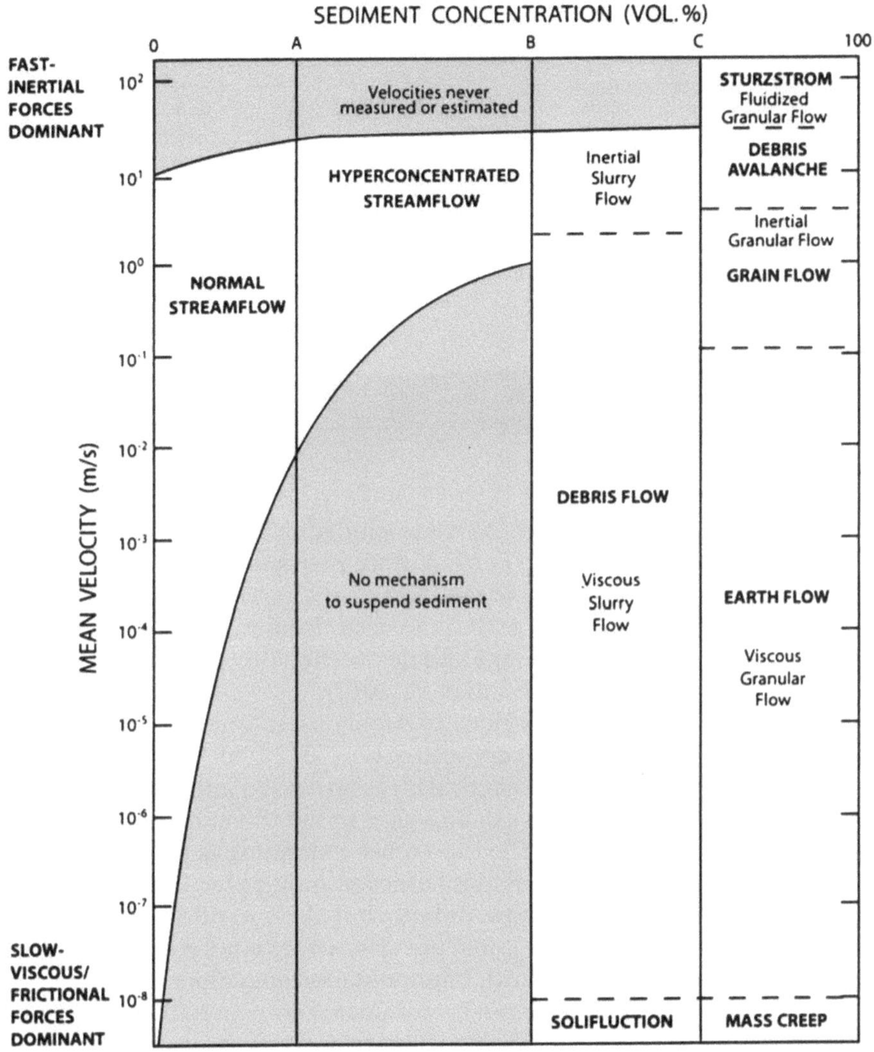


| | | | | |
|---|---|---|---|---|
| FLUID TYPE | NEWTONIAN | NON-NEWTONIAN | | |
| INTERSTIAL FLUID | WATER | WATER+FINES | | WATER+AIR +FINES |
| FLOW CATEGORY | STREAMFLOW | | SLURRY FLOW | GRANULAR FLOW |
| FLOW BEHAVIOR | LIQUID | PLASTIC | | |

FIG. 5.5 Diagram of various sediment concentrations mixed with water plotted against mean velocity showing the range of debris flows. *From Pierson and Costa, 1987; Shroder, J.F., 2014. Natural Resources of Afghanistan: Geographic & Geologic Perspectives on Centuries of Conflict. Elsevier, Amsterdam.*

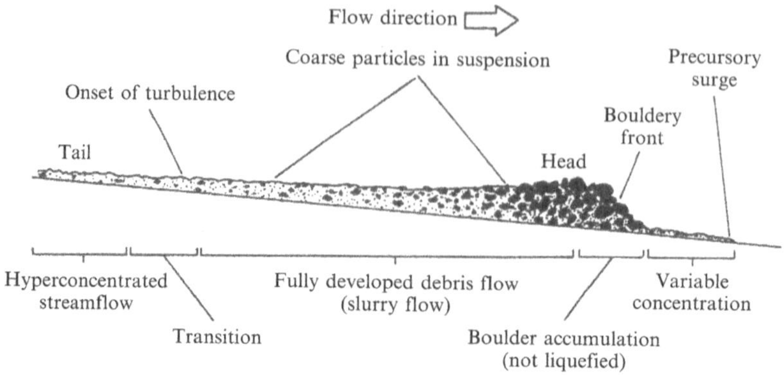


FIG. 5.6 Diagram of a debris-flow surge with a boulder front. *From Hungr, O., 2005. Classification and terminology [of debris flows]. In: Jakob, M., O. Hungr (Eds.), Debris-Flow Hazards and Related Phenomena. Springer/Praxis Publishing, Chichester, UK.*

and western Himalaya. In fact, Hewitt (2004) concluded that debris flows are the most widespread and frequent of the destructive Earth-surface processes in the mountainous regions of Afghanistan, Pakistan, and India. Part of the most dangerous aspects of debris flows are their rapid onset and the significant distances that can occur from where they initiate to where they may eventually flow to and cause damage and destruction down along channel sides and out on sediment (alluvial-colluvial) fans (Santi et al., 2010).

In the mountain villages of these regions of Afghanistan, some of the residents who must cope with the ever-present risk, know very well of how and where debris flows originate, how the flows behave, and they have quite practical ideas on how to mitigate the associated hazards and risks. Debris flows in these places can be a year-round threat because of the mountainous terrain and the regional climatology. The local relief commonly is thousands of meters above where people live, which not only provides sufficient orographic forcing of strong storms to cause heavy precipitation of rain or snow, or both, but also provides enough potential energy to generate far-traveling flow hazards. Most villages, arable land, roads, and other infrastructure are located on the arid and semiarid, intermontane valley floors and lower slopes of the sediment fans. The precipitation types can be dominated by snowfall, with debris flows being triggered by rapid snowmelt, or intense, monsoon-related, summer thunderstorms.

Santi et al. (2010) designed a mitigative approach for debris flows in parts of the southwest Asian mountains, which included a number of procedures that would help people to overcome their debris-flow hazards. The five general categories of possible debris-flow mitigation include education, avoidance and warning systems, channelization, interception, and slope treatment. Education would involve the holding of community meetings at the beginning of each monsoon season in order to talk over precursory warnings of incipient debris flows. Such things as heavy rains, ground shaking, roaring noises, changing stream colors; all can indicate the presence of the phenomena. The posting of information and the reviewing of these warning signs in mosques and schools can increase awareness and safety. This would include the establishment of alert and evacuation criteria, with clearly marked evacuation paths and safe locations.

Avoidance and warning of debris flows would include establishing of "safe" zones located above the expected and historical inundation zones that are based on topographic, geomorphic, and hydrologic conditions and mapped by trained geologists. If possible, and depending upon the level of local development, sirens or loud bells could be sounded in a simple and predetermined pattern that was known and understood by the whole populace of the village. Rainfall threshold values can be developed by trained professionals that indicate what level would occur before debris-flow warnings are to be activated.

Channelization can be done in certain areas on sediment fans where certain debris-flow tracks are preferred over others where people live. To the degree possible, the channel flow paths should be straightened out through villages, and berms built where necessary to constrain or divert flows. Straight, steep paths are needed to keep debris flows from slowing, depositing, and overflowing the banks of the channels. Rerouting, diverting, and deflecting any flow away from the village is ideal, but in most cases in Afghanistan this may require more resources than are available, even with typically inexpensive manual labor in profusion.

Interception, in which the debris flows are somehow caught and held in debris basins or behind dams, have been constructed in the more highly developed mountainous regions, but are not generally feasible in Afghanistan. Reasons for this can include not having enough room to develop such basins and dams, which are generally quite expensive as well, and therefore not feasible.

Slope treatments in Afghanistan have been done in a number of places, mainly in order to conserve soils and prevent their erosion away. Treatment consists of terracing in some places, as well as construction of small rock check dams in hillside swales and runnels. Revegetation and reforestation are also useful techniques to slow down and decrease overland water flow. If the surrounding slopes are already eroded down to a rocky substrate, revegetation generally will work if soil in pockets still exists. Although some terracing and check dams may help. In any case, it is obvious that only trained geoscientists who are familiar with the fundamental principles of water infiltration, overland flow, soil erosion, and debris-flow initiation can effectively do the analysis necessary to help villagers work out the means to reduce their vulnerability to such risk. Unfortunately in Afghanistan such expertise is rather limited and will have to await the proper training of such water- and slope-cognizant professionals in order to remedy the all too common situations of threatening debris flows where such procedures could be useful.

## 5.3 WATER FLOODS

Flooding by the rivers in Afghanistan has long been a severe problem in the country and it seems to have been getting much worse in recent years, which is either a fact due to climate change, a result of increasing environmental degradation, or could possibly be only an apparent effect from much better reporting. In fact the rate of flood-related deaths in Afghanistan is among the world's highest (Thomas and Eqrar, 2011) (Fig. 5.7). The likelihood of such flooding being only an apparent effect of better reporting is low, however, given all the prior information about environmental degradation and climate change that has been developed in recent years (Shroder, 2012, 2014). In the assessment of hazards in Afghanistan, Shroder (2014) detailed flooding in an extensively illustrated exposition of the phenomena. For example, in 2012 in eight provinces in the north of the country, flooding killed or injured 241 people, destroyed 8533 houses, as well as 1630 wells, canals, dams, small bridges, and roads. In terms of persons affected by flooding in that year alone, some 11,061 people became

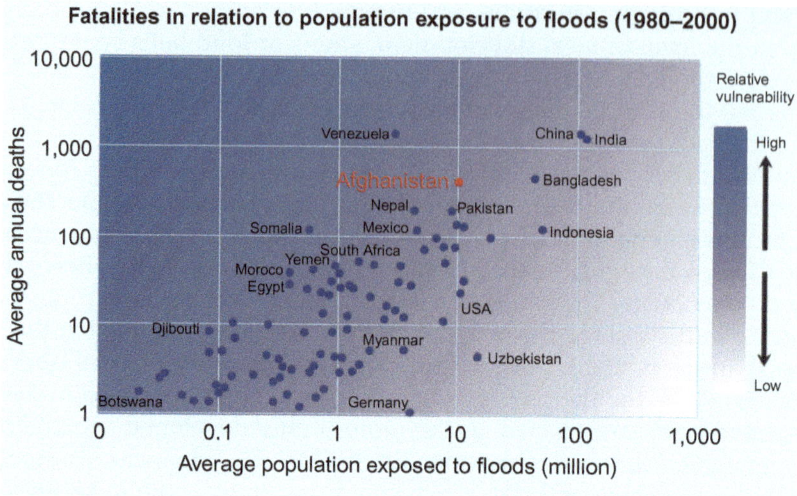


FIG. 5.7 Afghanistan is among the most vulnerable to floods in the world (Thomas and Eqrar, 2011). *Adapted from UNEP/GRID Geneva as quoted in ICIMOD presentation—Kabul (May 2008).*

internally displaced (Anon., 2013). In the first half of 2013, in the 54 total flood events that occurred, some 21,000 houses were destroyed (OCHA, 2013). Up to the present time, various divisions of the United Nations and the US Agency for International Development (USAID) continue to do their best to monitor all the hazardous situations that erupt in Afghanistan, and bring relief for affected survivors when and where they can. In any case, the flooding situation is obviously quite severe and needs further elaboration.

Such high levels of flood destruction in short periods of time, and on an annual basis are indicative of places where little to no flood protection is available, mainly because local people simply do not understand how such quantities of water come into existence, or how it spreads out to ruin their lives so profoundly (Fig. 5.8A and B). Not only do most Afghans

FIG. 5.8 Flooding in Afghanistan. (A) Flooding on the Helmand River (USAID); (B) Heavy rains in Kabul, 2 Apr. 2014 (Voice of America News).

not understand how floods occur, but they are mostly quite unaware of all the flood-control predictions and advances are made in the developed world, or how these control procedures might be used in their own country. Flooding in Afghanistan has long been such a serious problem in need of much new scientific description and mapping that people are desperate to avoid the hazard, if they can. But in general in Afghanistan, people also have only minimal concepts of flood danger and tend to build homes and infrastructure in very vulnerable places down on floodplains and too close to river banks. Attempts of limited capability at producing some sort of floodplain planning maps had been tried by such organizations as AIMS (Afghanistan Information Management Service), where their cartographers simply drew two lines alongside the existing river lines to indicate flood hazard close to the rivers, without regard to the actual topography on the ground. Normally in the developed Western countries, the common way to accomplish flood-hazard mapping is to assess the magnitude and frequency of known meteorological floods that have occurred in the past, and mark out on high-resolution maps exactly where the probabilities of floods of certain heights will likely fall in future. Fundamentally, the goal of all flood-frequency analysis is to determine, or at least realistically estimate, the probabilities that various flood-discharge levels or depths in any particular area will be exceeded in the future. To do this with statistical reliability requires a record of past flood levels at least as long as the projected recurrence interval (ie, prediction of the commonly mapped hundred-year floods requires at least a century of flood records at any particular place).

Calculation of recurrence interval or return period of the floods can be done analytically according to certain mathematical rules, or graphically on probability paper to estimate the frequency with which various heights of flood-discharge levels are exceeded. This is generally done using an annual flood series, which is a list of the largest flood that occurred in each water year for which records are kept. A long flood-series record of an area, coupled with high-resolution topographic contour maps with a small contour interval, enables accurate plotting of to where certain heights of floods will rise (10-year flood probability, 25-year flood, 100-year flood, etc.). Geological records of former high flood events in past centuries can sometimes be used to augment historical records.

In Afghanistan the lack of flood records for the past 30–40 years of war has effectively curtailed the standard methods of developing reliable flood-hazard maps because the statistical methodology is largely invalidated. Work by the U.S. Geological Survey has been done since 2001 to reinstall new river-flood gauging stations and to start keeping new records, as all developed countries must do to help understand and control their own flooding (see Fig. 2.2 and Chapter 2). In the interim, however, until these data can be augmented by years of new record keeping, other methods had to be developed (Hagen et al., 2010; Shroder, 2014). The parsimonious flood-hazard model (AFG-FHM) that was developed to overcome the shortcoming of flood prediction in Afghanistan relied on adequate available data from the Dartmouth Flood Observatory for calibration to reduce uncertainty in the virtual world of cellular automata flooding from one pixel to another that was developed to cope minimally with the problem (Fig. 5.9). The Afghanistan Flood Hazard Map was constructed for use on 1:100,000 digital terrain elevation data (DTED) of U.S. military specification with vertical resolutions of ≤30 m, which would not be an adequate scale for most developed nations, but will serve reasonably for largely rural Afghanistan. Unfortunately because the topographic data set of DTED is protected by a US government intelligence classification of "LimDis"

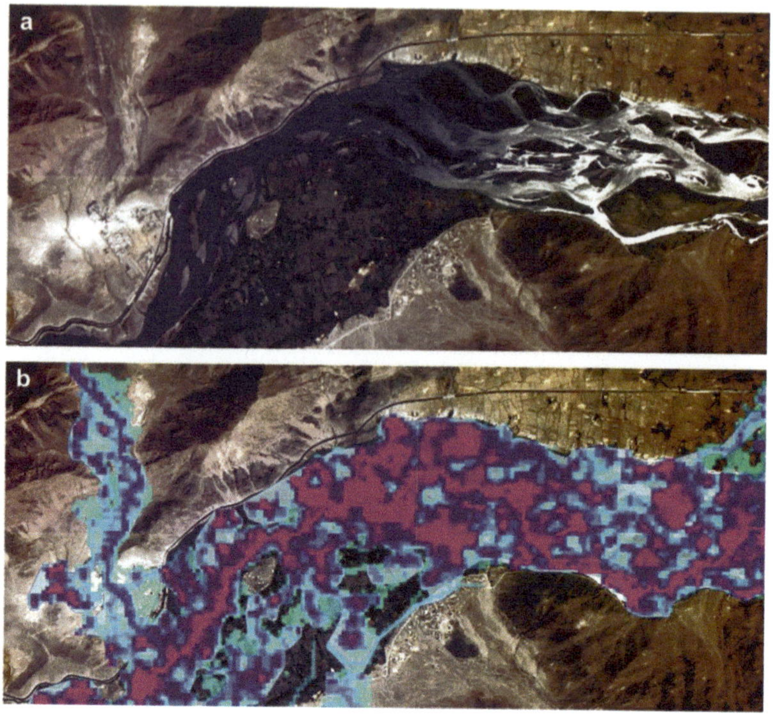


FIG. 5.9 Orthophotograph image (A); of lower Kunar River showing AFG-FHM cellular overlay with about an 8.7 m flood depth (B). *From Hagen, E., Shroder Jr., J.F., Lu, X.X., Teufert, J.F., 2010. Reverse engineered flood hazard mapping in Afghanistan: a parsimonious flood map model for developing countries. Quat. Int. 226 (1–2), 82–91.*

(Limited Distribution) to prohibit unauthorized use by the Taliban, it is not available for civilians, although this may change in future years following the withdrawal of so many of the Western Coalition troops by the end of 2014, and the accession to the head of government by new Afghan President Ashraf Ghani, who is a technocratic expert who can be expected to understand the necessity of security with such possibly predictive tools from the modern technology.

The ≤30 m resolution of the AFG-FHM proved to be adequate to categorize the general locations of floods in Afghanistan, but testing by our team did reveal some breakdowns in areas on the plentiful alluvial fans in the country. On those landforms, the common slurries of mud and rocks that constitute rapid, wet debris flows are known to avulse or jump out of the channels and seek new routes of travel that are only partly predictable from pre-existing topography (Hagen et al., 2010; Shroder, 2014) (Figs. 5.9 and 5.10). This adds to unpredictability. In terms of water flooding and climate change in future in Afghanistan (Shroder, 2014), predictions are that problematic flooding situations are unlikely to become ever able to be forecasted accurately because of high levels of inherent uncertainty. Instead impacts from flooding are expected to be amplified due to rapid snowmelt in spring because of predicted higher temperatures, in combination with further land degradation from advanced devegetation, soil erosion, and land mismanagement.

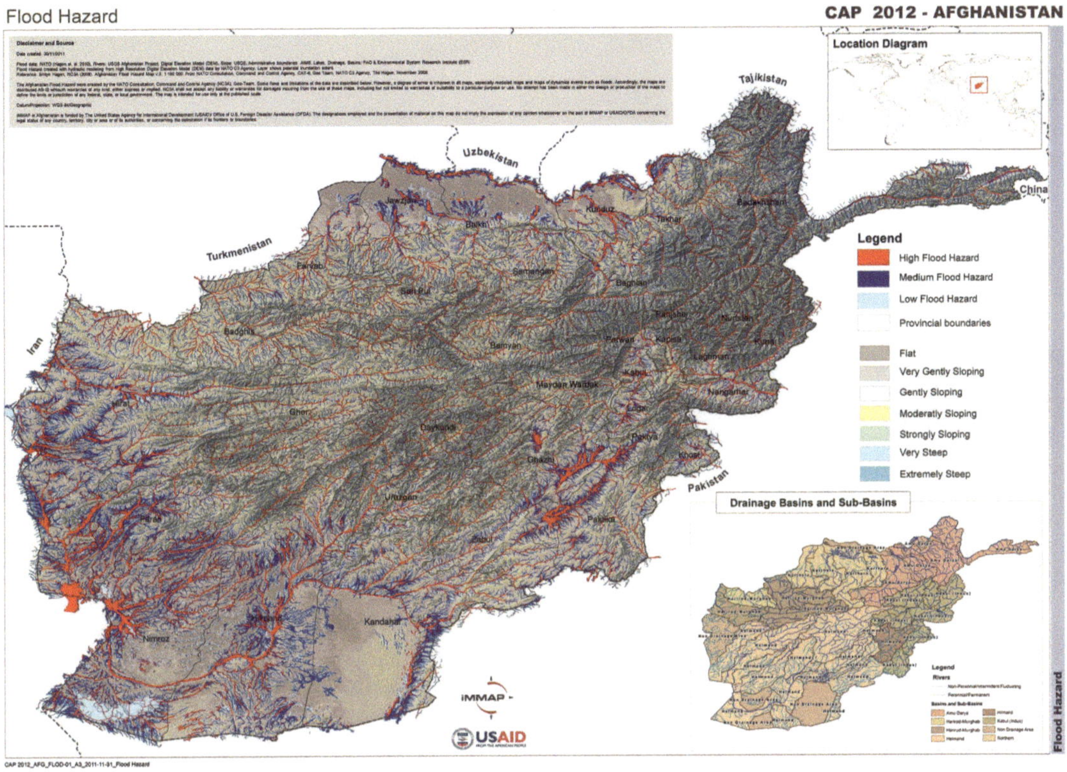


FIG. 5.10 Flood-hazard map of Afghanistan derived from the AFG-FHM of Hagen et al. (2010) (http://afghanag.ucdavis.edu/country-info/e-afghan-ag-maps-1/flood-risk-maps-immap/national_flood_hazard_jan_2011.pdf). The map on line is capable of considerable enlargement, although the original could generally be accurately enlarged to 1:100,000. *Red* indicates high flood hazard and *dark blue* is moderate flood hazard.

## 5.4 SNOW AND ICE HAZARDS

As Harris (2014) has noticed, snow and ice as hazards are quite diverse and their associated risks to people, infrastructure, and economic activity are undoubtedly becoming greater in a world of ever-expanding populations. Such a situation prevails in Afghanistan due to steep and high-altitude topography, the dynamics of snow and ice masses, and the distinctive cryogenic geological processes associated with phase changes from water to ice and conversely. As an example of one such hazard to deal with, when the Soviets were in Afghanistan in the 1980s, some of their scientists first began using remote sensing to map avalanche hazard areas to try to help alleviate a common hazard (Kravtsova and Bondareva, 1983).

Although the residents of Afghanistan really do want a great deal of snow in any given winter because of its great value as a meltwater resource the following spring and summer, the hazard from avalanches seems to have increased in recent years, which adds to their dilemmas. In fact, however, this increase might only be an apparent issue because of the lack of adequate records. In any case in strongly deforested Afghanistan (Shroder, 2014), the absence

FIG. 5.11 Snow avalanche at Salang Pass in 2010 showing wrecked cars downhill from snow sheds built by Soviet engineers to protect against avalanches; clearly the sheds were not built long enough. *Photo © S. Sabawoon, epa European Press Agency, permission sought but no answer to request.*

of woody vegetation to help retain deep snows on mountain slopes is part of the reason for the high mortality rates from increasingly pervasive large avalanches of snow and ice.

Snow avalanches and deaths, injury, and property damage in Afghanistan apparently have increased in magnitude and frequency in recent years in part from irregular deep snows, as well as the presence of villages on mountain sides and in valley bottoms, and major transportation routes on high steep slopes and ridges such as the Salang Pass. For example in the winter of 2010 at least 17 avalanches buried several kilometers of the main highway and killed 165 people (BBC, 2010; Norland, 2010) (Fig. 5.11). Thousands of people were trapped in their vehicles for days, some inside the Salang Tunnel, where the dangers of carbon monoxide poisoning from vehicle exhaust added to the hazard. Troops and helicopters from the Afghan National Army and the NATO International Security Assistance Force (ISAF) were brought in to search for survivors and drop food to them. Elsewhere in a number of provinces in the country at the same time numerous other people were killed in snow avalanches as well.

Then 2 years later in the winter of 2012, snow avalanches again devastated parts of the country (Bowley, 2012; BBC, 2012) when the whole village of Sherin Nazim with 24 families was wiped off a mountain side, killing at least 50 people at the outset. Snowfall accumulations of as much as 4 m curtailed travel, and rescue efforts were hampered for days before any relief at all was possible. Then 3 years later in Feb. 2015 a similar thing happened yet again with other avalanches in mountainous Badakhshan to the northeast killing people and cutting off access (Relief Web, 2015), as well as in the Panjshir Valley northeast of Kabul, where 196 people were killed as their houses were destroyed (AP, 2015).

## 5.5 DROUGHTS

Too much water in floods, too little water in droughts; Afghanistan has been subjected to both of these extremes for a very long time, but in recent years the hazard of far too little water has received progressively greater attention by the Government of Afghanistan and the many

FIG. 5.12 Drought conditions in Afghanistan in 2006. *From UNHCR and Albawaba.*

international nongovernmental agencies (NGOs) who try to bring some relief in that part of the world (Fig. 5.12). Archeological and environmental analysis has shown recently that extended periods of drought have caused the collapse of many ancient civilizations (Akkadian, Mayan, Anasazi, etc.; Fagen, 2003), with obvious severe implications for countries like Afghanistan with its dominantly semiarid to arid environments to begin with. Like so many other countries in the dominantly arid Middle East, those places with already deficient water supplies hover right at the edge of sustainability, given their razor-thin margins of ability to tolerate drought. Accordingly some assessments of drought character are necessary to comprehend the difficult situations that keeping cropping up with such high temporal frequency.

Drought is a difficult event to understand and define because it is commonly such a slow onset phenomenon. Any drought is some sort of departure from a statistical "normal" or expected precipitation, which can be considered as a long-term mean or average (WMO, 2006). Because of the creeping nature of droughts, the passage of time is a key controlling factor (Fig. 5.13). Although no exact time designation can be assigned, clearly the buildup of certain kinds of water deficit, stress factors can provide some guidance as to exactly what level of drought should be assigned, with four different drought types being possible: (1) meteorological; (2) agricultural; (3) hydrological; and (4) socioeconomic and political (Fig. 5.14).

### 5.5.1 Meteorological Drought

This first form of drought begins from within natural climate variability, first with a precipitation deficit in the water amount, intensity, or timing. This deficit can be exacerbated by high temperatures, high winds, low relative humidity, greater sunshine, or lesser cloud cover. These factors can further result in increased evaporation and transpiration, all of which reduce water infiltration into the soil, or runoff over it, as well as reduce deep percolation and ground-water recharge. Such meteorological drought is generally defined by a threshold precipitation deficiency in a certain period of time, but this varies by location all of the world, as well as inside Afghanistan.

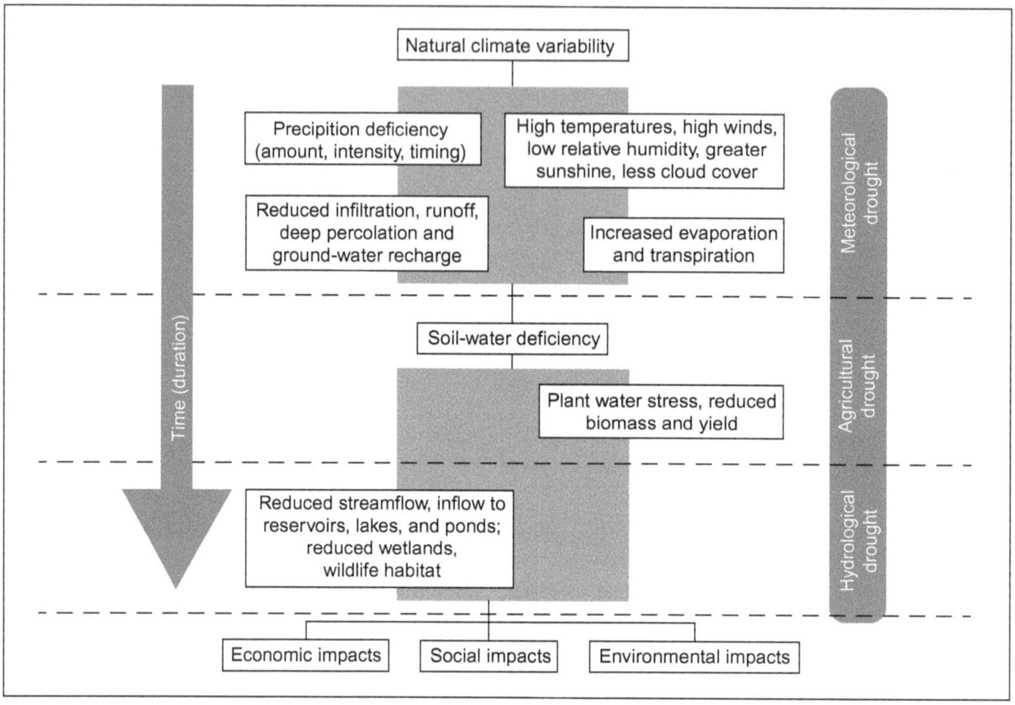


FIG. 5.13 Systems diagram showing sequence of drought occurrence and impacts for the most commonly accepted drought types. All droughts actually originate from the initial deficiency in precipitation, which is known as meteorological drought. The other forms of drought and the resulting impacts cascade through time from the initial deficiency. *With permission; from National Drought Mitigation Center, University of Nebraska—Lincoln, USA, in WMO, 2006. Drought Monitoring and Early Warning: Concepts, Progress and Future Challenges. World Meteorological Organization; WMO-No. 1006; 24pp.*

## 5.5.2 Agricultural Drought

This sort of water shortage is characterized by soil-water deficiency that results in plant stress (wilting), resulting in reduced biomass and agricultural yield.

## 5.5.3 Hydrological Drought

Such forms of water shortage are situations that develop when the precipitation deficit is sufficient to reduce streamflow to the point where inflow to lakes and reservoirs is curtailed so much that water levels drop dramatically, whereas wetlands dry up and wildlife habitat is adversely affected as well.

## 5.5.4 Socioeconomic Drought

This type of water deficit reflects the relationship between supply and demand for some commodity or economic benefit produced by water availability.

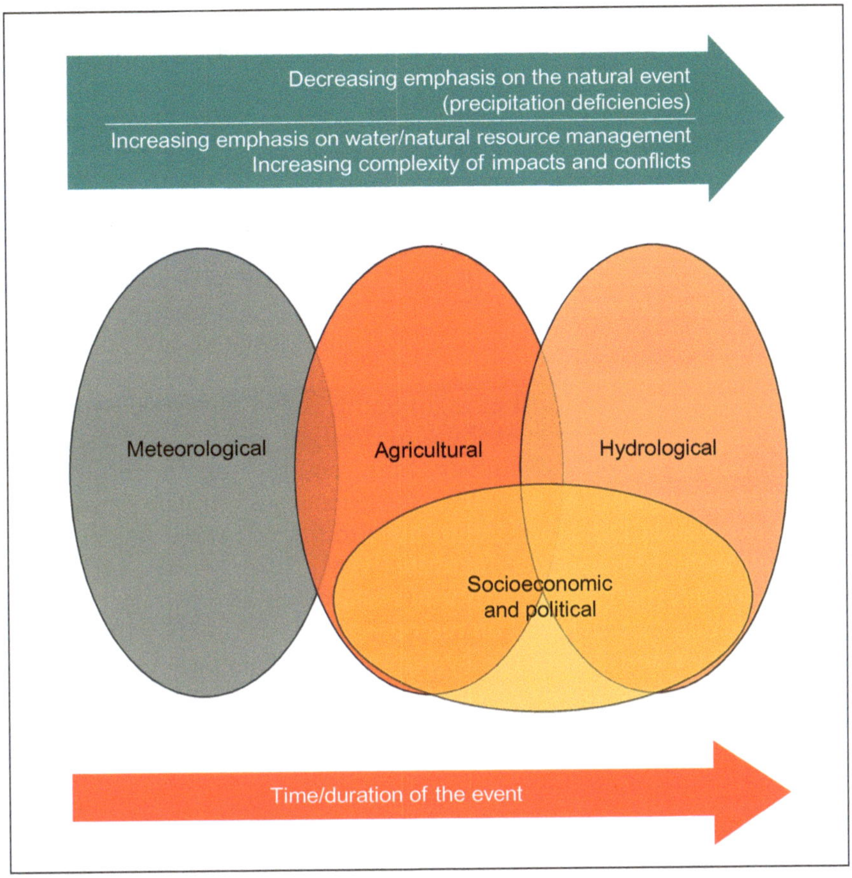


FIG. 5.14 Graphic of interrelationships between initial meteorological drought, followed by the cascade of agricultural, hydrological, and socioeconomic and political forms of drought. *With permission; from National Drought Mitigation Center, University of Nebraska—Lincoln, USA, in WMO, 2006. Drought Monitoring and Early Warning: Concepts, Progress and Future Challenges. World Meteorological Organization; WMO-No. 1006; 24pp.*

The interrelationships between these different types of events (Fig. 5.14) are time-bound in that over the passage of time a decrease in emphasis occurs with the natural event of the precipitation differences in the meteorological drought, whereas increasing emphases must occur in the natural-resource management of the water. This results in increasing complexity of impacts and conflicts over the water deficit from the first expression of meteorological drought through the later or following drought types over the passage of time.

Droughts in Afghanistan, as with most other drought-affected parts of the world, have three distinguishing features: (1) intensity; (2) duration, and (3) spatial extent. Such data are difficult to obtain in a country with few observation or monitoring stations or cognizant meteorological personnel. In the past all of the information about drought in Afghanistan has been collected by NGOs such as the UN (United Nations), the USAID, and other agencies. The only even marginal observation post in the country has been in the Civil

Aviation Authority mainly only in Kabul, but at the present time no drought early warning system exists in Afghanistan (GWP, 2014). Nonetheless, records of droughts were collected by the UN (ESCAP, 2012) and the University of Kansas (Denning et al., 2013); the three most severe droughts occurred in 1998–2003, 2006, 2008 and 2011. These droughts affected some 2,580,000, 1,900,000, and 1,750,000 people, respectively. The 2011 drought resulted in nutritional problems for roughly 2.6 million people, which included 210,350 pregnant and lactating women, and 523,876 children less than 5 years of age. Of those children who were the most affected, 73,622 were projected to be acutely malnourished and 25,242 were judged to be "severely acutely malnourished" (ESCAP, 2012). The food insecurity produced by these droughts was especially acute in different regions depending upon the depth of the drought; the 2011 drought was especially severe in the isolated communities of northern Afghanistan (Hall, 2012).

Afghanistan is at the top of the list of eight South Asian countries with ratings of very low, to no prediction capability to contribute to drought monitoring, or early warning of such a looming threat, as well as having very high requirements for technical and training support for drought monitoring. As a result of these facts, and similar needs in several other countries, a Southwest Asia Drought Monitoring Program (SADM) has been implemented by the International Water Management Institute (IWMI, 2015), which is a part of the CGIAR global consortium for international agricultural research. SADM is a near-real-time drought monitoring and reporting system based on remote-sensing data, set up to cover Afghanistan, Pakistan, and parts of western India. The system uses MODIS and AVHRR satellite data to produce drought-related indices. Every 16 days new images of the study area and individual districts and provinces are uploaded to SADM so that any user can examine the drought-related ground cover information at three levels:

- for the entire region
- at the district or province level
- at the pixel level (500 m by 500 m).

These satellite data are available back to 2002 and include three levels of analysis indices: (1) the normalized difference vegetation index (NDVI); (2) the NDVI deviation from the long-term mean; and (3) the vegetation condition index (VCI) (Weler and Herring, 2000; Quiring and Ganesh, 2010). These measures determine vegetation health, "normal" versus drought-affected plants, and factors that primarily measure agricultural drought, which is the most serious in Afghanistan, although of course, the other forms of drought can be just as devastating, but in somewhat different ways.

## 5.6 CONCLUSION

The hydro-meteorological hazards that Afghanistan undergoes are extreme with the most common being floods, droughts, ground-water contamination, and extreme winters (ADKN, 2009), although a number of other types also occur, such as the different types of water-induced slope failures discussed above. The top ten natural disasters in Afghanistan include the floods of 31 Sep. 1992 that killed 450 people and 6 Feb. 1991 that killed 415 (ANDMA, 2002). A summary of natural disasters in Afghanistan from 1954 to 2005 included 43 floods in total that killed 3368 people, with 483 injured, and 37,875 people made homeless. Those that were affected by the floods numbered as many as 966,199, an enormous number for what is generally a rather dry country. Similarly, in seven major droughts, over 7,383,000 people were affected (ANDMA, 2002).

The USAID in Afghanistan and the United Nations have also been maintaining data networks on hazards in the country (USAID, 2014; OCHA, 2012) (Fig. 5.15). Their maps and rapid assessment form (RAF) are done for every province, by each month, for nine major hazards, seven of which are directly water-related. These include heavy rain, flood, landslide, drought, avalanche, harsh winter (with avalanches), and river erosion. The numbers of individuals affected in each place are included in the data matrix, as well as the homes damaged and destroyed, and the locations are mapped by province and district.

Water is, indeed, essential to all life but perhaps not without considerable cost. Afghanistan has more than its share of water-related hazards that raise costs tremendously. Other countries in the region similarly share in these difficult situations; only continued data collection and study of the hazardous processes associated with water can potentially lead to greater understandings and lasting solutions.

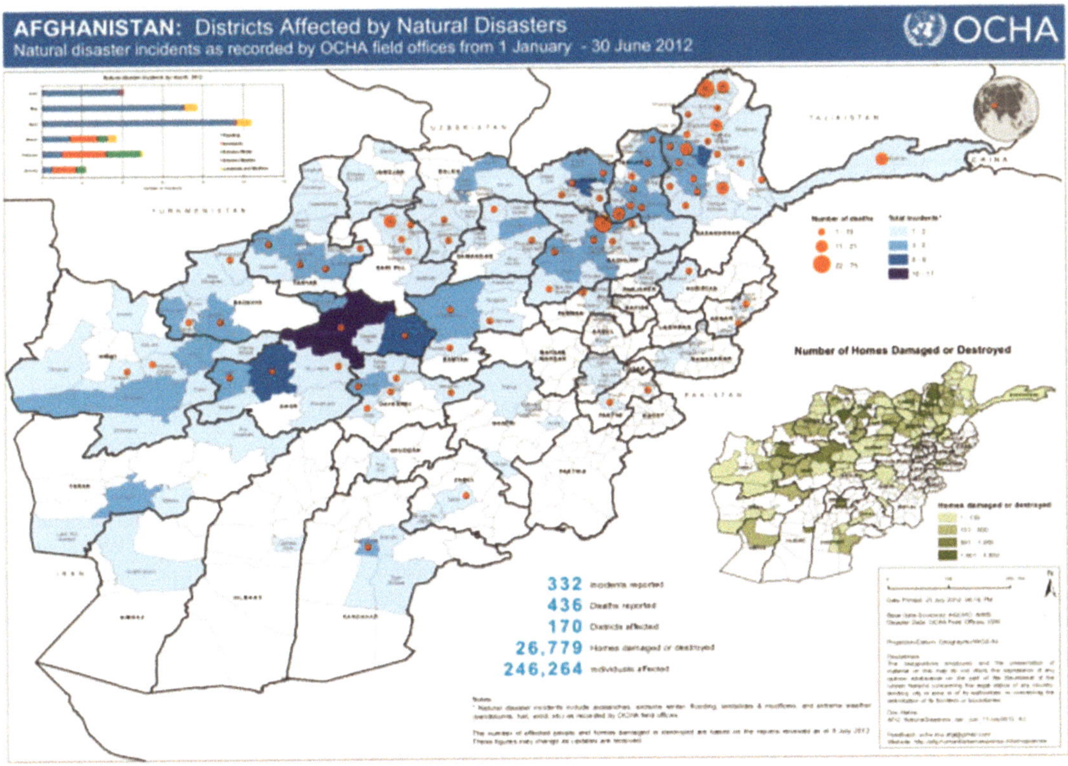


FIG. 5.15 Afghanistan districts affected by natural disasters, showing incidents and deaths reported, homes damaged or destroyed, and the individuals affected. *From OCHA, 2012. Afghanistan: Districts Affected by Natural Disasters. United Nations Office for the Coordination of Humanitarian Affairs. http://reliefweb.int/sites/reliefweb.int/files/resources/AFG_NaturalDisasters_Jan_Aug_10Sept2012_A3.pdf (accessed 21.02.15).*

# References


ADKN, 2009. Afghanistan Disaster Knowledge Network: Hazard Profile. http://www.saarc-sadkn.org/countries/afganistan/hazard_profile.aspx (accessed 21.02.15).

ANDMA, 2002. Natural Disaster Mitigation Policy in Afghanistan. Afghanistan National Disaster Management Authority. http://reliefweb.int/sites/reliefweb.int/files/resources/CFC_Afghanistan_Food-Insecurity_Jan2012.pdf (accessed 21.02.15).

Anon., 2013. Afghanistan IOM—Natural Disasters—Induced IDPs in 2012 (map). iMMAP, IOM, USAID. http://reliefweb.int/map/afghanistan/afghanistan-iom-%E2%80%93-natural-disasters-induced-idps-2012-10-february-2013 (accessed 16.10.14).

AP, 2015 Afghan avalanches cut off 3 northern districts. The Sun Herald, 3 March 2015. http://www.sunherald.com/2015/03/03/6099792/afghan-avalanches-bury-more-homes.html (accessed 14.03.15).

BBC, 2010. Afghanistan avalanches kill at least 165 in Salang Pass. BBC News; 10 February. http://news.bbc.co.uk/2/hi/south_asia/8506033.stm (accessed 21.02.15).

BBC, 2012. Afghanistan avalanche kills 42 in Badakshan. BBC News Asia; 6 March. http://www.bbc.com/news/world-asia-17273198 (accessed 20.02.15).

Beekma, J., Fiddes, J., 2011. Floods and droughts: the Afghan water paradox. Afghanistan Human Development Report 2011. CPHD (Centre for Policy and Human Development); Kabul University. http://www.researchgate.net/publication/263506848_Flood_and_drought_-_The_Afghan_water_paradox (accessed 14.07.15).

Bishop, M.P., Bush, A.B.G., Furfaro, R., Gillespie, A.R., Hall, D.K., Haritashya, U.K., Shroder Jr., J.F., 2014a. Theoretical foundations of remote sensing for glacier assessment and mapping. In: Kargel, J.S., Leonard, G.J., Bishop, M.P., Kääb, A., Raup, B. (Eds.), Global Land Ice Measurements from Space. Springer-Praxis, Berlin. ISBN-10: 354079817X, ISBN-13: 978-3540798170 (Chapter 2).

Bishop, M.P., Shroder Jr., J.F., Ali, G., Bush, A.B.G., Haritashya, U., Roohi, R., Sarikaya, M.A., Weihs, B.J., 2014b. Remote sensing of glaciers in Afghanistan and Pakistan. In: Kargel, J.S., Leonard, G.J., Bishop, M.P., Kääb, A., Raup, B. (Eds.), Global Land Ice Measurements from Space. Springer-Praxis, Berlin. ISBN-10: 354079817X, ISBN-13: 978–3540798170 (Chapter 23).

Bowley, G., 2012. Avalanche exposes Afghans' vulnerability to vagaries of climate and of aid. The New York Times, 7 March. http://www.nytimes.com/2012/03/08/world/asia/afghanistan-avalanche-death-toll-rises.html (accessed 20.02.15).

Bohannon, R.G., 2005. Geologic Map of Quadrangle 3166, Jaldak (701) and Maruf-Nawa (702) Quadrangles, Afghanistan. U.S. Geological Survey Open File Report 2005-1120-A.

Costa, J.E., Wieczorek, G.F. (Eds.), 1987. Debris Flows/Avalanches: Process, Recognition, and Mitigation. Reviews in Engineering Geology , vol. VII. Geological Society of America, Boulder, Co, USA.

Denning, G.M., Steele, W.B., Johnson, W.C., Chauvin, K.W., Omelicheva, M.Y., Dobson, J.E., Herlihy, P.H., Mandel, R.D., Hoopes, J.W., Macpherson, G.L., Haufler, M., 2013. Final Report of the University of Kansas Research and Educational Support for U.S. Army Programs and Initiatives at Fort Leavenworth, Kansas [Karez]. pp. 1–48, http://www.google.com/url?sa=t&rct=j&q=&esrc=s&source=web&cd=8&ved=0ahUKEwiij_2mnYTKAh-VC4CYKHaBCDSUQFghOMAc&url=http%3A%2F%2Fwww.dtic.mil%2Fcgi-bin%2FGetTRDoc%3FAD%-3DADA602982&usg=AFQjCNFOApMeFCrHFE3emxR6ic9RyDKMNg&sig2=lgJg2bPtDlNDXQZFMjmXUA (accessed 01.01.16).

ESCAP, 2012. Compendium on Water-Related Hazards and Extreme Weather Events in Central Asia and Neighboring Countries. United Nations Economic and Social Commission for Asia and the Pacific; Chapter II. Water-Related Hazards and Disaster Risk Management in Central Asia; 1. Afghanistan; ST/ESCAP/2640, pp. 5–10.

Fagen, B., 2003. The Long Summer: How Climate Changed Civilization. Basic Books, New York.

GWP, 2014. Drought Monitoring System for South Asia: Summary Report of the Need Assessment Survey on the Development of South Asian Drought Monitoring System (SA DMS). Global Water Partnership—South Asia (GWP—SAS). 18 pp. http://www.droughtmanagement.info/literature/GWP_SA_Summary_Report_Need_Assessment_Survey_2014.pdf (accessed 20.02.15).

Hagen, E., Shroder Jr., J.F., Lu, X.X., Teufert, J.F., 2010. Reverse engineered flood hazard mapping in Afghanistan: a parsimonious flood map model for developing countries. Quat. Int. 226 (1–2), 82–91.

Hall, M., 2012. Droughts and Food Insecurity in Afghanistan. Civil-Military Fusion Centre. http://reliefweb.int/sites/reliefweb.int/files/resources/CFC_Afghanistan_Food-Insecurity_Jan2012.pdf (accessed 21.02.15)..

Harris, C., 2014. Snow and Ice-related hazards, risks and disasters. Foreword. In: Haeberli, W., C. Whitman (Eds.), J.F. Shroder (Ser. Ed.), Snow and Ice-Related Hazards, Risks and Disasters, Amsterdam.

Hewitt, K., 2004. Geomorphic hazards in mountain environments. In: Slaymaker, O., Owens, P. (Eds.), Mountain Geomorphology. Hodder Scientific, London, pp. 187–218.

IWMI, 2015. About SWA Drought Monitor. http://dms.iwmi.org/about_swa_dm.asp (accessed 20.02.15).

Jakob, M., Hungr, O. (Eds.), 2005. Debris-Flow Hazards and Related Phenomena. Springer/Praxis Publishing, Chichester, UK.

Kravtsova, V.I., Bondareva, T.A., 1983. Mapping the avalanche hazard for mountainous areas of Afghanistan (original in Russian). Mapp. Sci. Remote. Sens. 30 (2), 137–150. http://www.tandfonline.com/doi/abs/10.1080/07493878.1993.10641928#preview (accessed 20.02.15).

Melosh, J., 1987. The mechanics of large rock avalanches. In: Costa, J.E., Wieczorek, G.F. (Eds.), Debris Flows/Avalanches: Process, Recognition, and Mitigation. Reviews in Engineering Geology, vol. VII. Geological Society of America, Boulder, CO, USA, pp. 41–49.

Norland, R., 2010. Avalanches kill dozens on mountain highway in Afghanistan. The New York Times, 9 February. http://www.nytimes.com/2010/02/10/world/asia/10avalanche.html?_r=0 (accessed 20.02.15).

OCHA, 2012. Afghanistan: Districts Affected by Natural Disasters. United Nations Office for the Coordination of Humanitarian Affairs. http://reliefweb.int/sites/reliefweb.int/files/resources/AFG_NaturalDisasters_Jan_Aug_10Sept2012_A3.pdf (accessed 21.02.15).

OCHA, 2013. Afghanistan: Overview of Natural Disasters. Natural Disaster Incidents as Recorded by OCHA Field Offices and IOM From 1 January 2013 to 30 June 2013. http://www.humanitarianresponse.info/operations/afghanistan/infographic/afghanistan-overview-natural-disasters-iom-and-ocha-1-january-30 (accessed 16.10.14).

Pierson, T.C., Costa, J.E., 1987. The rheological classification of subaerial sediment-water flows. Geol. Soc. Am. Rev. Eng. Geol. 7, 1–12.

Quiring, S.M., Ganesh, S., 2010. Evaluating the utility of the Vegetation Condition Index (VCI) for monitoring meteorological drought in Texas. Agric. For. Meteorol. 150 (3), 330–339.

Relief Web, 2015. Avalanches claim 6 lives NE Afghanistan. Kuwait News Agency (KUNA); 16 February.

Santi, P.M., Hewitt, K., vanDine, D.F., Barillas Cruz, E., 2010. Debris-flow impact, vulnerability, and response. Nat. Hazards 56 (1), 371–402. http://inside.mines.edu/~psanti/paper_pdfs/DF%20Impact%20Vulnerability%20and%20Response.pdf (accessed 14.10.14).

Shreve, R.L., 1968. The Blackhawk Landslide. Geological Society America Special Paper 507, 48 pp.

Shroder Jr., J.F., 1976. Mass movement on Nyika Plateau, Malawi. Z. Geomorphol. 2, 56–77.

Shroder Jr., J.F., 1989. Slope failure: extent and economic significance in Afghanistan and Pakistan. In: Brabb, E.E., Harrod, B.L. (Eds.), Landslides: Extent and Economic Significance in the World. A.A. Balkema, Rotterdam, Netherlands, pp. 325–341.

Shroder Jr., J.F., 2012. Afghanistan: rich resource base and existing environmental despoliation. Environ. Earth Sci. 67 (7), 1971–1986. http://dx.doi.org/10.1007/s12665-012-1638-7.

Shroder, J.F., 2014. Natural Resources of Afghanistan: Geographic & Geologic Perspectives on Centuries of Conflict. Elsevier, Amsterdam.

Shroder Jr., J.F., Bishop, M.P., 2010. Glaciers of Afghanistan. In: Williams Jr., R.S., Ferrigno, J.G. (Eds.), Satellite Image Atlas of Glaciers of the World—Asia. U.S. Geological Survey, Washington, DC, pp. 167–199 U.S. Geological Survey Professional Paper 1386-F3.

Shroder Jr., J.F., Giardino, J.R., 1978. Progress on rock glacier research. Trans. Nebr. Acad. Sci. VI, 51–54.

Shroder Jr., J.F., Weihs, B., Schettler, M., 2011a. Mass movement in northeast Afghanistan. J. Phys. Chem. Earth 36, 1267–1286. http://dx.doi.org/10.1016/j.pce.2011.03.003.

Shroder, J.F., Schettler, M.J., Weihs, B.J., 2011b. Loess failure in northeast Afghanistan. J. Phys. Chem. Earth 36, 1287–1293. http://dx.doi.org/10.1016/j.pce.2011.03.001.

Thomas, V., Eqrar, N., 2011. Managing water resources, scarcity and climate shocks. In: Kakar, K. (Ed.), Afghanistan Human Development Report 2011—The Forgotten Front: Water Security and the Crisis in Sanitation. Centre for Policy and Human Development, Kabul University. 238 pp. http://re.indiaenvironmentportal.org.in/files/Complete%20NHDR%202011%20final.pdf (accessed 25.07.15).

Unitar, 2014. The Ab Barak Landslide: Past and Future. UNOSAT. 5 pages, http://unosat-maps.web.cern.ch/unosat-maps/AF/FL20140430AFG/UNOSAT_Ab_Barak_landslide_report.pdf (accessed 14.03.15).

USAID, 2014. Afghanistan Rapid Assessment Form (RAF)—Country Overview January 2012 to August 2014. US Agency for International Development, iMMAP. http://reliefweb.int/sites/reliefweb.int/files/resources/iMMAP_AFG_A3_DRR_RAF_Country%20overview%20for%20the%20period%20January%20to%20August%202014.pdf (accessed 21.02.15).

Weler, J., Herring, D., 2000. Measuring Vegetation (NDVI & EVI). NASA Earth Observatory. http://earthobservatory.nasa.gov/Features/MeasuringVegetation/ (accessed 20.02.15).
Wheeler, R.L., Rukstales, K.S., 2007. Seismotectonic Map of Afghanistan, With Annotated Bibliography. U.S. Geological Survey Open-File Report 2007-1104.
Witze, A., 2014. Afghan landslide was "an accident waiting to happen.". Nature (Q&A). 4 pages, http://www.nature.com/news/afghan-landslide-was-an-accident-waiting-to-happen-1.15158 (accessed 14.03.15).
WMO, 2006. Drought Monitoring and Early Warning: Concepts, Progress and Future Challenges. World Meteorological Organization. WMO-No. 1006, 24 pp.
Zahori, H., Ahmed, A., 2014. Lack of orderly means to distribute aid is latest setback for Afghan village. The New York Times; May 6, 2014. http://www.nytimes.com/2014/05/06/world/asia/lack-of-orderly-means-to-distribute-aid-is-latest-setback-for-afghan-village.html?_r=0 (accessed 14.03.15).

CHAPTER

# 6

# Characteristics and Implications of Climate Change in Afghanistan and Surrounding Regions

*J. Shroder*

University of Nebraska at Omaha, Omaha, NE, United States

OUTLINE



## Abstract

Climate change in the Afghanistan region has become obvious because so much of the area is generally so close to the edge of habitability anyway. In recent years, desertification and water shortages have become quite obvious with ever recurrent droughts that seem to be increasing in frequency and magnitude. Water-related hazards (floods, water-induced landslides) seem to have increased as well, and water security issues have reverberated into intelligence and military circles as issues that must be attended to if the country is to survive. Possible solutions include reducing scientific uncertainty, taking measures for mitigation or adaptation, and increasing public awareness and active engagement in the problem.

**Keywords:** Desertification, Water security, Megadrought, Recurrent drought, Historical drought, Water-related hazards, Science-policy linkage, Mitigation, Adaptation, Public awareness, Integrated water-resources management (IWRM), Pleistocene Lake Seistan



http://dx.doi.org/10.1016/B978-0-12-801886-6.00006-9

## 6.1 INTRODUCTION

The fact that the world's climates are in a state of flux is growing ever more obvious as scientists study additional aspects to find further evidence, and as people begin to notice rather glaringly obvious weather problems and changes in their own regions. The changing climate in the whole of the Hindu Kush—Himalaya (HKH) of South Asia is very much on the minds of the scientists who have been working there on various aspects of the hydrological cycle for some time (Haritashya et al., 2009; Bishop et al., 2010; Erickson et al., 2009; Norgrove et al., 2008; Savage et al., 2008, 2009, and many others); now the evidence has accumulated to the point where few intelligent people will deny it. Unlike the eastern Himalaya, the summer monsoon precipitation dominates the climate of the western HKH only in July and August, and is dominantly of an orographic nature, which results in distinct variations with altitude, from the deep drier valleys, to greater moisture on the peaks. In many areas, a higher proportion than before of the total precipitation appears to be coming now as rain, with the result that snowmelt is earlier, with shorter winters. In fact, a decrease in monsoon precipitation, by up to 20%, is forecast by the end of the century in most parts of Pakistan and southeastern Afghanistan (Erickson et al., 2009).

## 6.2 REGIONAL CLIMATE CHANGE

It is thus now quite obvious that Afghanistan is indeed subject to serious climate change, and is clearly suffering from the early consequences of global warming. The persistent droughts and floods in Afghanistan that have resulted from serious land degradation and outright desertification (Shroder, 2012) ultimately increase the burden of poverty, disease, and mass-population displacement. The incidence and intensity of these disasters are seen to increase as climate change deteriorates further. Recently it has also been hypothesized by some people (Savage et al., 2008; Williams, 2009) that Afghanistan's recurrent droughts may actually be the forerunner warning signals of impending major climate change in the region, even while the water problems exacerbate civil unrest and make the country ever harder to rule effectively. In fact it is thought by some of us that the country of Afghanistan actually may be a bit of a bellwether to the rest of the world about the great potential for disruptive effects of climate change. This appears to be true, especially in countries already close to the edge from other stressors, including environmental disruptions of all kinds, religiously motivated upheavals, politically motivated insurgencies, and all the other aspects of failed states that Afghanistan is certainly heir to.

UNEP (2014) has noted that Afghanistan has been identified as one of the most vulnerable countries in the world to climate change. With all the indications of the need for adaptation to climate change, UNEP is attempting to strengthen institutional capacity to ensure effective adaptation planning and protection of communities, ecosystems, and development against deleterious effects. A number of UNEP programs are designed directly to help develop a national adaptation strategy for climate change at numerous levels in the government. Chapter 19 in Shroder (2014) on Afghanistan water and climate change points out the seriousness of the changes at hand, and all the consequences of not facing up to the challenges that have already occurred and that are coming. In fact, changes in transboundary water management and transparency in all dealings were viewed as essential to avoid predictable coming conflicts or new water wars. Unfortunately, as discussed in Chapter 17 on hydropolitics herein, the moratorium under President Karzai on discussing water within the Afghanistan government below ministerial or Presidential Cabinet level was exemplary of an understandable, rather long-term, head-in-the-sand approach to governance in Afghanistan, rather than to a more enlightened position of leadership in which concerted study and scientific conference discussion on water issues to find solutions is encouraged. Still, one can understand the current unease of the Government of Afghanistan (GoA) when all the countries around seem to want to have access to as much water from Afghanistan as they can get their hands on, at the same time that little to no help is offered for helping Afghanistan develop its own water resources for its own benefit. Under such conditions, it is quite understandable as to why the GoA must proceed most carefully until the water complexities are better understood. This book is very much a device to promote the education of Afghan officials on the need to learn more about their water resources.

Price et al. (2014) undertook interviews of people across South Asia, including Afghanistan, to explore attitudes towards domestic water management and transboundary water issues. In terms of understandings and attitudes toward climate change, their findings unearthed a number of rather pessimistic points of view. For example, climate change was seen by some respondents as a burgeoning problem, with apparent increases of related extreme weather events. The need to deconvolute these apparent increases as a result of climate change from the effects of progressive environmental despoliation (deforestation, soil erosion, etc.; Shroder, 2012) seemingly was not recognized, however. Nonetheless, respondents did view government approaches to climate change with a good deal of skepticism. For example, the new National Environmental Protection Agency (NEPA) in Afghanistan is the body responsible for setting government-wide policy and action on climate change, but the amount of actual or practical action on the issue was regarded as practically nonexistent (Price et al., 2014).

An additional issue was that a huge gulf was seen to exist between the national policy framework, which is wide-ranging, quite technical, and of an internationally recognized, high standard, with however, the measures built into the ground-level operations being underfunded, lacking oversight, and uncoordinated with national policy. Local programs were completely decentralized and set up with many commonly competing, decision-making centers (Price et al., 2014). Furthermore, a related issue is that Afghanistan's NEPA was regarded as a weak department, staffed by nepotism, lacking technical knowledge, and being underresourced. In fact international agencies provided most of the actual expertise as foreigners, who helped to write most of the regulations, but commonly did not translate the materials into local languages (Dari, Pashto) so that the Afghan people could even understand what was newly required for dealing with the effects of climate change. Some respondents even indicated that Afghanistan should not bother to be concerned about climate change because the country had contributed relatively little to the overall problem of world climate change and so they could safely ignore the problems. In addition, the greatest threats to water supply and water management were seen actually as stemming mainly from the neighboring countries, particularly Pakistan and Iran, rather than from climate change. The ongoing instability and violence in Afghanistan was seen additionally as being directly fomented to keep Afghanistan weak and undeveloped so that water flows out of Afghanistan into the neighboring countries would be kept at a maximum.

## 6.3 WATER SECURITY

Already the key judgments on global water security by the US intelligence community (Intelligence Community, 2012) have emphasized five major points of coming problems with water from a security point of view (Table 6.1), all with direct or indirect linkages to Afghanistan. In fact this intelligence community has also focused on two of only seven world major rivers, the Indus and the Amu Darya, both of which have significant tributary parts in Afghanistan (Table 6.2). They listed nine types of water issues in those river basins, all of which issues are directly applicable to Afghanistan, along with notably moderate to inadequate management capacities there, as well as in the surrounding countries. Any additional stressors, such as climate change in these river basins, only add additional pressures that likely will not be tolerable in Afghanistan, so that violence will increase in that already volatile place. Clearly, enhanced Afghan hydro-informatics (hydrographs, river discharges, ground-water levels, etc.) would add much to help overcome some of the lack of understanding about this absolutely vital resource in any country.

In addition to the above work of the intelligence community about climate change, the US Department of Defense (DOD) has recently also made significant plans to deal with the effects of climate change on security issues (DOD, 2014; Morton, 2015). DOD, of course, has essential commitments and responsibilities all over the world as a result of its great capabilities, which has resulted in considerable technical experience for operating in diverse climatic environments. Nonetheless, it is also quite likely that their experience that was gained over the 15 years by the US military operating in the difficult theater of Afghanistan at war, led many of the intelligence and operational officers who served in the country to appreciate the results of climate change writ large across the face of the nation as diverse climatic hazards have erupted. In any case and in spite of considerable climate-change denial in its conservative ranks, the Pentagon has recently released a new Climate Change Adaptation Roadmap (CCAR) detailing

TABLE 6.1 Key Judgments in Global Water Security, Based in Part on Climate-Change Projections, With Implications for Afghanistan.

A. During next decade water problems will contribute to instability in states important to US national security interests. Water shortages, poor water quality, and floods by themselves are unlikely to result in state failure, but water problems—combined with poverty, social tensions, environmental degradation, ineffectual leadership, and weak political institutions—contribute to social disruptions that can result in state failure.

B. Water-related, state-on-state conflict is unlikely during next decade. Historically, water tensions have led to more water-sharing agreements than violent conflicts. As water shortages become more acute beyond next decade, however, water in shared basins will increasingly be used as leverage, as a weapon, or to further terrorist objectives.

C. During the next decade, depletion of ground-water supplies in some agricultural areas because of poor management will pose risks to both US national and global food markets.

D. During the next 30 years water shortages and pollution probably will harm economic performance of important trading partners to USA.

E. During the next 30 years improved water management (eg, pricing, allocations, "virtual" water trade), and investments in water-related sectors (eg, agriculture, power, water treatment) will afford best solutions to water problems. Because agriculture uses approximately 70% of global fresh-water supply, greatest potential for relief from water scarcity will be through technology that reduces amounts of water needed for agriculture.

*From Intelligence Committee, 2012. Global Water Security. Intelligence Community Assessment; ICA 2012-08, 2 February 2012. Defense Intelligence Agency (DIA), NGA, CIA, State/INR, and DOE. http://fas.org/irp/nic/water.pdf (accessed 05.09.14).*

TABLE 6.2 Overview of Selected River Basins That are Part of Afghanistan

| River Basin | Type of Water Issue | Impact/Expected Time | River Basin Management Capacity |
|---|---|---|---|
| Indus | Poor water management | Degraded regional food security—now to 2040. | |
| | Inefficient agricultural practices | | Moderate |
| | Inadequate infrastructure | Reduced resiliency to floods and droughts—now to 2040. | |
| | Greater variability in water availability | | |
| | Water pollution | | |
| Amu Darya | Inadequate water agreements | Degraded regional food security—now to 2040. | Inadequate |
| | Degradation of water quality | Increased regional tensions over water—now to 2040. | |
| | Disruption of water flows in some states | Decrease of health of populations around Aral Sea. | |
| | Poor water management | | |

The two basins have a number of water issues that occur outside Afghanistan, but several parts are directly about Afghanistan. *From Intelligence Committee, 2012. Global Water Security. Intelligence Community Assessment; ICA 2012-08, 2 February 2012. Defense Intelligence Agency (DIA), NGA, CIA, State/INR, and DOE. http://fas.org/irp/nic/water.pdf (accessed 05.09.14).*

a comprehensive strategy against the advancing foe of climate change. These new recognitions will have some effect on the course of combatting terrorism in Afghanistan.

Recent moves by DOD and GoA to slow Western troop pullout in Afghanistan have developed in recognition of the advisability of helping keep the Taliban at bay as new President Ashraf Ghani's government attempts to stabilize and improve its performance. At the same time, the new DOD focus on the effects of climate change will no doubt be reflected in attention to CCAR guidelines that stress the increased frequency of extreme weather that could influence aviation, as well as intelligence, surveillance, and reconnaissance capabilities, especially in a country such Afghanistan. Recovery from extreme weather events will require increased intervention, something that the military does excel at worldwide. Drought was recognized as most likely impacting operations, creating water shortages and fire hazards. Dust storms, already a major problem in much of Afghanistan, will likely become more of a factor to be concerned about with training activities, security operations, sensitive equipment, and control measures to meet new laws on environmental compliance. The report also emphasizes the facts of global climate change as an international phenomenon that will cause instability by impairing access to food and water, damaging infrastructure, spreading disease, uprooting and displacing large numbers of people, which will compel mass migration, interrupt commercial activity, and restrict availability of electricity; all of these factors will obviously impede maintenance of security in Afghanistan. Furthermore, these predicted developments were also recognized as having the potential to undermine an already fragile government that are unable to respond effectively, or even challenge currently stable governments, as well as increasing competition and tension between countries vying for limited resources. These gaps in governance will create an avenue for extremist ideologies and conditions that foster terrorism (DOD, 2014; Morton, 2015); such factors will likely play to extremes in a country such as Afghanistan and must be carefully considered in the future of security control there.

In its consideration of the most probable situations prevailing in Afghanistan for the coming years of 2015–2019, the United Nations has endeavored to project the most likely futures, given the best available data (United Nations, 2014). In their results and resources framework for Afghanistan, their desired Output 8 specifically noted that increased community resilience to climate change would likely occur, which would thereby reduce vulnerability and sustain economic gains in the Afghanistan "Transformation Decade" that was projected to be coming. In their analysis, the indicators for success with this output were thought to be the number of provinces that had attained operational systems for early warning and data information management. The target would be ten provinces by the end of the campaign. Furthermore, an additional indicator of success would be the number of provincial development plans in which climate-change adaptation measures were explicitly addressed and budgeted for. Numerous other benchmark outputs were also listed by the United Nations (2014) as necessary for the benefit of the nation, and for measuring the success of UN programs in the country, but are not further relevant here concerning water.

## 6.4 WATER-RELATED HAZARDS IN AFGHANISTAN AND CLIMATE CHANGE

An additional stressor with water in Afghanistan that is not often considered in any detail when everyone is concerned most about climate change and not having enough water in future, is the problem of recurrent floods and water-related, mass movements, or landslides (Chapter 5). As Shroder (2014) has noted, likely changes in atmospheric circulation in South Asia and intensified monsoonal circulation have brought about increase in natural hazards, especially floods. Climate change involves most seriously, changes in the magnitude and frequency of extreme weather events, with high intensity rainfall leading to flashfloods, rapid wet debris flows, and a wide variety of other types of landslides. The recent (early May 2014) flow of wet loess over the village of Ab Barik situated largely on crystalline bedrock in Badakhshan Province, Afghanistan (see Chapter 5) was thought to be an indication of such apparently increased hazard, for example, although it should be noted that many other such flows have been occurring in the region for some time (Shroder et al., 2011). Aside from the direct impact of mass movement into towns where people live, other valuable infrastructure, such as hydropower plants, roads and bridges, and communication systems, will be increasingly at risk from these flooding events and other hazards as the climate changes.

Vulnerable water sectors in Afghanistan have been described by Norgrove et al. (2008) by using an analytical hierarchy process (AHP) to assess water vulnerabilities in Afghanistan. The multi-criteria, decision-making method of AHP is a structural technique for organizing and analyzing complex decisions (Saaty, 1977), based in this case on socioeconomic rankings (from 1 to 9), climate-induced parameters, and the sub-sectors of surface water, underground water, and hydropower generation (Table 6.3). The vulnerability index of −338.8 that resulted was the most extremely negative of any of the other six factors

TABLE 6.3 Weighting Matrix for Prioritizing Vulnerability Indices (VI) for Water-Resource Sectors in Afghanistan

| | Socioeconomic | Climatic Parameters | | | | Sub-Sectoral | |
|---|---|---|---|---|---|---|---|
| **Sub-Sector** | **Rank** | **T** | **ET** | **P** | **D** | **VI** | **VI** |
| Hydropower generation | 6 | −3 | −4 | −1 | −7 | −90 | |
| Underground water | 5.6 | −1 | −2 | −7 | −7 | −95.2 | −338.8 |
| Surface water | 6.4 | −6 | −7 | −3 | −8 | −153.6 | |

Socioeconomic ranks ranged 1–9; climatic parameters ranged 9 to −9, according to the Saaty (1977) approach in the analytic hierarchy processes (AHP). Minus values indicate severity of climate change. The sub-sectoral vulnerability indices are calculated by multiplying the socioeconomic rank to the summation of vulnerability ranks to climatic parameters. Sub-sectoral values sum to the vulnerability index (VI). T = temperature raise; ET = evapotranspiration and salinity; P = precipitation pattern change; D = drought. See text for explanation of terminology and calculations.
*From Norgrove, L., Bowling, B., Ahadi, M.S., Modaqiq, W., Sallari, S., ur Rahman, S.S., Haidari, G.H., Omar, M.O., Shams, H.K., Ahmadi, L.A., Jan, S., Shakir, S., Ali, R., Habib, M., Habib, G., Yunosi, A., Fakhiri, A., Haqbeen, M.A., Ahmadzai, A.B., Ibrahimi, M.W., Hamidullah, Sarwary, G.D., Yasin, M., 2008. Climate Change and Disaster Preparedness Working Group: Final Thematic Report. Global Environmental Facility; United Nations Environment Programme; Government of the Islamic Republic of Afghanistan; 83 pp. http://postconflict.unep.ch/publications/afg_tech/theme_02/afg_ccdp.pdf (accessed 24.02.15).*

(agriculture, forestry and rangeland, biodiversity, health, energy, water) that were measured, and probably one of the worst in the world; other such global data were not included in the study.

## 6.5 HISTORICAL DROUGHTS IN AFGHANISTAN

The country of Afghanistan is characterized by large, wind-blasted, desert areas with little precipitation where any water that does occur tends to have flowed down from snowmelt in the high mountains. Slightly more rain does fall along the border with Pakistan where the low pressures of the intertropical convergence zone (ITCZ) cause the in-rushing air masses of the summer monsoon. During these warm months the northerly winds that are drawn in to the low pressure regions cause the wind of 120 days (*badi sado bist roz*), which can blow day and night, rather like an electric hot-air blower set at 50°C (122°F) to create vast dust storms. When periodic droughts occur in this annual hot-wind cycle as well, the dust storms move vast quantities of clay- and silt-sized particles that form the extensive loess dust deposits of the country. The sand that is left behind piles up in vast seas of dunes in the southwest and northern lowlands of Afghanistan.

In the Pleistocene ice-age times of Afghanistan, when huge glaciers extended throughout the highlands, large lakes also existed in some of the lowlands, especially in the Seistan Desert of the southwest (Shroder, 2014). As the climate warmed at the end of the Pleistocene and the Holocene interglacial time started about 11,700 years ago, those lakes began to grow smaller through increased evaporation and reduced river inflow. Ancient pre-human hominids had walked out of Africa during the Pleistocene and had left their stone tools across South Asia in places that they found habitable (Fig. 8.3 in Shroder, 2014). Eventually thousands of years later their descendants began to build small cities at various places in Afghanistan, in some places around the lake shores or along the rivers where the water necessary for life was reasonably abundant. Overtime, however, periodic city-killer droughts would have occurred in that increasingly arid place, even if few historical records of those weather variations were preserved.

Recently, however, evidence of these droughts has been discovered in northwest India, close to the border with Pakistan and Afghanistan. This new evidence points to an abrupt weakening of the summer monsoon about 4100 years ago that is thought to be responsible for droughts that brought about the end of the Harappan Civilization of the Indus Valley (Dixit et al., 2014; Marris, 2014). In fact, the resulting drought lasted for some two centuries, which certainly constitutes a megadrought, or a dry cycle lasting more than two decades. Such megadroughts have been known to eradicate many other civilizations worldwide in the past (Anasazi, Mayan, Akkadian, Khemer, Yuan; Fagen, 2003), so that possibility exists for parts of Southwest Asia as well. The implications for Afghanistan's future are obvious (Shroder, 2014).

The most serious historical drought in Afghanistan was thought to be in 1970–72, but as Alim and Shobair (2002) have noted, other, but more minor droughts had occurred in 1948, 1955, 1961–63, and 1966–67 as well. Other less important droughts apparently also occurred in 1977, 1981, and 1992. The most serious recent drought cycles have been in 1998–2006, and again in 2008–09. UNEP (2013), with data from the European Commission Humanitarian Office (ECHO), noted that the period 1998–2005/6 marked the longest and most severe drought in Afghanistan's known climatic history. These observations on droughts and other climatic and atmospheric-emissions data have been assessed by technical analysts at the Stockholm Environment Institute, who were interested in future forecasting (Savage et al., 2008). With reservations appropriate to this 7-year-old analysis while so much new analysis is underway that is not yet available, still the indications of increased temperatures and decreased precipitation in Afghanistan, urge serious attention by the GoA to what will further reduce streamflow in the country. The Stockholm group thought they had evidence that much of the drying would be due to decreases in spring (March-April-May) precipitation, and winters were expected to be significantly drier in the south (Savage et al., 2008). They leveled rather severe criticism of the GoA that no sectoral or national plans had been made anywhere in the government to think about or address the coming climate change. They felt that no existing programs in the GoA were sufficient in scope or scale to meet the challenges indicated by the climate scenarios, so they needed to find solutions.

Approximately 73% of Afghanistan is shrubland, savannah, and grassland, whereas 15% is sparse or barren of vegetation, as well as snow and ice. These are fragile lands that have been subjected to severe negative impacts (overgrazing, deforestation, dearbification; Shroder, 2014) by people and their grazing animals for millennia. The result of this long-term land degradation is that no parts of the region, with the exception of the highest alpine areas, have escaped the human depredations. Nationwide, the balances between precipitation and primary agricultural production is rather precarious, and coupled with poor land management, conflict, insecurity, migration, poverty, and drought; all these factors have left Afghanistan exceptionally vulnerable to desertification (Fig. 6.1) (Norgrove et al., 2008).

## 6.6 POSSIBLE PARTIAL SOLUTIONS TO CLIMATE AND WATER PROBLEMS

The recommendations that the Stockholm group thought might work, at least in the shorter term (Savage et al., 2008), were to: (1) clean up the records of existing historical weather data and increase the level of computer-driven climate modeling of the region; (2) improve

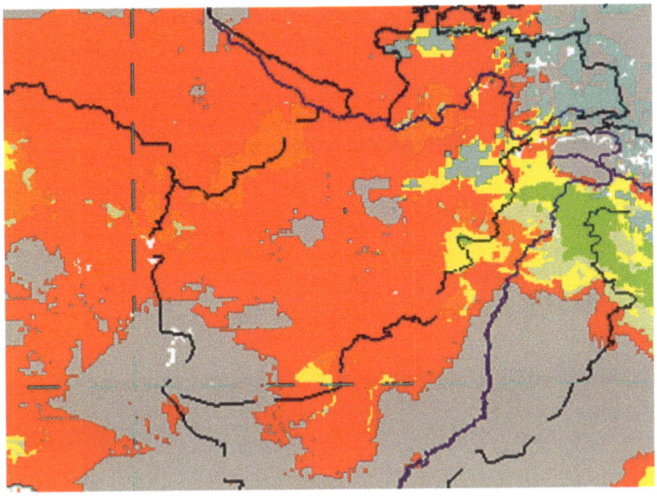

FIG. 6.1 Assessment of vulnerability of Afghanistan to desertification, where *red* indicates the highest vulnerability. As can be seen, most of the country is subject to full conversion to desert. *Light blue* and *light gray* are already full deserts. *From US Department of Agriculture, 1998; Norgrove, L., Bowling, B., Ahadi, M.S., Modaqiq, W., Sallari, S., ur Rahman, S.S., Haidari, G.H., Omar, M.O., Shams, H.K., Ahmadi, L.A., Jan, S., Shakir, S., Ali, R., Habib, M., Habib, G., Yunosi, A., Fakhiri, A., Haqbeen, M.A., Ahmadzai, A.B., Ibrahimi, M.W., Hamidullah, Sarwary, G.D., Yasin, M., 2008. Climate Change and Disaster Preparedness Working Group: Final Thematic Report. Global Environmental Facility; United Nations Environment Programme; Government of the Islamic Republic of Afghanistan; 83 pp. http://postconflict.unep.ch/publications/afg_tech/theme_02/afg_ccdp.pdf (accessed 24.02.15); UNEP, 2009. Afghanistan: National Capacity Needs Self-Assessment for Global Environmental Management (NCSA) and National Adaptation Programme of Action for Climate Change (NAPA). United Nations Environment Programme; 123 pp. http://www.unep.org/dgef/Portals/43/publications/Afghan_NCSA_and_NAPA_2009.pdf (accessed 23.02.15).*

the "climate awareness and resilience" of the development strategies of the Afghanistan Ministries; (3) coordinate between the Ministries to meet the cross-sectoral challenges of climate change by linking agriculture and livelihoods to water in a strategic plan; (4) the GoA should create priced-out strategies of adaptation-development investment towards applying to potential climate-adaptation, funding mechanisms in the international community; (5) GoA experts should continue engaging with larger professional entities to ensure best-practice climate assessments and adaptation approaches; (6) development impact should take precedence over world-emission standards in order to build adaptive capacity; and (7) small-scale, off-grid renewable energy technologies (small hydro, solar photovoltaic (PV), solar thermal, wind) could provide necessary enhanced access to energy. This report led directly to significant support from the United Nations of a $6 M gift to the GoA to combat the effects of climate change because of their current limited capacity to react to climate-change impacts. The initiatives intervention was designed to include: (1) improved water management and water-use efficiencies; (2) community-based watershed management; (3) improved terracing; (4) agro-forestry and agro-silvo-pastoral systems; (5) climate-related research and early warning systems; (6) improved food security; and (7) rangeland management (UN News, 2012). The whole package was designed to overcome the instability associated with climate change in Afghanistan as well as to increase the water-cognizant capacity of the GoA.

TABLE 6.4 Policy Change Recommendations in the World of Climate Change in the HKH. Although Developed for the entire region of high Asian Himalaya mountains, the schema can be applied usefully to Afghanistan alone

A. Reducing scientific uncertainty
   1. Develop scientific programs for climate-change monitoring.
   2. Application of regional climate models.
B. Mitigation measures
   1. Beyond the Kyoto protocol.
   2. Land-use management for carbon sinks and reduced emissions.
   3. Payment for ecosystem services (PES).
   4. Development of alternate technologies.
C. Adaptation measures
   1. Disaster-risk reduction and flood forecasting.
   2. Supporting community-led adaptation.
   3. National adaptation plan of action.
   4. Integrated water-resources management.
D. Public awareness and engagement
   1. Full disclosure and prior information for grass-root societies.
   2. Engagement of the media and academia.
   3. Facilitation of international policy dialogue and cooperation.

See text for explanations.
*From ICIMOD, 2009. Water Storage: A Strategy for climate change adaptation in the Himalayas. Sustainable Mountain Development; No. 56 Winter.*

The long-term climatic situation in Afghanistan may or may not be amenable to some remediation, but Erickson et al. (2009) from the International Centre for Integrated Mountain Development (ICIMOD) in Kathmandu, Nepal, have made a variety of policy recommendations to try to help mountain communities of the HKH to better cope. These recommendations are grouped into four major strategies: (1) reducing scientific uncertainty; (2) developing mitigation measures; (3) taking adaptation measures; and (4) increasing public awareness and engagement (Table 6.4). Within the context of climate change and water, their main points involved developing scientific programs for climate-change monitoring and development of computer-driven, regional climate models (RCMs) that are downscaled enough to apply to impact assessments for watersheds and subcatchments. In addition, integrated water-resources management (IWRM) was seen as needing scaling up from watersheds to river basins, with water allocation for households, agriculture, and ecosystems deserving particular attention. The characteristics of water storage that were based on local practices were seen as needing encouragement in the mountain regions.

A severe gap exists in the knowledge of both short- and long-term implications of the impact of climate change on water and water-related hazards in the HKH and their downstream river basins (Erickson et al., 2009). All too many studies have excluded the HKH because of its extreme and complex topography, and lack of the typical rain-gauge data that is so common in the nation-states of the West. An urgent need exists for closing the knowledge gap by establishing accessible monitoring schemes for snow, ice, and water, as well as for both downscaling and upscaling climate models for the complex topography. Some such possible solutions are presented below in the final chapter on the future of water management in Afghanistan. Developing basin-wide scenarios that take into account water demand and socioeconomic development, while also applying hydrological models to predict water availability, would be tremendously useful applications.

In addition to these recommendations, Erickson et al. (2009) have also noted a need for better land use to encourage carbon sinks into tree farms and reduce emissions. Reforestation or transitions into forest recovery have been emphasized in a number of other HKH countries but not yet in Afghanistan, or only in a minor way, and then generally at the behest of the foreigners in ISAF and the good works of their PRTs and ADTs (the military provincial reconstruction and agricultural development teams). Aside from the beneficial increase in infiltration of precipitation into the soils of reforested areas, the diminution of soil erosion that would accrue is also highly useful. These factors are all highly important in the world of climate change because such offsets are necessary to overcome all the other detriments.

The idea of payments for ecosystem services (PES) in the headwater regions of Southwest Asia would be a major game changer if it were to come to pass. The mountains of Afghanistan, in addition to all the other highland areas of the region, provide abundant services to the burgeoning downstream populations in terms of water for household purposes, agriculture, hydropower, tourism, spiritual values, and transport. Where they exist, land and water managers in the upstream regions must shoulder the responsibility of ensuring reliable provision of good quality water downstream; where they don't exist, as in much of Afghanistan, they should be established with the encouragement of downstream stakeholders because of the obvious benefits that would accrue. Various PES schemes can be developed at different scales, from local to national and even regional, and can involve local communities, various levels of the GoA, and the private sector. At the present time these ideas of payment for ecosystem services have not been explored much in the region, and would require a great deal of realistic discussion at government levels to effect. Policy and decision makers, in addition to land and water managers should be encouraged to look for win-win solutions in this context because it is clear that any sort of payment schemes from the huge populations of the lowlands, upstream to the countries of the watershed sources would make an enormous difference in Southwest Asia in the face of climate change. In addition, the setting of precedents of modest PES for delivery of good water supplies downstream would establish whole new regional dynamic that would be so unusual in the region as to establish entirely new ways of thinking that would reverberate greatly to the benefit of people and ecosystems, as well as shift hydro-hegemonic thinking in various national capitals in whole new directions that would be potentially quite useful (see Chapter 14 for explanations on hydro-hegemony).

The development of alternative technologies in the face of climate change in this fragile region would be an excellent development, because the diffusion of micro- and macro-hydropower, solar power, wind power, geothermal energy, and biogas manufacture would make tremendous use of now largely wasted resources in Afghanistan. Any sort of emphasis of these neglected resources could help greatly to take advantage of renewable sources and reduce emission of the problematic greenhouse gases $CO_2$ (carbon dioxide) and $CH_4$ (methane) into the atmosphere. Admittedly Afghanistan is not responsible at the present time for much of a carbon footprint onto the environment, but in any kind of carbon-footprint accounting system in South Asia, these schemes would be multiply beneficial and potential remunerative in PES developments.

Adaptation measures to climate change in Afghanistan and neighboring countries will require much better methods of forecasting floods and reduction of risks from disasters. Technical advances in flood forecasting have been developed specifically for Afghanistan and other third-world countries with spotty hydrologic records (Hagen et al., 2010), which can offer an opportunity for regional cooperation in disaster management, should neighboring governments have enough foresight to avail themselves of the opportunity. Recent rapprochements by the Government of Pakistan (GoP) to the GoA (Masood, 2015), after many years of Pakistani support of Taliban, show that it might actually be possible to work closely with Afghanistan in future to try to control floods sweeping down out of the HKH. This should become a political agenda between the two countries (Erickson et al., 2009). Preparedness in climate change is obviously essential in order to reduce the potential destruction from floods and other natural disasters.

Support for community-led adaptation to climate change in Afghanistan would be an obvious reinforcement of grass-roots activities that would be a most worthy action. "Bottom-up" community processes built on local knowledge, innovations, and practices are an excellent approach to vulnerability and local-level adaptation that can be encouraged if recognized and focused upon. Empowering communities to base their adaptations to changing climates and environments on their own decision-making processes and participatory technology development should be encouraged with support from outsiders. Equally important would be giving priority to the most vulnerable groups of the poor, women, and those living in fragile environments such as steep slopes and riverbanks.

Under the initiative of the UN Framework Convention on Climate Change, national adaptation plans of action (NAPAs) were prepared for many countries, and that for Afghanistan also assessed global environmental management. Afghanistan identified 51 potential activity options for adapting to climate change, and through a series of evaluation exercises two adaptation options were selected and worked into summary priority proposals entitled, *Improved Water Management and Use Efficiency*, as well as *Land and Water Management at the Watershed Level* (UNEP, 2009).

IWRM was seen as an integral procedure for risk reduction in disaster preparedness, and included future climate-change scenarios to be scaled up from watersheds to river basins. Water allocations for individual households, to more pervasive agriculture, and up to entire ecosystems were seen to be deserving of particular attention. In mountain regions, water storages that are based on local practices are obviously to be encouraged.

In the course of bringing these factors into greater public awareness and gaining their more robust engagement, a number of procedures were recommended in the ICIMOD report (Erickson et al., 2009). For example, grass-roots societies should be supplied with full disclosure and prior information about the effects of coming climate change in their appropriate languages. The various media and academia need to be engaged with awareness and knowledge to play a significant role in public education, awareness building, and projection of trends. International policy dialogues and cooperation were seen as being needed to address the implication of climate change in Afghanistan on ecology, socioeconomics, and cultures. Cooperative regional ventures require the involvement of the international community, including donors, decision makers, and the private and public sectors. Such would be required for achieving sustainable and efficient management of the transboundary rivers of Afghanistan.

## 6.7 SCIENCE AND POLICY LINKAGES IN CLIMATE CHANGE

Formations of good polices within the context of climate change are mainly based upon good science that is founded upon credible, salient, legitimate knowledge. Such credibility is attained from field observations and actually tested by local communities. Whether or not any knowledge is actually salient at the time depends upon whether or not the information is immediately relevant and useful to policy-makers. Legitimacy of information is founded upon whether or not it is unbiased in its origins or manner of creation, and is both fair and reasonably comprehensive in the manner in which it deals with pro and con interests and views. The policies that result from this process are derived from the formulas or manners of utilizing the power deriving from the application of the knowledge gained.

Large questions arise from this sort of logical analysis, as to how realistic would be these methods of assessment and what could the differences be that would weigh local knowledge more favorably than scientific knowledge? This brings up the issues of achieving some sort of pragmatic balance between local understandings and knowledge, and the more scientific certainties. Who might be more correct in any given situation? In fact scientific knowledge is generally always useful, but because of the complex scale issues of the HKH, it may be more limited and laden with sufficient uncertainties to suggest that considerable caution should be leveled with its unquestioned use. Lebel et al. (2004) have studied alternate perspectives in Southeast Asia and concluded that as many as four contrasting perspectives might arise in enigmatic situations, with different points of view from the state, the market, the environmentalists (scientists), and the locals. In some cases then, the perspective that "nobody seems to know best" might be the safest course of action in some complex situations in Afghanistan until greater certainty might be derived at a later time. In the absence of unequivocal or firm scientific evidence to the contrary, great care would have to be exerted to not make situations worse by blind faith in any particular method, just because it seemed so "scientific" or otherwise seemingly so unimpeachable.

Multiple alternate perspectives carry different values and perceptions about where the best knowledge lies to guide decision making and exactly who should be making the rules, as well as what additional scientific studies are needed to produce new knowledge. Table 6.5 shows five different locations or actors involving public engagement that can be used to help resolve issues of scientific uncertainty, adaptation, and mitigation for developing policy where considerable uncertainty exists. This process of engaging multiple actors allows development of the most consensual policy by using essential international cooperation to transfer technology from the outside to the locals, as well as building regional cooperation and developing capacity to downscale important results to more scales.

## 6.8 CONCLUSION

The HKH region, of which Afghanistan is an important part, holds one of the most complex and dynamic mountain systems in the world, which is exceptionally vulnerable to global warming and reduction of water resources (Bandyopadhay and Gyawali, 1994). They note especially that without substantial rethinking, the possibilities for sustainable utilization of this great natural resource of plentiful water is remote, and the possibilities for dissipation of the vast resource from the highlands, and even irreparable damage, are considerable. Erickson et al. (2009) have written in particular that in relation to water, the climate change at hand is gradually and powerfully changing the ecological and socioeconomic landscapes of the region. Imperative calls have been made for environmental decision makers and managers in government to revisit and redesign their research agendas, policies of development, conservation and management practices, and even to develop appropriate new technologies.

TABLE 6.5 Policy Matrix to Cope With Climate-Change Uncertainty in Afghanistan

| Location | Scientific Uncertainty | Adaptation | Mitigation |
|---|---|---|---|
| State | Engage in new research | Collaboration between sectors | Develop good new policies |
| | Support long-term climate research | Support environmental control and alleviation of poverty | Commit to new international water treaties |
| | Develop regional cooperation | | |
| Market | Corporate/academic/GoA research partnerships | Apply new technologies | GoA environmental regulation without rampant corruption |
| | | Support relevant education and community development | |
| Civil Society | Participatory analysis of vulnerability, linking local to global, facilitating knowledge learning | Community preparedness, facilitating local learning and adaptation | Social auditing, green watch, and monitoring |
| Local Community | Local indicators and monitoring, local knowledge, innovations, and practices | Improved land/resource management | Help renewable energy, alternative livelihoods, and mitigation |
| | | Preparedness for surprises | |
| NGO Roles | Impact assessments, knowledge synthesis, regional database, forecasting, and monitoring | Capacity building | Facilitating the clean development mechanism (CDM[a]) and C marketplace, designing payments for water and environmental services |
| | | Support for new water policies | |
| | | Pilot demonstrations | |
| | | Optimizing land-use for patterns and livelihoods | |

[a]*The Clean Development Mechanism (CDM) is an arrangement made under the Kyoto protocol of 1997 that creates emissions-reduction credits through emissions-reduction projects in developing countries.*
*From Erickson, M., Xu, J., Shrestha, A.B., Vaidya, R.A., Nepal, S., Sandström, K., 2009. The Changing Himalaya—Impact of Climate Change on Water Resources and Livelihoods in the Greater Himalayas. International Centre for Integrated Mountain Development (ICIMOD), 28 pp. http://www.preventionweb.net/files/11621_icimodthechanginghimalayas1.pdf (accessed 07.09.14); ICIMOD guidelines.*

The uncertainty inherent in the little studied and the remote and difficult-to-access areas of Afghanistan and surrounding regions means that great uncertainty exists about the rates, and even directions, of all the expected environmental changes. It will certainly get dryer in some places, and it may even get wetter in a few places in Afghanistan, although no one is at all certain that this will occur. The reason for this is, of course, that far too little is known or understood about the dynamics of the regional topo-climates and hydrologic processes. The computer-driven, global circulation models used to predict mathematically changing climates are too broad scale to capture local details of climate change, especially the delivery of precipitation into watersheds. Reduction of this uncertainty requires well-equipped baseline stations, long-term modeling, networking and open sharing of data, and extensive cooperation between neighboring countries of the HKH, factors that are quite limited at the present time with Afghanistan. Several NGOs do specialize in overcoming various kinds of the physical and political restrictions that greatly hinder such activities, including several agencies of the United Nations, ICIMOD, GLORIA (Global Observation Research Initiative in Alpine Environments), GMBA (Global Mountain Biodiversity Assessment), MAIRS (Monsoon Asia Integrated Regional Studies), and MRI (Mountain Research Initiative).

Climate change in this region of Southwest Asia has been a fact of history (Shroder, 2014), even while its relentless nature has been so temporally slow that most ordinary people do not recognize this fact. The long-term nature of climate change in Afghanistan is most exemplified by the millennia of reduction of Lake Seistan in the Helmand Basin (Tate, 1910–1912; Pias, 1972; Smith, 1974; Whitney, 2006), and this fact can be utilized, together with the existing and highly obvious, recent drought sequences in Afghanistan, to help people understand the need to adapt to more coming environmental and climate change. In addition, these people have lived with flash floods, rapid, wet debris flows, snow avalanches, and debilitating droughts and horrendous dust storms for millennia, so they are no strangers to the hazardous phenomena; thus it is not generally an issue to convince them of coming effects. It is altogether another matter, however, in getting people in Afghanistan to change behaviors to either avoid such phenomena, or especially to stop certain behaviors to help alleviate the causative factors leading to such phenomena in the first place. In fact this willful avoidance of such behavior change is commonly more a matter of lack of choice, because people in these societies are generally already far too poor, as well as too uneducated to have much ability to make such changes. Only extensive and repeated efforts at education and reeducation, coupled with plentiful help for outside donors are ever likely to have much effect. Erickson et al. (2009) noted that only adaptive policies and major efforts to reduce human drivers of climate change will have to be incorporated into all sectors to make any difference; land use, water management, disaster management, energy consumption, and human health. In fact a very useful scientific procedure that has not received much attention would be for detailed hazard mapping to help local communities and decision makers. Such a concerted effort could help everyone to understand the current situations that are developing through climate change, which would thereby enable people to anticipate or to assess their flexibility to adapt to coming changes through concerted planning and new technical designs.

## References

Alim, A.K., Shobair, S.S., 2002. Drought and human suffering in Afghanistan. In: TAEAS2002 Proceedings, Agricultural and Forestry Research Center, University of Tsukuba, Japan, pp. 119–127.

Bandyopadhay, J., Gyawali, D., 1994. Himalayan water resources: ecological and political aspects of management. Mt. Res. Dev. 14 (1), 1–24.

Bishop, M.P., Bush, A., Copland, L., Kamp, U., Owen, L.A., Seong, Y.B., Shroder, J.F., 2010. Climate change and mountain topographic evolution in the Central Karakoram, Pakistan, special issue on climate change. Ann. Assoc. Am. Geogr. 4, 772–793.

Dixit, Y., Hodell, D.A., Petrie, C.A., 2014. Abrupt weakening of the summer monsoon in northwest India ~4100 yr ago. Geology. http://geology.gsapubs.org/content/early/2014/02/24/G35236.1.abstract.

DOD, 2014. Climate Change Adaptation Roadmap. U.S. Department of Defense. http://www.acq.osd.mil/ie/download/CCARprint_wForeword_c.pdf (accessed 22.02.15).

Erickson, M., Xu, J., Shrestha, A.B., Vaidya, R.A., Nepal, S., Sandström, K., 2009. The Changing Himalaya—Impact of Climate Change on Water Resources and Livelihoods in the Greater Himalayas. International Centre for Integrated Mountain Development (ICIMOD). 28 pp. http://www.preventionweb.net/files/11621_icimodthechanginghimalayas1.pdf (accessed 07.09.14).

Fagen, B., 2003. The Long Summer: How Climate Changed Civilization. Basic Books, New York, NY.

Hagen, E., Shroder Jr., J.F., Lu, X.X., Teufert, J.F., 2010. Reverse engineered flood hazard mapping in Afghanistan: a parsimonious flood map model for developing countries. Quat. Int. 226 (1–2), 82–91.

Haritashya, U.K., Bishop, M.P., Shroder, J.F., Bush, A.B.G., Bulley, H.N.N., 2009. Space-based assessment of glacier fluctuations in the Wakhan Pamir, Afghanistan. Clim. Chang. 94, 5–18.

Intelligence Committee, 2012. Global Water Security. Intelligence Community Assessment; ICA 2012-08, 2 February 2012, Defense Intelligence Agency (DIA), NGA, CIA, State/INR, and DOE. http://fas.org/irp/nic/water.pdf (accessed 05.09.14).

Lebel, L., Contreras, A., Pasong, S., Garden, P., 2004. Nobody knows best: alternative perspectives on forest management and governance in Southeast Asia. In: International Environmental Agreements: Politics, Law and Economics, 4. Kluwer Academic Publishers, Dordrecht, Netherlands, pp. 111–127.

Marris, E., 2014. Two-hundred-year drought doomed Indus Valley civilization. Nature News. September, http://www.nature.com/news/two-hundred-year-drought-doomed-indus-valley-civilization-1.14800 (accessed 18.09.14).

Masood, S., 2015. In a shift, Pakistan pats Afghanistan on the back. The New York Times. 13 February, http://www.nytimes.com/2015/02/13/world/asia/pakistan-credits-afghanistan-in-arrest-of-militants-in-peshawar-attack.html?_r=0 (accessed 22.02.15).

Morton, M.C., 2015. Pentagon report calls for military to prepare for climate change. Earth. February, http://www.earthmagazine.org/article/pentagon-report-calls-military-prepare-climate-change (accessed 22.02.15).

Norgrove, L., Bowling, B., Ahadi, M.S., Modaqiq, W., Sallari, S., ur Rahman, S.S., Haidari, G.H., Omar, M.O., Shams, H.K., Ahmadi, L.A., Jan, S., Shakir, S., Ali, R., Habib, M., Habib, G., Yunosi, A., Fakhiri, A., Haqbeen, M.A., Ahmadzai, A.B., Ibrahimi, M.W., Hamidullah, Sarwary, G.D., Yasin, M., 2008. Climate Change and Disaster Preparedness Working Group: Final Thematic Report. Global Environmental Facility; United Nations Environment Programme; Government of the Islamic Republic of Afghanistan. 83 pp. http://postconflict.unep.ch/publications/afg_tech/theme_02/afg_ccdp.pdf (accessed 24.02.15).

Pias, J., 1972. Signification geologique pedologique et paleoclimatique de formations paleolacustres et deltaiques au Seistan (Afghanistan meridional). C. R. Acad. Sci. 274, Paris.

Price, G., Alam, R., Hasan, S., Humayun, F., Kabir, M.H., Karki, C.S., Mittra, S., Saad, T., Saleem, M., Saran, S., Shakya, P.R., Snow, C., Tuladhar, S., 2014. Attitudes to Water in South Asia. Chatham House Report. The Royal Institute of International Affairs. 114 pp. http://www.chathamhouse.org/sites/files/chathamhouse/field/field_document/20140627WaterSouthAsia.pdf (accessed 18.09.14).

Saaty, T., 1977. A scaling method for priorities in hierarchical structures. J. Math. Psychol. 15 (3), 234–281.

Savage, M., Dougherty, B., Hamza, M., Butterfield, R., Bharwani, S., 2008. Socio-Economic Impacts of Climate Change in Afghanistan. Department for International Development (DFID CNTR 08 8507); Stockholm Environmental Institute (SEI). 7 pp. http://www.necsi.edu/afghanistan/pdf_data/2007447_AfghanCC_ExS_09MAR09.pdf (accessed 24.02.15).

Savage, M., Dougherty, B., Hamza, M., Butterfield, R., Bharwani, S., 2009. Socio-Economic Impacts of Climate Change in Afghanistan. A Report to the Department for International Development. Project Report (SEI) Stockholm Environment Institute. 44 pp. https://static.weadapt.org/placemarks/files/1092/5345354491559sei-dfid-afghanistan-report-1-.pdf (accessed 24.02.15).

Shroder Jr., J.F., 2012. Afghanistan: rich resource base and existing environmental despoliation. Environ. Earth Sci. 67 (7), 1971–1986. http://dx.doi.org/10.1007/s12665-012-1638-7.

Shroder, J., 2014. Natural Resources in Afghanistan: Geographic and Geologic Perspectives on Centuries of Conflict. Elsevier, Amsterdam.

Shroder, J.F., Schettler, M.J., Weihs, B.J., 2011. Loess failure in northeast Afghanistan. J. Phys. Chem. Earth 36, 1287–1293. http://dx.doi.org/10.1016/j.pce.2011.03.001.

Smith, G.I., 1974. Quaternary deposits in southwestern Afghanistan. Quat. Res. 4 (1), 39–42.

Tate, G.P., 1910–1912. Seistan: A Memoir on the History, Topography, Ruins and People. 2 vols. Government Printing Office, Calcutta.

UNEP, 2009. Afghanistan: National Capacity Needs Self-Assessment for Global Environmental Management (NCSA) and National Adaptation Programme of Action for Climate Change (NAPA). United Nations Environment Programme. 123 pp. http://www.unep.org/dgef/Portals/43/publications/Afghan_NCSA_and_NAPA_2009.pdf (accessed 23.02.15).

UNEP, 2013. Afghanistan's National Biodiversity Strategy & Action Plan. Framework for Implementation 2013–2017. http://www.unep.org/disastersandconflicts/portals/155/countries/Afghanistan/pdf/AfghanistanActionPlan.pdf (accessed 06.09.14).

UNEP, 2014. Climate Change Adaptation. Disasters and Conflicts. United Nations Environment Programme. 1 pp. http://www.unep.org/disastersandconflicts/CountryOperations/Afghanistan/ClimateChangeAdaptation/tabid/133225/Default.aspx (accessed 11.09.14).

United Nations, 2014. Draft Country Programme Document for Afghanistan (2015–2019). Executive Board of the United Nations Development Programme, the United Nations Population Fund and the United Nations Office

for Project Services. DP/DCP/AFG/3; 18 pp. http://papersmart.unmeetings.org/media2/3816313/dpdcpafg3.pdf (accessed 17.10.14).
United Nations News, 2012. UN Environmental Agency to Help Afghanistan Combat Effects of Climate Change. http://www.un.org/apps/news/story.asp?NewsID=43268#.VAt36U3QfIU (accessed 06.09.14).
U.S. Department of Agriculture, 1998. Global Desertification Vulnerability Map. http://www.nrcs.usda.gov/wps/portal/nrcs/detail/soils/use/?cid=nrcs142p2_054003 (accessed 27.03.16).
Whitney, J.W., 2006. Geology, Water, and Wind in the Lower Helmand Basin, Southern Afghanistan. Scientific Investigations Report 2006-5182, US Geological Survey.
Williams, D.B., 2009. Finding water in the heart of darkness: Afghanistan's ongoing water challenges. Earth. July; 4 pp. http://www.earthmagazine.org/article/finding-water-heart-darkness-afghanistans-ongoing-water-challenges (accessed 05.09.14).

PART II

# WATER-MANAGEMENT ISSUES IN CENTRAL AND SOUTH ASIA



Consideration of the administration of all things hydrological in Central and South Asia is a daunting task because of the great depths of knowledge required for understanding all the many complexities involved. At the very outset is coping with all the many indigenous or traditional means of water management by the local people, which have a tendency to trump some issues of alternate usages suggested by modern engineering, especially where ethnic cards are played or corrupt practices unduly influence water apportionments. A major issue at hand is the fact that Afghanistan has a dramatically diminished usage of the water that rises within its borders, because of the proportionally low and non-consumptive use of water to generate hydroelectricity because of lack of dams and generators. This paucity of dams has resulted in plans for a wide variety of new structures, some of which are presently under construction to overcome the diminished capacity. An additional factor of great importance is the atrocious treatment of the water resources by people who fail to appreciate the need to protect the quality of the surface or the ground water. Contamination of all sorts includes human waste, chemicals, dead animals, medical waste, and trash of all kinds; these are the pollutants thrown into rivers without regard for protection of the resource. Part of the reason for this mindless bad behavior is the ignorant belief that flowing water naturally "purifies" itself. While this is true to a limited extent in terms of providing greater aeration, education about the resource is critical to its future protection.

Collection of information about water resources in Central and Southwest Asia is an essential scientific exercise to enable proper prediction and control. Under the normal scientific conditions of free and unfettered access to data, academic scientists and government agencies assess and measure precipitation and river discharges, collect velocities, water depths and channel widths, among other measures. Flood heights are determined and inundation extents are mapped. Depths to the water table are investigated and hydraulic conductivities are measured for aquifers. In these water-short Asian regions, however, a number of conditions inimical to good data collection occur: (1) lack of access to remote and backward areas where proper measuring equipment on the ground is not available; (2) a perception by most of the governments that outsider scientists and others with interests in water in the region are mostly rather unwelcome because their motives and viewpoints are unknown or misunderstood; and (3) the treatment of water data by the governments as various levels of state secrets. In the meantime, many of the outsider scientists and foreign governments (the United States, Germany, etc.) can see various forms of problems with water looming progressively larger and in need of outsider skills and experience to alleviate the difficulties. One of the solutions to some of these problems has evolved to use various categories of remote sensing to make appropriate measurements.

In recent years, in particular as a result of widespread recognition that in a world of burgeoning human populations, awareness has peaked that free access to water is becoming an ever increasingly contentious issue. This has led to development of much new thinking about international water law, with special notice of the significant differences between upstream, headwater states such as Afghanistan, and downstream nations such as the surrounding countries of Pakistan, Iran, Turkmenistan, Uzbekistan, and Tajikistan. These concepts are particularly poignant in terms of ideas of absolute territorial integrity, wherein cases are made that upstream riparian states can do as they like with water that rises inside their boundaries, which point of view is directly countered by equally valid ideas that downstream states are certainly also entitled to the uninterrupted flows of fixed quantities of water, just as they had always done.

Another considerable problem in the region is the absence of up-to-date, transboundary water treaties with the surrounding countries into which the waters of Afghanistan flow. In all cases, the original treaties are many decades old, were made with governments that no longer exist, and were written under conditions that have completely changed in the intervening years. In some cases, the terms of the original treaties were routinely invalidated or ignored. Instead of seeking to revise or renew the old treaties, since 2001 the new governments in Kabul have been studiously avoiding any sort of treaty revisions, reportedly in the belief that the time gained by lack of negotiation will enable the Afghan government to become more hydro-cognizant and build hydro-capacity among its government agents.

This leads directly into concepts of hydro-hegemony, which are the recent theoretical attempts to assess and equate national strengths in dealing with all aspects of the control of water. For example, the geographical position of any riparian nation, whether upstream, midstream, or downstream, very much determines its ability to exploit the resource. This is countered however, by the material power of the riparian nations in terms of their ability to develop the water, their other infrastructure, their military strength, and their population size, as well as their bargaining strength or diplomatic powers to attain favorable terms in their water usage. In the end, however, the power of ideas, or the ability to think in new ways and conceive novel ways to solve water problems may account for a very great deal in working with water needs. This latter concept of ideational power is very much in the nature of the modern academic environment where new ideas are the coin of the intellectual realm, and success with these endeavors may be able to solve hosts of problems.

The result of much of the new thinking about water in Afghanistan is the recognition that a new awareness about water must be built to enable the population and the government to be more intelligent about their critical resource. This so-called "capacity-building" exercise in training about all aspects of water, involves new training in the considerable variety of capacity types, as well as the need to develop a suite of techniques, methodologies, and institutions to collect, assess, and store hydrologic data, even while local officials, technicians, and scientists learn how to better understand and use these data to benefit the country.

In the course of assessing the transboundary water resources of Afghanistan, three main workshops were set up to learn a maximum about how water is perceived: in Dushanbe, Tajikistan; Islamabad, Pakistan; and Kabul, Afghanistan. Financial and political limitations precluded inclusion of water workshops in the neighboring countries of Iran, Turkmenistan, or Uzbekistan, but such may occur in future years. In any case, the problems and possibilities of equitable division and apportionment of transboundary water resources will be assessed, along with issues dealing with ethnic divisions, landowner controls, and access to water according to one's gender, particularly where women are prohibited from public participation in such activities. Finally, the strategic and security environments concerning water in Central and Southwest Asia must be assessed as well, even as we consider the future of water management in the region.

CHAPTER

# 7

# Modern Water Management Issues in Central and Southwest Asia

*J. Shroder*

**University of Nebraska at Omaha, Omaha, NE, United States**

OUTLINE



## Abstract

The water crisis in Afghanistan is acute and in need of serious effort to overcome even partially. Integrated water-resource management (IWRM) is a preferred course of action in dealing with the complex and difficult problems of all the varieties of water management in Afghanistan. Water management in any place is all about knowing as precisely as possible what water you actually have and apportioning it out to the many water users in some sort of reasonably equitable fashion. Accordingly, water management systems are designed to account for all sorts of waters, especially the so-called "blue" waters of precipitation and river flow, the "green" waters of evapotranspiration through plants, and the "gray" waters that have been utilized in agriculture, industry, and domestic uses where they become variously polluted. The new Water Accounting Plus (WA+) system that has been developed for enhanced water management in an increasingly water-short world utilizes four detailed lists with all the various water processes and products that are organized into the resource-base sheet, evapotranspiration sheet, productivity sheet, and withdrawal sheet. The WA+ system enables rigorous water management utilizing these accounting systems.



http://dx.doi.org/10.1016/B978-0-12-801886-6.00007-0

**Keywords:** Water governance, Integrated water-resource management (IWRM), Water Accounting Plus (WA+) system, Resource-base sheet, Gross inflow, Net inflow, Exploitable water, Landscape ET (evapotranspiration), Land use, Managed water use, Reserved outflow, WA+ evapotranspiration sheet, WA+ productivity sheet, WA+ withdrawal sheet, Water footprint, Blue water, Green water, Gray water

## 7.1 INTRODUCTION

A terrible water crisis exists in Afghanistan that is defined by a number of unpleasant facts (Kuonqui et al., 2011).

- About 16.8 million people, or over half of the total population, drink unsafe or contaminated water.
- Some 23 million people in the country have inadequate access to improved toilets and waste-disposal facilities.
- Increased population growth rates produces ever more competition for less water per capita.
- About 3 million hectares were irrigated in the 1970s, but that number has been reduced to about 1.8 million hectares, or only 60% of the pre-war days, because of destructive losses of the irrigation infrastructure during the war, salinization, and other land degradation.
- Environmental degradation is exacerbating water scarcity.
- Flooding and droughts are continuing to cause premature mortality, injury, destruction of property, food shortages, and lost earnings.
- The dangers of national and international disputes over water resources are growing ever more acute.

In general, so many Afghans lack access to good water because they are also so very poor. Their infrastructure is crude, they are unable to mobilize effective institutions, and their general lack of capacity, weakened institutions, and inequitable power relationships thus determine who is entitled to tap into irrigation systems (Kuonqui et al., 2011). The result is that the water crisis affects the most vulnerable people, such as female-headed homes, children who must fetch water, squatter's homes in unplanned urban areas, destitute farmers in downstream canal areas, and the Kuchi and other nomads.

The water crisis in Afghanistan is actually largely an issue of governance (Amarkhail and Kakar, 2011). Shortages of water in the country are driven primarily by inefficient services, rather than by insufficient water resources, especially in terms of drinking water and disposal of waste or simple sanitation. Thus the manner in which individuals, communities, and institutions select the governance of their water resources has great control over their livelihoods, development, and environmental sustainability.

Water management and governance structures commonly depend upon the level of development in any given river basin (Table 7.1). For example, three stages of development of river basins can be formulated (Amarkhail and Kakar, 2011), which can be thought of as an initial development stage where the priority is generally to increase access to water-supply services. This is followed by an intermediate utilization stage wherein sharply increasing water consumption occurs as improved supply meets increasing demand. The third and final stage occurs as water resources become scarcer relative to demand so that reallocation of waters from one user to another begins to occur. If this rise in demand can no longer be met by creating more supply, the river basin is referred to as "closing." In these cases as competition to obtain ever more water share increases, the need for improved water governance is clear.

TABLE 7.1 The Development of River Basins in Afghanistan

| | Stage 1: Development | Stage 2: Utilization | Stage 3: Reallocation |
|---|---|---|---|
| Allocation flow, % | Low 0–40 | Moderate 40–70 | High 70–100 |
| Dominant activity | Construction: supply and storage infrastructure | Managing supply | Managing demand |
| Value of water | Low | Increasing | High |
| Ground water | Development | Conjunctive use | Regulation |
| Conflicts | Few | Within the subsector | Cross-sectoral |
| Typical institutional tasks | Construction: planning and implementation | Operation, maintenance expansion and rehabilitation | Intersectoral planning |
| Based by allocated flow | Panj-Amu Darya Kabul to Indus | Hari Rud-Murghab Helmand | Northern |

*From Amarkhail, M.A., Kakar, K., 2011. The need for governance in the water sector. In: Kakar (Ed.), Afghanistan Human Development Report 2011—The Forgotten Front: Water Security and the Crisis in Sanitation. Centre for Policy and Human Development, Kabul University, pp. 169–193. http://re.indiaenvironmentportal.org.in/files/Complete%20NHDR%202011%20final.pdf (accessed 23.07.15).*

## 7.1.1 Integrated Water-Resource Management

The overarching concept of integrated water-resources management (IWRM) has arisen from the development of the new Water Law in Afghanistan and the new water-sector strategy (Amarkhail and Kakar, 2011). These new concepts make decentralization and participation of stakeholders in water resources a high priority. Bringing IWRM into a practical reality in Afghanistan is, however, replete with difficulties because it is not supported entirely by the stakeholders, in part because some diminished capacities in understanding. Furthermore, the relatively diminutive funding available for the water sector in Afghanistan by foreign donors and local means additionally restricts progress. It is also clear that greater commitment to water aid is critical, just as it needs to be quite in line with the principles of the Paris Declaration on Aid Effectiveness (OECD, 2005), which include taking ownership, alignment to national goals, harmonization between donor and recipient, managing for results, and mutual accountability between donor and recipient government.

Governance of the water sector is complex and requires focus on the four pillars of human development, such as the social development, economics, political dimension, and environmental activities so that usages are equitable, efficient, equally democratic, and sustainable (Fig. 7.1; Amarkhail and Kakar, 2011). In Afghanistan, after the governmental collapse from the aftermath of the Soviet invasion, the civil war, and the incompetence of the Taliban, in 2002 in the Kabul Understanding, an international conference laid the foundation for a wholly new formulation of a Strategic Policy Framework for the Water Sector, including: (1) revision of Water Law of 1991; (2) water-resources management policy; (3) institutional structure for water resources management; (4) irrigation policy and regulations; (5) regulations for water-user associations (WUA); (6) national urban and rural water supply and sanitation policies and institutional development; (7) ground-water policy; (8) hydropower development policy; and (9) environmental law.

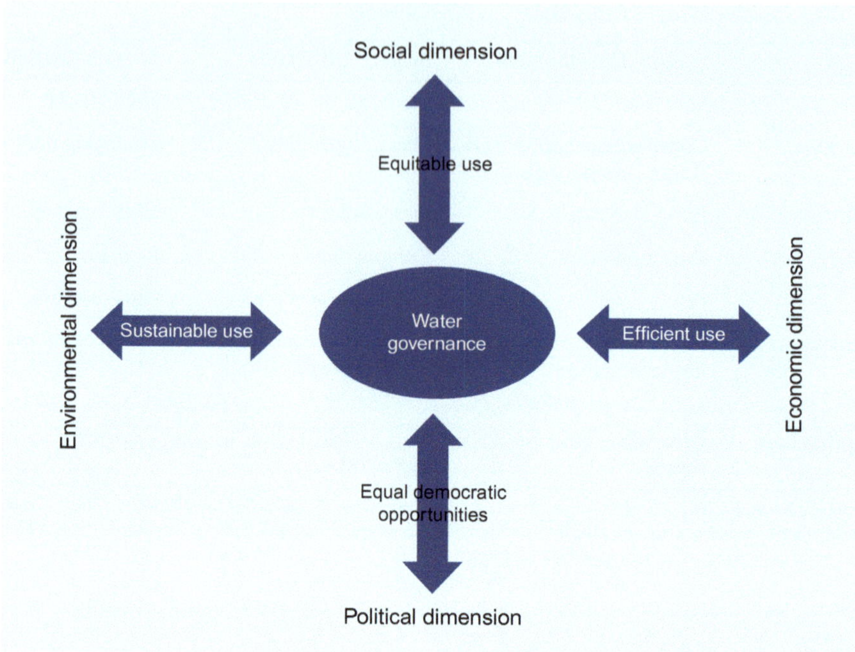


FIG. 7.1 Four dimensions of water governance are linked with human development concerns such as equity, sustainable development, equal demographic opportunities, and efficient use. *From Tropp, H., 2007. Water governance: trends and needs for new capacity development. Water Policy 9 (Suppl. 2), 9–30 http://www.watergovernance.org/documents/WGF/Reports/Water_Governance_trends_and_needs_paper.pdf (accessed 23.07.15); Kakar, K., 2011. Afghanistan Human Development Report 2011—The Forgotten Front: Water Security and the Crisis in Sanitation. Centre for Policy and Human Development, Kabul University, 238 pp. http://re.indiaenvironmentportal.org.in/files/Complete%20NHDR%202011%20final.pdf (accessed 23.07.15).*

The vision and goal of the water-sector strategy is to facilitate management of the nation's water resources in order to reduce poverty, sustainably increase social and economic development, improve the lives of Afghans, and ensure an adequate water supply at the present and in the future. To this end in particular, the vision of IWRM in Afghanistan is the key to planning and development of water resources. The goal is for a fully functional and effective set of procedures involving a variety of stakeholders who would mutually help to implement IWRM throughout the country (Fig. 7.2). This is the process that promotes coordinated development and management of water, land, and related resources so as to maximize resultant economic and social welfare equitably without compromising sustainability of vital ecosystems (Amarkhail and Kakar, 2011). The foundation of the IWRM is based on the well-known four Dublin Principles (Table 7.2), whose simplicity and equitability are marks of real civilization where truly implemented.

The river-basin approach in IWRM is a decentralization process, moving from the existing, rather purely administrative purview to the more scientifically logical approach of that of a hydrological basin. Much has been written about the great advisability of these changes, but the implementation of them in reality is quite another matter. Perhaps the complexity and all the feedbacks and meeting required between all the stakeholders in Fig. 7.2 is illustrative of a measure of the difficulty in making an IWRM for different drainage basins in Afghanistan

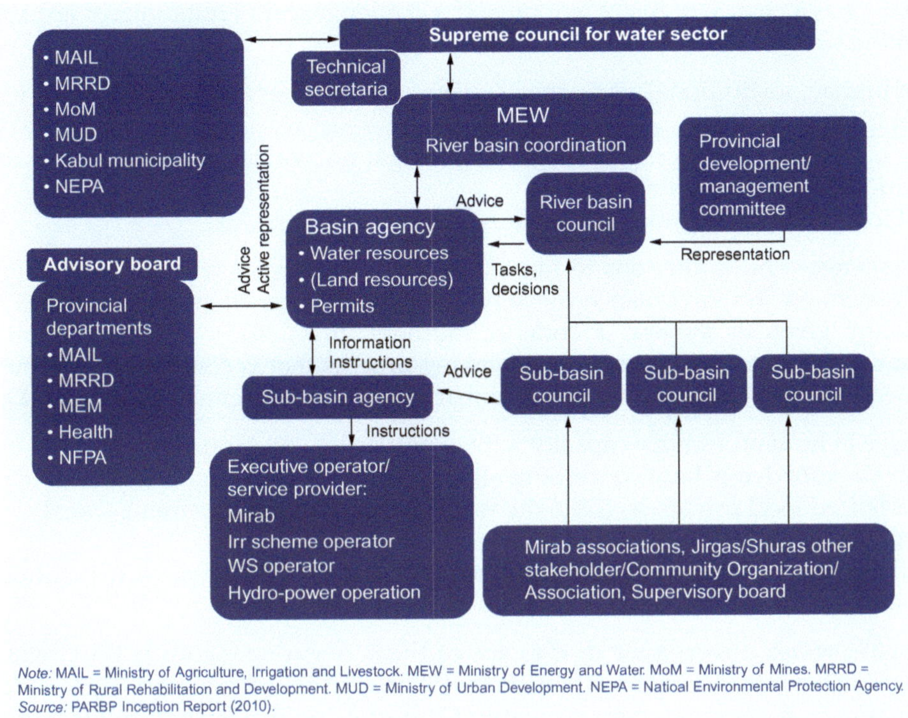


FIG. 7.2 The complex organizational structure in Afghanistan for implementing integrated water-resource management (IWRM) (From Kakar, 2011). Failures in attempts to make IWRM work properly in the country have occurred in part because of the complexity of this structure, and the wide variety of opinions and points-of-view of all the different stakeholders, some of whom are in conflict with each other or between organizations.

TABLE 7.2 The Dublin Principles Were Generated at a Meeting of Experts on Water-Related Problems at the International Conference on Water and the Environment (ICWE), Dublin, Ireland in Jan. 1992

| | |
|---|---|
| Principle 1 | Fresh water is a finite and vulnerable resource, essential to sustain life, development and the environment |
| Principle 2 | Water development and management should be based on a participatory approach, involving users, planners and policy-makers at all levels |
| Principle 3 | Women play a central part in the provision, management and safeguarding of water |
| Principle 4 | Water has an economic value in all its competing uses and should be recognized as an economic good |

The Dublin Statement on Water and Sustainable Development recognizes the increasing scarcity of water as a result of the different conflicting uses and overuses of water, which applies absolutely to Afghanistan.

that actually has a chance of really working. The challenges and limitations of implementing reform in the water sector are manifold and include:

- lack of human, institutional, or financial resources to deliver adequate water services to the population;
- lack of mechanisms to regulate water use for irrigation, water supply, hydropower generation, and basic sanitation; and
- lack of integrated governance of the water sector.

The challenges and limitations to implementing water sector reform are plentiful, even while improving water governance is essential for the country to ever recover and develop appropriately, given its mineral abundance (Shroder, 2014, 2015). Implementation of new laws about water in Afghanistan is a painstaking process that is confounded by inadequate enforcement. The gaps to ensure better administration and enforcement include (Kakar, 2011):

- shortages in human resource capacity with expertise in water issues;
- gaps in the knowledge base concerning all aspects of water in Afghanistan;
- inexperienced local private sector in the water sector (drilling, damming, canal construction, etc.); and
- weak enforcement capacity to adjudicate water disputes in an effective, expeditious, and transparent manner.

Improving water governance in the face of these negative issues is thus regarded as critical, and can be fixed only by attention to: (1) filling the information gaps, and sharing and disseminating knowledge about water; (2) providing infrastructure for better water use; (3) developing skills and enhancing capacity; (4) focusing on conflict management and rules enforcement; and (5) developing self-financing procedures. Kakar (2011) and her collaborators have developed a comprehensive list of priority recommendations that would address most of these issues, providing the new government of President Ashraf Ghani can be induced to bring them into effective existence (Table 7.3). Such will be a

TABLE 7.3 Priority Recommendations Designed by Kakar (2011) and Her Collaborators to Deal With the Forgotten Front in Afghanistan on Water Security and the Crisis in Sanitation

1. Filling the data and information gap about all aspects of water in Afghanistan
2. Stimulating public awareness and developing the capacity for action
   - Developing community awareness and capacity
   - Supporting water-user associations and other local institutions
   - Developing the capacity of government staff
   - Developing capacity in the private sector
   - Developing capacity in higher education
3. Improving supply and promoting soil and water conservation
   - Filling the infrastructure gap
   - Adopting soil and water conservation measures
4. Supporting water-governance reform
   - Introducing institutional change and increasing international aid
   - Transforming integrated water-resources management into a practical reality
   - Developing regulations to support the implementation of the new Water Law
   - Ensuring community participation in design, implementation and monitoring
   - Operation and maintenance by the government
5. Enhancing aid to the water sector

difficult prospect at best, given the strong overhaul of society and government that is implied, and all of that required in a culture that is collectively known for a certain resistance to change.

## 7.2 WATER MEASUREMENT

Worldwide, the intelligent management of vital water supplies has been long regarded as an absolutely essential matter for various government agencies to focus infrastructure and personnel resources upon. The World Bank has noted (2014) that practically every development challenge of the 21st century—food security, managing rapid urbanization, energy security, environmental protection, and adapting to climate change, most definitely will require urgent attention to management of the water resources. In fact, the observed meteorological and hydrological data that are collected by most (but not all) competent governments are the basis upon which almost all progress in the science of hydrology, as well as the practice of its uses are founded. Few will dispute this vital necessity, yet severe issues occur, especially in Central and South Asia. Most notably it has been realized that you cannot manage what you do not measure! Thus if Afghanistan does not, or cannot, measure its water resources, the country will have little hope in managing its hydrological treasure, especially as it tries to deal with its neighbors, who want as much of the precious fluid as they can obtain.

It is not that water cannot be measured in any particular place, it is instead that improvement in the management of water resources anywhere will not come with better models alone (Smakhtin and Wichelns, n.d.). Data collection and sharing information has been progressing slowly over the past decades in the late 20th century and many countries and regions of the world remain quite poorly gaged, or are actually decreasing in capability after many years of success. Even in the developed world (the USA, Canada, the UK, China, South Africa, etc.), the old observational networks have been declining in scope and number, and in the developing world (Thailand, Nepal, Bosnia-Herzegovina, Sri Lanka, and Afghanistan) the situation has been growing steadily worse, or measurement has stopped entirely. In spite of over six major resolutions at international levels in various divisions of the United Nations to the effect that nations should provide hydrologic data on a free and unrestricted basis, most hydrologists in the world apparently do not even know about these resolutions, and only 20 member states out of the 170 in the World Meteorological Organization share their information, in any case. What is at stake is a whole host of essential aspects to improve lives and livelihoods, without which much human misery and environmental damage will go on increasing with fraught circumstances (Tables 7.4 and 7.5).

In recent years, in a move that augers very well for two regional powers who are seeking to recover from the twin but quite different traumas of their national formation, Afghanistan and Tajikistan are both seeking means to work on much better transboundary hydrological and environmental cooperation (Omar, 2013). Accordingly they have had a series of cooperative workshops under UN auspices on the upper Amu Darya River Basin on: (1) environment and security river basins on Sep. 20–21, 2007 (ENVSEC, 2007); (2) severe floods, vulnerable ecosystems and climate-change impacts on Mar. 27–29, 2012 (UNECE, 2012); (3) environment and hydrology from Mar. 25–27, 2013 (UNECE, 2013); and (4) water and the environment on Nov. 24–25, 2014 (UNECE, 2014) (See the last Chapter 20 herein on Concluding Transboundary Water Activities.)

TABLE 7.4 Problems With Lack of Hydrological Measurements That Will Make the Following Difficult or Impossible

| | |
|---|---|
| 1. | Monitoring of impacts of climate changes on hydrological regimes |
| 2. | Design of adaptation measures on hydrological regimes undergoing the impacts of climate changes |
| 3. | Meeting the Millennium Development Goals of the United Nations on water and sanitation to improve the design of small-scale rural water supplies |
| 4. | Protection of the aquatic environments through implementation of proper environmental-flow practices |
| 5. | Improved trans-boundary water management versus water conflicts or water-sharing agreements that are impossible to implement |
| 6. | Improved accuracy of global hydrological models, in turn, providing data for data-poor regions, which is a form of humanitarian aid |

*From Smakhtin, V., Wichelns, D., n.d. Hydrometeorological Data Collection and Sharing: The Global Challenge. International Water Management Institute, Colombo, Sri Lanka. http://www.oecd.org/tad/sustainable-agriculture/45066665.pdf (accessed 31.12.14).*

TABLE 7.5 Reasons for the Poor Status of Hydrologic Measurements in Central and South Asia

| | |
|---|---|
| 1. | Benefits of meteorological and hydrologic data collection are not understood by general public or politicians |
| 2. | Benefits of meteorological and hydrologic data collection are regarded negatively by all too many of those responsible for data collection and storage, because they are thought of only as basic inventory work without scientific excitement |
| 3. | The detriments because of lack of funding as donors have other priorities |
| 4. | The lack of trained staff, no resources for training, and low salaries |
| 5. | The unfortunate over-estimation of the ability of new methods of data collection to replace traditional methods |
| 6. | No defined mechanisms for sharing data exist |
| 7. | No measurable goals occur |
| 8. | Hydrologic data are seen as a national security issue, which makes them sacrosanct and unable to be utilized scientifically to help people |
| 9. | Few incentives exist to share data |

*From Smakhtin, V., Wichelns, D., n.d. Hydrometeorological Data Collection and Sharing: The Global Challenge. International Water Management Institute, Colombo, Sri Lanka. http://www.oecd.org/tad/sustainable-agriculture/45066665.pdf (accessed 31.12.14).*

After the meetings in 2013, Counselor Azizullah Omar of the Afghanistan Mission to the UN in Geneva (Omar, 2013) produced a document on Hydrology and Environment Cooperation in the Upper Amu Darya River Basin that introduced all the basic geographic and socio-economic features of the two countries, as well as information on: (1) the growing economic importance of the shared region in terms of international roads and the shared friendship bridges (6 bridges); (2) existing hydroelectric facilities (16 in Tajikistan at present); (3) proposed hydroelectric facilities (14 in Tajikistan; 10 located on the border, 8 in northern Afghanistan); (4) locations of glaciers; (5) relative discharge amounts; (6) areas of irrigation; (7) freshwater withdrawals from the Amu Darya per country; (8) environmental, hydrological, and hydrometeorological observations; (9) differences and similarities between water issues and environment in the two countries; and (10) possible sharing of hydrological and environmental activities that the two countries can share jointly (Tables 7.6 and 7.7).

Economic development and natural-resource management are the top priorities for cooperation between the Islamic Republic of Afghanistan and the Republic of Tajikistan. The recommended joint activities between the two countries shown in Tables 7.6 and 7.7 are clearly designed to allow a variety of scientists and lower-level bureaucrats to work separately on some occasions and later meet together to share relevant information. Afghan-Tajik hydrological- and environmental-collaboration activities have only just started and they need to be strengthened by adequate donor inputs and cross-border coordination, as well as by dedicated support in the initial steps, coupled with facilitation by international organizations. In addition, Afghanistan and Tajikistan are on record (Omar, 2013) as welcoming exchange of experience and collaboration with other international river basins. To further this sort of exchange, the two countries have produced the new *Visual Atlas of Cooperation—Afghanistan and Tajikistan: Environment and Hydrology in the Upper Amu Darya Basin*, which contains over 50 maps and graphs, with around 100 photographs (Zoï Environment Network, 2013). The new book contributes to the UN International Year of Water Cooperation 2013 and the Afghanistan-Tajikistan (AFG-TJK) environment and cooperation process. It presents information at the level of the river basin as a whole, rather than at a national level, and it presents challenges from a regional context, rather than the country perspective. The book aims to support governmental decision making, as well as local and international awareness of the Upper Amu Darya basin and the two riparian countries from which it emanates. The book builds upon existing and new knowledge, and takes a cross-sectoral approach; it was produced by the United Nation's Zoï Environment Network, in association with the UN Economic Commission for Europe (UNECE) and with funding from the Russian Federation. This international cooperation between adjoining riparian nations can serve as a beginning standard and source of fruitful ideas to enhance similar cooperative relationships that are in some critical need between Afghanistan and Pakistan on the Kabul River and its tributaries,

TABLE 7.6 Joint Hydrology Activities Recommended to be Shared Between Afghanistan and Tajikistan

| | |
|---|---|
| 1. | Compile fact sheets listing hydrological and data transfer equipment, locations and status of hydrologic stations, and needs and plans for modernization (with a focus on the Panj River international border) |
| 2. | Establish points of contact, and draft procedures and agreements on sharing and exchange of hydrological data: routine, historical, and data for floods and forecasts |
| 3. | Facilitate installation of new automated hydrological station on the Sherhan-Bandar Bridge across Panj River |
| 4. | Try to pilot a data exchange |
| 5. | Catalyze support for a new hydrological station at Ayvaj |
| 6. | Establish snow-cover monitoring, especially in the range of altitudes of 2000–4500 m, using common approaches and collect and exchange data |
| 7. | Conduct joint glacier survey and assessment, and collect and exchange data |

*From Omar, A., 2013. Hydrology and environment cooperation in the Upper Amu Darya River Basin. United Nations Economic Commission for Europe (UNECE). http://www.unece.org/fileadmin/DAM/env/water/meetings/joint_bodies/presentations/3.3AFG-TJK-presentation-unece-sep2013.pdf (accessed 31.12.14).*

TABLE 7.7 Joint Environmental Activities Recommended to be Shared Between Afghanistan and Tajikistan

| | |
|---|---|
| 1. | Establish experience and data exchange on national communications and activities on climate change mitigation, adaptation, and resilience |
| 2. | Establish experience and data exchange for agrometeorological services, and conduct joint work on early warning and mitigation of extreme climate-related events such as locust infestations, droughts, and floods |
| 3. | Conduct joint work on conservation and monitoring of globally significant biodiversity in near-border protected areas, wetlands, migratory species habitats and ecological corridors |
| 4. | Map genetic resources and important ecosystem services in the Upper Amu Darya Basin |
| 5. | Facilitate afforestation and reforestation programs and sustainable land management, including documentation of good practices and their replication |
| 6. | Conduct joint survey and assessment of environmental quality and the state of the environment in the Panj/Amu basin, common environmental indicators and priorities |
| 7. | Provide capacity—building and training on environmental reporting and shared environment information systems in the countries and provinces of the Amu Darya Basin |
| 8. | Work jointly on raising awareness, success stories, and good practices; training for students |

*From Omar, A., 2013. Hydrology and environment cooperation in the Upper Amu Darya River Basin. United Nations Economic Commission for Europe (UNECE). http://www.unece.org/fileadmin/DAM/env/water/meetings/joint_bodies/presentations/3.3AFG-TJK-presentation-unece-sep2013.pdf (accessed 31.12.14).*

Afghanistan and Iran on the Helmand and other smaller rivers that cross the international border, between Afghanistan and Turkmenistan with the Tenjen (Hari Rud) and Murghab rivers that cross the international boundary between them, and between Afghanistan with Uzbekistan and Turkmenistan and the Amu Darya border.

## 7.3 WATER-BALANCE ACCOUNTING SYSTEMS

Any river basin in the world can be assessed in terms of a simple water balance in terms of total incoming precipitation to the basin and outgoing evapotranspiration and river discharge from the basin. According to the equation:

$$P = Q + \text{ET} + \Delta S \tag{1.1}$$

where $P$ is precipitation, $Q$ is river-water discharge, ET is evapotranspiration, and $\Delta S$ is any change in water storage in the soil, on the surface, or in the ground water (Wisler and Brater, 1959; Karim and Veizer, 2002).

Many other water-balance accounting systems have been tried or newly developed over the years, with varying limitations of one kind or another. *Aquastat* is the global information system on water and agriculture of the UN's Food and Agriculture Organization (FAO). This system focuses on water withdrawals but does not distinguish between consumptive use, such as E, and nonconsumptive use, such as return flows. The UN Statistics Division also has a proposed system *SEEA-Water* of environmental economic accounting for water (United Nations, 2012). SEEAW accounts for precipitation, evapotranspiration, and soil water, as well as differentiating consumptive uses from nonconsumptive, but its applicability to account for every water resource was compromised in many cases because key required data could not be obtained (Perry, 2012). Nonetheless, SEEA-Water does provide a conceptual framework for organizing hydrological and economic information in a coherent and consistent manner (Table 7.8). SEEAW also did not especially differentiate between a number of water pathways, or between blue- and green-water types (Falkenmark and Rockström, 2006), where blue water is fresh surface water and ground water, and green water is that precipitation that does not run off or go into the ground, but that is stored in the soil and used by plants, where it is transpired into the atmosphere as water vapor (Fig. 7.3). This sort of thinking in terms of blue- and green-water thinking has been assessed also in terms of time since humankind has been involved in manipulating and using water in all its forms (Fig. 7.4), not necessarily to much good affect.

The Australian water-accounting system was designed for semiarid areas so river discharge is much less than all the other water movements in such basins. It was based upon SEEA-Water guidelines, and it accounts mainly for water withdrawals rather than consumptive use, and it does not provide any information on rainfed agriculture. The International Water Management Institute (IWMI) developed a water-accounting procedure that was originally designed for irrigation schemes within a basin, and was later used

TABLE 7.8 Topics, Information Sources, and Assistance to Policy Makers Covered in the SEEA—Water Effort by the United Nations to Harmonize Approaches, Concepts, and Definitions in Dealing With Water Supplies in Any Location (United Nations, 2012)

| | |
|---|---|
| 1. | Stocks and flows of water resources within the environment |
| 2. | Pressures imposed on the environment by the economy in terms of water abstraction and emissions added to wastewater and released into the environment or removed from wastewater |
| 3. | The supply of water and its use as an input in the production process and by households |
| 4. | The reuse of water within the economy |
| 5. | The costs of collection, purification, distribution, and treatment of water, as well as the service charges paid by its users |
| 6. | The changing of these costs, that is, who is to pay for the water supply and sanitation services |
| 7. | The payment of permits for access to abstract water or use it as a sink for the discharge of wastewater |
| 8. | The hydraulic stock in place, as well as investments in hydraulic infrastructure made during the accounting period |
| 9. | Allocating water resources efficiently |
| 10. | Improving water efficiency |
| 11. | Understanding the impacts of water management on all users |
| 12. | Getting the most value for the money for investing in infrastructure |
| 13. | Linking water availability and use |
| 14. | Making available a standardized information system, which is capable of harmonizing information from different sources, is accepted by the stakeholders, and is used for the derivation of indicators |
| 15. | Getting stakeholders involved in decision-making |

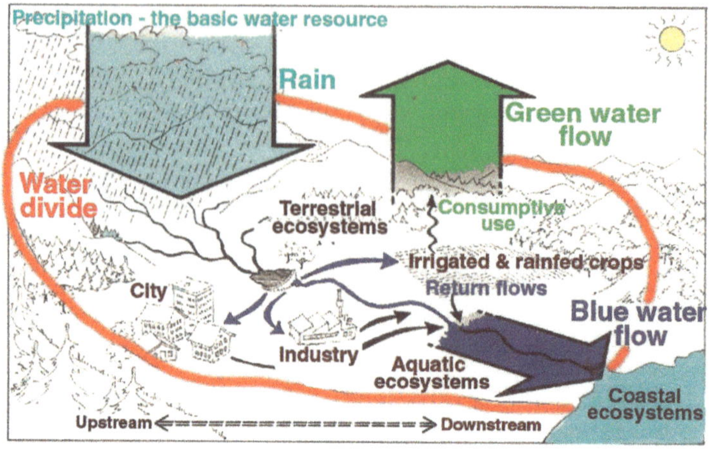


FIG. 7.3 Conceptual cartoon model of flows of blue, green, and gray water in a drainage basin. Precipitation over an area (large, light-blue arrow) provides the runoff (blue-water flow, BWF), which is partitioned to feed cities, industry, agriculture, and rivers (From Falkenmark and Lannerstadt, 2005). Part of this BWF is consumed in plant production, with plentiful transpiration from leaves and evaporation from wet surfaces (green water or vapor flow), which are consumptive uses. Gray-water flow can be return flow of polluted water. Return flow can also be clean.

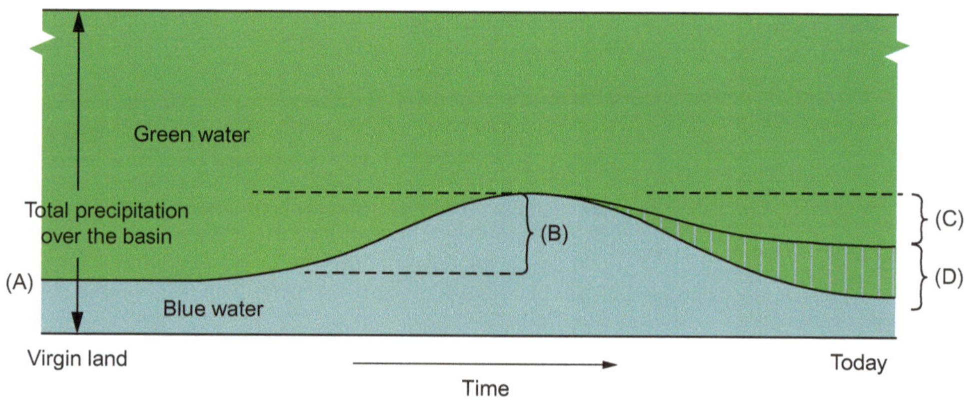


FIG. 7.4 Blue- and green-water flow changes in world drainage basins over the past ~10 K yr (From Falkenmark and Lannerstadt, 2005). The left (A) begins with undisturbed virgin lands back in antiquity as humanity began to take control of its environment. (B) represents the major forest clearance world-wide as humanity began wide-spread agriculture. (C) is the time of multiple general improvements of crops and agriculture, which resulted in decreased blue-water flows as plant growth increased. (D) with the shading lines is the time of increased consumptive water use, with increased irrigation from rivers and ground water, as well as increased evaporation from reservoirs (diagram is not to scale).

for basin analysis, but some of its components were too generic for basin-level studies (Karimi et al., 2013a).

Studies of the 1994–95 water balance between precipitation, evapotranspiration, and river discharge of the Indus River Basin (Karim and Veizer, 2002) show that of the ~398 $km^3$ that falls as rain or snow over the basin, some 83% returns to the atmosphere by evapotranspiration with only about an eighth of the flux, or ~53 $km^3$ is discharged by the Indus. Furthermore, analysis of hydrogen and oxygen isotopes in the Indus near its mouth reveal that up to ~72% of its water is derived from Mediterranean (Westerly) sources, rather than the Indian monsoon, which means that the bulk of the Indus drainage is the result of delayed runoff from the melt of snow and ice sources in the Karakoram and Himalaya.

More advanced water-balance accounting systems, however, can be used in river basins where exceptionally high appropriations of water are the norm, and more specialized accounting may be required to sort out details of the varied water uses. Accordingly, Karimi et al. (2013a,b) have designed an advanced new water-balance accounting system that is designated as *Water Accounting Plus* (WA+). Such detailed assessments are required in order to more clearly understand river basin hydrological processes, manageable and unmanageable water flows, interactions of water with highly variable water uses, and opportunities to mitigate negative effects and help overcome the effects of water depletion on societies. In fact, WA+ is a new intellectual framework that is designed to provide explicit spatial information on water depletion and net withdrawal processes in complex river basins (Table 7.9).

At the present time in much of South and Central Asia, water-cognizant professionals do not have a common framework with which to deal with net water withdrawal processes, nor to link beneficial and nonbeneficial water depletions to different user groups. One of the reasons for this is because of the unwise, even ignorant treatment of most of the river-flow data as state secrets so that the nonavailability of so much information causes so much misunderstanding, imprecision, and confused thinking that leads to underestimation and overexploitation of water resources (aquifers, flow discharges, etc.) (Karimi et al., 2013a,b).

The WA+ system of water accounting includes what are referred to as *sheets*, which are designed as paper or electronic accounting procedures to keep track of the details of four basic accounting elements that control the assessments of water supplies in any basin: (1) resource-base sheet; (2) evapotranspiration sheet; (3) productivity sheet; and (4) withdrawal sheet. Each of the so-called "sheets" is comprised of multiple sets of indicators to summarize the overall situation affecting water resources in any given area, and in aggregate for the entire basin. Impacts of external controls such as climate change, and internal influences such as natural or anthropogenic diversions or infrastructure construction such as dams or irrigation canals, allow estimations of effects qualitatively or even quantitatively where the data are good enough. In many cases satellite assessments of different phenomena allow adequate measure of specific or broad-brush, water-related phenomena to use as inputs into the WA+ system. Water-accounting systems such as this are very rigorous in that attention must be paid to making accurate estimates of many rather esoteric divisions of the myriad aqueous phenomena. The effort appears justified where the need is very great in the face of severely declining supplies, as in Afghanistan and many others of the drier parts of the world.

### 7.3.1 WA+ Resource-Base Sheet

The resource-base sheet (Fig. 7.5) shows how the water incoming to any particular basin is accounted for in terms of which water fractions go where, and how each fraction should be considered (Table 7.8). The *Gross inflow* is the gross precipitation, plus any surface water flow ($Q_{in}^{SW}$), as well as any ground-water flow ($Q_{in}^{GW}$) into a basin. The *Net inflow* includes all of the above, plus any water-storage changes ($\Delta S_{f,sw}$ = surface water storage change; $\Delta S_{f,GW}$ = ground water including soil moisture storage change in the vadose zone;

TABLE 7.9 Water Definitions in WA+ and Elsewhere Arranged Alphabetically Rather Than in Order of Receipt of the Water Into a Basin

| | |
|---|---|
| Available water | Exploitable water minus reserved outflows and nonutilizable outflows, representing water that can be used in the basin domain |
| Beneficial use | Beneficial water use refers to a reasonable quantity of water applied to a nonwasteful use, such as water used for domestic purposes, livestock watering, industry, commerce, irrigation, mining, fisheries, wildlife, recreation, fire, dust suppression, and energy production |
| Conserved land use | Relates to environmentally sensitive land uses and natural ecosystems that are set aside for protection, including protection from the sea |
| Depleted water | Is the total Landscape ET plus the Utilized flow (managed water use plus flows in the sinks) |
| Exploitable water | Water present in rivers, reservoirs, lakes, and ground water that is used for utilized, utilizable, nonutilizable, and reserved outflows |
| Gross inflow | Total amount of water that flows into the basin domain, including precipitation, plus any inflow from surface and ground-water sources |
| Landscape ET | Water that evapotranspires directly from the natural water cycle without artificial supply |
| Managed water use | Represents landscape elements that receive withdrawals from utilized flows |
| Modified land use | Relates to replacement of the original vegetation for increased utilization of land resources |
| Net inflow | Gross inflow after correction of the storage change ($\Delta S$), which represents water for landscape evapotranspiration (ET) and exploitable water |
| Nonbeneficial water | Wasteful water use and water-logging from over-irrigation, use of flood irrigation where drip irrigation would work better; premature evaporative upflux loss, and so forth |
| Nonutilizable outflow | Wasteful water use mostly arises during and after flooding events when excess water threatens to innundate large areas that need to be evacuated |
| Outflow water | Is all the flows out of the basin as surface water bodies or as ground water |
| Reserved outflow | Water that has to be reserved to meet prior commitments for certain outflow, navigational flows and environmental-maintenance flows |
| Utilized flow | That portion of available water that is depleted for different uses |
| Utilizable outflow | Water available for resources development |
| Utilized land use | Represents a low to moderate resource utilization, such as savannah, woodland, and mixed pastures |

*From Karimi, P., Bastiaanssen, W.G.M., Molden, D., 2013a. Water Accounting Plus (WA+)—a water accounting procedure for complex river basins based on satellite measurements. Hydrol. Earth Syst. Sci. 17, 2459–2472.*

$\Delta S_{f,sm}$ = snow and glacier melt storage change). The *Net inflow* partitions into *Landscape ET* and *Exploitable Water* (green water, or evapotranspiration). The Landscape ET is a consequence of rainfall across all sorts of composite terrain with mixed land use caused by such variables as geology, soils, slopes, altitudes, and stream drainages. The exploitable water is the net inflow minus the Landscape ET.

Resource base sheet

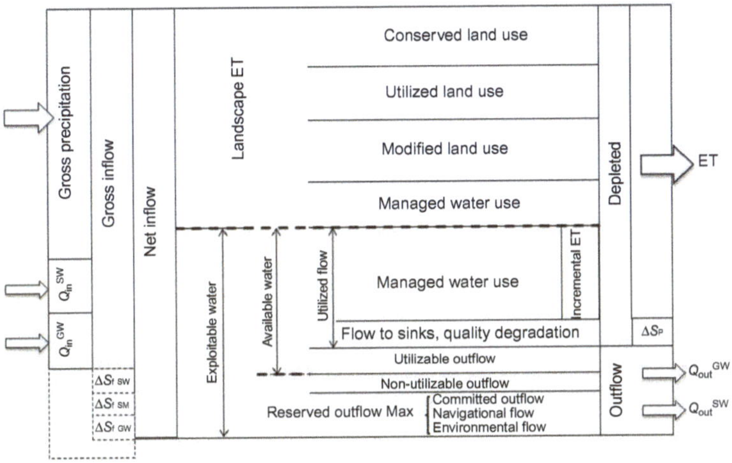


| Indicators | Definition |
|---|---|
| Exploitable water fraction | Explitable water divded by the net inflow |
| Storage change fraction | Fresh water storage change divided by exploitable water |
| Available water fraction | Available water divided by exploitable water |
| Basin closure fraction | Utilized flow divided by available water |
| Reserved flow fraction | Reserved outflows divided by the total outflow |

FIG. 7.5 The resources base sheet is a schematic presentation that provides information on water volumes. The water inflows occur on the left of the diagram, the middle part provides information on how and through what processes the water is depleted in any particular domain, and on the right information on exploitable water and outflow reports are summarized. *From Karimi, P., Bastiaanssen, W.G.M., Molden, D., 2013a. Water Accounting Plus (WA+)—a water accounting procedure for complex river basins based on satellite measurements. Hydrol. Earth Syst. Sci. 17, 2459–2472; Karimi, P., Bastiaanssen, W.G.M., Molden, D., Cheema, M.J.M., 2013b. Basin-wide water accounting based on remote sensing data: an application to the Indus Basin. Hydrol. Earth Syst. Sci. 17, 2473–2486.*

The Landscape ET is subdivided into three types of land use (*Conserved*, *Utilized*, and *Modified Land Use*), and one type of *Managed Water Use*. These four types of land and water use classes have similarities in ecosystem services, provisioning services, human interaction, and interventions in the hydrological cycle. The four classes form the basis for the different management options in the WA+ system (Table 7.10).

The *exploitable water* fraction of the net inflow into a basin can be subdivided into the available water and the *Reserved outflow*, which itself is that which has to be reserved to meet the requirements of various water rights downstream, such as committed outflows, navigational flows, and environmental flows that are required to meet certain nonagricultural needs, or other nonutilizable outflows from the basin. The *Available water* then is exploitable water that can be allocated to the various water-use sectors in the domain of the basin (Table 7.8). In the course of use of the available water in a basin, some of the utilized flow is depleted, mainly through incremental ET, as well as that which flows into sinks where it is lost due to contamination, salinity increase, or movement into areas

TABLE 7.10 Categories of Land-Use Classes That Have Similar Characteristics for Management

| Land Use | | | |
|---|---|---|---|
| **Conserved** | **Utilized** | **Modified** | **Water Managed** |
| Reserved, national parks | Closed natural forests | Plantation trees | Irrigated pastures |
| Conservation set-asides | Tropical rain forest | Rainfed pastures | Irrigated crops |
| Glaciers | Open natural forest | Rainfed crops | Irrigated fruits |
| Coastal protection sites | Woody savanna | Rainfed fruit | Irrigated biofuels |
| | Open savanna | Rainfed biofuels | Reservoirs, canals |
| | Sparse savanna | Rainfed parks | Greenhouses |
| | Shrub land | Fallow land | Aquaculture |
| | Natural pastures | Dump sites | Residents |
| | Deserts | Oases, wadis | Industrial areas |
| | Mountains | Roads, lanes | Irrigated parks |
| | Rocks | Peri-urban areas | Managed wetlands |
| | Floodplains | | Inundation areas |
| | Tidal flats | | Mining |
| | Bare land | | Evaporation ponds |
| | Wasteland | | Wastewater treatment beds |
| | Moore fields | | Power plants |
| | Wetlands and swamps | | |
| | Alien invasive species | | |
| | Permafrosts | | |

*From Karimi, P., Bastiaanssen, W.G.M., Molden, D., 2013a. Water Accounting Plus (WA+)—a water accounting procedure for complex river basins based on satellite measurements. Hydrol. Earth Syst. Sci. 17, 2459–2472; Karimi, P., Bastiaanssen, W.G.M., Molden, D., Cheema, M.J.M., 2013b. Basin-wide water accounting based on remote sensing data: an application to the Indus Basin. Hydrol. Earth Syst. Sci. 17, 2473–2486.*

where it cannot be recovered. *Outflow water* is all of that that exits the basin either as surface outflow or ground-water flow.

Karimi et al. (2013a,b) have also shown performance indicators as fractions wherein various water and flow types are divided into each other to obtain diverse quantitative assessments of water management in a basin. Such measures can be quite instructive in judging impacts of water policy interventions on water scarcity and benefits from water-management decisions (Table 7.10).

## 7.3.2 WA+ Evapotranspiration Sheet

The evapotranspiration sheet (Fig. 7.6) is an accounting of managed, manageable, of nonmanageable ET, so it shows some of the same subdivision parts of the landscape ET, but

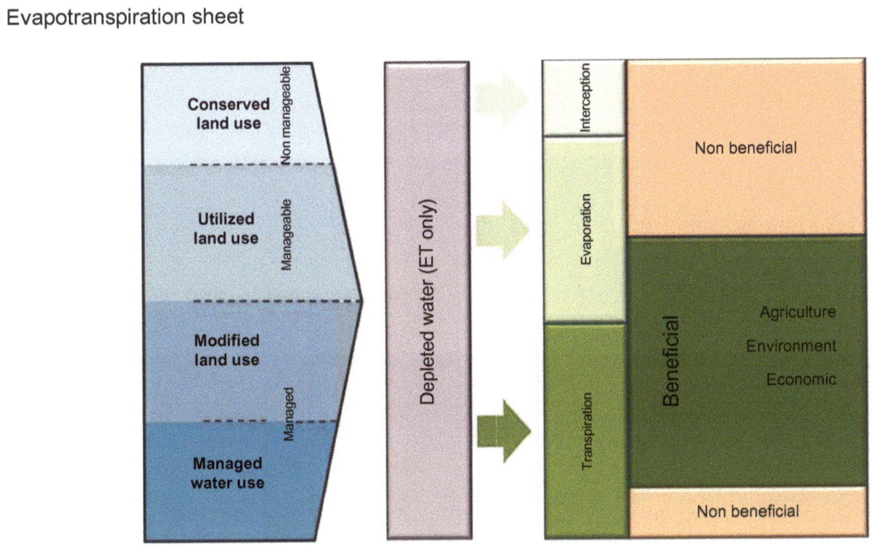


| Indicators | Definition |
|---|---|
| T fraction | Total T divided by the total ET |
| Beneficial fraction | Beneficial E and T divided by the total ET |
| Managed fraction | Managed ET divided by the total ET |
| Agri. ET fraction | Agricultural ET divided by the total ET |
| Irri. ET fraction | Irrigated agricultural ET divided by the agricultural ET |

FIG. 7.6 The evapotranspiration (ET) sheet describes the different parts of the ET processes that are managed, manageable capability, or nonmanageable, and their beneficial and nonbeneficial characteristics as the parts are subdivided into E, T, and interception. *From Karimi, P., Bastiaanssen, W.G.M., Molden, D., 2013a. Water Accounting Plus (WA+)—a water accounting procedure for complex river basins based on satellite measurements. Hydrol. Earth Syst. Sci. 17, 2459–2472; Karimi, P., Bastiaanssen, W.G.M., Molden, D., Cheema, M.J.M., 2013b. Basin-wide water accounting based on remote sensing data: an application to the Indus Basin. Hydrol. Earth Syst. Sci. 17, 2473–2486.*

it shows far more detail about the beneficial and nonbeneficial aspects of the ET processes. For example, transpiration from agricultural crops is obviously deemed beneficial, but that from weeds or alien invasives is not. Evaporation from farm fish ponds would be considered beneficial, whereas that from fallow wet soils would not. Performance indicators for evapotranspiration provide a number of measures of the magnitudes of benefits from ET in a basin (Table 7.10).

## 7.3.3 WA+ Productivity Sheet

The agricultural or biomass production per unit of water is shown in Fig. 7.7. The depletion of water by ET, in addition to general biomass production, is also involved in specific biomass production directly from water itself, sequestration of carbon dioxide, or equivalent

Productivity sheet

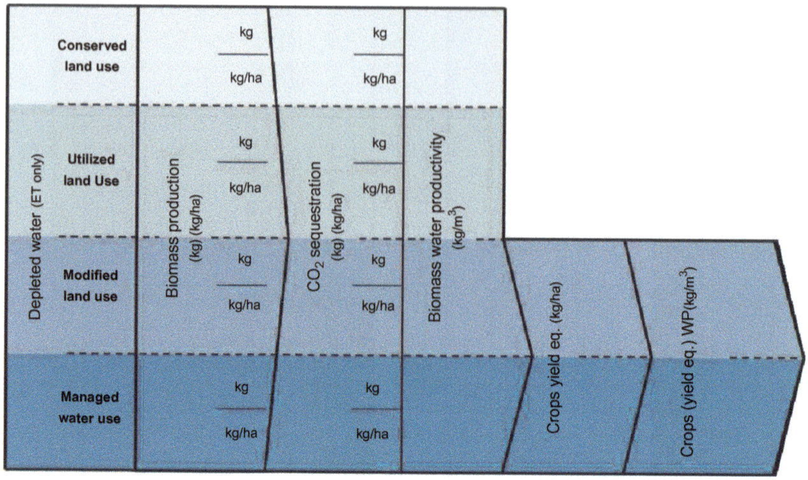


| Indicators | Definition |
|---|---|
| Land productivity$_{crops}$ | Crop biomass times harvest index divided by cropped area |
| Land productivity$_{pastures}$ | Pastures biomass times harvest index divided by pasture area |
| Water productivity$_{crops\ rainfed}$ | Rainfed crops biomass times harvest index divided by rainfed crops ET |
| Water productivity$_{crops\ irrigated}$ | Irrigated crops biomass times harvest index divided by Irrigated crops ET |
| Food Irri. dependency | Irrigated food production divided by total food production |

FIG. 7.7 The productivity sheet describes the agricultural production per unit of water by reporting on the bio-physical land productivity and water productivity in the four land categories. *From Karimi, P., Bastiaanssen, W.G.M., Molden, D., 2013a. Water Accounting Plus (WA+)—a water accounting procedure for complex river basins based on satellite measurements. Hydrol. Earth Syst. Sci. 17, 2459–2472; Karimi, P., Bastiaanssen, W.G.M., Molden, D., Cheema, M.J.M., 2013b. Basin-wide water accounting based on remote sensing data: an application to the Indus Basin. Hydrol. Earth Syst. Sci. 17, 2473–2486.*

crop yields. These measures can be quantified into performance measures of the various types of land and water productivities, as well as a measure of food to irrigation dependency (Table 7.10).

## 7.4 WA+ WITHDRAWAL SHEET

The withdrawals sheet is a more detailed accounting of managed water use in terms of surface and ground-water uses (Fig. 7.8), where the management options of those uses are quite different because the surface water can be so easily subdivided into reservoirs, agriculture/aquaculture, domestic, and industrial uses. Ground water obviously cannot be so treated. The performance indicators for the withdrawal sheet are in part quite classical measures of efficiency in irrigation, as well as measures of fractions of ground-water withdrawal and recoverable water (Table 7.10).

Withdrawals sheet

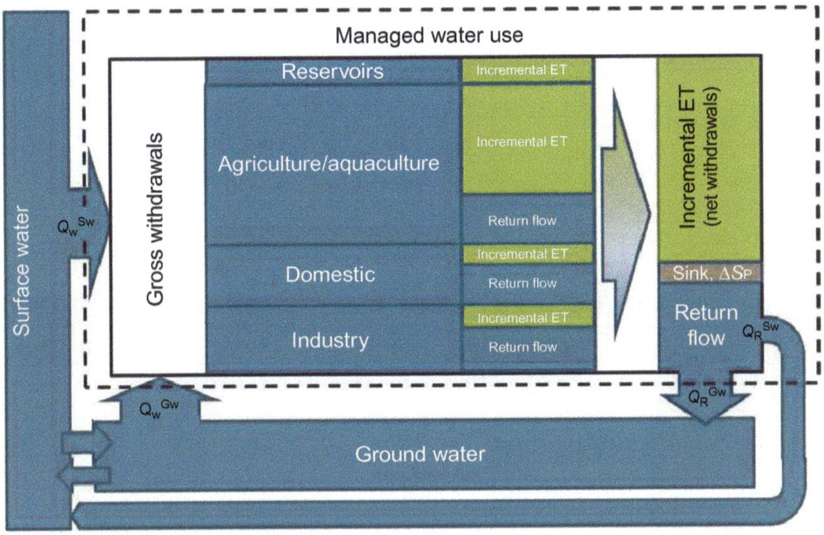


| Indicators | Definition |
|---|---|
| GW withdrawal fraction | Groundwater withdrawals divided by total withdrawals |
| Classical irrigation efficiency | Incremental ET of agriculture divided by withdrawals for agriculture |
| Recoverable fraction | Return flow divided by total withdrawals |

FIG. 7.8 The withdrawal sheet differentiates between surface and underground water and the different types of managed water use. *From Karimi, P., Bastiaanssen, W.G.M., Molden, D., 2013a. Water Accounting Plus (WA+)—a water accounting procedure for complex river basins based on satellite measurements. Hydrol. Earth Syst. Sci. 17, 2459–2472; Karimi, P., Bastiaanssen, W.G.M., Molden, D., Cheema, M.J.M., 2013b. Basin-wide water accounting based on remote sensing data: an application to the Indus Basin. Hydrol. Earth Syst. Sci. 17, 2473–2486.*

## 7.5 APPLICATIONS OF THE WA+ SYSTEM IN CENTRAL AND SOUTH ASIA

The WA+ system uses various accessible measures from new generations of satellites as a main input data source, rather than detailed hydro-meteorological measurements that are so hard or impossible to obtain in Central and South Asia. In fact, as a proof of concept, Karimi et al. (2013b) applied the methodology to the particularly complex and "messy" data environment of the Indus Basin. The result is that the methodology can be used basically everywhere, with the resulting possibility of a much more rigorous and pervasive accounting system to make water management much better in water-short regions (Table 7.11). Even so, the collection of data from the various sources and institutes to help validate the satellite data is the largest obstacle for water accounting in the operational context (Karimi et al., 2013a). Only the withdrawal sheet requires more

**TABLE 7.11** Performance Indicators for the Various Sheets in the WA+ System of Water Accounting

**Resource-base sheet**

$$\text{Exploitable water}_{\text{fraction}} = \frac{\text{Exploitable water}}{\text{Net inflow}}$$

$$\text{Storage change}_{\text{fraction}} = \frac{\Delta S_{\text{fw}}}{\text{Exploitable water}}$$

$$\text{Available water}_{\text{fraction}} = \frac{\text{Available water}}{\text{Exploitable water}}$$

$$\text{Basin closure}_{\text{fraction}} = \frac{\text{Utilized flow}}{\text{Available water}}$$

$$\text{Reserved outflow}_{\text{fraction}} = \frac{\text{Reserved outflows}}{Q_{\text{out}}^{\text{SW}} + Q_{\text{out}}^{\text{GW}}}$$

**Evapotranspiration sheet**

$$\text{Transpiration}_{\text{fraction}} = \frac{T}{\text{ET}}$$

$$\text{Beneficial}_{\text{fraction}} = \frac{E\text{beneficial} + T\text{beneficial}}{\text{ET}}$$

$$\text{Managed}_{\text{fraction}} = \frac{\text{ET managed}}{\text{ET}}$$

$$\text{Agricultural ET}_{\text{fraction}} = \frac{\text{Agricultural ET}}{\text{ET}}$$

$$\text{Irrigated ET}_{\text{fraction}} = \frac{\text{Irrigated agricultural ET}}{\text{Agricultural ET}}$$

**Productivity sheet**

$$\text{Land productivity}_{\text{crops}} = \frac{\text{Crops biomass production} \times \text{Harvest index}}{\text{Cropped area}}$$

$$\text{Land productivity}_{\text{pasture}} = \frac{\text{Pasture biomass production} \times \text{Harvest index}}{\text{Pasture area}}$$

$$\text{Water productivity}_{\text{crops rainfed}} = \frac{\text{Rainfed crops biomass production} \times \text{Harvest index}}{\text{Rainfed crops ET}}$$

$$\text{Water productivity}_{\text{crops irrigated}} = \frac{\text{Irrigated crops biomass production} \times \text{Harvest index}}{\text{Irrigated crops ET}}$$

$$\text{Food} - \text{irrigation dependency} = \frac{\text{Irrigated flood production}}{\text{Total food production}}$$

**Withdrawal sheet**

$$\text{Ground} - \text{water withdrawal fraction} = \frac{Q_{\text{w}}^{\text{GW}}}{Q_{\text{w}}^{\text{SW}} + Q_{\text{w}}^{\text{GW}}}$$

$$\text{Classical irrigation efficiency CE} = \frac{\text{incremental ET irrigated agriculture}\left(\text{ET}_Q\right)}{Q_{\text{w}}\ (\text{Irrigated agriculture})}$$

$$\text{Recoverable fraction} = \frac{Q_{\text{R}}}{Q_{\text{W}}}$$

*From Karimi, P., Bastiaanssen, W.G.M., Molden, D., 2013a. Water Accounting Plus (WA+)—a water accounting procedure for complex river basins based on satellite measurements. Hydrol. Earth Syst. Sci. 17, 2459–2472; Karimi, P., Bastiaanssen, W.G.M., Molden, D., Cheema, M.J.M., 2013b. Basin-wide water accounting based on remote sensing data: an application to the Indus Basin. Hydrol. Earth Syst. Sci. 17, 2473–2486.*

classical, water-accounting procedures where field data must be accumulated. Even in such cases some of the data must be very broad-brush or low resolution, such as measures of ground-water depletion from spotty measures of decline of water-table levels, versus gravitational satellites, such as the Gravity Recovery and Climate Experiment (GRACE). Data from hydrological models and water-allocation models can also be utilized as input data in WA+.

The use of the WA+ system allows access to thinking about water accounting to go far beyond that required in most traditional hydrologic basins, or by the national governments. The availability of standard sets of indicators based on data-collection procedures that are quite transparent and not subject to government classifications of secret is a great boon to the people of the arid regions. This seeking of workable solutions in the decision-making processes of water-resource management is greatly improved, once the steep learning curves of remote- sensing methodologies and formulation of new water policies are undertaken in this fashion.

## 7.6 BLUE-WATER, GREEN-WATER, AND GRAY-WATER PARADIGMS

In the course of working with difficult water-related problems in the mountains of South and Central Asia, we know that we must, if possible, develop and apply many other new methods. Thus the thinking about the blue- and green-water paradigms has possible useful applications in working with a declining and progressively more fraught resource in the region (Falkenmark and Rockström, 2006) (Figs. 7.3 and 7.4). The blue-water resource, or the fresh surface water in lakes, rivers, and ground water in aquifers, is contrasted with the green water that is precipitation on land that is stored as moisture in the soil, or that stays temporarily on top of the soil or vegetation (interception). Eventually this portion of the precipitation evaporates or transpires through the plants. Green water can be made productive for the growth of crops, although not all of it, because some will always evaporate from the soil, and not all periods of the year are generally suitable for the growth of crops. In addition to blue and green water, gray water is that portion of blue water that becomes contaminated with various compounds that pollute the water.

Concomitantly with the blue-green water paradigm, the concept of a "water footprint" also has emerged. Thus the water footprint is an indicator of the use of freshwater that can be both direct and indirect water use as a consumer or producer. Water use is measured in terms of water volume consumed (evaporated or incorporated into the product), and/or polluted per unit of time. The direct water-use footprint is the freshwater consumption and pollution that is used in a straightforward fashion to produce the product to be used by the producer or consumer. It is the volume of freshwater appropriated to produce the product, taking into account the volumes of water consumed and polluted in the different steps of the supply chain. The direct water footprint is equal to the sum of the water footprints of all products consumed by the consumer or of all (non-water) inputs used by the producer. This direct water footprint is quite distinct from the indirect-water footprint, which is the water consumption and pollution that is "behind" or can be associated with the production of the goods and services consumed by the consumer, or the inputs used by the producer.

Thus the water footprint in any given region is the amount of water used in and around the home, school, office, or farm field throughout the day. It includes all the water obtained directly from the well or the tap, as well as the water that it took to produce the food that was eaten, the products that might be purchased, or even the energy that was consumed. A person might not actually drink, feel, or see this virtual water, but in fact, it may make up the majority of the individual's actual water footprint. It may seem a hopeless endeavor, or an incalculable venture, but actually a water footprint can be computed for an individual human, a whole household, a business, or an entire nation. Calculation of a water footprint may be a valuable enterprise because it will reveal water-use patterns at many different levels; all the way from individuals to entire countries. This may be a valuable exercise because it can enable intelligent behavior modification to conserve water, or to redirect energies to promote efficiencies of conservation. In the course of doing this, of course, the water use will be split into the three corresponding footprint elements of: (1) the blue-water footprint volume of surface and ground water consumed (evaporated or incorporated in the product) during the production process; (2) the green-water footprint volume of the rainwater consumed (evaporated or incorporated in the product) in making the product; and (3) the gray-water footprint volume, which is the freshwater required to mix and dilute pollutants and maintain water quality according to water-quality standards.

These concepts are used to formulate a shared global standard on definitions and calculation methods, which is crucial given the rapidly growing interest in the world to use water footprint accounting systems as a basis for formulating sustainable water strategies and policies for companies and nation states. Such water-accounting tools are effective only when used in a strict or rather rigid manner, not as a limp metaphor that people can shrug off as a nice idea but perhaps unsuitable in their weak or failed-state status, such as a country like Afghanistan.

In fact, the water footprints for all the world's nations have been worked out in considerable detail, and are available on line, although a complete look at Afghanistan had to be left out in part because consumption data were too limited to work with (Mekonnen and Hoekstra, 2011a,b). The global water footprint (1996–2005) was 9087 km$^3$/yr, with 74% green, 11% blue, and 15% gray. Agricultural production contributes 92% to this total water footprint. In addition, about one-fifth of the global water footprint relates to production for export. A total world volume of 2320 km$^3$/yr (68% green, 13% blue, and 19% gray) of international virtual water flows was related to trade in agricultural and industrial products. The water footprint of the global average consumer from 1996 to 2005 was 1385 m$^3$/yr, with ~92% of the water footprint related to the consumption of agricultural products, 5% to consumption of industrial goods, and 4% to water use domestically.

When Mekonnen and Hoekstra (2011a,b) looked at the total international blue-water footprint per capita (excluding Afghanistan because of missing consumption data), countries in Central and Southwest Asia appeared on the top of the list. Consumers in Turkmenistan appear to have the largest blue-water footprint of all countries, namely 740 m$^3$/yr/cap

TABLE 7.12 The WA+ Framework Can Evaluate the Impact of Interventions in Water Management and Some Land Mismanagement

| | |
|---|---|
| 1. Water re-allocations | 10. Urban expansion |
| 2. Reduced ground-water withdrawals | 11. Deforestation/dearbification |
| 3. Deficit irrigation | 12. Introduction of biofuel crops |
| 4. Water logging | 13. Cropping pattern change |
| 5. Irrigation modernization (flood to drip, etc.) | 14. Altered cultivation practice |
| 6. Artificial recharge | 15. Soil salinization/desalinization |
| 7. Water retention and storage | 16. Overgrazing |
| 8. Waste-water treatment | 17. Soil erosion |
| 9. Water-productivity improvement | |

*From Karimi, P., Bastiaanssen, W.G.M., Molden, D., 2013a. Water Accounting Plus (WA+)—a water accounting procedure for complex river basins based on satellite measurements. Hydrol. Earth Syst. Sci. 17, 2459–2472.*

on average. Other countries with a large blue-water footprint are (in descending order): Iran (589 $m^3$/yr/cap), the United Arab Emirates (571), Egypt (527), Libya (511), Tajikistan (474), Saudi Arabia (447), and Pakistan (422). The global average blue-water footprint of consumption for each person is 153 $m^3$/yr/cap, which is 11% of the total water footprint. Afghanistan, however, could be compared in some limited ways (national production, virtual water flows, national water savings related to trade) with its neighboring countries, so that some water footprint data on national production in Afghanistan have become available (Tables 7.12 and 7.13).

## 7.7 CONCLUSION

Water management can be viewed as a complex human endeavor to measure and control water supplies that becomes ever more essential as water supplies decline in quantity and quality as human withdrawals and uses increase. The rather extreme situation concerning water in Afghanistan is one wherein the basic principles of IWRM clearly must be followed to ensure a better future for the people of the country, but bringing people and organizations into these new ways of thinking has been difficult. Thinking about the so-called blue, green, and gray water types is one way of subdividing water types and uses that can classify the complex subject. More recently, however, the latest procedures include the WA+ system of rigorous measures of all water factors that must be considered in detail and accounted for in the resource base, the various forms of evapotranspiration, the different types of productivity, and the withdrawals of all the water types. Collectively these rigorous details about water can enable the most efficient and useful system of managing water that are available at the present time.

TABLE 7.13 Water Footprints of National Production in Afghanistan and Surrounding Countries in $Mm^3/yr$ for the Period 1996–2005

| | Water Footprint Crop Production | | | Water Footprint Grazing | Water Footprint Animal Supply | Water Footprint Industry | | Water Footprint Domestic Supply | | Total Water Footprint | | |
|---|---|---|---|---|---|---|---|---|---|---|---|---|
| **Country** | **Green** | **Blue** | **Gray** | **Green** | **Blue** | **Blue** | **Gray** | **Blue** | **Gray** | **Green** | **Blue** | **Gray** |
| Afghanistan | 7586 | 7664 | 149 | 4043 | 132 | 0.1 | 1 | 42 | 378 | 11,899 | 7838 | 528 |
| Iran | 43,027 | 39,823 | 8826 | 13,212 | 415 | 55 | 1045 | 620 | 5580 | 56,239 | 40,913 | 15,451 |
| Pakistan | 40,561 | 74,272 | 21,805 | 34,113 | 907 | 173.5 | 3296.5 | 327 | 2943 | 74,674 | 75,679 | 28,044 |
| Tajikistan | 1834 | 2865 | 222 | 4243 | 32 | 28 | 532 | 44 | 396 | 6077 | 2969 | 1150 |
| Turkmenistan | 3961 | 7086 | 1 | 3558 | 61 | 9.5 | 180.5 | 42 | 378 | 7519 | 7199 | 559 |
| Uzbekistan | 9083 | 22,236 | 0 | 6881 | 143 | 60 | 1140 | 277 | 2493 | 15,964 | 22,716 | 3633 |

*From Mekonnen, M.M., Hoekstra, A.Y., 2011b. National Water Footprint Accounts: The Green, Blue and Gray Water Footprint of Production and Consumption. Volume 2: Appendices. UNESCO-IHE; Institute for Water education; Value of Water; Research Report Series No. 50, 94 pp. http://www.waterfootprint.org/Reports/Report50-NationalWaterFootprints-Vol2.pdf (accessed 01.12.15); Appendix I.*

## References

Amarkhail, M.A., Kakar, K., 2011. The need for governance in the water sector. In: Kakar (Ed.), Afghanistan Human Development Report 2011—The Forgotten Front: Water Security and the Crisis in Sanitation. Centre for Policy and Human Development, Kabul University, pp. 169–193. http://re.indiaenvironmentportal.org.in/files/Complete%20NHDR%202011%20final.pdf (accessed 23.07.15).

ENVSEC, 2007. Environment and Security Initiative ENVSEC. Summary of the Regional Meeting on Environment and Security in the Upper Amu Darya River Basin Between Afghanistan and Tajikistan, 6pp. http://www.envsec.org/publications/Summary%20of%20the%20Regional%20Meeting%20on%20Environment%20and%20Security%20in%20the%20upper%20Amu%20Darya%20River%20basin%20between%20Afg.pdf (accessed 01.01.15).

Falkenmark, M., Lannerstadt, M., 2005. Consuptive water use to feed humanity - curing a blind spot. Hydrol. Earth Syst. Sci. 9, 15–28. https://www.diva-portal.org/smash/get/diva2:219522/FULLTEXT01.pdf (accessed 31.03.16).

Falkenmark, M., Rockström, J., 2006. The new blue and green water paradigm: Breaking new ground for water resources planning and management. J. Water Resour. Plan. Manag. 132, 129–133.

Kakar, K., 2011. Afghanistan Human Development Report 2011—The Forgotten Front: Water Security and the Crisis in Sanitation. Centre for Policy and Human Development, Kabul University, 238pp. http://re.indiaenvironmentportal.org.in/files/Complete%20NHDR%202011%20final.pdf (accessed 23.07.15).

Karim, A., Veizer, J., 2002. Water balance of the Indus River Basin and moisture source in the Karakoram and western Himalayas: implications from hydrogen and oxygen isotopes in river water. J. Geophys. Res. 107 (D18), 4362. http://dx.doi.org/10.1029/2000JD000253. ACH 9, 1–12.

Karimi, P., Bastiaanssen, W.G.M., Molden, D., 2013a. Water Accounting Plus (WA+)—a water accounting procedure for complex river basins based on satellite measurements. Hydrol. Earth Syst. Sci. 17, 2459–2472.

Karimi, P., Bastiaanssen, W.G.M., Molden, D., Cheema, M.J.M., 2013b. Basin-wide water accounting based on remote sensing data: an application to the Indus Basin. Hydrol. Earth Syst. Sci. 17, 2473–2486.

Kuonqui, C., Thomas, V., Ahmazai, A., 2011. Water security and human development in Afghanistan. In: Kakar, K. (Ed.), Afghanistan Human Development Report 2011—The Forgotten Front: Water Security and the Crisis in Sanitation. Centre for Policy and Human Development, Kabul University, 238pp. http://re.indiaenvironmentportal.org.in/files/Complete%20NHDR%202011%20final.pdf (accessed 25.07.15).

Mekonnen, M.M., Hoekstra, A.Y., 2011a. National Water Footprint Accounts: The Green, Blue and Grey Water Footprint of Production and Consumption. Volume 1: Main Report. UNESCO-IHE; Institute for Water education; Value of Water; Research Report Series No. 50, 50pp. http://www.waterfootprint.org/Reports/Report50-NationalWaterFootprints-Vol1.pdf (accessed 01.12.15).

Mekonnen, M.M., Hoekstra, A.Y., 2011b. National Water Footprint Accounts: The Green, Blue and Grey Water Footprint of Production and Consumption. Volume 2: Appendices. UNESCO-IHE; Institute for Water education; Value of Water; Research Report Series No. 50, 94pp. http://www.waterfootprint.org/Reports/Report50-NationalWaterFootprints-Vol2.pdf (accessed 01.12.15).

OECD, 2005. The Paris Declaration on Aid Effectiveness: Five Principles for Smart Aid. Organization for Economic Cooperation and Development (OECD), 2pp. http://www.oecd.org/dac/effectiveness/45827300.pdf (accessed 22.07.15).

Omar, A., 2013. Hydrology and environment cooperation in the Upper Amu Darya River Basin. United Nations Economic Commission for Europe (UNECE). http://www.unece.org/fileadmin/DAM/env/water/meetings/joint_bodies/presentations/3.3AFG-TJK-presentation-unece-sep2013.pdf (accessed 31.12.14).

Perry, C.J., 2012. Accounting for water; stocks, flows, and values. Inclusive Wealth Report 2012: Measuring Progress Toward Sustainability. UNU-IHPD and UNEP. Cambridge University Press, Cambridge, UK, pp. 215–230. http://isp.unu.edu/events/2012/files/Inclusive_Wealth_Report-2012.pdf (accessed 01.07.15).

Shroder, J., 2014. Natural Resources in Afghanistan: Geographic and Geologic Perspectives on Centuries of Conflict. Elsevier, Amsterdam.

Shroder, J.F., 2015. Progress with Afghanistan extractive industries: will the country know resource success or failure evermore? Extractive Ind. Soc. 2, 264–275.

Smakhtin, V., Wichelns, D., n.d. Hydrometeorological Data Collection and Sharing: The Global Challenge. International Water Management Institute, Colombo, Sri Lanka. http://www.oecd.org/tad/sustainable-agriculture/45066665.pdf (accessed 31.12.14).

United Nations, 2012. SEEA—Water System of Environmental Economic Accounting for Water. United Nations Statistics Division; Economic and Social Affairs, Geneva. 216 pp. http://unstats.un.org/unsd/publication/seriesf/Seriesf_100e.pdf (accessed 01.06.15).

UNECE, 2012. UNECE Supports Afghan-Tajik Cooperation to Address Severe Floods, Vulnerable Ecosystems and Climate Change Impacts in Amu Darya Basin, 2 pp. http://www.unece.org/?id=29474 (accessed 01.01.15).

UNECE, 2013. UNECE Supports Afghan-Tajik Cooperation on Environment and Hydrology in the Amu Darya Basin, 2 pp. http://www.unece.org/index.php?id=32619 (accessed 01.01.15).

UNECE, 2014. UNECE Supports Afghan-Tajik Cooperation on Water and the Environment in the Amu Darya Basin, 2 pp. http://www.unece.org/unece-main/info-resources/unece-weekly/articles/environment/2014/unece-supports-afghan-tajik-cooperation-on-water-and-the-environment-in-the-amu-darya-basin/unece-supports-afghan-tajik-cooperation-on-water-and-the-environment-in-the-amu-darya-basin.html (accessed 01.01.15).

Wisler, C.O., Brater, E.F., 1959. Hydrology. John Wiley & Sons, New York.

World Bank, 2014. Water Resources Management Overview. http://www.worldbank.org/en/topic/waterresources-management/overview#1 (accessed 31.12.14).

Zoï Environment Network, 2013. Visual Atlas of Cooperation—Afghanistan and Tajikistan: Environment and Hydrology in the Upper Amu Darya Basin, 47 pp. http://www.zoinet.org/web/sites/default/files/publications/AFG-TJK-atlas-web.pdf (accessed 01.01.15).

CHAPTER

# 8

# Introduction to the Indigenous Water-Management System in Afghanistan

*S.J. Ahmadzai*

Center for Afghanistan Studies, University of Nebraska at Omaha, Omaha, NE, United States

OUTLINE



## Abstract

Afghans, like many other members of agrarian societies, have developed a system of water management to address their needs to ensure fair distribution of water to their fields or lands. The Mir-Aab system is an indigenous water-management system that has been used in Afghanistan for centuries. This system is the result of community engagement and cooperation to sustain their lives and manage their water dependency through a collective decision of all stakeholders in a village. One person, a Mir Aab is entrusted with making vital and fair decisions on the distribution of the most precious natural resource. This chapter informs the reader of the water-management system that Afghans have been using and how it could be made more useful with the help of the government and the introduction of the latest water-management techniques that supplement the already accepted norms and systems of water distribution. The traditional wells, underground karezes, and constructed canals are explained, along with the best insights developed from the irrigation systems, as well as the historical development phases of irrigation in the country, and the historical events and factors that have affected the traditional Mir-Aab systems.



*Transboundary Water Resources in Afghanistan*
http://dx.doi.org/10.1016/B978-0-12-801886-6.00008-2

**Keywords:** *Mir Aab* water master (mirab), *Mir Aab Bashi* supervisor, *Khada Dar* supervisor, *Kok Bashi* assistant, *Chack-bashi* assistant, Irrigation, Traditional well, Karez, *Hashar* volunteer work, *Sarband* head dam, *Nahr* canal, *Juie* canal, *Arhat* (*rahat*) "Persian" wheel

## 8.1 MIR AAB, THE WATER MASTER

As an agrarian society with 74% of the people living in rural areas (Sadat, 2013) and struggling with poor irrigation systems, dry weather, and diminishing but not easily accessed water resources, Afghans have been involved with issues concerning water usage for centuries. Lacking other means to sustain their livelihoods, their dependency on water underlines the importance of water in their lives. Water usages for agriculture and drinking purposes have been the source of social, economic, and political cohesion and also problems in Afghanistan. Irregular or unfair use of water for any purpose can cause conflict among the stakeholders of any water resource, such as canals or wells. Most of rural Afghanistan is inhabited along rivers, canals, or around lakes, shallow wells, springs, and or sources which underscore the reliance of the people on easy access to water sources (Fig. 8.1).

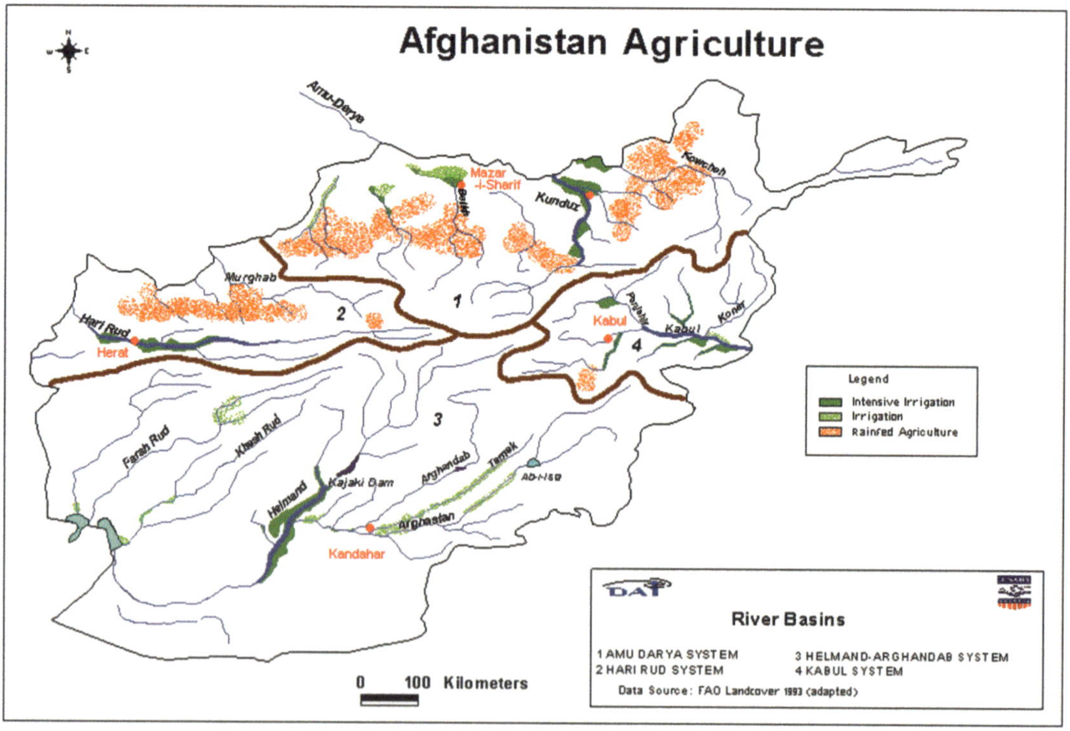


FIG. 8.1 Agriculture in Afghanistan showing intensive irrigation along the major river courses, irrigation of a less intense nature and areas of rainfed agriculture (which tend to occur on loess-covered (wind-blown dust) uplands. *Map from Garner et al. (2002).*

Dependency on dwindling water resources would naturally force people to look for ways to better manage water-flow regulations to certain fields and at specific times, and mitigate any wastage or unauthorized usage of water. In rural Afghanistan where people and livestock depend on proper distribution of this scarce resource, Afghans have traditionally acquired local knowledge and expertise to address their issues related to water and its usage.

Traditionally, communities with the consensus of all would appoint a local elder, preferably from the downstream end of the canal, as the manager of water who would be responsible to fairly distribute water among the landowners in a village for irrigation purposes. This person is called "Mir Aab," or Water Master. The reason to have a Mir Aab from the people down the stream is to make sure that water reaches all of the fields along the canal without any favors or discriminations. The Mir-Aab's position can also be inherited from a father to his son (Mohmand, 2011), since they would have institutional memory of how things are supposed to be done. It is important to understand that any Mir-Aab's implementation power is only as strong as his political and social influence, or his stance in his area.

The residents of the villages along canals can belong to different ethnic or tribal groups. The selection of a Mir Aab transcends all ethnic and tribal lines for the sake of fair distribution of water among all farmers (Ward et al., 2012). Anybody sticking to ethnic or tribal lines undermines the authority of the Mir Aab and can easily spur conflict among the people of the villages. Characteristically such conflicts are resolved through *jirgas* or village councils where elders of the villages come together to discuss the problem and find solutions that are acceptable to all sides involved.

Depending on the size of land being irrigated and the number of canals to be managed [up to 20 km of the river (Sexton, 2012)], a council of Mir Aabs manage water flow to the lands along the canals or canal. A Mir Aab, typically responsible for one canal, can have assistants supervising his activities. The Mir Aab or his assistants usually spends most of his time patrolling the canals and ditches to stop any violations by farmers and ensure fair and timely distribution of water. Patrols can be at any time of the day or night. In some areas, if and when deemed necessary, Mir Aabs can carry weapons, especially during the nighttime.

Typologies of designation of Mir Aabs can be different area to area in Afghanistan. A Mir-Aab's supervisors can be called *Mir Aab Bashi* in the north, west and east, and a *Khada Dar* in the areas to the south of Kabul. His assistants can be called *Kok Bashi* as they are referred to in Balkh Province (Wegerich, 2009) or *Chackbashi* in some other parts of the country. In Kandahar Province's Daman district, landowners or farmers also assist the Mir Aab. According to Mr. Saber Ali Amin, who hails from Kandahar, at times when water is directed to their fields, farmers, individually or in groups, monitor the flow of water from river intake to their fields. This job of monitoring water flow is called Ab Mali or as colloquially pronounced Ow Mali (Amin, 2015) (Fig. 8.2).

## 8.2 CONCEPTUAL UNDERSTANDING OF THE MIR-AAB SYSTEM IN AFGHANISTAN

The interlinked components, which are not necessarily exhaustive in Fig. 8.3, of the Mir Aab systems in Afghanistan exist in a milieu of interactions between the local nodes of the agro-ecological environment, the socio-economic and political environments, the available

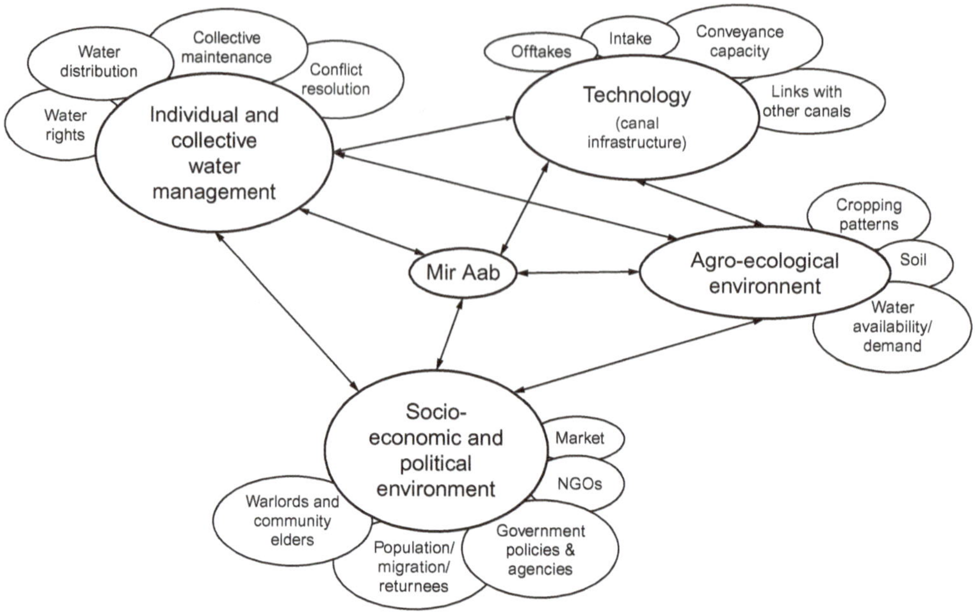


FIG. 8.2 The elements of the Mir-Aab system of the Jangharoq Canal in Baghlan Province that is located in the mid-stream section of the Lower Kunduz sub-basin. The Mir Aab is the person or agent (service provider) that operates within these structural nodes (technology, different environments, and individual and collective water management), all of which are influenced by a variety of factors (nonexhaustive) associated with each node. *Diagram courtesy of Thomas, V., Ahmad, M., 2009. A Historical Perspective on the Mir Aab System: A Case Study of the Jangharoq Canal, Baghlan. Afghanistan Research and Evaluation Unit, Kabul. Retrieved from: http://www.areu.org.af/EditionDetails.aspx?EditionId=214&ContentId=7&ParentId=7.*

hydro-technology, and the individual and collective water-management styles (Thomas and Ahmad, 2009). Each of these nodal components has myriads of associated factors to be considered as well. Understanding the Mir-Aab system and its local culture means understanding how the key individual and collective practices of water management have been managed down at the local canal level since its origin some time (centuries) back. The emphasis is on the factors and actors that might have influenced or actually directly shaped the practices of the local Mir Aab and the related collected actions. In most areas in Afghanistan, the individual and collective water-management practices are generally limited to water distribution (overlain by the water rights attached to it), as well as the collective maintenance of the water-distribution structure, and especially any conflict resolution required.

The social capital of any water-distribution system is the institutions, relationships, and norms that shape the quality and quantity of a society's social interactions (Thomas and Ahmad, 2009). On an operational level in this regard to the Mir-Aab system, the concept of social capital to draw on, would include:

- groups and networks: the collections of individuals who promote and protect personal relationships that improve social welfare;
- trust and solidarity: the elements of interpersonal behavior that can foster increased cohesion more robust collective action;

FIG. 8.3 Group of Afghan villagers who have turned out as collective volunteer labor or *hashar* to dig out a canal outtake for the main irrigation canal of Chata (photograph courtesy of USAID). Note that some of the volunteer workers have rubber boots to keep their feet dry and some do not. Also the shovels seem largely imported from the USA rather than locally made because those tend to have a wooden cross bar on which to stand to add leverage.

- collective action and cooperation: the capability of people to work together in order to resolve communal issues;
- social cohesion and inclusion: the mitigation of conflict risk and promotion of equitable access to development benefits by enhancing participation by marginalized people; and
- information and communication: the breaking down of negative social capital and the enabling of positive social capital by improving access to reliable information.

With these factors as background, Thomas and Ahmad (2009) have assessed the historic events and various complex interactions that have so affected the evolution of the Mir-Aab system among the multiple problems that exist in Afghanistan (Table 8.1).

### 8.2.1 Factors Impacting Local Water-Management Systems

Water distribution is determined by the size of irrigated land, type of crop and time of year. This traditional system of water management employs "a combination of proportional and rotational allocations, [and]… has flexibility in adjusting irrigated area and cropping intensity to match water supply levels" (Rout, 2008). The share of water (based on the size of land) also determines the responsibilities of the landowner and his share of payments to the Mir Aab and labor provision when irrigation canals need maintenance. At the times when severe sediment and flood damages canal banks and impacts water flow, community members or farmers contribute to canal-maintenance services either by taking part or by hiring a

TABLE 8.1 Historical Events and Factors in Afghanistan Affecting Traditional Mir-Aab Systems of Water Management Over Past Century to Greater or Lesser Effect

1. Technological changes have occurred in water management and cropping practices
2. Dramatically shifting politics (monarchy, revolution, communism, totalitarian rule, democracy, theocratic anarchy, etc.)
3. Mercurial government policies
4. Unremediated canal siltation
5. Immigration in of new ethnicities
6. Emigration out (displacement) of old ethnicities
7. Rising water demands from increased population during stable times
8. Climate change has reduced water flow in drought periods
9. Controversy over water usages has occurred between head and tail uses on canals
10. Historical increase in need for Mir-Aab arbitration over time as irrigation development has grown in complexity
11. Arbitrary and/or government-enforced changes have occurred to remuneration amounts from farmers to Mir Aabs
12. Enforcement has varied over required collective volunteer labor from each family to dig out or desilt canals
13. Different warlord command areas occurred during war with Soviet Union caused collapse of regular water deliveries, and ultimately the collapse of the Mir Aab system as Mir Aabs became refugees out of the country
14. Natural river course shifts have caused changing water demands
15. Collapse of state during war years meant collapse of the Mir-Aab system
16. State collapse caused collapse of Mir-Aab water-management systems at canal level because state was main enforcer of rules, and also the decision maker on water distribution and crop production types (water intensive or not)
17. Unregulated construction of irrigation canal outtakes occurred during mujahedeen period
18. No regulation or curtailment occurred of high water-use crops (rice) during times of collapsed Mir-Aab arbitration, which adversely affected downstream water users
19. Problems of water access between canal head-end and tail-end farmers were not adjudicated during times of state collapse and mujahedeen takeover
20. Mir-Aab successes over the years have been shown to be largely due to personal reputation and ability to work with different people
21. Defaulting on canal desiltation maintenance duties was seen to be greatly reduced during Taliban times because of their harsh punishments for infractions
22. Bribing Mir Aabs to get water became a factor on post-2001 years, with some farmers who refused to pay arbitrary assessments getting no water
23. By 2007, water sharing and water rights were far from being clear or transparently defined or negotiated
24. Free-for-alls developed in some upstream areas with downstream farmers only being given water by Mir Aabs if they paid significant bribes
25. Shifts between early farmer-managed water-management systems to more Mir-Aab management systems have been noted over the years

*In part courtesy of Thomas, V., Ahmad, M., 2009. A Historical Perspective on the Mir Aab System: A Case Study of the Jangharoq Canal, Baghlan. Afghanistan Research and Evaluation Unit, Kabul. Retrieved from: http://www.areu.org.af/EditionDetails.aspx?EditionId=214&ContentId=7&ParentId=7.*

replacement to do the job for them (Fig. 8.3). "Under the system each water user contributes to the system's maintenance in proportion to that user's land area, and the land's location on the canal" (Ward et al., 2012). For example, in Kunduz Province, "per 40 jerib [8 ha of land] one labor day has to be contributed per day of collective maintenance" (ter Steege, 2007). This collective action of the landowners and farmers is called "*hashar*." *Hashar* can also be called for other community projects such as road building, mosque building, or other public works-related activities in a village.

Payments to the Mir Aab are done annually in kind. Farmers pay the Mir Aab a specific portion of their harvest at the end of the year as agreed upon. For example, in some areas where wheat is grown, farmers pay 4 kg of wheat per jerib (0.2 ha) to the Mir Aab (Mohmand, 2011). Some farmers might opt for cash payments also if deemed appropriate or when the crop grown by the farmer is not of any interest or benefit to Mir Aab.

More than three decades of war since the late 1970s in Afghanistan have tremendously impacted the power structure of rural Afghanistan and its tribal and social dynamics. Traditional players have been replaced with contemporary and nontraditional power structures and new people who hold power that was not available to them previously. Weak government, and emerging local leaders, usually with different and sometimes opposing interests, have had negative impacts on the system.

Prior to the Soviet invasion of Afghanistan in 1979, government agencies played a vital role as an enforcer of the laws or rules in assisting farmers to manage local water irrigation systems, when deemed necessary. Government would not be involved in the micromanagement of the irrigation systems. Specifics of the agreements would be managed by the Mir Aabs or community leaders. Government involvement was also related to the crop the government wanted the farmers to grow (Thomas and Ahmad, 2009). For example, in the northeast (Baghlan and Kunduz provinces) and in the South (Helmand River Valley) the government would encourage farmers to grow certain crops. These crops would require specific water regimes that would affect water distribution in an area watered by a canal managed by a local Mir Aab. The local agriculture directorate would provide technical assistance and on occasion, material assistance to the farmers in maintaining and repairing their irrigation systems.

After the Soviet invasion and with the start of resistance against the occupation, during which the central government lost control of many agricultural areas, local Mujahedeen commanders started to control everything in their areas of operations. They would collect taxes, in the form of the particular harvest, from the farmers. To have more tax collected in their areas of operations, they would permit their farmers to have more water and or allow the construction of illegal off-takes (Fig. 8.4) at the cost of the farmers downstream. Mujahedeen commanders continued their practices of controlling water resources until they were removed by the Taliban who were also not that different and favored their own loyal residents in their

FIG. 8.4 Illegal outtake built during mujahedeen times to take water from the Jangharoq Canal in Baghlan Province that is located in the mid-stream section of the Lower Kunduz sub-basin.

affairs and collected taxes. The only advantage of the Taliban rule was uniformity in their approach towards dealing with local farmers, because they had removed all other power brokers in their areas of operation.

After 2001 and the removal of the Taliban from power, the gap previously filled by the Taliban government was once again filled by the local power brokers and warlords. Most of the time, warlords would use water as a political tool to gain more power in their areas. For example, in Samangan Province, " control of irrigation water from the Khulm and Balkhab river basin was used as leverage by a warlord to attempt to dominate downstream areas" (Sexton, 2012). Such examples are plentiful across Afghanistan, where water is used for political power.

Climate change resulting in a rise in temperature, uneven precipitation, and lesser snow accumulation, which affects water reservoirs and water flow and availability, has further impacted water management and the behavior of the stakeholders and power brokers in rural communities in Afghanistan. Climate change is anticipated to severely impact Afghanistan and add to its already distressed environmental situation (Savage et al., 2009; Shroder, 2014). More droughts, floods, and deforestation are expected to adversely affect agriculture and irrigation. Droughts, in particular, will reduce water availability affecting millions of people who depend on water for their livelihoods. Then shortage of water in different seasons can be one of the reasons local power brokers, generally warlords, would try to influence the system of water distribution in their favor or to the benefit of their supporters. The influence can come in various modes such as threats, bribes, or if deemed necessary, the application of force.

More than 80% of the population of Afghanistan is dependent on agriculture and natural resources. The booming population and the return of refugees from Iran and Pakistan has put the already weak agriculture system under further stress that can potentially add to conflict and instability in Afghanistan (Ministry of Agriculture and Irrigation, 2006).

## 8.3 EVOLUTION OF MIR-AAB CONCEPTS IN THE PRESENT-DAY

Because of the many changes long underway in past times in Afghanistan, the ill-defined functions of the Mir-Aab system have had to evolve to meet ever-fluctuating pressures and needs. In general, however, the Mir Aab was a person or group of people who had traditional knowledge of an irrigation canal, that was elected by the land-owning farmers generally from a downstream location, and who would be responsible for proper distribution of the allocated water among users, as well as handling routine operation and maintenance along the canal. The Mir Aabs would mobilize the collective volunteer *hashar* work (Fig. 8.3) and coordinate with government authorities when necessary, especially if illegal diversion of the water needed to be cut off, or if *hashar* avoidance or other defaulters needed to be identified to the communities and authorities (Ziaee et al., 2015). Thus at the present time the foundations of the existing Mir-Aab culture, was never clearly defined in great detail because the concept by necessity has diffused in many directions through the years. Accordingly since 2002 the government of Afghanistan has planned and carried out major institutional reforms in the water sector.

Integrated water resources management (IWRM) is the new approach to river basin management, in which the government is now fostering the formation of Water User Associations (WUA) at the local level; this is the first step at the present time of changing the older and rather illusive Mir-Aab approach to water management. New water law has noted that the existing Mir-Aab institutions are to be strengthened towards these more formalized WUA. The existing Mir-Aab culture is to be preserved, but is to be structured so as to improve its management effectiveness and enhance its legitimacy and power. Thus the old unwritten Mir-Aab understandings would be converted into a more management-oriented and accountable institution. This new transparency would be developed democratically, and in manners that are acceptable to local norms, customs, and values (Ziaee et al., 2015).

## 8.4 TRADITIONAL IRRIGATION INFRASTRUCTURE

### 8.4.1 Canals

These structures of the surface-water distribution system, locally known as *nahr*, or *juie*, form most of the means for distribution of surface waters in Afghanistan (Fig. 8.5). With five major rivers basins across the country that are dependent upon the level rainfall and snowmelt in spring and early summer, the sources of water in canals across Afghanistan have been rivers and dams irrigating more than 3 million ha of land (Savage et al., 2009) These five rivers basins are the Amu Darya River Basin to the north, Kabul River Basin to the east, Hari Rud River Basin to the west, Helmand River Basin to the southwest, and Kunduz-Kokcha River Basin to the northeast.

Wherever possible, water is diverted from dams and rivers to arable lands. The distribution of water from dams for irrigation purposes is part of the formal irrigation system in Afghanistan developed and managed by the government. Most of the canals managed in this fashion feed from dams constructed on various rivers in Afghanistan, such as the Dala Dam in Kandahar Province on Arghandab River, Darunta Dam in Nangarhar Province on the Kabul River, and the Pul-e-Khumri Dam on the Kunduz River in Baghlan Province. The flow of the water from the dam into canals is managed by government in these canals but when the water reaches farmland, it is managed by local farmers under the supervision of a Mir Aab. The main canals feeding from dams are then distributed into regulated secondary or smaller canals that reach different arable lands in different villages (Rout, 2008).

FIG. 8.5 A typical secondary canal in Afghanistan of medium size that distributes water to the fields in the background. *Photograph courtesy of B. Rout, Afghanistan Research and Evaluation Unit.*

On the other hand, the informal system has been in place in Afghanistan for centuries and is managed and maintained by local communities. Water is diverted into canals by a vulnerable infrastructure locally called the *sarband* or head dam (Fig. 8.6). Sarbands are constructed using any available resources such as timber, sand bags, tree branches or rocks and at times, in fact rarely, when and if the community can afford it, a concrete structure is built to divert water from the river into canals (Fig. 8.7). Many times, high levels of water in the river or floods have usually damaged a sarband, thus forcing the villagers to completely reconstruct the structure. However, this can generally be avoided if the structures are made of concrete (Fig. 8.8).

FIG. 8.6 A sarband or head dam built of timber and branches to divert water from the river into side canals. *Photograph courtesy of B. Rout, Afghanistan Research and Evaluation Unit.*

FIG. 8.7 A canal offtake from the Panshir River that has been constructed of stones set in mortar cement, which makes it far more permanent and less liable to degradation over time. Afghans commonly do not know how to utilize the correct proportions of water, sand, and mortar to ensure longevity of their concrete cemented structures, which can easily lead to long-term degradation of their built structures such as this. In general, however, the use of such expensive construction techniques is commendable in any case because it argues for longer-term survival of such vital canals.

FIG. 8.8 A well-built canal offtake of concrete cement on the Kunduz River. *Photograph by A. Roe in Rout, B., 2008. How the Water Flows: A Typology of Irrigation System in Afghanistan. Afghanistan Research and Evaluation Unit (AREU), Kabul. Retrieved from: http://www.areu.org.af/Uploads/EditionPdfs/811E-Typology%20of%20Irrigation%20Systems.pdf.*

## 8.4.2 Karezes

These underground tunnels are one of the oldest irrigation systems in Afghanistan and have been used for many centuries (Fig. 8.9). With its dry arid topography in the southeast, south and west of Afghanistan, the only source of water in some areas are karezes, which are dug in the shape of an sloping narrow tunnel underground connecting multiple vertical shafts or wells. Traditionally, karezes are dug by hand using very basic tools (Fig. 8.10). Their length can vary and could go underground for kilometers until the tunnel reaches the ground-water sources or the aquifer, and then they feed by gravity so that the water flows outside to a reservoir for drinking and or irrigation purposes.

The first shaft of the tunnel at the source is the mother well (*sarchah*) and the end is the eye of the karez (*owkura*) leading to a pond or reservoir from where water is then channeled to fields. In between are long lines of spoil craters that show the course of the karez underground

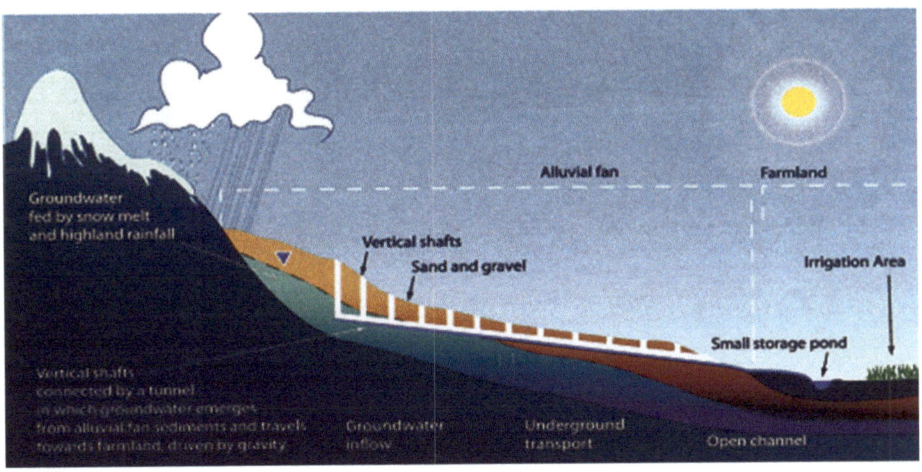


FIG. 8.9 Diagram of a karez in Afghanistan wherein the uppermost vertical shaft is commonly excavated down through the sands and gravels of an alluvial fan to the water table, at which point the karez construction team knows about what slope would be needed to allow the water to flow along a gentle gradient through a tunnel they dig to reach the outside world lower down the fan. Accordingly by means of lateral tunneling and removal of the spoil dirt up through additional vertical shafts, the karez is constructed. *Diagram courtesy of* Earth *magazine, by permission.*

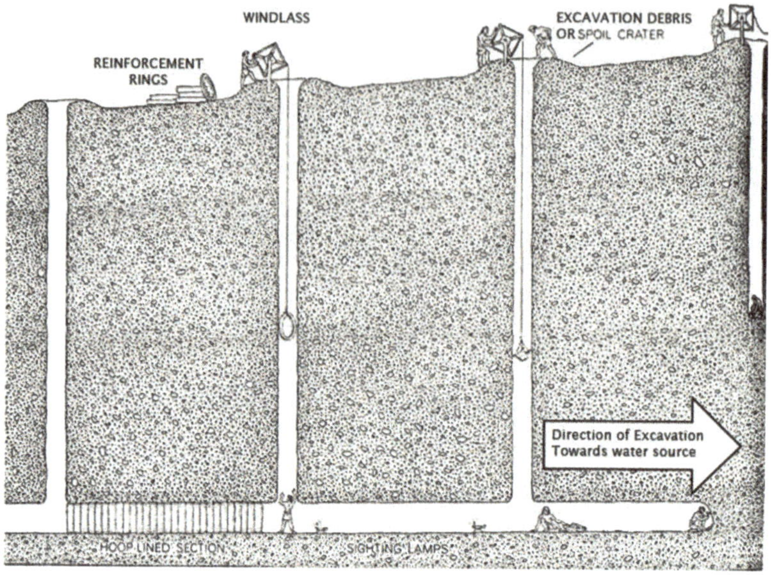


FIG. 8.10 Cartoon cross section of a karez under construction by excavating with pick and shovel at depth and removing sand and gravel up vertical shafts to surface by windlass. The vertical shafts have rings of excavation debris around them to form a spoil crater. The karez may need support of the tunnels in some places, which can be accomplished by addition of reinforcement rings of baked clay.

FIG. 8.11 Aerial view of karez spoil craters of varying ages, where some very old ones are completely filled in where likely collapsing of the karez roof at depth forced offset excavation around the collapse zone to keep the karez functioning. *Photograph courtesy of the US military.*

(Fig. 8.11). Either climate change that lowers the water table, or boreholes and tubewells with diesel pumps that do the same, can dry up karezes; digging another mother well uphill from the first can sometimes solve the problem.

People can also dig up underneath a major river to find similar karez water sources. In some areas, and depending on the ability of the people using karez water, karez tunnels are surfaced with slabs or tiles to ensure un-interrupted flow of water and avoid water loses due to permeable gravels in the tunnel (Rout, 2008). This system is also used in other countries such as China, Iran, Central Asian, Middle Eastern, and North African countries, where they are called *qanat*, *fogara*, *falaj*, *khitara*, or *kanerjing*.

Just as with any other irrigation system, karezes also require proper maintenance. The flow of water in the tunnel erodes the soil along the surface and walls of the tunnel, resulting in sediment that needs to be cleaned out periodically (Fig. 8.12A and B). In other cases, maintenance and cleaning is required due to the formation of subsurface collapse of the tunnel affecting water flow to the reservoir (Fig. 8.13). Professional workers/karez diggers are hired to clean out a karez by entering the tunnels through vertical shafts or wells and pulling the dirt out to the surface. Cleaning costs are shared by all who benefit from the water of a karez in a village, and water distribution is usually administrated by a community-appointed Mir Aab or Water Master, who makes sure that all farmers receive their fair share of water for their crops. According to some estimates, at least 6761 karezes in Afghanistan have been irrigating about 163,000 ha of land (Pedersen, 2009).

### 8.4.3 Wells

Compared to canals and karezes, the amount of land irrigated by wells is nominal (Fig. 8.14). According to the estimates form late 1960, wells in Afghanistan were used to irrigate only 1% of land, thus making only about 12,000 ha of land (Rout, 2008). The availability of water in the wells, most of the time depends on the subsurface water level, which, however, has been fluctuating a lot recently due to drought and unauthorized wells installed by people (Fig. 8.15). For example, the water level in Kabul city has gone down due to many unauthorized wells installed everywhere around the city (see previous Chapter 4). During the civil war, and under the Taliban regime when government services were nonexistent, people

(A) (B)

FIG. 8.12 (A) Photograph of traditional windlass for removal of spoil from a karez. Note the forked tree trunk at the right that allows the windlass to be turned and used directly over the vertical shaft. (B) Provincial Reconstruction Team (PRT) from the Czech Republic helping Afghan villagers clean and refurbish their karez with a less well-built tripod over the vertical shaft and an offset windlass set in tree forks to haul up the dirt. *Photographs courtesy of C. Greplova, Czech Republic Development Cooperation, Ministry of Foreign Affairs of the Czech Republic.*

FIG. 8.13 In Baraki Barak District, Logar Province, the local community Development Council repaired a dilapidated karez system by moving dirt out of the lower karez channel by using a grant from USAID. *Photograph by A. Salarzi and R. McGovern, courtesy of USAID.*

FIG. 8.14 Photograph of traditional hand dug well with mud plastered brick topmost portion and sandbags to hold pulley from rope to draw water with. Such a water acquisition device is not capable of drawing much water for irrigation, but is generally only used for drinking water and such. *Photograph courtesy B. Routt, AREU.*

FIG. 8.15 One of thousands of recently installed wells with hand-pumps that abound in the Kabul basin and elsewhere in the country. Such wells are an essential tool to enable access to water, but are commonly shallow and likely to be contaminated unless built and maintained properly.

started to install wells at their backyards and pumped out water for their daily needs. The population increase (return of refugees from abroad) in the last 14 years after 9/11, and lower precipitation levels in the last decade, have put water sources under further stress, not only in Kabul city but also in other populated areas of Afghanistan. The number of wells increased when many NGOs tried to provide basic services in the absence of government. According to an article published online in springerlink.com, "between 1997 and 2005, approximately 1500 shallow-supply wells, with a median depth of 22m, were installed by NGOs in the Kabul Basin" (Mack et al., 2013). Traditionally, Afghans have used well water for drinking purposes but on occasion they have also use them to irrigate their lands, especially orchards and small plots of land inside or around their housing compounds.

In some areas of Afghanistan where water cannot be diverted from low rivers to higher ground, and where the water table is not too deep, people have used wells to irrigate their patches of land. However, since the initial purpose of digging a well in the middle of a desert, or in areas located far away from any nearby river was irrigation, people have acquired modern technology to pump water out of the wells. They use diesel-powered engines, which is financially not a very efficient way to water their lands, and has drawn down the water tables in many places to cause the drying up of many karezes, springs, and wells.

Traditionally, wells are dug by hand using shovels and pickaxes. Digging wells is a fairly laborious work that can take days or weeks if the deeper ground is hard and resistant. The sides of the well are reinforced by precast concrete rings and wherever the water table goes down, the wells can be dug deeper to reach the water table. However, with new technology, people have also started drilling wells for irrigation purposes wherever possible. Drilled wells or tube wells generally go deeper than the traditional wells. Tube wells are mostly used in more arid areas where the water level is very deep.

(A)

(B)

FIG. 8.16 Drawing and photograph of the well-known Persian wheel or *rahad* (*rahat*) method of ancient irrigation wherein a pair of draft animals walks in an endless circle dragging a long pole behind them that turns the central horizontal wheel suspended on a vertical axis, which in turn with gear teeth turns a vertical wheel with buckets that lifts ground water out of a shallow well and dumps it into a nearby irrigation ditch. (A) 19th century drawing of irrigation wheel obscures the means by which the bullocks are turning the central horizontal wheel. (B) Photograph of animals dragging white log (center) that turns the central toothed wheel. A buried axle from the central wheel turns the vertically mounted bucket wheel in the right, which causes the suspended buckets to pick up water and thereby fill the irrigation ditch. *Photograph by Sreeya Sen, by permission from ICIMOD, Kathmandu, Nepal.*

In some areas where the water level is high, people also use an ancient system of irrigation, the water wheels, to lift up water to the surface with the buckets attached to the wheel. In these shallow wells, a so-called "Persian" wheel, or *arhat* as locally called in Afghanistan dialect (*rahat* in many other places) (Fig. 8.16), the wheel is located right in the middle of the shallow well and rotated with a spindle attached to another wheel on the surface away from the well which is rotated by an animal (a bull, water buffalo, or a camel) walking around it. This system, however, is not common throughout Afghanistan. Some places in eastern Afghanistan have water wheels that are used to lift up water from wide shallow wells.

## 8.5 HISTORICAL PHASES OF IRRIGATION DEVELOPMENT

The historical development of irrigation in Afghanistan shows a number of significant changes beginning with simple community initiatives, to government assistance, to international aid; all interwoven with various phases of collapse during wartime, and restoration during the intermittent times of peace (Table 8.2). The informal traditional irrigation systems can be subdivided into small, medium, and large-scale systems depending upon their size, which is also partly dependent upon their geographical location in Afghanistan (Ziaee et al., 2015). The formal and modern irrigation systems are those built in the past 70 years or so that were constructed using modern engineering methods (Table 8.3).

In general, the large formal irrigation projects in Afghanistan have been underwritten financially and undertaken with plentiful assistance from foreign donors because of the necessity for foreign aid and the use of foreign-trained engineers (see Chapter 9 on dams for a review of a number of these projects). Although the use of foreign funding and outsider engineering has been regarded as essential by a succession of governments in Afghanistan,

TABLE 8.2 Development Phases of Irrigation in Afghanistan

1. *Community-Led Development Era* (~5000 years ago–1879). The largely historic time of farming in antiquity up to the late 19th century, wherein local farmers through their own efforts developed indigenous irrigation systems using locally available materials by mobilizing their own resources, and using social norms for allocation and distribution of water to form a traditional irrigation structure as an integral part of Afghan society
2. *State-Supported Development Era* (1880–1979). Under the new reign of Abdur Rahman Khan ("The Iron Amir") support was extended from the state for expansion of irrigation facilities and development of hydropower. Funding came from the state and foreign sources
3. *Destruction Era* (1980–89). The communist takeover and Soviet invasion led to the outright destruction or severe degradation of the developed irrigation infrastructure, along with the total collapse of governance and institutional systems, and exodus and death of competent human resources
4. *Restoration Era* (1990–2001). Post-war recovery efforts of traditional irrigation systems with a few unplanned rehabilitation efforts through the United Nations (UN) and some nongovernmental agencies (NGOs) from foreign aid sources
5. *Post-Conflict Rehabilitation Era* (2002–the present time). Establishment of the transitional new government has led to more formal and planned rehabilitation of irrigation with plentiful outside assistance from UN and NGO sources, along with collateral objectives of poverty alleviation, agricultural intensification, and/or diversification and capacity development among both government professionals and local farming communities

*Courtesy of Ziaee, F., Suman S., SayedSharif, S., 2015. Water wisdom and sustainability: insights from irrigation systems in Afghanistan. In: International Commission on Irrigation and Drainage; 26th Euro-Mediterranean Regional Conference and Workshops; Innovate to Improve Irrigation Performances; 12–15 October 2015, Montpellier, France.*

TABLE 8.3 Insights From Afghanistan Irrigation Systems

1. Traditional irrigation systems more sustainable than modern, so that key traditional elements (resource mobilization, conflict resolution, equitable water distribution) inherent in Mir-Aab system should be preserved and built upon in the modern irrigation systems
2. Institutional strengthening is essential component of any irrigation development endeavor and must be included in all subsequent irrigation development works
3. Systematic technical assessment of all problems and consultation with Mir Aabs and farmers is best approach for identifying and planning irrigation projects, on the basis of river basin and sub-basin planning units, with systematic survey and assessment prior to priority selection and formulation of specific projects in order to ensure preservation of traditional water rights and allocations, as well as minimization and mitigation of downstream impacts and conflicts so that proposed solutions are sustainable
4. Cost-effective and labor-intensive methods for designs of irrigation works are most suitable in rural Afghanistan context, with every opportunity utilized to provide skills training to villagers during construction to improve community capacity and enable improvement and sustainability of preventive maintenance
5. Rehabilitation projects are best formulated, designed and implemented in consultation and participation with communities, local farmers, and Mir Aabs, and best implemented, supervised, and monitored at provincial and local levels
6. Government and donors should not pay for work traditionally done by farmers and *hashar* cleaning and maintenance of canals and karezes, because otherwise the traditional methods are undermined and become unsustainable in future

*Courtesy of Ziaee, F., Suman S., SayedSharif, S., 2015. Water wisdom and sustainability: insights from irrigation systems in Afghanistan. In: International Commission on Irrigation and Drainage; 26th Euro-Mediterranean Regional Conference and Workshops; Innovate to Improve Irrigation Performances; 12–15 October 2015, Montpellier, France.*

the result in the traditional and most conservative agrarian society has also been an all-to-common failure when normal maintenance has been dismally omitted through ignorance or lack of continued funding. In a number of cases, with the Helmand Valley Project being the most glaring example, decades of American foreign aid have largely come to no good end as cultivation of opium poppy is rampant, and electricity production at Kajaki Dam has been curtailed.

## 8.6 CONCLUSION

Currently, Afghanistan has 7.5 million ha of cultivatable land, of which 3.8 million ha are irrigated. Currently about 65% of the irrigation network in the country still follow the Mir-Aab system of water management, although problems with corruption and mismanagement of the traditional system have emerged in the post-war years. New WUA are being developed in the country as a part of the new IWRM in an attempt to augment traditional methods with more transparent and formalized systems to improve water management in the country of Afghanistan.

## References

Amin, S.A., 2015. S.J. Ahmadzai, Interviewer.

Garner, D.A., Svendsen, M., Assifi, T., Mickelwait, D.R., Viala, E., Aini, A., Waiss, H.A., 2002. Assessment of the Irrigation Sector in Afghanistan and Strategy for Rehabilitation. Development Alternatives, Inc, Bethesda. Retrieved 2015, from http://pdf.usaid.gov/pdf_docs/Pnadg628.pdf.

Mack, T.J., Chornack, M.P., Taher, M.R., 2013. Groundwater-level trends and implications for sustainable water use in Kabul Basin, Afghanistan. Environ. Syst. Decis. 33, 457–467. Retrieved from Springer: springerlink.com.

Ministry of Agriculture and Irrigation, 2006. National Report of Islamic Republic of Afghanistan on the Implementation of United Nations Convention to Combat Desertification (UNCCD). MoAg, Kabul.

Mohmand, R., 2011. An Overview of Current Water Distribution System of Khawajazai Canal (Intake) in Goshta District of Nangarhar Province, Afghanistan. Van Hall Larenstein University of Applied Sciences, Leeuwarden, Netherlands.

Pedersen, F.S., 2009. Sustainable agricultural production: providing an alternative to opium in Afghanistan. Master Thesis, Aalborg University.

Rout, B., 2008. How the Water Flows: A Typology of Irrigation System in Afghanistan. Afghanistan Research and Evaluation Unit (AREU), Kabul. Retrieved from: http://www.areu.org.af/Uploads/EditionPdfs/811E-Typology%20of%20Irrigation%20Systems.pdf.

Sadat, S.H., 2013. Spate Irrigation in Afghanistan. Spate Irrigation Network. Retrieved from: http://www.spate-irrigation.org/wordpress/wp-content/uploads/OP10_SI_Afghanistan_SF.pdf.

Savage, M., Dougherty, B., Hamza, M., Butterfield, R., Bharwani, S., 2009. Socio-Economic Impacts of Climate Change in Afghanistan. Stockholm Environment Institute, Stockholm. Retrieved from: https://static.weadapt.org/placemarks/files/1092/5345354491559sei-dfid-afghanistan-report-1-.pdf.

Sexton, R., 2012. Natural Resources and Conflict in Afghanistan: Seven Case Studies, Major Trends and Implications for the Transition. Afghanistan Watch, Kabul. Retrieved from: http://www.watchafghanistan.org/files/Natural_Resources_and_Conflict_in_Afghanistan/Natural_Resources_and_Conflict_in_Afghanistan_Full_Report_English.pdf.

Shroder, J.F., 2014. Earth Magazine. American Geosciences Institute, Virginia, USA.

ter Steege, B., 2007. Infrastructure and Water Distribution in teh Asqalan and Sufi-Qarayateem Canal Irrigation System in the Kunduz River Basin. The Open Access Publication Server of the ZBW-Leibniz Information Center for Economics. Retrieved from: http://hdl.handle.net/10419/88342.

Thomas, V., Ahmad, M., 2009. A Historical Perspective on the Mir Aab System: A Case Study of the Jangharoq Canal, Baghlan. Afghanistan Research and Evaluation Unit, Kabul. Retrieved from: http://www.areu.org.af/EditionDetails.aspx?EditionId=214&ContentId=7&ParentId=7.

Ward, F.A., Amer, S.A., Ziaee, F., 2012. Water Allocation Rules in Afghanistan for Improved Food Security. Retrieved from: http://agecon.nmsu.edu/fward/water/Food%20security%20afghan%20water%20allocation.pdf.

Wegerich, K., 2009. Water Strategy Meets Local Reality. Afghanistan Research and Evaluation Unit, Issues Paper Series. Kabul, Afghanistan Research and Evaluation Unit (AREU).

Ziaee, F., Suman, S., SayedSharif, S., 2015. Water wisdom and sustainability: insights from irrigation systems in Afghanistan. In: International Commission on Irrigation and Drainage; 26th Euro-Mediterranean Regional Conference and Workshops; Innovate to Improve Irrigation Performances; 12–15 October 2015, Montpellier, France.

CHAPTER

# 9

# Dams in Afghanistan

A.K. Stewart

St. Lawrence University, Canton, NY, United States

OUTLINE





*Transboundary Water Resources in Afghanistan*
http://dx.doi.org/10.1016/B978-0-12-801886-6.00009-4

## Abstract

Afghanistan is on the cusp of renewal having just completed its first free and democratic transfer of power in 2014. Over 40 years of war and strife, however, have left infrastructure and human capital in serious need of support. Since the 2001 defeat of the Taliban by the US-led coalition, partner nations and non-governmental agencies have focused their efforts on humanitarian and infrastructure development. One key infrastructure development opportunity, appearing to provide large-scale and long-term benefits is water-resources infrastructure. The infrastructure of choice in Afghanistan and all developing nations is centered on building dams. In an effort to speed up development and provide a more stable water resource, large dams, from a western perspective, appear to be a panacea for kick-starting development. In order to best assess Afghanistan's ability to leverage such large and expensive structures, baseline data of water infrastructure are needed. Due to the lack of institutional records and record keeping, it is unclear exactly how much water infrastructure exists in Afghanistan, so in order to provide independent authentication of available data, a three-tiered effort was employed to gather as much data as possible regarding dams and diversions in Afghanistan: (1) Google-Internet search; (2) library research at the University of Nebraska, Omaha's Arthur Paul Collection of Afghan materials (in English); and (3) satellite imagery analysis using either Google Earth/Maps or TerraServer. From this extensive survey, 62 existing dams spanning >9000 m of combined crest length, 109 major diversions totaling over 500 m in length diverting some 23% of affected rivers and 104 dams proposed by Afghanistan and partner nations were identified. Every province in Afghanistan is represented with Ghazni Province having the most existing dams ($n=11$), Paktia and Wardak provinces have the greatest proposed dams ($n=9$ each) and Kabul Province had the most diversions ($n=23$). This document focuses on four main existing dams, Bandi Sultan, Dahla, Kajaki, and Salma Dams, which represent current large-dam infrastructure and associated problems. Additionally, the proposed large dams of Almar, Baghdara, Bakhshabad, and Kamal Khan are presented as representative of the future of dam building in Afghanistan. No dam occurs in Afghanistan, existing or planned, that is not currently plagued with serious issues of maintenance, operation and/or funding, and that is not part of the build-neglect-rebuild cycle that prevents any of these structures from actually providing as planned.


**Keywords:** Bandi Sultan Dam, Dahla Dam, Kajaki Dam, Salma Dam, Tangi Azao Dam, Tangi Shah Dam, Water diversions, Dam-crest length, Dam-maintenance issues, Water infrastructure

## 9.1 INTRODUCTION

After nearly 40 years of warfare and strife, Afghanistan is finally on the cusp of renewal, reform, and prosperous growth by having undergone her first democratic and peaceful transfer of power (IROA, 2014). As a result of 13 years of nation building supported by the US-led coalition and other international partners, Afghanistan must begin caring for itself and preparing to care for future generations. Thanks to increased security and quality of life in Afghanistan, the population is growing significantly; projected to be the 29th most populous country, surpassing even Canada and Italy, by 2050 (USCB, 2015). More specifically, the population of Afghanistan has grown from about 11 million in the late 1980s to over 32 million in 2015, with projections to over 63 million by 2050 (USCB, 2015) representing over 160 people/km$^2$ (Roser, 2014). Overall, population hotspots will center on the major urban areas, such as Kabul, which currently retains some 26% of the population of Afghanistan (54% of the urban population, WBG, 2015). Despite increased urban migration, Afghanistan will likely remain mostly rural, with currently approximately 74% living rural, subsistence life ways with nearly 60% of the population economically active in agriculture (FAO, 2015).

Having just begun to emerge from decades of conflict, along with a rapidly increasing population and quality of life, water resources in Afghanistan will only become more stressed with expected water availability dropping some 50% by 2050 under current usage. Currently, and likely well into the future, agriculture is the prime consumer of available freshwater in Afghanistan (98%, cf. 70% world average; Alim, 2006) with a full 85% of this water sourced from the major rivers and streams irrigating only half of the 12% arable lands (Alim, 2006).

Afghanistan is an arid to semi-arid country with a wide distribution of precipitation from a low of 75 mm/year in Farah Province to over 1100 mm/year in Parwan Province (Favre and Kamal, 2004). Precipitation is temporally limited to the winter and spring months with summers and falls seeing little to no precipitation. Overall water resources are limited by the amount of winter snows in the high mountains and the ensuing spring melt. Little snow and fast break-up leads to a very limited runoff time; however, lots of snow with a slower spring break-up can lead to extended spring-early summer runoff. On average, Afghanistan receives approximately 327 mm rain/year generating more than 200 billion $m^3$ of freshwater leading to about 75 billion $m^3$ (75 $km^3$) of practical, available freshwater (Qureshi, 2002; FAO, 2013); providing over 2000 $m^3$/person/year (cf. 1400 $m^3$/person/year for Iran or 1200 $m^3$/person/year for Pakistan, Qureshi, 2002). Afghanistan is currently underusing its water resources with 40% of this surface water potential flowing out of the country unused (Saffi and Kohistani, 2013). Once water enters the agricultural system of irrigation ditches, 75% of irrigation water is lost due to inefficiencies in poor quality and/or improperly maintained systems (Alim, 2006). In sum, over 85% of Afghanistan's surface-water resources either flow beyond her borders or are lost to evaporation in poorly maintained water infrastructure making Afghanistan a water-insecure nation, despite its relative abundance of freshwater (Habib et al., 2013).

In order for Afghanistan to overcome both the rapid increase in population and the spatially and temporally erratic precipitation and irrigation losses, infrastructure development is key (eg, Brown and Lall, 2006; Bosshard, 2012; Tortajada and Biswas, 2014). Because Afghanistan's problem with water is not necessarily volume, but spatial and temporal distribution, retention of spring run-off via water-control structures (eg, dams and diversions) is necessary to promise development (Brown and Lall, 2006). Afghanistan is not without water-control infrastructure, however; in fact, it underwent significant, first-rate development projects from the 1930s to the 1970s country wide. By the late 1970s, three large-scale water-control systems had been implemented: Helmand-Arghandab valleys in the southwest (ie, Kajaki and Dahla Dams and the Boghra Canal); the Jalalabad area in the east; and the Kunduz-Khanabad in the northeast (Favre and Kamal, 2004). Just about the time these irrigation networks were fully online, the Soviet Red Army invaded. As a result of the Soviet-Afghan War (1979–89), over 3000 irrigation schemes were destroyed by the Russians (27–36%, Formoli, 1995; Yusufi, 1988) with the remaining falling into prolonged neglect (Favre and Kamal, 2004). The intervening years, from 1989 to the 2001 defeat of the Taliban by the US-led Coalition, continued to leave this infrastructure in disrepair. It is likely, despite the negative effects from this time, that after 30–40 years, irrigation schemes in Afghanistan would be in severe need of refurbishment or replacement due to lack of long-term, maintenance funding (Ward, 2010). Since 2001, significant international expertise and monetary support were provided to help reconstruct Afghanistan and bring these, and new, water-infrastructure programs to working order (see Lashkaripour and Hussaini, 2008; Habib et al., 2013; Ahlers et al., 2014).

Of all these infrastructure efforts implemented in Afghanistan over the past half century only dams (large dams, especially) are contentious at all scales (cf. WWF, 2003; Tullos et al., 2009; Ansar et al., 2014). Because Afghanistan sits atop the regional water sheds, its relatively few dams ($n = 62$, approximately 20 large dams; cf. 500 Iranian dams; Zafarnejad, 2009) have been causing transboundary water issues since the damming of the Helmand River at Kamal Khan some 1000 years ago. Transboundary water issues will only multiply as Afghanistan attempts to develop her water infrastructure through the building of more than 100 new dams with at least 20 falling into the large dam category; many affecting currently unaffected/virgin stretches of river entering her northern neighbors (eg, Uzbekistan, Tajikistan). Baseline water-resource infrastructure data are needed in order to best evaluate the impact of what could be characterized as the indiscriminate damming of Afghan rivers. This chapter begins to provide information on existing dams, diversion schemes and proposed dams.

## 9.2 BACKGROUND

### 9.2.1 Water-Resources Planning and Development

Managing human water needs by means of infrastructure likely began many thousands of years ago when humans first realized they could "control" water to increase availability. As small tribes began coalescing for mutual support and security, in nascent community centers, a more formalized need of water management arouse. Likely, for thousands of years, until about the mid-20th century, water-resource planning was about engineering solutions (Goodman, 1984), providing maximum access to water at a particular site. Since the enactment of the Flood Control Act of 1936 in the United States of America, inclusive of a cost-benefit analysis requirement, water-resources planning has become holistic in its approach (Goodman, 1984). This holistic approach recognizes that water needs are not only a community-level concern, but one that impacts both the riparian "owner" of a particular stretch of water, but also those affected by their actions downstream (or upstream, in the case of reservoirs). The inclusion of economists, planners, ecologists, sociologists and other environmental and social scientists, to the core engineering concerns, has made water-resource planning much more complex. This complexity, however, results in a "plan" that is cognizant of four-dimensional space, inclusive and understanding of community social and physical needs and expectations and the environment's needs (Linsley et al., 1991). This holistic plan is an effort to maximize water allocation to the community in question and to "guarantee" continued access to water based on projected needs. Water-resources planning and development is an essential element to human development (cf. Tortajada, 2014) and is a required tool for developing nations (eg, WWF, 2003; Grey et al., 2013; Troell and Weinthal, 2013).

The goals of water-resources planning and development are met by structural (eg, dams) and/or nonstructural (eg, restrictive zoning) measures. Typically, projects may consist of one or more structural and nonstructural measures, which depend on, not only technical effectiveness, but also other standpoints (eg, economic or financial) (Goodman, 1984). The goal too, is to not focused on one project for one concern, but to broaden the functionality of the effort to the basin level and beyond (eg, electrification via hydropower beyond the basin's extent). The purposes and functions of water-resources planning and development are many, such as: (1) water supply for municipal, industrial and agricultural needs; (2) flood control and damage prevention; (3) hydroelectric power; (4) commercial fishing; and (5) recreation (Goodman, 1984). In addition to these "economically driven" purposes and functions, environmental concerns are also warranted, such as: (1) retention of natural water and flows with a focus on esthetics; (2) archeological, historical, biological and geological resources; and (3) water, land and air quality (Goodman, 1984). At the broadest, most holistic scale, the promotion of general welfare may also be used, such as: (1) regional economic development and quality of life issues; (2) health and safety; and (3) emergency preparedness.

One facet of this complex and holistic system is the implementation of dams and diversions to meet the structural needs of water management and development (eg, Tortajada and Biswas, 2014). Dams and diversion schemes have made significant contributions to human welfare and development (WCD, 2000). In an inverse relationship, their cost is high and is associated with a point-type project, which likely has the most profound reach in water-development efforts.

## 9.2.2 Dams

Dams have been part of humans' effort at controlling the environment for over 6000 years. Since this time dam-engineering technology has advanced to allow the safe operation of gargantuan earth-and-rock structures such as the Tarbela Dam in Pakistan (approximately $125 \times 10^6\,m^3$) and huge concrete-and-steel structures, such as the Three-Gorges Dam in China (approximately $27 \times 10^6\,m^3$). Despite the technological know-how to build these monuments to modernization (cf. Kaika, 2006), their occurrence is inversely related to their size, so the most prevalent and cheapest can be as tiny as rock-filled basket check dams (maybe $5\,m^3$ or less). Regardless of size, these structures attempt to facilitate both human and economic development by provision of dependable water supply for domestic, agricultural and industrial purposes, flood mitigation and/or to generate electrical power (de Wrachien and Mambretti, 2009). In order to provide these benefits, dams, especially large ones, are expected to be reasonably water tight and stable, built with the consideration of topography of the dam/reservoir area, strength and variability of the foundation, and availability and suitability of construction materials (de Wrachien and Mambretti, 2009). To promise the best structure is to incur the most expense; however, in reality, dams are built in an economically feasible way while attempting to maximize stability and safety. Despite significant advances in dam engineering and best of intentions, large dams still require huge capital investments with social, environmental, financial and safety costs mostly left underestimated (WWF, 2003; Scudder, 2005; Tilt et al., 2009).

### 9.2.2.1 *Large Dams*

The International Commission on Large Dams (ICOLD) defines large dams as those rising 15 m or more from their foundation or, if between 10 and 15 m in height, meeting at least one of the following criteria: (1) having a storage capacity of over $3 \times 10^6\,m^3$; (2) a crest length $\geq$500 m; (3) with a spillway proportional to a flood $\geq 2000\,m^3\,s^{-1}$; (4) with particularly difficult geotechnical site conditions; and/or (5) is of an unusual design (de Wrachien and Mambretti, 2009). At the turn of the 20th century, several hundred large dams occurred around the globe; however, by 1949, that number was about 5000. Since this time, the number of large dams has risen to well over 45,000 in over 140 countries; peaking with about 5000 being built between 1970 and 1975 (WCD, 2000). Despite their cosmopolitan occurrence, nearly 80% of large dams are in five countries: China (~22,000), US (6390), India (>4000), Spain and Japan (between 1000 and 1200 each) (Scudder, 2005). Of these, China has built over 99% of all their 20,000 large dams since 1949 and has recently begun exporting its large-dam expertise (McDonald et al., 2009). This rapid, large-dam development in these countries has fundamentally used all the prime locations, leading these countries to either begin developing lesser quality sites (ie, China) or to focus more on rehabilitation, renovation and optimizing function for multiple operations.

Large dams have been promoted as instruments of development to meet water and energy needs while supporting economic development. As a result, nearly two-thirds of current large-dam development is occurring in developing nations. With an average construction time of 5–10 years, current development is between 160 and 320 large dams per year (WCD, 2000). Estimated annual investments in support of these dams is estimated at $22–31 billion per year with a focus on hydropower development ($12–18 billion) and irrigation ($8–11 billion), with individual dams representing the largest infrastructure projects ever implemented in developing nations (Scudder, 2005). The promotional aspects of large-dams in developing nations is based on a development strategy by a powerful coalition of politicians, banks, engineering firms and others (Scudder, 2005). A major component of US President Roosevelt's New Deal (1933–38) was the Tennessee Valley Authority (TVA), which was a basin-scale water management program implemented as a series of large dams to raise the living standards in one of US' least developed regions (Scudder, 2005; Baghel and Nüsser, 2010). The TVA model was enthusiastically promoted as a cure for "backwardness" all over the world, leveraging its inclusion of social dimensions and public legitimacy as a grassroots-styled effort (Baghel and Nüsser, 2010). Large dams, however, are not without their problems, with researchers in the early 1960s criticizing the undesirable and unnecessary environmental and social impacts (Scudder, 2005) leading to the current large-dam debate (Scudder, 2005 and Korman, 2014).

### 9.2.3 Dam Safety

Large dams require both stability and reasonable water tightness; however, both of these are dependent upon the underlying geology of the site. When balancing cost versus stability, the location and design of a dam are carefully suited to the particular structure/geology of the site. Along these lines, the final decision for dam emplacement is founded on three general factors: (1) the topography of the dam site and reservoir area; (2) the strength and variability of the foundation; and (3) the availability of suitable construction materials (de Wrachien and Mambretti, 2009). Despite the best intentions to maximize stability and minimize cost, movement and deformation of the dam structure and foundation cannot be ignored and must be recognized or predicted and implemented in the overall design (de Wrachien and Mambretti, 2009).

Dam failure is defined by ICOLD (2015) as a "collapse or movement of part of a dam or its foundation, so that the dam cannot retain water. In general, a failure results in a release of large quantities of water imposing risks on the people or property downstream." Failures can occur when designs and reality fail to meet and/or when natural disasters strike, such as flash floods, earthquakes and mass movement, or due to intentional man-made actions such as wartime attacks or terrorism (de Wrachien and Mambretti, 2009). For example, during the period from 1900 to 1965, approximately 1% of the 9000 large dams worldwide failed, with another 2% having suffered serious accidents, which did not lead to a failure (de Wrachien and Mambretti, 2009). In Afghanistan, the dam-failure rate is 0.07%; based on one failure of 27 dams in 50 years (ie, Bandi Sultan in 2005, see below, Sadat, 2012).

Generally, dams are most likely to fail during their "infant" stage (0–5 years) when they have to adapt to stresses acting on them during a period of operational adjustment (He et al., 2008). Based on He et al. (2008) data, for example, infant dam failure rate for Chinese dams is about 72% and approximately 30% for other countries. Once a dam is beyond this "infant" stage, likelihood of failure is reduced. Because dam failures can result in significant disasters with loss of human life and property, dam safety and maintenance has come to a critical point in those countries having significant large-dam infrastructure. After the peak in dam building some 40 years ago, many dams in the Europe, Asia and North America are reaching middle age. Although data suggest that middle-aged dams are more stable than "infant" dams, their performance has decreased significantly (He et al., 2008) and the general focus of dam operations has become maintenance, refurbishment and/or removal (Lane, 2008). Even though the focus has shifted from dam building to preservation/removal of dam infrastructure, even the USA has no clear strategy to deal with these dams; let alone the funding mechanisms for such projects (Lane, 2008). As a result of the USA and Europe beginning to realize the maintenance needs of this aging infrastructure, they are beginning to promote a longer view of dam building, at all scales, in developing nations, which is beginning to include financing for long-term maintenance (eg, UN, 2007; Pisaniello et al., 2011).

## 9.3 METHODS

The ability to present the "entire" dam infrastructure in Afghanistan, or any other definable variable from agriculture and irrigation networks to roadways and bridges is, essentially, impossible. Before the westernized modernization of Afghan infrastructure and concomitant data management that supports these efforts began in the 1930s and 1940s, Afghanistan was a "graveyard of empires," (Shroder, 2014, p. 5) which underwent centuries of warfare punctuated by brief periods of quiet (Shroder, 2014). In addition to warfare, the approach to living in Afghanistan before the middle of the 20th century was one of survival based on subsistence farming, with little to no formal management of infrastructure beyond the local merab (mirab), or water manager (Wegerich, 2009; Torell and Ward, 2010).

Formal, government-sanctioned infrastructure development formally began in the early-mid 20th century with help from Westernized nations (eg, Germany, US), which resulted in a handful of large, formal dams and diversion schemes in support of water management for irrigation, power and flood control. Data on these large, externally supported dams and diversions are, more or less, readily available through general internet searches, which easily uncovers scanned documents from the original efforts. Since this mid-20th century infrastructure boom, the Afghan government has refurbished or built numerous dams, diversions, and irrigation schemes with the support from a variety of non-governmental organizations (eg, Danish Committee for Aid to Afghan Refugees, DACAAR; Aini, 2007). The near 40 years of war since this renewed development phase has seen chaos, destruction and the lack of any preservation of documentation associated with Afghanistan (cf. Yusufi, 1988; Formoli, 1995; Shroder, 2007, 2014). Not only are baseline data unavailable due to confusion and chaos, the loss of human capital due to killing and/or expatriation also severely limits "institutional knowledge," all resulting in "information" based on best guesses or estimates (Yusufi, 1988; Formoli, 1995).

In order to circumvent this "best guessing" and provide independent authentication of available data, a three-tiered effort was employed to gather as much data as possible regarding dams and diversions in Afghanistan: (1) Google-internet search for webpages, documents, and books; (2) library research at the University of Nebraska at Omaha's Arthur Paul Afghanistan Collection of materials (in English); and (3) satellite imagery analysis using either Google Earth/Maps or TerraServer.

## 9.3.1 Google-Internet Search

Using only Google Web Search, data were gathered by searching terms such as "Afghanistan dam" or "Afghanistan water" or "Afghanistan resources." Once dam or diversion names were evident, they were tallied and used as additional search words using Google (eg, "Kajaki Afghanistan dam") to increase the likelihood of finding dam-specific information. Some primary sources of information retrieved from Google Search were: (1) United Nations Environmental Program documents (eg, UNEP, 2015); (2) Food and Agriculture Organization of the United Nations data (eg, FAO, 2015); (3) Afghanistan Energy Information Center maps and documents (AEIC, 2015); (4) Islamic Republic of Afghanistan documents (eg, IROA, 2014); (5) United States Agency for International Development documents (eg, USAID, 2015); (6) United States Army Corps of Engineers documents (eg, USACE, 2009); (7) United States Geological Survey documents (eg, Perkins and Culbertson, 1968); (8) Scott's Helmand Valley Archives (SHVA, 2015); and (9) the Afghan War Diary for security-related issues (AWD, 2015). In addition to general searching via Google Search, Google Books was critical in finding out-of-print "classic" books, which provided insight into the history of particular regions and structures (eg, Vigne, 1840); Google Scholar, too, was productive, albeit in a limited way due to the general lack of scholarly literature citing "Afghanistan dams" (eg, Cullather, 2002; Ahlers et al., 2014).

## 9.3.2 Library Research at the Arthur Paul Afghanistan Collection, University of Nebraska at Omaha

With support from Dr. J.F. Shroder and Ms. Shaista Wahab, curator, complete access to the UNO's Criss Libraries, Arthur Paul Afghanistan Collection was essential to uncovering unpublished and inaccessible documents, which are only available in this collection (eg, MK, 1951a; TEC, 1956). Working with this collection, moreover, provided personal support, which is not available using the internet.

### 9.3.3 Satellite Imagery Analysis

In order to confirm published documentation and to generate a "complete" data set, a river-by-river search was completed using Google Maps with sites confirmed using TerraServer. The Google Map server is a simple-to-use interface that allows the viewer not only to view the landscape, at times to 1 m or better resolution per pixel, but also to annotate and manage site data. When imagery resolution was poor (some 2 m or more per pixel), the general area under consideration was analyzed using TerraServer, a pay-to-use site (currently $45 USD/month), providing aerial and satellite imagery generally superior (mostly 0.75 m per pixel for Afghanistan) to Google Map imagery, although at times cloud-covered.

#### *9.3.3.1 Published Sites*

Published dams, proposed dams and diversion sites were confirmed using the Google Earth (GE) platform, which allowed published maps to be used as layers (co-ordinates were rarely available for these point objects). This layering property in GE allows published maps to be georectified onto the GE image of Afghanistan and made partially transparent. Once the published site location was determined, usually by locating a map symbol such as a dot or triangle (eg, PISU, 2005), a regional search of the non-layered imagery was completed. This regional search was, at a minimum, some 50 $km^2$ due to the size of the projected map symbol; including the imperfections in georectifying the image in GE, the search area likely increased to 100 $km^2$ or more, depending on the riverine density in the area of concern.

#### *9.3.3.2 Unpublished Sites*

By searching thousands of kilometers of rivers and streams in Afghanistan using Google Map (and TerraServer) "holes" in the published data were filled. This intensive search was not comprehensive, but representative following a logical regression from large, main-trunk streams (eg, Helmand) to minor tributaries (eg, Kaj). The focus of efforts in the tributary streams was centered on rivers that had good imagery and were associated with human settlement. By manually searching imagery, the dam and diversion data set was "completed"; however, because of the lack of known published information about these "found" sites, site names are after the closest, map-named community and metadata were restricted to measurements made with the imagery available (eg, better quality imagery resulted in better quality measurements). No effort was made to re-focus attention to "fill in" blank spots on the map (eg, the Wakhan Corridor of Badakhshan Province) and, for this reason, the distribution of data across Afghanistan is not skewed, but, more likely, representative of the country.

### 9.3.4 Data Management

In order to provide a spatial view of the data, they were imported into Google Maps, Google Earth and ArcGIS using.csv files of metadata generated using Microsoft Excel. Metadata for dam sites included, for example, location, dimensions, province, river basin, river and even a shortened link to imagery stored on the TerraServer server. Diversion metadata also included approximate percent of the river blocked by the diversion scheme and approximate water diversion (based on a 2-D view).

## 9.4 HELMAND RIVER WATER DISPUTES BETWEEN AFGHANISTAN AND IRAN (TABLE 9.1)

The Helmand River drains approximately 262,000 km$^2$ or 41% of Afghanistan. The top of the watershed begins in the Kohi-Baba Range about 50 km west of Kabul and flows in a southwesterly direction for approximately 1300 km toward the Afghan-Iranian border where it branches as distributaries into the landlocked Sistan (Seistan) depression (USACE, 2004). The river flow is disrupted along its journey by two major dams, the Kajaki and Grishk hydrodams with increased disruption likely due to the remaking of the ancient dam at Kamal Khan in Nimruz Province near the Iranian border. In addition, its flow is diverted by more than 15 major irrigation diversions with the Dahla Dam disrupting its major tributary, the Arghandab River.

TABLE 9.1 Generalized Chronology of Afghanistan-Iran Transboundary Agreements Associated With the Helmand River

| Date | Agreement | Comments |
|---|---|---|
| Mar. 4, 1857 | Treaty of Peace between Afghanistan and Persia | After the 1856 capture of Herat by the Persians over the sovereignty of the Sistan Basin (and waters), this treaty proposed that all future conflicts be mediated by the British government |
| 1872 | Goldsmid Arbitration | This arbitration dealt with the division of Helmand waters by adjusting the frontier boundary between Afghanistan and Persia with a statement that neither country would interfere with irrigation-water supply by developing works on the Helmand. No specific mention of water allocation was made |
| 1903 | McMahon Commission | To ameliorate current disputes and provide more time for in-depth analyses to water allocation, McMahon proposed border adjustments, which were accepted by both countries |
| 1905 | McMahon Award | Based on 35 years of hydrological, meteorological and water-use data (1870–1905), this award allotted 1/3 (435,000 acre-feet) of Helmand water entering at Kamal Khan Dam to Iran during a "normal" year |
| 1931 | Afghan adjustment of the McMahon award | After Iran's eager development in its portion of the Sistan in the 1920s, a series of droughts prompted additional water concerns. In order to ameliorate this problem, Afghanistan kindly adjusted the McMahon award by increasing the Iranian share to 50% of the Helmand waters reaching Kamal Khan Dam |
| 1931–39 | Failed attempts at international agreements | Despite Afghanistan's neighborly increase of water share in 1931, Iran continued to push the issue by expecting water allocation to be pushed upstream from the Kamal Khan Dam *and* for Afghanistan to restrict diversion development upstream. Lack of agreement internally in Afghanistan prevented any formal agreements |

TABLE 9.1 Generalized Chronology of Afghanistan-Iran Transboundary Agreements Associated With the Helmand River—cont'd

| Date | Agreement | Comments |
|---|---|---|
| 1948 | Delegation to Washington, DC | Iranian concerns over the development of the Boghra Diversion and the Kajaki and Dahla Dams upstream of Kamal Khan Dam led them to seek assistance from the USA to mediate. An Afghan–Iranian delegation met in Washington, DC in early 1948, but negotiations were futile because the Iranian delegate lacked any authority to make agreements |
| 1949 | Creation of the Neutral Technical Commission (later known as the Helmand River Delta Commission) | Delegates from Afghanistan and Iran met again in Washington, DC, which led to the creation of a Neutral Technical Commission to make field surveys in each country to provide facts, conclusions and recommendations as to the water supply issue. One field excursion per country was completed to better understand the problem using scientific and engineering expertise |
| Sep. 7, 1950 | Neutral Technical Commission terms of reference | The terms of reference and interpretive statement from the commission was submitted and agreed upon by both parties and included recurring assessments to monitor changes in the system. Agreement never ratified |
| 1959 | Delegation to Washington, DC | With the completion of the Kajaki and Dahla Dams, tensions were revived, which lead to another delegation to Washington, DC for mediation. These negotiations failed |
| 1973 | Sistan region Draft Agreement | For over 10 years, disputes continued until both countries prepared a draft agreement involving water use in the Sistan region. This treaty, too, was never ratified and failed |
| | | According to this aborted treaty, Iran was to receive 22 $m^3/s$ of Helmand waters as its share with the option to purchase an additional 4 $m^3/s$ from the Afghan government, which has become the accepted allotment for Iran |
| 2003–05 | United Nations Environment Program support | Afghanistan and Iran exchanged information regarding the Sistan region in each country |
| 2010–13(?) | Afghan-Iranian Helmand River Commission | Both countries met quarterly to share information on flood control of the Helmand River |

*Data from Sulaiman, M.S., 1957. The Irano-Afghan dispute over the Helmand River: a study of its legal aspects and implications. Unpublished Master's Thesis, George Washington University, 195 pp.; Favre, R., Kamal, G.M., 2004. Watershed Atlas of Afghanistan: 1st Edition—Working Document for Planners: FAO and Afghanistan Information Management Systems (AIMS), 242 pp.; UNEP (United Nations Environment Program), 2013. Natural resource management and peace building in Afghanistan: United Nations Country Team Brief, 85 pp. http://www.unep.org/disastersandconflicts/portals/155/countries/Afghanistan/pdf/UNEP_Afghanistan_NRM.pdf.*

The history of the Helmand Valley goes back some 3000 years to when it was called *Haetumant* (abounding in dams) during the Avestan age. Since this time, medieval geographers called the region the "garden spot of Asia" because the waters of the Helmand were harnessed by a system of diversion dams and canals, thus providing water much beyond its riparian extent (Sulaiman, 1957). During the Mongol invasions, (AD 1218–21), much of the irrigation infrastructure was destroyed and the "garden of Asia" quickly returned to its austere, desert conditions (Sulaiman, 1957). The complex history of the basin resurged during the Perso-Afghan conflict, which was partially fought over access to Helmand River waters by Persia (called Iran after March of 1935). In 1856, Persia incurred war with England by seizing the city of Herat over the dispute of the sovereignty of the Sistan (Seistan) basin. As the result of the treaty of peace signed on Mar. 4, 1857, all future conflicts between Afghanistan and Persia (Iran) were to be mediated by the British government. A first effort at resolving the water allocation of the Helmand between Afghanistan and Iran was a failure, despite a tripartite commission of the British, Afghans and Iranians (Sulaiman, 1957).

Because of this failure, the foreign office of Great Britain sent General Goldsmid to find a solution as early as 1870. The Goldsmid arbitration of 1872 primarily dealt with boundary delineation between the two countries and a statement that no works were to be carried out on either side of the border, which would interfere with the irrigation-water supply of the Helmand River. This was the first international regulation of boundary waters in the Middle East (Hirsch, 1956). Because this solution was boundary related and not associated with water allotment, no real solution to the water needs of either country was found.

From 1872 to 1895, problems arose, but were settled locally by both Afghan and Iranian provincial governors who were well aware of each other's needs (Sulaiman, 1957). This all changed in 1896 when the Helmand River changed its course and water supplies became contentious again. At this time, Iran again asked for help from Great Britain to help arbitrate the current disputes. Great Britain appointed Sir Henry McMahon as the commissioner to settle the Perso-Afghan dispute over the apportionment of Helmand River waters (Sulaiman, 1957). As a relatively quick fix, he proposed border adjustments, which were accepted by both countries in 1903. This provided time for a more in-depth analysis of the water-resource problem in the Sistan. Extensive investigations were completed by the Perso-Afghan Boundary Commission dealing with agriculture, economics, meteorology and ethnological information relative to the Iran-Afghan Sistan (Sulaiman, 1957). The report was based on scientific evidence associated with water and cultivated and uncultivated lands on both sides of the border. The compromise by McMahon was presented in 1905 and was to follow Goldsmid's arbitration regarding boundaries; however, in addition it specifically provided allocation percentages of the Helmand River waters to both Afghanistan and Iran. Based on 35 years of data (1870–1905), evidence suggested that the Iranian Sistan was amply supplied with water with only 3 years being in deficit. As a result, the McMahon award provided one-third to Iran and two-thirds to Afghanistan of all water reaching the Kamal Khan Dam (Sulaiman, 1957). This award, specifically, allocated 435,000 acre-feet of water to Iran in a "normal" year distributed throughout the year. The arbitration was binding and both countries, generally, accepted the decision.

During the 1920s, Persia (Iran) proactively pushed irrigation projects in their region of the Sistan (Seistan) but a series of droughts in the early 1930s prompted additional concerns about water distribution. In 1931, an effort to ameliorate the problem, the Government of

Afghanistan broke from the McMahon agreement of one-third, two-thirds apportionment to provide additional waters to the Iranians in a new, unofficial and rescindable, 50:50 split of waters arriving at the Kamal Khan Dam. Despite this neighborly offer of additional waters, Iran continued to push the issue, asking for the 50% rule to be started upstream of Kamal Khan Dam along with restrictions on Afghan diversions upstream of Kamal Khan. No agreement was formally accepted at this time either (Sulaiman, 1957). Up to 1939 additional efforts were attempted, but no treaty was formally ratified because Afghans could not come together, refusing to sign the treaty (Favre and Kamal, 2004). In 1939, with the onset of WWII, water issues were of lessened priority.

After WWII, water access issues returned to the forefront because of a once-in-a-lifetime drought in 1947 after a few years of general drought, which had the city of Kandahar without water for a few mid-year months. This drying of the Arghandab River, the primary tributary to the Helmand, significantly reduced water flow into the Iranian Sistan. As a result of recent, water-management efforts upstream of Kamal Khan Dam, the Iranian government blamed their lack of water not on meteorological concerns, but on Helmand Valley development (Table 9.2). In August of 1947, Iran gave an ultimatum

TABLE 9.2 Generalized Chronology of the Helmand Valley Development Authority

| Dates | Involvement | Comments |
|---|---|---|
| ~1000 to ~800BP | n/a | Helmand Valley archeological evidence of first settlement of the region. During the Avestan Age, it was called *Haetumant* or "abounding in dams" |
| 1218–21 | n/a | Mongol invasions destroyed much of the irrigation infrastructure and the "garden of Asia" fell into disrepair |
| 1920–40 | USA, Germany, and Japan | Multiple times until 1934, the Afghan government reached out to the USA for development support; each time, it was turned down. Work began on the Boghra canal under Afghan authority using Afghan funds |
| | | In the 1930s both Germany and Japan provided technical expertise in support of the region. The start of WWII ended this support |
| 1946–49 | USA | US involvement during this phase was unofficial and only related to the hiring of the US engineering firm, Morrison-Knudsen by the Afghan government. The primary reason for this development was to settle nomads and landless farmers through continued construction of irrigation works, especially the Boghra canal |
| 1949–53 | USA | Second phase of Morrison-Knudsen involvement, which initiated the idea, design and completion of both the Kajaki and Dahla Dams. In addition, significant maintenance was performed on previously unseen problems from earlier canal work |
| | | As part of the Point Four Program of development, in 1951, the USA became officially associated with HAVDA efforts |

Continued

TABLE 9.2 Generalized Chronology of the Helmand Valley Development Authority—cont'd

| Dates | Involvement | Comments |
|---|---|---|
| 1953–60 | USA | Despite the completion of two large dams, unseen maintenance costs were piling up due to poor planning. To improve the image of the project Morrison-Knudsen proposed a package plan to include education, health, power generation and rural development in support of the agriculture and irrigation efforts completed thus far. Despite these efforts, the project was undergoing set backs including farmers abandoning their lands |
| | | The Afghan Construction Unit was established and given the responsibility for the operation and maintenance of the canals and road projects |
| 1960–70 | USA | This was a period of study, planning, training and development, which was headed by the Bureau of Reclamation. Many scientific and feasibility studies were completed within the region. Included was the authorization for the installation of hydro-electric turbines in Kajaki Dam. By the mid-1960s, the project was beginning to make headway |
| 1970–78 | USA | US financial aid dwindle to less than $500,000 in 1973 as the result of reduced Cold War stress in the region. The deposition of King Zahir by Daoud Khan (supported by pro-Moscow communists) in 1973 prevented any real US efforts along with the end of the Bureau of Reclamation contract (and inability to draft a new one, due to the coup). By mid-1979, the USA had fully pulled out of Afghanistan after the Soviet-backed, communist takeover in 1978 |
| 1979–2001 | n/a | After nearly 25 years of warfare, chaos and turmoil, HAVDA infrastructure was either destroyed or fell into disrepair. In conjunction, prompted by Soviet soldiers, poppies (for opium production) became the crop of choice in the Helmand |
| 2001–today | US, UK, and others | After the US-led invasion of Afghanistan and the defeat of the Taliban, reconstruction efforts began in order to increase agricultural output and reduce poppy growth in the Helmand |
| | | Coalition (and NGO) efforts centered on refurbishing or replacing or repairing irrigation infrastructure and dams |

*Modified from Baron, L.Z., 1975. The water supply constraint: an evaluation of irrigation projects and their role in the development of Afghanistan. Ph.D. Dissertation, McGill, University, Montreal, 331 pp.; Azam, 1999.*

to Afghanistan to cease and desist, or else (Sulaiman, 1957). At this time, the Iranian government asked for assistance from the United States of America to mediate this water supply issue. A tripartite delegation met in Washington, D.C. in early 1948; however, negotiations were futile because of the lack of authorizations on the Iranian delegate's part. Negotiations were taken up again in the U.S. in 1949 with both parties ready to negotiate. The U.S. Department of State developed a Neutral Technical Commission (later known as the Helmand River Delta Commission) to make field surveys in each country to provide facts, conclusions and recommendations as to the water supply issue. Two field excursions were made, 1949 and 1950, pertaining to streamflow data, irrigation needs and previous arbitrations of Goldsmid and McMahon for the apportionment of the Helmand River to Afghanistan and Iran. This report reaffirmed the fact of natural variability in the system and that droughts not only happen, but also exacerbate the problems associated with water sharing. The terms of reference of this commission and interpretive statement were submitted to and agreed upon by both Afghanistan and Iran with proposed recurring assessments as the situation changes (Sep. 7, 1950; Favre and Kamal, 2004). The completion of Kajaki and Dahla Dams and the Boghra Diversion in 1950s revived tensions, which led to sending more representatives to the U.S. for mediation in 1959 (Favre and Kamal, 2004). These negotiations, too, failed with disputes continuing until 1973 when both countries prepared a draft agreement involving their water use in the Sistan region; however, this agreement, too, failed to be ratified (Favre and Kamal, 2004). According to this aborted treaty, Iran was to receive 22 $m^3/s$ of Helmand waters as its share with an additional 4 $m^3/s$ purchasable from the Afghan government (Favre and Kamal, 2004; Frenken, 2012; UNEP, 2013). This quantity of water allotted to Iran was actually less than the amount presented in the McMahon's arbitration (1/3 flow from Kamal Khan Dam). Just subsequent to this aborted treaty, the Iranian monarchy was overthrown in 1979 and Afghanistan was occupied by the Soviet Union preventing any further efforts to settle this dispute (Favre and Kamal, 2004). Recently, with the support of the United Nations Environment Program, between 2003 and 2005, Iran and Afghanistan had exchanged information on the Sistan; most recently, from at least 2010 to 2013, a joint Afghan-Iranian Helmand River Commission has met quarterly to share information on flood control of the Helmand River (UNEP, 2013).

## 9.5 HELMAND RIVER AND THE DEVELOPMENT OF THE HELMAND-ARGHANDAB DEVELOPMENT AUTHORITY (HAVDA) (TABLE 9.2)

During the Mongol invasions of Genghis Khan and Tamerlane (13th and 14th centuries), much of the irrigation infrastructure in the Helmand Valley (Fig. 9.1) was destroyed and the "garden of Asia" quickly returned to its austere conditions (Sulaiman, 1957). For centuries, this once-breadbasket of Asia was sparsely populated, and affected by varying degrees of poor soils and erosion (Bradsher, 1985). After the McMahon arbitration of 1905, Afghanistan became sensitive to the idea of having to provide Helmand waters to Iran who was beginning to out-develop its section of the Sistan Basin. As a result, Afghans approached the US government repeatedly in the 1920s for support in developing the region (Azam, 1999). The Afghan government under Nadir Shah approached the USA again in 1930 and 1933, again, to no avail (Poullada and Poullada, 1995). Despite the lack of support from the USA, the Afghan government moved forward with redeveloping the Helmand Valley with technical support from Germany (1934) and, later, Japan (1937) (PAS, 1959; Baron, 1975; Ansari, 1978). With the advent of WWII, this technical support ceased, but the Afghans continued developing the region through the war in an effort to improve living

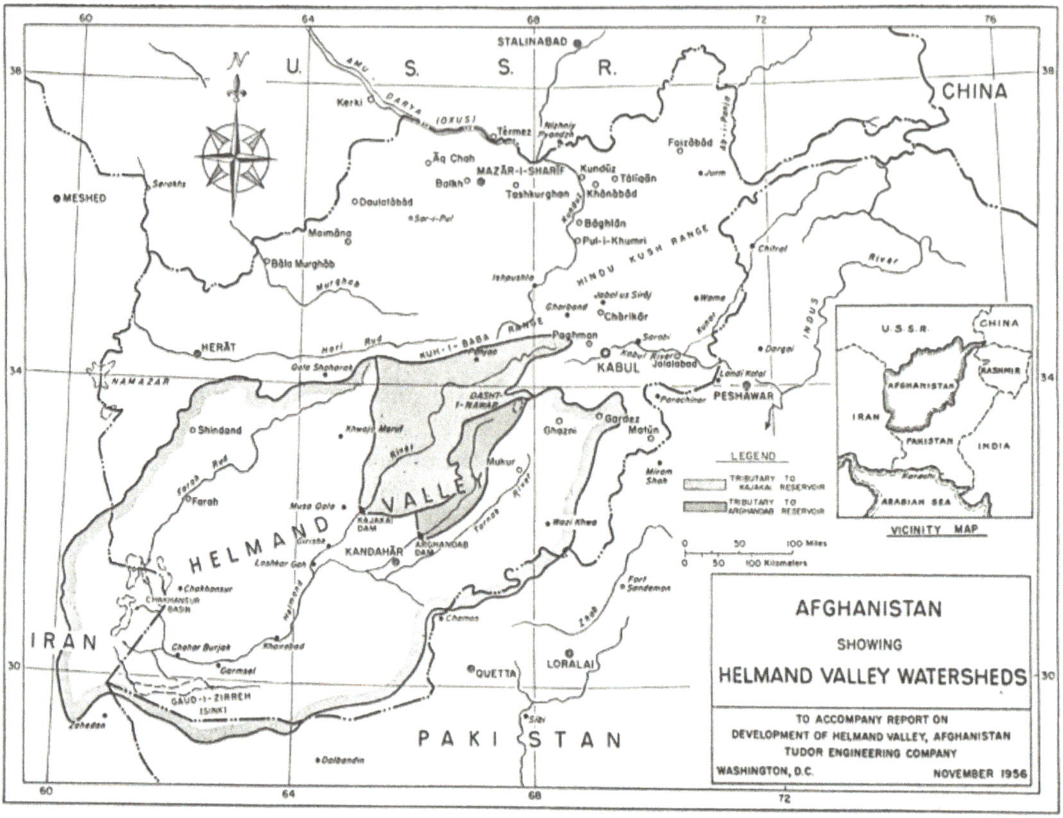


FIG. 9.1 Map of Afghanistan showing the Helmand Valley watershed (TEC, 1956).

conditions in this part of Afghanistan and, strategically, to begin centralizing power under the new king (Dupree, 1973; Chandrasekaran, 2012). Most of the money supporting this project during the war and just after, was thanks to their supplying foodstuffs for allied armies in India (Baron, 1975) and the recent diaspora of German-Jewish furriers to the USA where they found a new source for their pelts—Afghanistan and its karakul pelts (Baron, 1975; Chandrasekaran, 2012). Becoming a new source of pelts for American furriers brought in $100 million by the end of the war, which helped fund initial efforts at Helmand recovery (Chandrasekaran, 2012). After the war, the continued impression by the Afghans that Iran was out-developing their region of the Sistan (and probable need of additional Helmand waters) brought the Afghans back the USA for skills, technology and equipment to assist in construction.

## 9.5.1 Phase I (1946–49), the Preliminary Phase

In 1946, despite Afghanistan having a balanced budget and no internal or external debt, the US government, again declined Afghanistan's request for assistance (Azam, 1999). To keep the connection with the USA open, the Government of Afghanistan (GOA) hired the

US firm of Morrison-Knudsen (MK), who had good connections with the US government through regional wartime projects, to make a proposal for seven infrastructure projects (ie, major roads, electrification and irrigation) across the country, including the Helmand Valley (Gopalakrishnan, 1980). After the proposal was submitted, GOA decided to move forward with MK with a $17 million contract for the continued construction of the Boghra Canal and extension to irrigate the Nadi Ali and Marja terrace tracts and Shamalan Spur of the Boghra to begin irrigating the Helmand Delta region (Azam, 1999). The primary reason for this development project in the Helmand Valley was to settle nomads and landless farmers and increase agriculture production through the development of modern irrigation networks and land development including provisions for flood control and electric power generation (TEC, 1956). As these projects were well underway, GOA continued to request additional funds for projects to be completed by MK. During this early survey-and-construction phase, MK and GOA began to realize that the proposed irrigated land was shallow, relatively poor soils with limited drainage (Franck, 1954), and implemented cost-saving measures at the expense of the long-term efficiency and functionality of the project (USAID, 1983). After much of the $17 million was spent on American salaries, machinery and transportation costs, MK suggested that they would assist GOA in applying for a loan from the US government. It was at this point, that MK proposed that continued efforts in the Helmand Valley would be most positively affected by storing winter/spring runoff; thereby, providing for year-round water supplies. Furthermore MK proposed integrated development in the valley with dams to be constructed on the Helmand (Kajaki Dam) and the Arghandab (Dahla Dam) rivers (Baron, 1975). To this, the GOA agreed because its projected an increase in fruit-and-vegetable needs for the newly independent Pakistan and India, to bring tangible results to a huge investment already made and a recent (1946–47) drought in the region, which left many canals dry, killing vines and fruit trees (Azam, 1999).

In August of 1949, GOA formally presented the Export-Import Bank with a loan proposal for $55 million, supported by MK, which took nearly 2 years to be dispersed. During this delay, MK curtailed some if its efforts, laid off Afghan employees with only about 1/3 of the projects completed (57 km of canal) for $20 million; $11 million for US and Pakistan machinery and another $4 million of which went to work camps and equipment (Baron, 1975). The formal approval of a $21 million loan for the construction of dams occurred in November of 1949, but without funding for the needed studies on regional groundwater and soil surveys where problems were apparent (Azam, 1999). After 1949, US involvement in the region changed when the communists took over China, with the US international aid budget being used to check the progression of communism in Asia. Thus with the profits made by MK and the growing influence of communism and the Cold War in the region, the USA became formally involved with Afghanistan (Fraser-Tytler and Gillett, 1967).

## 9.5.2 Phase II (1949–53), MK's Second Contract

The first efforts of MK under this second phase, financed through the US Export-Import Bank, were to correct previously unseen problems from earlier canal work. Canal leakages, poor irrigation practices, and previously unrecognized hard pans, for example, began ruining lands in the Nadi-Ali area by raising the ground-water table from some 5 m below the surface to just 10 cm This elevated ground water reduced drainage and allowed salt to accumulate

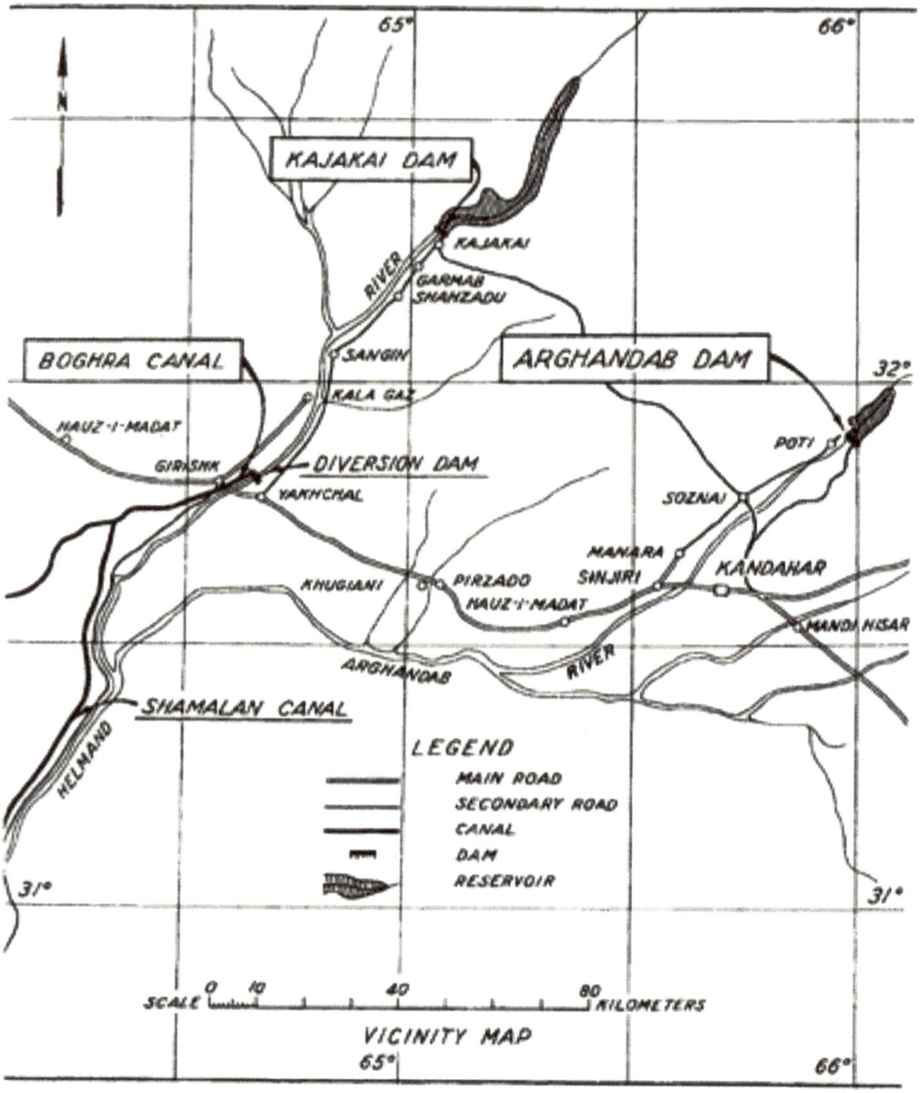


FIG. 9.2 Vicinity map of select projects within the Helmand Valley Development Authority's regional development (RGA, 1956). Kajaki and Arghandab (Dahla) Dams are mentioned in text.

at/near the surface (Maierhofer, 1961); problems that may have been recognized and mitigated earlier, if not for cost-saving measures by the Export-Import Bank (Baron, 1975).

It was also during this phase that two the two largest dams, at the time, in Afghanistan were started and completed (Fig. 9.2). Both the Kajaki Dam and reservoir on the Helmand River and the Dahla Dam and reservoir on the Arghandab River were started in 1950 and completed in 1952 for a sum of $20.48 million (Azam, 1999). Part of the drive to complete two massive dam projects simultaneously was associated with the Cold-War intentions of the USA to provide a TVA for the Hindu Kush (Cullather, 2002). These dam-building projects hoped to expand and intensify the authority exercised by the central government in Kabul (Cullather, 2002), helping to extend the reach of a more centralized government. As for the USA, these projects were a "weapon, which, if properly employed, might outbid all the social ruthlessness of the Communists for the support of the peoples of Asia" (Hamby, 1985, pp. 72–73). Monetarily, the building of Kajaki Dam was justified at the time by the future inclusion of electric-power generation intended for the establishment of a fertilizer plant (nitrogenous) where such a plant would derive hydrogen from the electrolysis of water and nitrogen from liquid air (TEC, 1956). Projected needs of fertilizer in the basin (35,000 tons) would need some 50,000 kW of electricity; the problem was, the farmers would need to be convinced to use the fertilizer (TEC, 1956). Combined, dams, reservoirs, canals and the future prospect for power in the region and continued funding from the Export-Import bank required an autonomous Helmand Valley Development Authority (HAVDA/HAVA/HVA), which was established at this time (Azam, 1999). As the idea of the TVA was implemented in Afghanistan, the US Point Four Program was established to provide technical assistance in agricultural matters in February of 1951, which formally connected the US government with the Helmand Valley Project (Azam, 1999). The Point Four Program was administered by the newly established Technical Co-operation Administration (TCA, later USAID); basically, to continue to thwart communist propaganda that "free nations are incapable of providing a decent standard of living for the millions of people in under-developed areas of the earth. The Point Four Program will be one of our principal ways of demonstrating the complete falsity of that charge" (President Truman quoted in the NYT, 1950).

## 9.5.3 Phase III (1953–60), MK's Third and Final Contract

By the middle of 1953, the $21 million loan from the US Export-Import Bank had been expended on two dams, a diversion, two long feeder canals resulting in approximately 10,000 ha of partially developed, sparsely settled lands in the Nadi Ali area (Michel, 1959). As a result of this construction, many unseen maintenance costs were becoming apparent because of poor planning. In addition to maintenance costs, the local farmers were less-than-impressed by their new farmlands and many abandoned them (Azam, 1999). To improve the image of the project, MK proposed a package plan that would add education, health, industries, power generation and rural development to the irrigation and agriculture efforts (Azam, 1999). The inclusion of these projects prompted the formal name change to the Helmand Valley Development Authority (HAVDA), which was used by the Afghans to promote the image of the project.

After these disappointing setbacks and lack of financing, the Afghan government, again, approached the US Export-Import Bank for a non-restricted loan of $36 million (Dupree, 1973). This loan was formally approved on Jun. 21st 1954 and was used for: (1) major drainage construction in the Marja, Nadi Ali and Shamalan areas; (2) diversion dams and canals in the Darwishan, South Arghandab and Tarnak areas; (3) a small hydro-electric plant (3 MW) on the Boghra Canal with distribution lines to serve Grishk and Lashkar Gah (ie, "Little America"); (4) the establishment of the Afghan Construction Unit (ACU), which was given responsibility for the operation and maintenance of the canals and road projects (and subsequent replacement of MK); (5) land development and preparation continued in Marja and

Shamalan areas; and (6) the Tudor Report (TEC, 1956), which was published as an evaluation in preparation for the US Technical Co-operation Administration (Azam, 1999).

Despite these efforts, farmers were still abandoning their lands and the monetary return on investment was only 30% of the expected (cf. $29/acre vs. projected $95/acre; Franck, 1954). Because of the lack of actual quality-of-life and social gains in a vast project meant to use modern technology to settle nomads with no farming experience, the USA was blamed for the entire project, but was unable to recuse itself because of losing prestige in the face of the Soviet Union and the on-going Cold War. The situation became so dire, that the Assistant Director of the Export-Import Bank wrote to the US Secretary of the National Advisory Council on International Monetary and Financial Problems that "... there was considerable danger that the Helmand Valley Project would fail completely" (Roberts, 1992, p. 341).

### 9.5.4 Period of Study, Planning, Training and Development (1960–73), So Many Problems

At the peak of the Cold War, the USA continued to recognize the strategic importance of the Helmand Valley for the region. In an effort to garner additional support from the GOA, the TCA assigned the Bureau of Reclamation to work in Helmand and address the land development issues (Azam, 1999). With the end of MK efforts in 1959, the project was now completely in the US government's hands. The recently established Afghan Construction Unit replaced MK with American technicians acting as advisors and trainers (Azam, 1999). As early as 1961, the American advisory role expanded to include the formal training of Afghans at the behest of the GOA in all manner of agriculture, design engineering, maintenance and operation and health and administration (Poullada and Poullada, 1995).

Headed by the US Bureau of Reclamation, many scientific and feasibility studies (eg, DOI, 1970) were completed within the HAVDA region. The United States Geological Survey undertook a hydrographic and sedimentation survey of the Kajaki reservoir attempting to understand a "forgotten" part of dam reservoir age limits; with the results of this study suggesting that in only 40–45 years (2008–13) the delta front would reach the irrigation outlet (Perkins and Culbertson, 1968). An updated study by the USGS (Vining and Vecchia, 2007a,b), extended the life of the reservoir an additional 45 years to 2057. Also associated with the Kajaki Dam was the US government's capital assistance paper supporting the proposed hydroelectric turbine additions to generate 33 MW of power (USAID, 1967). The result of the assistance paper was to authorize a $12 million loan (additional $3.4 million from GOA) to install 2 × 16.5 MW turbines, as well as associated facilities and technology and supervisory and technical training. Along with efforts at improving infrastructure and understanding the long-term viability of these irrigation structures, Owen (1970) completed a "farm economic survey" to evaluate the current state of agriculture and social improvements. As to be expected, the implementation of improved seed varieties, along with increased (or beginning) use of fertilizer showed improvement in the region. Despite the temporary gains, the net return on investment was some 30–50 cents/dollar invested, which resulted in reduced social improvements.

In the mid-1960s the HAVDA project was making headway both agriculturally and politically, with Afghanistan gradually distancing itself from the Soviet Union. Despite evidence that political efforts were making progress in Afghanistan, the USA became ambivalent and began reducing funding (Azam, 1999). This loss of funding prompted the GOA to accept a

Soviet-sponsored project to develop the Nangarhar irrigation project in eastern Afghanistan to produce olives and citrus (Azam, 1999). This infusion of Soviet influence into eastern Afghanistan prompted the USA to increase funds for HAVDA efforts. After this resurgence of interest and funding, the threat of a Soviet global offensive was no longer considered so much of a problem due to détente negotiations in the early 1970s. This reduction in Cold War hostilities reduced US interest, so, incrementally, US financial aid to Afghanistan dwindled to less than $500,000 in 1973 (Azam, 1999).

### 9.5.5 Period of Building, Planning and Change (1970–76)

The deposition of King Zahir by Daoud Khan, supported by pro-Moscow communists, in 1973, made the USA pull out fully from the HAVDA project in July of 1974. Changing regional political tensions made the USA rethink its withdrawal and within 18 months, personnel were back in the Helmand Valley again (Dalton, 1981). In 1975, the USA resumed operations in Afghanistan, but in a much reduced capacity lacking the motivation and drive seen during the 1950s and 1960s. After a 1976 farm economic survey, USAID and the Soil Conservation Service focused on a drainage project under the food-for-work program (Dalton, 1981).

With the Soviet-backed, communist takeover of Afghanistan in 1978, the HAVDA project was viewed as a strong base of American imperialism (Azam, 1999). Despite the interest of Americans to remain in Helmand during the occupation, the communists jailed or killed most US-trained Afghans who had served on the project, accusing them of being agents of imperialism (Azam, 1999); more importantly, Musa Shafiq, the last Prime Minister of Afghanistan under King Zahir Shah, was executed by the communists because of his support of the monarchy and being the signatory to the Afghan-Iran water treaty on the Helmand River. In mid-1979, the USA pulled out of Afghanistan. During the Soviet occupation, agricultural production dropped 70% due to the destruction and negligence of over 3000 irrigation works (Formoli, 1995). In this poorly irrigated landscape, along with encouragement from self-serving Soviet soldiers, opium-poppy production became a prevalent way of supporting the family (Azam, 1999).

## 9.6 RESULTS

The country of Afghanistan has at least 62 dams (approximately 20 large dams), at least 109 diversion schemes and proposals for at least another 104 dams (Tables 9.3–9.5; Fig. 9.3). Existing dam infrastructure totals over 9000 m of dam crest and is focused primarily on supporting Kabul's water and hydropower needs (Table 9.3). The Kabul River Basin has 20 existing dams totaling over 1500 m of crest length (Table 9.6) focused in the Ghazni and Kabul provinces (Table 9.3). Second to the Kabul River Basin, the Helmand Valley has 19 dams totaling over 2300 m of crest length. Thirdly, the Amu Darya Basin retains 16 dams totaling over 1400 m of crest length spread out across Badakhshan, Baghlan, Balkh, Bamiyan, and Jowzan provinces (Table 9.3).

In addition to existing dam infrastructure, at least 109 diversion schemes total over 500 m in infrastructure (Fig. 9.3). These diversions reduce overall river flow for irrigation purposes by some 22% (Table 9.5). Diversion schemes, such as dams, are primarily focused on supporting Kabul's irrigation needs with 40 existing in the Kabul basin; 23 in Kabul Province,

TABLE 9.3 List of Provinces With Existing Dams

| Province | No. | Crest-Length Sum (m) | Basin(s) |
|---|---|---|---|
| Badakhshan | 4 | 110 | Amu Darya |
| Baghlan | 4 | 278 | Amu Darya |
| Balkh | 2 | 29 | Amu Darya, Kabul |
| Bamiyan | 2 | 435 | Amu Darya |
| Faryab | 1 | 211 | Amu Darya |
| Ghazni | 11 | 1190 | Helmand, Kabul |
| Ghowr | 4 | 102 | Hari Rud |
| Helmand | 6 | 618 | Helmand, Kabul |
| Herat | 1 | 119 | Hari Rud |
| Jowzan | 3 | 113 | Amu Darya |
| Kabul | 7 | 837 | Kabul |
| Kandahar | 1 | 503 | Helmand |
| Khowst | 1 | 45 | Helmand |
| Konar | 2 | 75 | Kabul |
| Kunduz | 1 | 165 | Amu Darya |
| Logar | 4 | 272 | Kabul, Helmand |
| Nangahar | 1 | 180 | Kabul |
| Nimruz | 1 | 3500 | Hari Rud |
| Nuristan | 1 | 34 | Kabul |
| Paktia | 1 | 80 | Helmand |
| Pansjir | 1 | 11 | Kabul |
| Parwan | 1 | 10 | Kabul |
| Sar-e Pol | 1 | 144 | Amu Darya |
| Wardak | 1 | 24 | Kabul |
| | 62 | 9085 | |

alone (Tables 9.5 and 9.6). Second to the Kabul River Basin, the Helmand Basin retains 34 diversions with the Amu Darya ($n=27$) and Hari River ($n=8$) following.

Proposed dams for the country total 104 projects (Fig. 9.3; Table 9.4), which, like dams and diversions, are primarily focused on the Kabul River Basin ($n=39$) with the Helmand ($n=35$), Amu Darya ($n=19$) and Hari River ($n=14$) completing these proposals (Fig. 9.4). Unlike dams and diversions, proposed dams are better represented across the country and in provinces with no evidence of either existing dams or diversions (ie, Baghdis, Daykundi, Farah, Ghor, Kapisa, Laghman, Samangan, Takhar, Uruzgan, and Zabul; Fig. 9.3).

TABLE 9.4 List of Provinces With Proposed Dams

| Province | No. | Province | No. |
|---|---|---|---|
| Badakhshan | 2 | Khowst | 3 |
| Badghis | 2 | Lagham | 2 |
| Baghlan | 1 | Nangrahar | 1 |
| Balkh | 1 | Nimruz | 2 |
| Bamiyan | 2 | Nuristan | 2 |
| Daykundi | 2 | Paktia | 9 |
| Farah | 5 | Paktika | 7 |
| Faryab | 2 | Panjshir | 3 |
| Ghazni | 8 | Parwan | 7 |
| Ghor | 3 | Samangan | 1 |
| Ghowr | 5 | Sar-e Pol | 2 |
| Helmand | 1 | Takhar | 4 |
| Herat | 5 | Uruzgan | 2 |
| Jowzjan | 1 | Wardak | 11 |
| Kabul | 3 | Zabul | 1 |
| Kandahar | 1 | | 104 |
| Kapisa | 3 | | |

TABLE 9.5 List of Provinces With River Diversions

| Province | No. | Width Sum (m) | Flow Reduction (%) |
|---|---|---|---|
| Balkh | 3 | 11 | 50 |
| Daykondi | 4 | 35 | 22 |
| Faryab | 9 | 14 | 18 |
| Ghazni | 4 | 7 | 38 |
| Ghowr | 3 | 12 | 25 |
| Helmand | 11 | 72 | 8 |
| Herat | 5 | 33 | 12 |
| Jowzan | 8 | 31 | 20 |
| Jowzjan | 2 | 4 | 50 |
| Kabul | 23 | 99 | 13 |
| Kandahar | 4 | 35 | 30 |

*Continued*

TABLE 9.5 List of Provinces With River Diversions—cont'd

| Province | No. | Width Sum (m) | Flow Reduction (%) |
|---|---|---|---|
| Khowst | 4 | 16 | 15 |
| Kunar | 2 | 9 | 5 |
| Lowgar | 12 | 58 | 18 |
| Nimruz | 3 | 46 | 0 |
| Nuristan | 2 | 7 | 30 |
| Paktia | 1 | 4 | 0 |
| Paktika | 3 | 5 | 50 |
| Sar-e Pol | 5 | 10 | 36 |
| Wardak | 1 | 1 | 10 |
| | 109 | 509 | 22.5 |

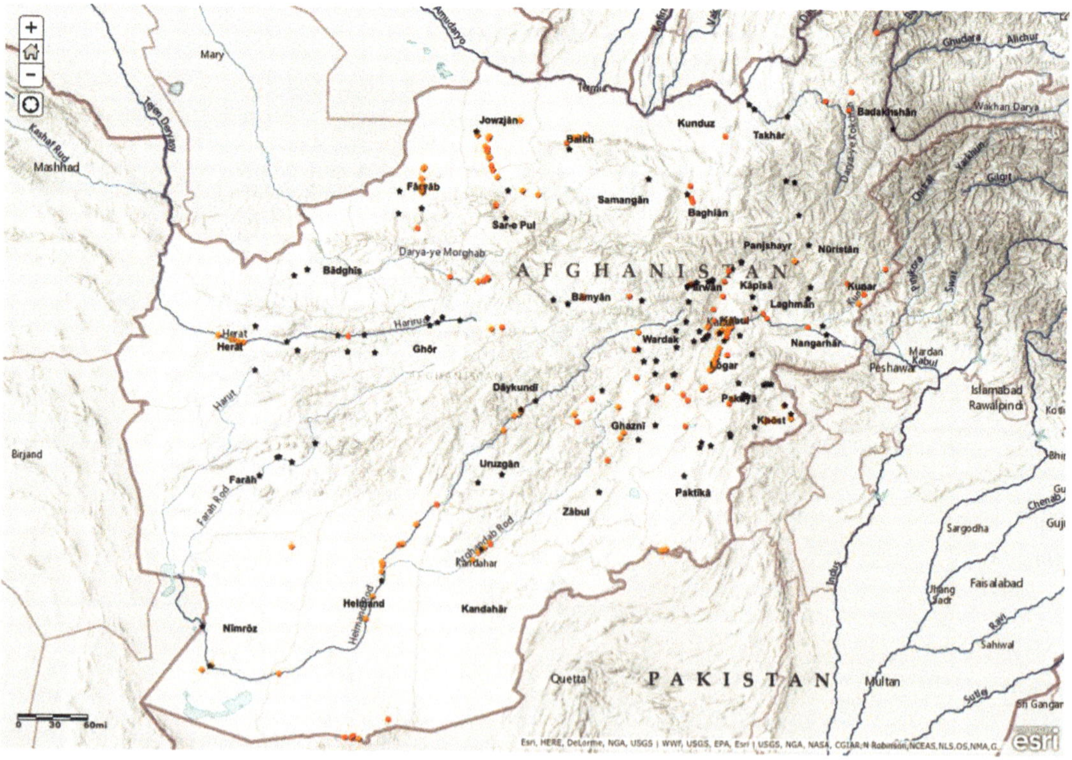


FIG. 9.3 Map of Afghanistan showing existing dams (red dots), proposed dams (black stars) and diversions (orange diamonds).

TABLE 9.6 List of Existing Dams, Proposed Dams, and Diversions by Basin

| | Dams | | Proposed Dams | Diversions |
|---|---|---|---|---|
| Basin | No. | Crest-Length Sum (m) | No. | No. |
| Amu Darya | 16 | 1479 | 19 | 27 |
| Hari Rud | 6 | 3721 | 14 | 8 |
| Helmand | 19 | 2323 | 35 | 34 |
| Kabul | 20 | 1562 | 39 | 40 |

FIG. 9.4 Map of Afghanistan with existing dams (red stars) and proposed dams (black stars) mentioned in the chapter.

Overall, Afghanistan has at least 171 dams and diversions totaling nearly 10 km of maintenance-requiring infrastructure focused on supporting the capital city of Kabul and the Helmand Basin. In order to best represent the current infrastructure and proposed infrastructure, four existing dams and four proposed dams were selected (Tables 9.7 and 9.8). These dams represent the largest and most important structures ever made (or proposed to be made) in the history of Afghanistan.

TABLE 9.7 List of Existing Dams Mentioned in the Chapter

| Name | Band-e Soltan | Dahla | Kajaki | Salma |
|---|---|---|---|---|
| Latitude | 33.757029 | 31.8496204 | 32.321944 | 34.330833 |
| Longitude | 68.380598 | 65.8884648 | 65.118889 | 63.825278 |
| Satellite image | http://bit.ly/15m5snh | http://bit.ly/15lVB0N | http://bit.ly/15mmjpX | http://bit.ly/15tUIUn |
| Length (m) | 200 | 535 | 290 | 119 |
| Width (m) | 8 | 44 | 242 | 150 |
| Height (m) | 32 | 55 | 100 | 107.5 |
| Province | Ghazni | Kandahar | Helmand | Herat |
| Basin | Helmand | Helmand | Helmand | Hari Rud |
| River | Ghazni | Arghandab | Helmand | Hari Rud |
| Elevation | 2362 | 1117 | 1016 | 1745 |
| Project Type | Storage | Storage | Hydropower | Hydropower |
| Commission date | 1901 | 1952 | 1952 | 1976 |

TABLE 9.8 List of Proposed Dams Mentioned in Chapter

| Name | Almar | Baghdara | Bakhshabad | Kamal Khan |
|---|---|---|---|---|
| Latitude | 35.777401 | 34.838111 | 32.880677 | 30.322952 |
| Longitude | 64.535254 | 69.571039 | 62.815209 | 61.881345 |
| Satellite image | http://bit.ly/195mo41 | http://bit.ly/195mYyu | http://bit.ly/195nzAi | http://bit.ly/195qwRu |
| Province | Faryab | Kapisa | Farah | Nimruz |
| Basin | Amu Darya | Kabul | Helmand | Helmand |
| Height (m) | 64 | 152 | 60 | 10 |
| Length (m) | 240 | 400 | 265 | 6700 |
| Project Type | Storage | Hydropower, 210MW | Hydropower, 24MW | Hydropower, 8.5MW |
| Estimated cost ($) | 51 million | 600 million | 500 million | 100 million |

## 9.6.1 Kajaki Dam

The Kajaki Dam is a 100-m tall, rockfill (with impervious clay core) hydropower/irrigation dam (290-m top length) impounding the Helmand River with an active capacity of 1.134 billion $m^3$ and is located in Kajaki District, Helmand Province (Tables 9.7 and 9.9; Figs. 9.5–9.7). The dam comprises a powerhouse, an ungated spillway channel (incomplete) in a saddle about 700 m from the main dam, and a 7-m diameter, 100-m-long penstock

**TABLE 9.9** Summary of News and Security Reports Published and Available Online

| | News Stories (No.) | Coverage (Years) | Security Issues (No.) | Coverage (Years) | Summary |
|---|---|---|---|---|---|
| *Existing* | | | | | |
| Kajaki | 10 | 2006–15 | 108 | 2006–09 | The New York Times (Rohde, 2006) presented the Kajaki Dam as an "Afghan symbol for change" as becoming a failure in the Helmand region. In 2011, the British Broadcasting Corporation presented the British-led military mission to provide security for the USAID effort of transporting the needed third turbine and what went wrong (Urban, 2011). This initiated a series of articles (and a British movie, "Kajaki") that claimed the Kajaki Dam effort as a "white elephant" (AP, 2013), which represented misguided efforts by the west to counter insurgency in the Helmand region (Druzin, 2015) |
| Dahla | 13 | 2009–15 | 135 | 2004–10 | Just a year after project commencement by the Canadian government, The Toronto Star began citing security concerns at the site and corresponding project failures (Potter, 2010). From this point until 2015, the project has failed to be completed and as coalition forces were removed in 2014, so was the funding. As a result, this project is on hold until the Afghan government can afford continuation (Glasse, 2015) |
| Bandi Sultan | 14 | 2005–14 | 1 | 2007 | Exclusive of century-old (and older; eg, Vigne, 1840) writings on the Bandi Sultan, the Mar. 2005 failure of the dam initiated a decade-long series of new stories centered on the failure and its failed reconstruction (eg, IRIN, 2005; Sayed, 2014) |
| Salma | 13 | 2004–15 | 10 | 2007–09 | Reconstruction efforts on the Salma Dam were lauded by the Afghan government in 2004 (EOA, 2004) with the promise of a 2009 completion. In 2013, evidence of a possible plot to blow up the dam (PAN, 2013) along with additional security concerns, initiated additional news regarding the failure of Salma to be completed (eg, Zaland, 2015) and the larger issues of transboundary water issues between Afghanistan, Iran, Pakistan, and Turkmenistan (eg, Peterson, 2013) |
| *Proposed* | | | | | |
| Almar | 12 | 2009–15 | 13 | 2008–09 | In 2010, Pajhwok News mentioned that the Almar site had been surveyed and was awaiting funding from the Afghan government for a 2011 start (PAN, 2010). From this point forward, news of the dam was related to slow progress due to security and technical issues and lack of funding (eg, Kohi, 2015) |

*Continued*

TABLE 9.9 Summary of News and Security Reports Published and Available Online—cont'd

| | News Stories (No.) | Coverage (Years) | Security Issues (No.) | Coverage (Years) | Summary |
|---|---|---|---|---|---|
| Baghdara | 4 | 2009–11 | 3 | 2006–08 | The primary news article associated with the Baghdara Dam project centers on the transboundary dispute between Afghanistan and Pakistan with regard to Afghanistan's proposed development (supported by India) of the shared water way of the Kabul and tributaries (Mustafa, 2011) |
| Bakhshabad | 15 | 2005–14 | 4 | 2007–13 | Most news is centered on security issues and transboundary problems with Iran (eg, Tabish, 2011) along with funding and delay issues (eg, Wadsam, 2012) |
| Kamal Khan | 8 | 1995–2012 | 1 | 2008 | Because of the proximal location of the Kamal Khan to Iran and its placement in water agreements between Afghanistan and Iran makes most news transboundary water related (eg, Mogelson, 2012) |

*Security data primarily sourced from WarDiaries, 2015. Iraq & Afghan War Diaries Explorer. https://wardiaries.wikileaks.org (accessed August, 2015).*

feeding a 33 MW power station (USACE, 2007). The reservoir irrigates approximately 1800 km$^2$ up to approximately 500 km downstream.

#### 9.6.1.1 History

Final studies for the dam began in 1946 and a preliminary design was created in 1950 (Figs. 9.8 and 9.9). Construction of the Kajaki Dam was commenced by the Morrison-Knudsen firm in 1951 (Figs. 9.10 and 9.11) and completed in November of 1952 with gate closing in January of 1953 (Perkins and Culbertson, 1968) at a cost of $13 million (Whitney, 2008). Water was primarily stored for irrigation, but plans were completed for future use as a hydroelectric facility for Helmand Valley development (projected at 50 MW). Original capacity of the reservoir was 1.6 million m$^3$ at a spillway elevation of 1033.5 m and a capacity of 2.3 million m$^3$ at an elevation of 1045 m (inclusive of the never-installed, 12-m gates). Tailwater level is at 968 m providing a 65.5 m maximum head.

Prior to the hydroelectric improvements in 1975, the USGS completed a hydrographic and sedimentation survey of the reservoir (Perkins and Culbertson, 1968). The result of this study suggested that the mean annual sediment influx was ~7800 acre-feet or about 8,500,000 tons per year (basin net loss of 7% volume/year). Based on these values, they projected the serviceable life of the reservoir for future irrigation and/or hydroelectric reasons would be about 2008–13 when the distal edge of the delta would reach the irrigation outlet. A more recent study by the USGS (Vining and Vecchia, 2007a,b) reassessed the sedimentation problem at Kajaki and their determination was a 22% reduction in reservoir volume (at the 1033.5-m crest) between 2006 and 2057. They also estimate that even with the planned, updated 1045-m crest, a severe drought could result in large multiyear irrigation-supply deficits where the water level would remain below 1022 m for more than 4 years; a 47% chance exists that during

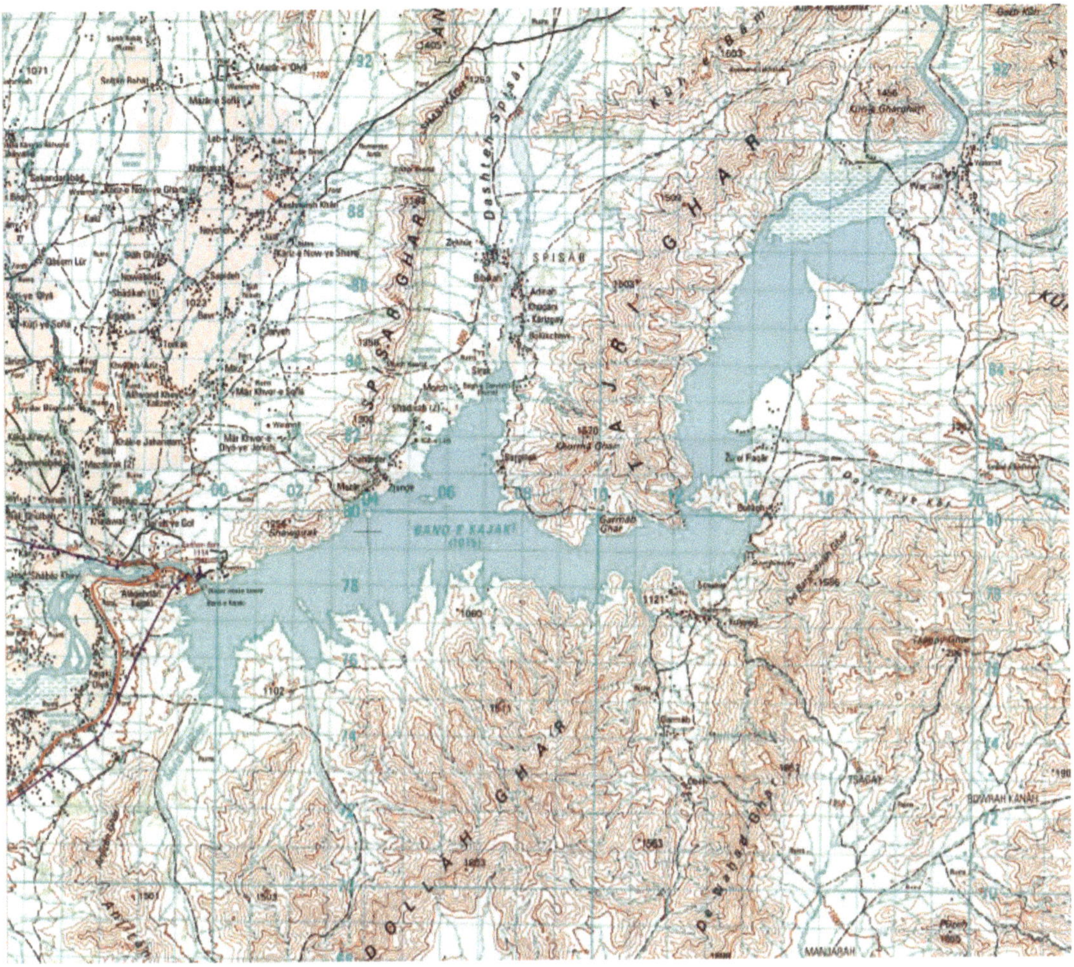

FIG. 9.5 Topographic map (1:100k) of the Kajaki Reservoir (NGIA, 2002).

FIG. 9.6 Oblique photographic view of the Kajaki Dam showing overflow spillway (bottom of image) and powerhouse release (USMC, 2011).

FIG. 9.7 Satellite image of the Kajaki Dam (Terraserver, 2015). [I can provide one of these for each selected dam site.]

the worst drought month, there would be sufficient waters. In addition, a 75% chance exists that the reservoir will be able to fill in 2 years (if the additional 12-m gates are added) under current irrigation needs; double/projected needs place this at a 50% chance of filling.

The power-generating facilities of the dam were planned by the R.W. Beck and Associates who recommended 150MW (USACE, 2007) and were funded by USAID in 1975 with two vertical Francis turbines installed in units 1 and 3 (leaving the second spot for future use). Each turbine has a rated capacity of 17.5MW based on a final net head of 65.5m. In 1977, the Water and Power Consultancy Services (WAPCOS; India) completed a geotechnical report on the proposed extension of the power station with a focus on the site geology and current powerhouse foundation.

During the Soviet invasion in 1979, contractors left the site to the engineers, where it was maintained to the best of their abilities until it was bombed by US forces in Oct. 31, 2001 making the Kajaki Dam the American connection to Afghanistan through the war and beyond. In 1998, the Taliban government closed the Kajaki sluices, which blocked the flow to

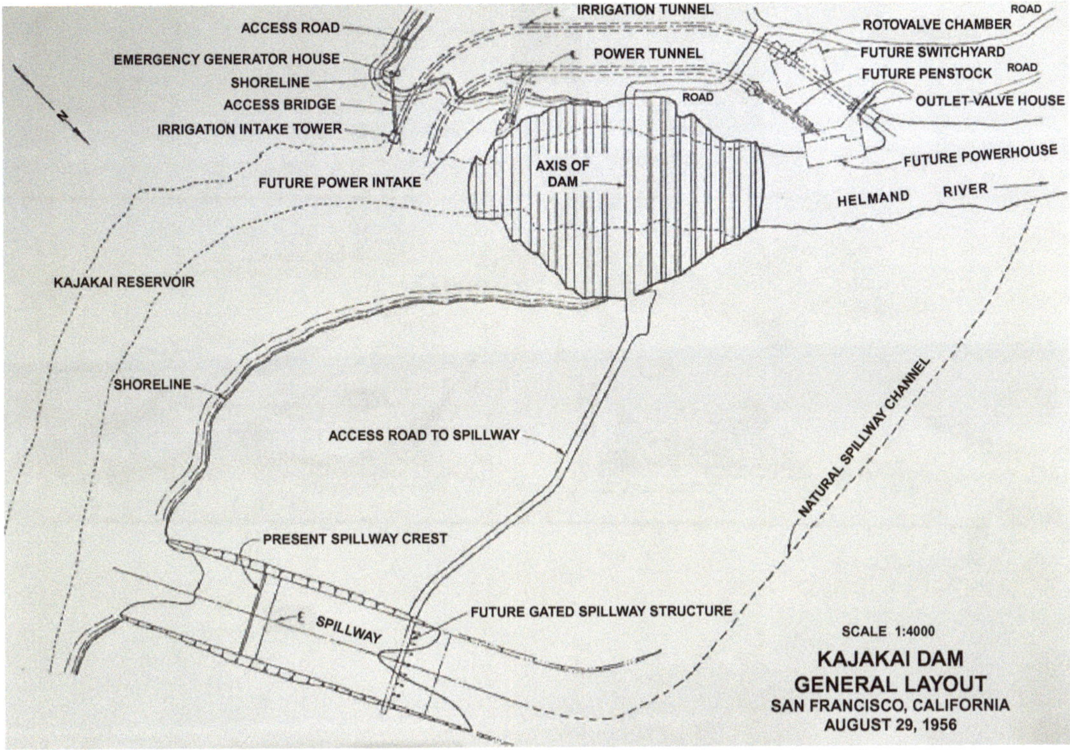


FIG. 9.8 Proposed, general layout for the Kajaki Dam (TEC, 1956).

the much-needed Helmand River to Iran's Sistan Baluchistan Province, which lasted until 2002 and coincided with one of the worst droughts on record. This action catastrophically impacted the region's ecosystem and led to extensive migrations out of the basin.

Two months before the US-led invasion of Afghanistan in 2001, the Afghanistan Ministry of Water and Power drafted a contract with the China Machine-Building International Corporation (2001) for $4 million ($400,000 for the turbine) to install the third turbine to be completed by 2007. In 2003, the Louis Berger Group completed an assessment of structures and facilities at Kajaki, recommending that turbine repair/maintenance be top priority, followed by transmission lines, the emplacement of a third turbine, and the dam refurbishment being last priority. In Nov. 2003, the Louis Berger Group projected cost estimates for the repair of turbines 1 and 3 at $8.1 million and the emplacement of turbine 3 at $18.2 million. With support from USAID, World Bank and others, unit 1 was operational again in September 2005 and unit 3 operational in Oct, 2009.

In 2004, USAID had revived the idea of installing the third turbine (#2) in an effort to counter the insurgency in the basin, hoping to double the electric output for about $500 million (Urban, 2011). It was not until 2007 that USAID had obtained the extra turbine. The problem was, the region became inundated with Taliban fighters and became impassable for contractors, so most development ceased. At this time the British were in charge of this area of operation and were slow to develop a plan to move this 210-ton turbine 150 km through the

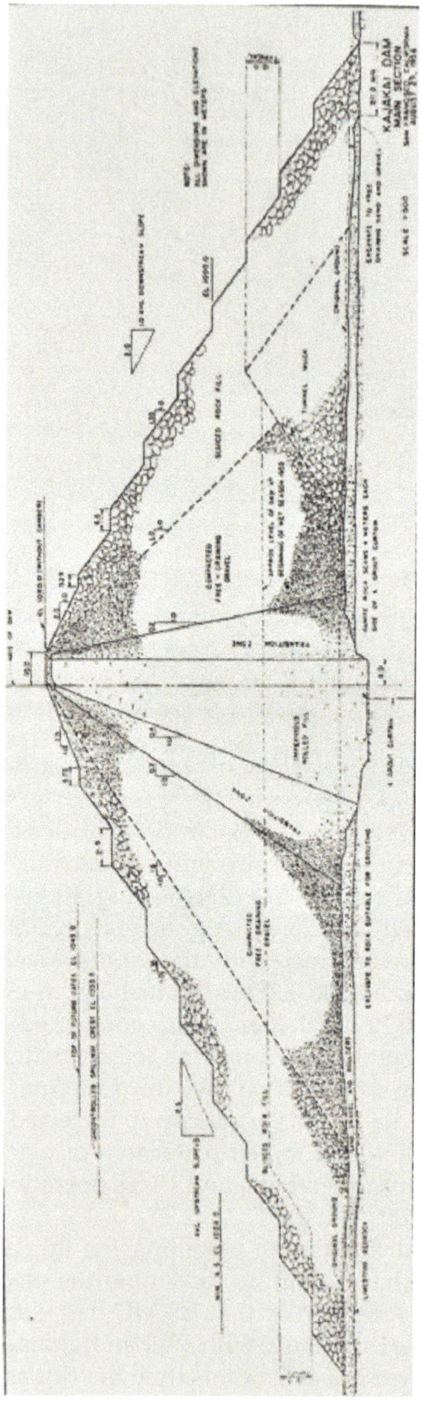

FIG. 9.9 Proposed, cross-sectional view of the Kajaki Dam (TEC, 1956).

FIG. 9.10 Ungated spillway during construction, six, 40×40' gates had yet to be installed (MK, 1951a).

FIG. 9.11 The old and new ships of the desert during the construction of Kajaki (MK, 1951a).

heart of the insurgency. In 2008, the mission "Eagle Summit" was launched, involving thousands of NATO troops and 100s of vehicles. The turbine was dismantled and transported on multiple "low-boys." Despite significant fighting in Kajaki Sofla (200 Taliban killed), the turbine reached the dam. Unfortunately, as of 2015, the turbine is still sitting in the same spot for a lack of cement for the foundation. Some 500 tons of cement needs to be trucked into where fighting is occurring daily. With the removal of allied troops (Brody, 2014), the project is now officially stagnant. Based on a formal response to the Special Inspector General for Afghanistan Reconstruction, USAID claims the third turbine will be in operation by Dec. 2015 (USAID, 2014); however, most recent news suggests that USAID pulled the plug on the project in 2011, handing it over to the Afghan government (re-contracting due to the flight of the Chinese contractors over security concerns). The best case scenario will be a 2018 working date. Most recent press (Druzin, 2015) extols the problems of aid development in a place without security. The prospects are poor for the completion of this project any time soon. In addition to turbine-emplacement problems and overall operation and maintenance, if and when the planned 12.5-m gates were installed, up to 50,000 villagers (16,000 USAID estimate, 2005) would need to be relocated and compensated for some 26,000 ha of property (8320 ha USAID estimate) (Campbell, 2005).

## 9.6.2 Dahla Dam

The Dahla Dam is a 55 m (soon to be 63 m) earth-and-rock fill embankment, irrigation dam 535 m long impounding the Arghandab reservoir (modeled at 388,000 acre-feet) along the Arghandab River in Kandahar Province (Tables 9.7 and 9.9; Figs. 9.12 and 9.13). The dam has six saddle dikes, extending 2040 m at a height of 15–25 m, two concrete service spillways and a low-level outlet works controlled by two Howell-Bunger energy-dissipation bypass valves located near the downstream toe of the dam (FBO, 2012). The dam is the largest in Kandahar Province and the second largest in Afghanistan situated 34 km northeast of Kandahar City and is operated by the Helmand and Arghandab Valley Authority (HAVDA). The dam was constructed by the USA (Morrison-Knudsen Afghanistan Incorporated) from 1950 to 1952 during the reign of Zahir Shah, the last king of Afghanistan (Figs. 9.14–9.16).

### *9.6.2.1 History*

The lands of the Central Arghandab Valley have been cultivated for centuries. Before construction of the Dahla Dam and reservoir, the ancient *juis* (canals) diverted seasonal flows of the Arghandab River for 4 months of the year. Local farmers along the canals provided the annual (8-month low-flow period) maintenance and rehabilitation necessary to keep the canals operational (as opposed to the current 30–40-day downtime) (USAID, 1973).

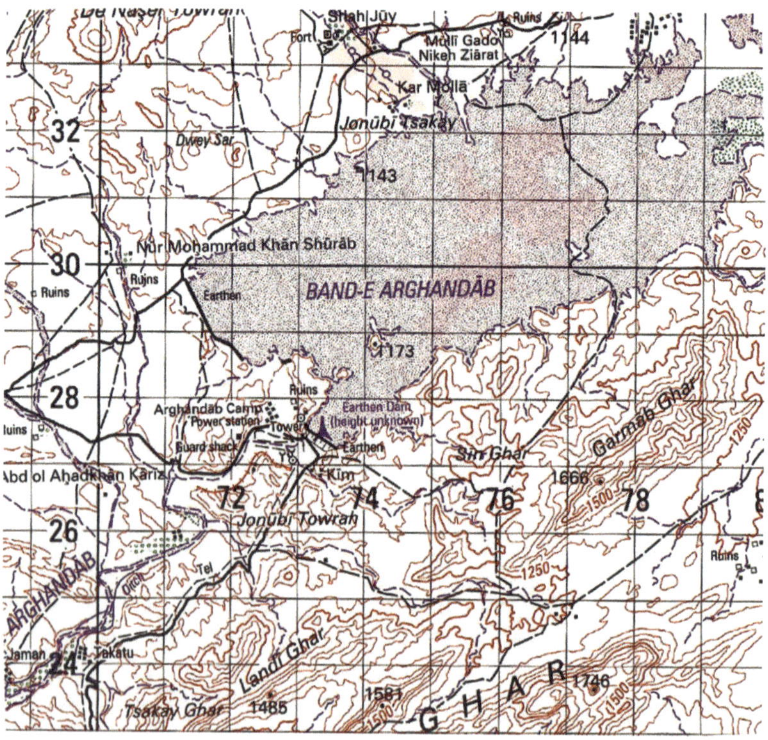


FIG. 9.12 Topographic map (1:100k) of the Arghandab/Dahla Reservoir (NGIA, 2001a).

FIG. 9.13 Oblique, aerial view of the Dahla Dam and Arghandab Reservoir (USACE, 2012).

The construction of the dam and major South Canal altered the water supply and made irrigation water available almost year round. This, in turn, altered the agricultural practices and maintenance required to keep the *juis* in fair operating condition. In some instances, maintenance is neglected until it becomes mandatory (hence the Canadian project, below). These conditions also adversely affected the regional groundwater table (USAID, 1973).

The current irrigation distribution system comprises six main *juis*: Shah, Patow, Rhorabad, Nushjan, Meon, and Punjab. These *juis* allow irrigation of water-intensive crops such as orchards, field crops and vegetables. During the evaluation by USAID (1970), evidence occurred of the lack of proper water-control structures in the *juis* system, coupled with a lack of proper operation and maintenance procedures, inefficient irrigation practices, and seepage losses (up to 70%); moreover, inadequate water supply to the distal reaches of the system was a problem. This caused saline-sodic soils to develop due to lack of water-flushing capacity. As a result of these deficiencies, the report suggested: (1) a new distribution system; (2) improvement of major drainages; (3) increased and/or improved farm-to-market roads; and (4) one, sole governing agency that is apolitical and inclusive of local involvement and farmer training. Just after this report was completed, a 2-year drought occurred (1971–72), as well as a non-violent coupe (1973), a violent communist coup (1978), and the Soviet invasion (1979). As a result, an additional 30 years of neglect were added to the above observations; hence, the Coalition effort at restoring water resources associated with the Dahla Dam and canals. Regarding the reservoir, the silting-up process has taken up some 7% of the reservoir (up to 1970) with projections of 0.7%/annum (to about a 43% loss by 2014; 30–50%, CAN, 2008; Sher Mohammad Atai quoted in Watson, 2012).

#### 9.6.2.1.1 COALITION EFFORTS

As part of Canada's role in development of Afghanistan, in 2007 the Canadian government created an Independent Panel to examine their overall mission in Afghanistan, which was to make recommendations on the future of Canada's efforts in Afghanistan. This panel, grounded in the Afghanistan National Development Strategy and Afghan compact,

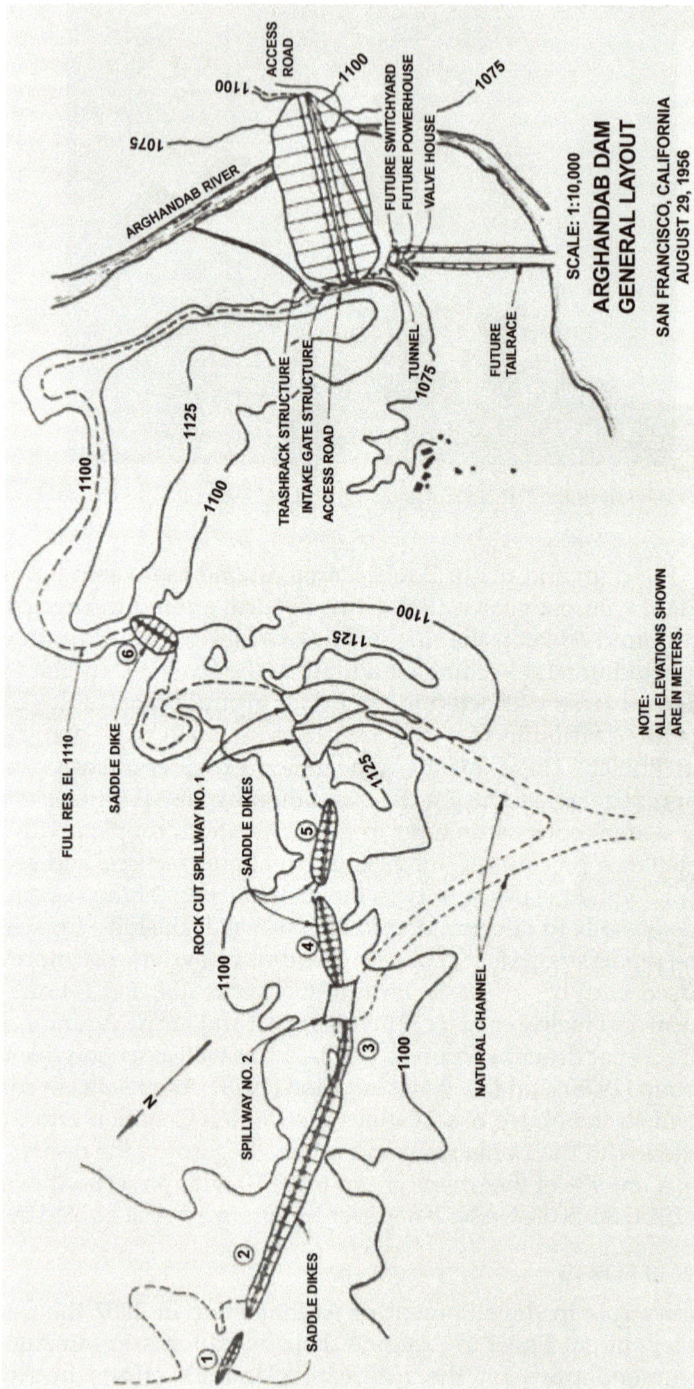


FIG. 9.14 Original, general dam layout for the Arghandab/Dahla Dam (TEC, 1956).

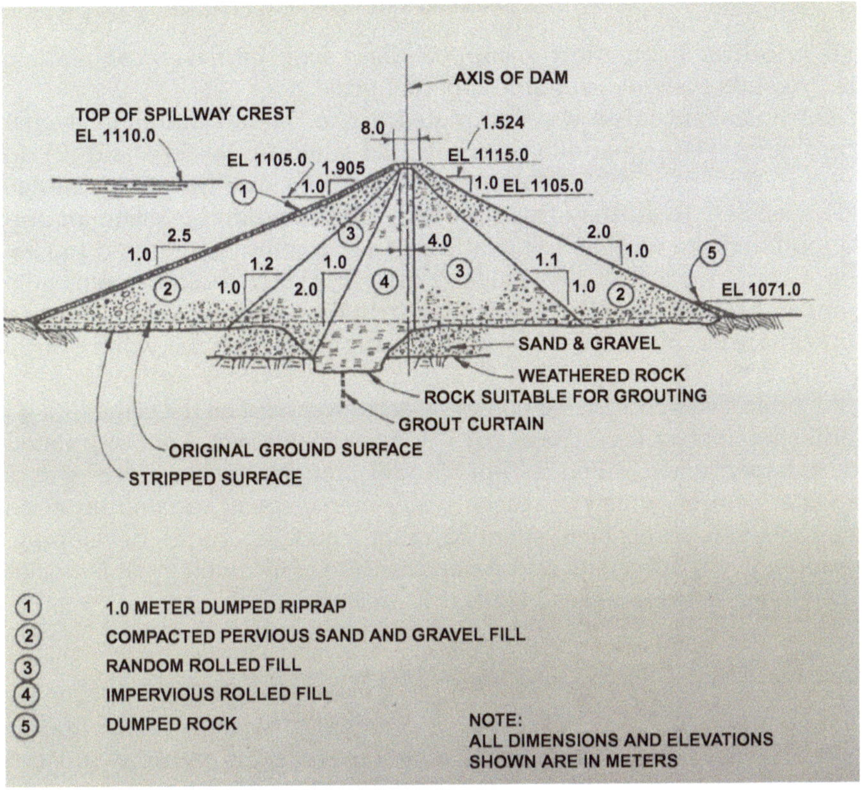


FIG. 9.15 Original, planned cross-section of the Arghandab/Dahla Dam (TEC, 1956).

FIG. 9.16 First-stage of construction of the Arghandab/Dahla Dam, with equipment placing impervious fill in the core trench (MK, 1951b).

suggested six priorities. In an effort to support these six priorities, rehabilitating the Dahla Dam and its irrigation system was 1 of 3 signature projects.

In 2008, a technical appraisal was completed of the Dahla Dam and system and noted two major problems: (1) Arghandab reservoir had silted up 30–50% and (2) downstream irrigation was in disrepair, with an estimated 70% loss due to infiltration/evaporation and other deficiencies. To address this problem, the Canadian government approved $50 million in a joint venture with SNC-Lavalin and Hydrosult (transferred to Central Asian Development Group due to security issues in 2011) to: (1) rehabilitate main canals, secondary canals, and water control of the Arghandab irrigation system; (2) build irrigation water management capacities; and (3) build capacities for farmers at the water-control community level (CAN, 2008; Harvey, 2009).

By 2012, the project was considered complete and successful by the Canadian government; however, conflicting reports (eg, Glasse, 2015) suggest the project is not completed. The project repaired, reshaped, and desilted 500 km of canals, repaired water valves and fixed gating to improve water control, which increased water-management capabilities at all levels. In addition, the Canadian government spent $2 million clearing mines, developing a partnership between Kandahar University and Asian Institute of Technology; it is unclear whether long-term maintenance funding was part of this project.

## 9.6.3 Salma Dam

The Salma Dam is a 107.5-m tall, earth-and-rockfill, hydropower/irrigation dam impounding the Hari River (Hari Rud) with an active reservoir capacity of 560 million $m^3$, and is located in Chishti Sharif District, Herat Province (Tables 9.7 and 9.9; Figs. 9.17 and 9.18). The dam comprises a power house, chute spillway, and switch yard with the following design specifications for 42 MW energy generation: a $3.25 \times 6.72$ m (63 $m^3/s$ discharge) power sluice, a 4.25-m-diameter penstock (2.46 m after trifurcation), a 15 $m^3/s$ discharge irrigation sluice, a 630-m-long, 8.5-m-diameter diversion tunnel (horseshoe shape) and a power house tunnel/pressure shaft (4.25-m-diameter, 457 m long) (Kumar et al., 2014). The dam will be the largest in Afghanistan, once entirely completed, and is located ~150 km east of Herat City.

The dam was originally constructed in 1976 and was damaged early during the Afghan civil war. Reconstruction was first started in 1988 by the Indian government through their Water and Power Consultancy Services (WAPCOS; India), but was left incomplete for nearly 20 years. In 2004, India returned with $79 million to begin reconstruction, which is yet to be completed.

### *9.6.3.1 History (FAO, 1965)*

From 1960 to 1964, FAO investigated, under a Special Fund Project, water control of river flows, using storage reservoirs, for the improvement and extension of irrigated agriculture within certain river basins in Afghanistan. This work was intended to help irrigate some 131,000 ha needing some 1865 $m^3$/annum of renewable freshwater.

In an effort to maintain over-year storage of waters in the Hari River Basin for continued/improved agriculture, four dams were proposed: Assarasum and Salma (on the Hari River) and Tangi Shah and Tangi Azao (on the Kawgan tributary). In an effort to begin water

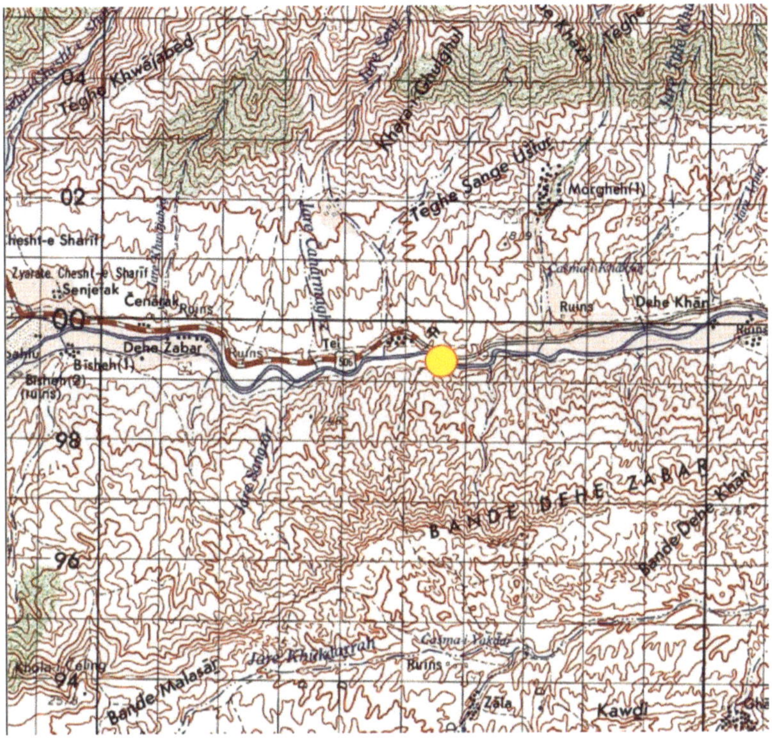


FIG. 9.17 Topographic map of the Salma Dam area (DMA, 1986a). Yellow dot represents current location of the dam; despite being extant at the time of mapping, it is not annotated.

FIG. 9.18 Continued construction efforts on the Salma Dam (AZ, 2015).

control in this basin, these four plausible sites were evaluated on multiple levels. The net result of the Food and Agriculture Organization study in the Hari River basin was that the Assarasum Dam had the optimum economic index, that is, the lowest capital investment per $m^3$ of available storage with nearly all storage used annually. Regardless, the authors suggest that the Tangi Shah Dam would likely be the first project to begin irrigation development (never started?). The Salma Dam, at the time, ranked the most expensive (6x Assarasum and 4x Tangi Shah). Design specifications in this report proposed a 60-m concrete dam and spillway for the Salma site, which would impound 220 million $m^3$ for a net 160 million $m^3$ storage (after a 100-year silting period, estimated with a 0.6 million $m^3$/annum silt rate). The dam was proposed to have a concrete spillway, irrigated outlets, hydropower station (rated at 24 MW) using three inlets and a diversion tunnel for an estimated cost of $11.5 million.

Despite the recommendations of this special project, the Assarasum and Tangi Shah (and Azao) Dams were never begun. Only the Salma Dam was nominally completed some 10 years later. The current dam specifications, however, deviate from those proposed by the UN Food and Agriculture Organization in order to increase the hydropower output to 42 MW and to develop additional lands for irrigation.

## 9.6.4 Bandi Sultan Dam

The Bandi Sultan Dam is a stone-masonry, gravity dam 23 km north of Ghazni City along the Ghazni River in the Jaghatu District of Ghazni Province (cited as Wardak Province by most references) (Tables 9.7 and 9.9; Figs. 9.19–9.21), which was built upon the footing of an 11th century dam built by Sultan Mahmood Ghaznavi (Fig. 9.22). The modern dam, built by

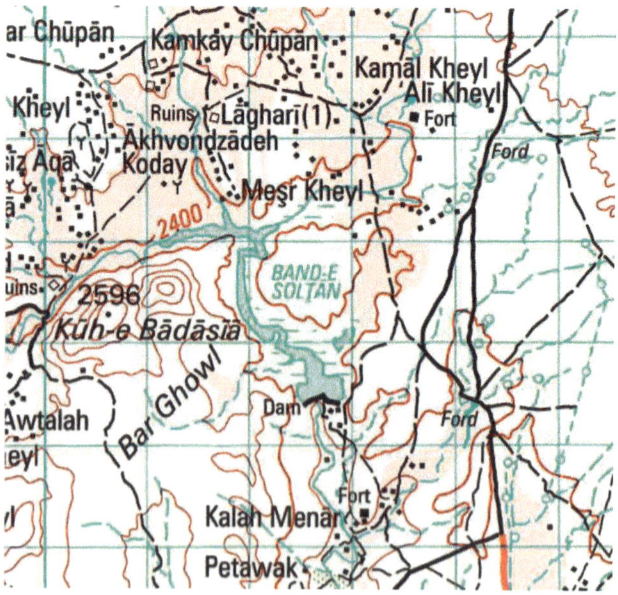


FIG. 9.19 Topographic (1:100k) map of the Bandi Sultan area and reservoir (NGIA, 2001b).

FIG. 9.20 Oblique view of the Bandi Sultan Dam (Stewart, 2009).

FIG. 9.21 Perspective view along the top deck of the Bandi Sultan Dam (Stewart, 2009).

the British in 1901, is 32m high, 200m long with a 32-m long side spillway on natural rock with no downstream stilling basin and a 200-m long shoulder dam; drained by two 1-m pipe outlets and an approximately 12 million $m^3$ reservoir (dgMarket, 2007) able to irrigate some 15,000ha and support 25,000 families.

#### 9.6.4.1 *History*

During the 1990s, the dam was in serious need of repair, so the Danish Committee for Afghan Aid to Refugees (DACAAR) worked at refurbishing the pointing on the downstream side of the main dam where it was beginning to leak from 2000 to 2002 ($40,000; DACAAR, 2005) along with an 80cm addition to crest height (contested by DACAAR; Stewart, pers. obsv.; Sadat, 2010).

FIG. 9.22 Original, 10th century portion of the Bandi Sultan Dam, currently submerged by the reservoir. This dam is made of bricks and mortar (Stewart, 2009).

In early spring 2005, after four days of crest overtopping (Sadat, 2010) due to heavy rains and snowmelt the reservoir broke through the second wing wall causing significant flooding downstream exacerbated by lodged shipping containers under bridges killing six, displacing thousands and washing away thousands of hectares of land and multiple shops in Ghazni City (IRIN, 2005). As the result of President Karzai's recognition of the disaster, multiple Afghan and international agencies were deployed to the region: Afghanistan's Rural Rehabilitation and Development Minister and the Public Works Minister were sent by the President to oversee operations while the Ministry of Defense sent helicopters to better understand the problem; food and tents were supplied and the Ministry of Water and Power and the Afghan Red Crescent Society lent tents, kitchen kits and blankets along with some 70 volunteers; US-led Coalition forces provided helicopters, truck support along with Coalition engineers to make initial dam assessments (IRIN, 2005). DACAAR responded immediately (and defensively to the thoughts that their recent repair helped cause the failure) by sending a water and sanitation team to the area to test drinking water and, with other NGOs, provided 20 laborers to help clean up (DACAAR, 2005).

The actual repair of the failed wing wall was begun shortly thereafter, with phase 1 completed by Jul. 2006 for $500,000 (or $2.1 million depending on reference) by Koh-e-Alborz Construction company (Koh-e-Albroz, 2012). This first phase was the emplacement of a new wing wall, which was only partially completed as of 2009 (Stewart, 2009) and, possibly, even 2014 (Fig. 9.23). Based on an on-site assessment of the repairs (Stewart, 2009) strength calculations of the replaced wing wall rated it at approximately 45% stability for a recommended water safety-level of 6.5 m below the crest or at the 4.8 m mark. Phase 2 efforts were designed, reviewed and submitted with a 2007 Request for Expression of Interest submitted by the World Bank for an Emergency Irrigation Rehabilitation Project (dgMarket, 2007) and its unclear exactly what they entail(ed). Phase 2 (or completion of Phase 1?) efforts were hampered by security concerns at the dam. In July of 2007, two German engineers and six Afghans were kidnapped (or handed over) to the Taliban (Ariana, 2007) with additional attacks through at least 2009 (Stewart, pers. obsv.).

FIG. 9.23 Repaired wing-wall section of the Bandi Sultan Dam (Stewart, 2009). This represents 4 years of progress in repairing the dam after the Mar. 2005 failure.

#### *9.6.4.2 Historical Dam*

The original Bandi Sultan Dam, still partially extant and one of the older in Afghanistan, is a curved gravity dam made of rough stone masonry (Vigne, 1840; Murray, 1886). Its first mention was by Babar in his *Memoirs* (1505) as 40–50 *qari* (yards?) high and some 300 long with stored waters let out as required. He recognized that it was destroyed by Alauddin *Jahan-soz Ghuri* (burner of worlds) when he conquered the country in 1152. Babar recognized the dam was in ruins during his visit and, as a result, felt there might be hope for it, sent money to make repairs in 1526. Nearly 300 years later, Vigne (1840), on his visit to Ghazni, described the dam as "rough stone-work" some 200 yards long and 25 yards above the reservoir level (Fig. 9.24). He recognized two flood gates, one at the top and other at the foot for regulating flow. At this time, the dam was "out of repair" and the reservoir was nearly empty. Some 40 years later a British expedition mentioned the dam in their report (Murray, 1886) as a wall of masonry some 300 yards long with height varying from 8 to 20 ft and some 6–7 ft thick. The 1902 edition of the Encyclopedia Britannica mentioned the dam, along with the 1932 book *Life and Times of Sultan Mahmud of Ghazna*, as much out of repair but still in use. In 1901 a new dam was built upon the same footing on the eastern bank, but on a line slightly downstream, so that a portion of the original, 11th century dam is visible (and accessible) from the main dam.

### 9.6.5 Almar Dam (Proposed)

The Almar dam is a $51 million, concrete gravity dam (64 m high × 240 m crest length) for irrigation, which was proposed by the Ministry of Energy and Water and, supposedly, to be funded by Afghanistan. The dam site is located in Faryab Province on the Almar River in the Amu Darya basin (Tables 9.8 and 9.9; Figs. 9.25 and 9.26). The dam is to be completed in four phases covering all the requisite infrastructure (10 km road, etc.), dam, canals and sluices.

The project got off the ground in 2010 when the Hi-Tech Engineering Firm completed a feasibility assessment (with social impacts) with a Tajik firm, Tajik Joint-Stock Company

FIG. 9.24 Wood-cut engraving of the Bandi Sultan Dam, early 19th century (Vigne, 1840).

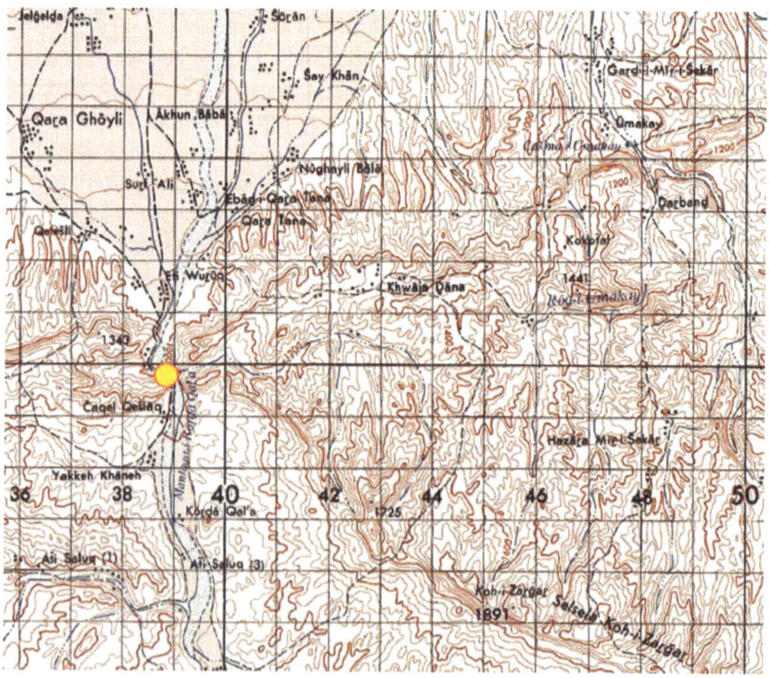


FIG. 9.25 Topographic map (1:100k) of proposed location for the Almar Dam site (yellow dot) (DMA, 1986b).

FIG. 9.26 Perspective view looking upstream approximately 300 m north of the proposed Almar Dam site (Opsingh, 2011).

(TADES), winning the international tender with support from an Afghan firm. Since its inception, the project has suffered from monetary issues (lack of allocation/provisions) and security concerns. As of late 2014, the project is indefinitely postponed.

## 9.6.6 Baghdara Dam(s) (Proposed)

Originally planned in the 1970s by the Afghanistan government, the project for the Baghdara Dam(s), named for the village in Kohi Safi District, Parwan Province, represented a multi-dam series to provide hydropower to the Kabul City area for an estimated $600 million. As the result of feasibility work, contracted in 2005, by Fichtner (MEW, 2006), an alternate dam site "D" was proposed, which subsequently has come to represent this project (Fig. 9.27). The Baghdara "D" option is proposed as a 152 m (or 142 m; 190-m gross head) high dam, with an approximately 400-m crest length and a 400 (200?) million m$^3$ reservoir at 1420 m providing 640–954 MW annually (Fichtner, 2007). The benefit of the "D" model with a 1420-m reservoir level, originally unplanned, is the virtual lack of resettlement and rehabilitation of villagers/properties and only needing a 1.2 km penstock (cf. 10–15 km for other options).

This "D" site is located 60 km northeast of Kabul city in the Panjshir Gorge and will dam the Panjshir River, which is a typical glacier-snowmelt stream that originates near the Anjuman Pass (~4500 m) in the Hindu Kush to the Naghlu Reservoir. This dam, as planned, is a hydropower-only dam, which will likely affect the Naghlu and Surobi Dams/reservoirs downstream; moreover, the Logar/Panjshir Rivers (Upper Kabul River basin) provide (or are expected to provide with pending population growth) significant water to the Kabul City area.

Feasibility reports (MEW, 2007) are comprehensive and cover all the modern aspects of dam emplacement (eg, feasibility, ecology, social and hydro/power). Of particular note, it was presented that gas-fired power plants would cost less than 50% (kWh/installed) of the dam and

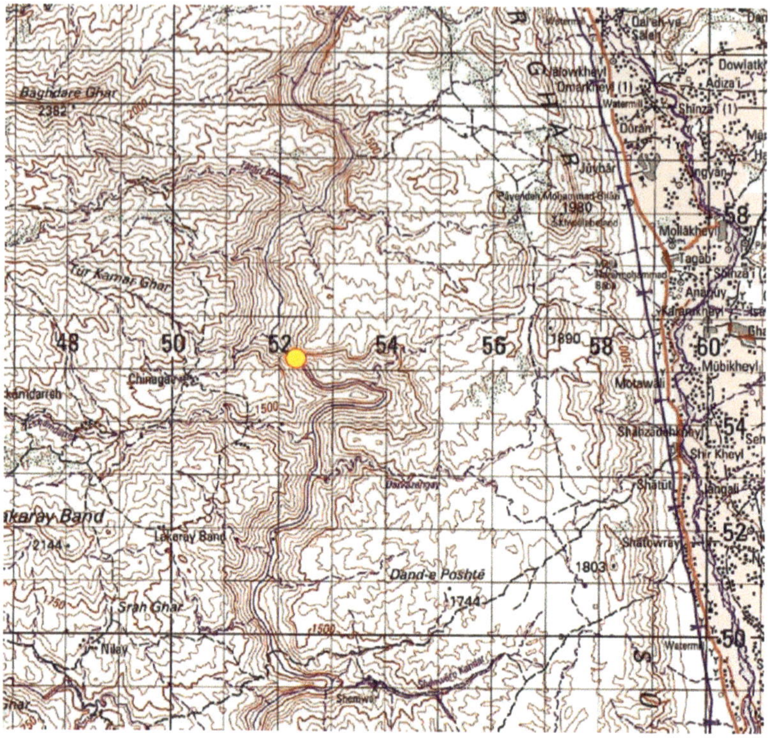


FIG. 9.27 Topographic map (1:100k) of the Baghdara Dam (yellow dot) area (NGIA, 2001c).

would be operating in about ½ the time. Due to logistical concerns about diesel supply and projects' longevities they stated, it was up to the government of Afghanistan to decide.

Because of lack of legal/policy framework with which to make particular decisions (ie, social and hydrological), Fichtner (2007) used the World Bank Policies for dam development. Because the project will affect an international waterway (ie, Kabul/Indus), World Bank policy OP7.50 (WBG, 2012) was used and *requires* an agreement or an arrangement between concerned riparian neighbors (ie, Pakistan). Regarding resettlement and rehabilitation, AF Official Gazette no. 639, 1987 (governs land compensation) is available; however, no recent use can be cited as a way forward. World Bank policies OP/BP 4.12 (involuntary resettlement) and OP 4.20 (indigenous people) were used to aspire to providing displaced persons improved or similar quality of life (the point becomes moot with the "D" option, due to lack of resettlement needs).

## 9.6.7 Bakhshabad Dam(s) (Proposed)

The Bakhshabad Dam locality was originally scouted by Tim de Jong in 1950, but no storage possibilities were visible to him at that time (FAO, 1965). The Food and Agriculture Organization (1965) returned to this area for their country-wide study on possible dam locations and irrigation in Afghanistan. As a result of significant field work (geological and

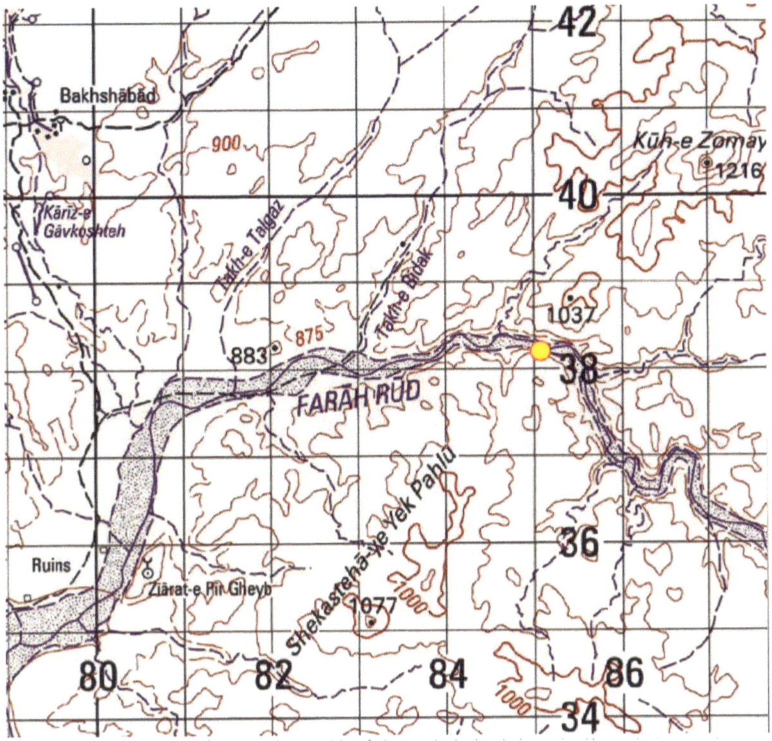


FIG. 9.28 Topographic map (1:100k) of the Bakhshabad Dam (yellow dot) area (NGIA, 2001d).

engineering), the Bakhshabad Dam site was the primary selection in the Helmand Valley (competing with the Alikini and Lashkagar sites) because of its proximity to proposed irrigated areas and was the least expensive option planned at $15.5 million (FAO, 1965) (Fig. 9.28). Despite being planned as providing the least hydroelectric power, 24 MW (cf. 26 MW for Alikini and 43 MW for Lashkagar), at the time considerations for power needs were secondary to irrigation and agricultural development. Also despite the increased need for electrification across Afghanistan, this project has been "grandfathered" into the current list of over 100 proposed dam projects without any evidence of reviving the partner dams in Alikini or Lashkagar, which would provide up to twice the electricity.

The Bakhshabad Dam (proposed) is a 60 m high, 265 m crest length, concrete gravity dam, as well as a power station, spillway, and irrigation outlet in the Bala Baluk District of Farah Province along the Farah River. When completed, the dam will provide 24 MW of electricity and irrigate some 69,200 ha using a 540 million $m^3$ reservoir.

## 9.6.8 Kamal Khan Dam(s) (Proposed)

The Kamal Khan Dam (KK) in the Charborjak District, Nimroz Province, was built as early as the 11th century by an unknown industrious and visionary leader, with baked bricks and an ancient lime mortar; supplying for hundreds of years a complex canal system that irrigated

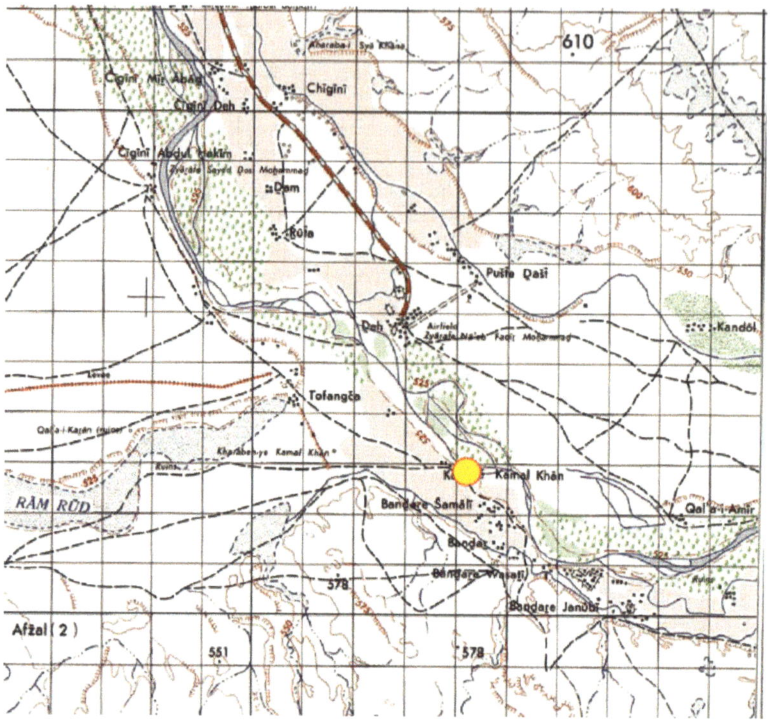


FIG. 9.29 Topographic map (1:100k) of the Kamal Khan Dam (yellow dot) area (DMA, 1986c).

what was then fertile wheat-and-barley country (Mogelson, 2012) (Fig. 9.29). In the late 1300s the Turkish conqueror Tamerlane punished Nimroz by destroying KK and it has returned to the desert ever since.

Not until the early 1970s did President Sardar Muhammad Daoud Khan set about rebuilding KK as part of his 5-year master plan for Nimroz (RRERS, 2013). During this time, the Asian Development Bank, in conjunction with the International Engineering Company (San Francisco) completed a feasibility study of the Chakhansur Region including flood control and irrigation development (ADB, 1973). This project proposed two major irrigation dams (KK and Qala Afzal), two minor irrigation dams (Rud-i-Seistan and Sikhsar) along with associated irrigation works/canals and road infrastructure totaling some $20.9 million. This proposal was appropriately sensitive to the transboundary water problems that would arise as the result of this project (referencing Kajaki, also) with a focus on the agricultural development of the region. This proposal suggested a zoned earth-embankment diversion dam leading water, via a dike, to the associated Qala Afzal Dam. The KK would need to be some 6700 m long, 15 m high, and 8 m wide (at top). This size would allow for a 1.5 m freeboard modeled to the greatest known flood. The dam, originally designed to *not* store water, was to be an impervious-core with pervious/semi-pervious random fill zones up/downstream with a gravel outer shell.

When Daoud was then ousted in a coup, invasion by the Soviets and Taliban rule left the KK project impractical until 10AUG11 when the MEW signed a contract with the Tajik Joint-Stock

Company (TADES) to begin a $10 million, phase-one, site preparation. At this time, the KK project, appearing to be KK only (no Qala Afzal Dam), was estimated at $100 million and intended to irrigate some 80,000 ha and generate some 8.5 MW of electricity (Yuldashev, 2011). This initial work included the construction of a 3400-m wall having 32 m width (base) and a 6–11-m height. The second phase, yet to be assigned or completed, would be to build a similar wall on the other side of the river and cementation of the riverbed (and 1 or 2 years to be fully operational). The third stage would be the emplacement of the turbines (PAN, 2011). As recent as 2014, the dam project is on hold due to security concerns and monetary issues. MEW states that KK, along with other dam projects, are on hold due to lack of funds (KK needing some $80 million, alone, for phase 3); however, Ministry of Finance claims money has been allocated (but none is available).

## 9.7 SUMMARY AND CONCLUSIONS

Afghanistan is a semi-arid country; however, with its placement atop the regional watershed in Central/South Asia, the country is relatively water rich. Afghanistan receives some 75 billion m³ of renewable, freshwater resources annually through winter snowfall and spring rains and melt. At her current level of agricultural water use (some 98% of all freshwater used in the country), the country is barely able to maintain subsistence agriculture, let alone a national economic agriculture focused on exports. Despite this key positioning to controlling her water resources, Afghanistan still loses some 85% of its surface-water resources by means of riverine runoff beyond her borders and poor construction and maintenance of her thousands of irrigation schemes.

Because these water resources are spatially and chronologically erratic, water-resource infrastructure has been a key part of development in Afghanistan for at least 1000 years, with the building of the Bandi Sultan and the Kamal Khan Dams as examples. With the advent of modernized water infrastructure developed by the west during the turn of the 20th century, Afghanistan was at the fore in accepting and using this technology. In 1901, in an effort to begin modernizing their water infrastructure (Ahlers et al., 2014), for example, the British helped Afghanistan control Ghazni River waters for irrigation and flood control by (re)building the Bandi Sultan Dam as a formal, 32-m-high large dam. Although unclear from the records, it is likely that additional water infrastructure was emplaced during the early parts of the 20th century; however, it was not until the reign of Mohammad Nadir Shah (1929–33) that modernizing Afghanistan became a national effort. As Iran began "out developing" its region of the Siestan Basin in the Helmand Valley, the Afghan government sought technical and monetary support from the USA to begin (re)developing its part of the Helmand. Without US assistance, the Afghan government accepted support from Germany and Japan, which renewed efforts to settle the nomads and farmers of the Helmand Valley and to centralize and extend the new Afghan government. With the advent of WWII and the diaspora of Jewish furriers to the USA, Afghan wool merchants found a new customer for their karakul pelts. This significant export opportunity provided over $100 million to the Afghan coffers. With this money, King Mohammad Zahir Shah, son of Nadir Shah, also recognized the importance of modernizing Afghanistan and used this money to begin funding water-resource efforts in the Helmand Valley and elsewhere. Interestingly, at this time, Afghanistan was not considered "behind the curve," but, possibly, on the cutting edge for non-western nations. Water-resource development was centered on the Helmand Valley from about 1946 to 1978 supported primarily by US private contractors and later the US government.

In 1973 with the deposition of King Zahir by Daoud Khan, supported by pro-Moscow communists, the overall positive trend of infrastructure development in Afghanistan began to unravel. In 1979, the USA formally left Afghanistan due to the 1978 Soviet-backed communist takeover of the country. After nearly 40 subsequent years of chaos, war, and destruction, Afghanistan was, more or less, returned to its 19th century capabilities. Unfortunately, war and strife continue to ravage the country despite this most recent free and democratic transfer of power in 2014. Primary losses from these tumultuous years were centered on infrastructure and human capital, which has only, slowly, begun to return to the country.

In all, over 60 dams (approximately 20 large dams) with a combined capacity of over 2 $km^3$ (FAO, 2015) representing over 9000 m of crest, along with 109 diversions schemes, exist across Afghanistan, with a focus on supporting the water-infrastructure needs of the capital, Kabul and the Helmand Basin. Proposed water-infrastructure projects, in the form of dams, total over 100 (>20 are planned to be large dams) and are spread out relatively evenly across the country and into areas without much previous government support and are projected to cost much greater than $10 billion (cited for only 31 major water projects and 15 storage dams; UNEP, 2013). This very low estimate represents building costs with little to no long-term maintenance or safety planning.

The sum of water-infrastructure efforts in Afghanistan over the past century has entered the "vicious cycle of build-neglect-rebuild" (Tortajada and Biswas, 2014, p. 7) where even the modest amount of dams and diversions in Afghanistan are in need of repair, refurbishment and technical and operational expertise. In order to break this cycle, long-term maintenance and human capital investments need to be included with these, already, most expensive undertakings in Afghanistan (eg, Meinzen-Dick, 2007; Ward, 2010; Pisaniello et al., 2011; Hall et al., 2014; Hewitson, 2015). With a full 99% of water-infrastructure schemes being government run, and 100% non-governmental organization maintained, in a country that can only provide about 10% of public sector funding is a recipe for projected failure of good intentions (Qureshi, 2002; UNEP, 2013; Ahlers et al., 2014). With the withdrawal of most Western Coalition forces in 2014 and the significant drop in funding that came with them (WBG, 2015; FTS, 2015), present and future security concerns in Afghanistan remain the number one obstacle to functional development and forward momentum onto the world stage.

Of the four existing large dams and the four proposed large dams presented here, without exception, all projects are mired in fundamental problems. Independent of the failure of Afghanistan, or donor nations, to provide monetary support for these projects to be completed, these projects exist in, likely the most insecure country in the world (eg, IEP, 2015) with a lack of infrastructure, roads, or human capital. Despite the free and democratic transfer of power in 2014 from Karzai to Ghani, overall the water-infrastructure outlook for Afghanistan is not likely to improve in the near future. Generally, until security and governance are significantly stabilized over the long term, these projects will remain in disrepair and/or unfinished or unstarted.

## Acknowledgments

I thank Richard B. Scott (retired, USAID) and his webpage (http://scottshelmandvalleyarchives.org) for helpful insight into the Helmand Valley and U.S. operations there since the middle of the 20th century. Thanks also to Shaista Wahab (UNO Criss Library, Arthur Paul Collection on Afghanistan) for document support and finding additional materials, and Melanie Swick (St. Lawrence University) for help in manuscript preparation. Thanks also to St. Lawrence University and the William B. Bradbury, Jr. Faculty Support Award for travel support.

## References

ADB (Asian Development Bank), 1973. Chakansur Flood Control and Irrigation Project in Afghanistan, a Feasibility Report. International Engineering Company, San Francisco, CA, 158.

AEIC (Afghan Energy Information Center), 2015. Afghan Energy Information Center, variety of links, maps and documents. http://home.dabs.af/web/aeic/index.php (accessed May, 2015).

Ahlers, R., Brandimarte, L., Kleemans, I., Sadat, S.H., 2014. Ambitious development on fragile foundations: criticalities of current and large dam construction in Afghanistan. Geoforum 54, 49–58.

Aini, A., 2007. Water conservation in Afghanistan. J. Dev. Sustain. Agric. 2, 51–58.

Alim, A.K., 2006. Sustainability of water resources in Afghanistan. J. Dev. Sustain. Agric. 1, 53–66.

Ansar, A., Flyvbjerg, B., Budzier, A., Lunn, D., 2014. Should we build more large dams?: the actual costs of hydropower megaproject development. Energy Policy 69, 43–56.

Ansari, M.T., 1978. Helmand Valley Authority: Making the Desert Bloom. The Asian Student Newspaper, United States.

AP (Associated Press), 2013. Kajaki Dam: the great white elephant of Afghanistan Kabul: the Independent. http://www.independent.co.uk/news/world/asia/kajaki-dam-the-great-white-elephant-of-afghanistan-8439742.html (accessed August, 2015).

Ariana, 2007. Police handed us to Taliban, says Afghan freed with German. http://e-ariana.com/ariana/eariana.nsf/allDocs/dda40a624c0991bf8725737100529f97!OpenDocument&Click= (accessed August, 2015).

AWD (Afghanistan War Diary), 2015. War diaries: Iran and Afghanistan War Diaries explorer. https://wardiaries.wikileaks.org (accessed May, 2015).

AZ (Afghan Zaria), 2015. The construction work on Salma Dam delayed due to security concerns. http://www.afghanzariza.com/2015/01/10/the-construction-work-on-salma-dam-delayed-due-to-security-concerns#sthash.K6cXProN.dpuf (accessed May, 2015).

Azam, G.F., 1999. The effects of local, regional and global politics on the development of the Helmand-Arghandab Valley of Afghanistan. Ph.D. Dissertation, University of London, School of Oriental and African Studies, 255 pp.

Baghel, R., Nüsser, M., 2010. Discussing large dams in Asia after the World Commission on Dams: is a political ecology approach the way forward? Water Altern. 3 (2), 231–248.

Baron, L.Z., 1975. The water supply constraint: an evaluation of irrigation projects and their role in the development of Afghanistan. Ph.D. Dissertation, McGill, University, Montreal, 331 pp.

Bosshard, P., 2012. Infrastructure for Whom?: A Critique of the Infrastructure Strategies of the Group of 20 and the World Bank. International Rivers, 24 pp..

Bradsher, H.S., 1985. Afghanistan and the Soviet Union. Duke University Press, NC, USA, 336, pp.

Brody, B., 2014. USAID's $500 million dam project circling the drain in Afghanistan: Global Post, Groundtruth Blog. http://www.globalpost.com/dispatches/globalpost-blogs/groundtruth/usaid-kajaki-dam-afghanistan-failure (accessed August, 2015).

Brown, C., Lall, U., 2006. Water and economic development: the role of variability and a framework for resilience. Nat. Resour. Forum 30, 306–317.

Campbell, S., 2005. Kajakai [sic] Environmental & Sociological Assessment—Project Inception and First Monthly Report. Report submitted to United Nations Office for Project Services, Mausell Aecom, 27 pp.

CAN (Canada), 2008. Project profile: Arghandab Irrigation System Rehabilitation. International Development Project. http://www.acdi-cida.gc.ca/cidaweb/cpo.nsf/vWebCSAZEn/CF769135A8B240A185257507003C938C (accessed August, 2015).

Chandrasekaran, R., 2012. Little America. Knopf, New York, 368, pp.

Cullather, N., 2002. Damming Afghanistan: modernization in a buffer state. J. Am. Hist. 89 (2), 512–537.

DACAAR (Danish Committee for Aid to Afghan Refugees), 2005. Facts about DACAARs work on the Band-e-Sultan Dam. Kabul.

Dalton, J.J., 1981. Area Development Projects—Evaluation Criteria. USAID, Kabul.

de Wrachien, D., Mambretti, S., 2009. Dam-Break Problems, Solutions and Case Studies. WIT Press, MA, p334.

dgMarket, 2007. Dam design consultant for review of the study, design, tender documents & reports for Band-e-Sultan Dam repair works. Request for expression of interest, Kabul, 2 pp.

DMA (Defense Mapping Agency), 1986a. Chesht-e Sharif, Afghanistan: sheet 1785, series U611, edition 2-DMA, scale 1:100,000, 1 sheet.

DMA (Defense Mapping Agency), 1986b. Meymaneh, Afghanistan: sheet 1988, series U611, edition 3-DMA, scale 1:100,000, 1 sheet.

DMA (Defense Mapping Agency), 1986c. Bandar, Afghanistan: sheet 1377, series U611, edition 2-DMA, scale 1:100,000, 1 sheet.

DOI (U.S. Department of the Interior), 1970. Central Arghandab Valley unit feasibility report (Kabul: December, 1970), 210 pp.

Druzin, H., 2015. US agencies adjust to shifting landscape in Afghanistan. Stars and Stripes. http://www.stripes.com/news/us-agencies-adjust-to-shifting-landscape-in-afghanistan-1.329562 (accessed August, 2015).

Dupree, L., 1973. Afghanistan. Princeton University Press, USA, p. 804.

EOA (Embassy of Afghanistan-Tokyo), 2004. Reconstruction of Salma Dam begins with Indian aid: News Archive. http://www.afghanembassyjp.com/en/news/947 (accessed August, 2015).

FAO (Food and Agriculture Organization of the United Nations), 1965. Report on the land and water resources: Afghanistan: 6 volumes.

FAO (Food and Agriculture Organization), 2013. Irrigation in Central Asia in figures: AQUASTAT survey-2012: FAO Water Reports #39, 246 pp.

FAO (Food and Agriculture Organization), 2015. Country profile: Afghanistan. http://www.fao.org/countryprofiles/index/en/?iso3=AFG (accessed May, 2015).

Favre, R., Kamal, G.M., 2004. Watershed Atlas of Afghanistan: 1st Edition—Working Document for Planners: FAO and Afghanistan Information Management Systems (AIMS). 242 pp.

FBO (Federal Business Opportunities), 2012. Dahla Dam improvements project: Solicitation Number: W5J9LE-12-SS-0003, 3 pp.

Fichtner, 2007. Feasability study for Baghdara hydropower project, preliminary environmental and social impact report (volume 1). Stuttgart, Germany, 83 pp.

Formoli, T.A., 1995. Impacts of the Afghan-Soviet War on Afghanistan's environment. Environ. Conserv. 22 (1), 66–69.

Franck, P.G., 1954. Obtaining Financial Aid for a Development Plan: The Export-Import Bank of Washington Loan to Afghanistan. United States Government Printing Office, Washington, DC, p. 55.

Fraser-Tytler, W.K., Gillett, M.C., 1967. Afghanistan: a study of political developments in Central and Southern Asia, 3rd ed. Oxford University Press, London, p. 362.

Frenken, K., 2012. Irrigation in Central Asia in figures: AQUASTAT survey—2012. FAO Water Rep. 39, 246.

FTS (Financial Tracking Service), 2015. Trend analysis, Afghanistan. https://fts.unocha.org/pageloader.aspx?page=emergemergencyCountryDetails&cc=AFG&yr=2015 (accessed August, 2015).

Glasse, J., 2015. Ice cream and dreams melt as Kandahar's electricity falters: Al Jazeera America. http://america.aljazeera.com/watch/shows/compass/articles/2015/6/23/kandahar-electricity.html (accessed August, 2015).

Goodman, A.S., 1984. Principles of Water Resource Planning. Prentice Hall Publishers, NJ, USA, 563 pp.

Gopalakrishnan, R., 1980. The Geography and Politics of Afghanistan. Concept Publishing Company, New Delhi, p. 280.

Grey, D., Garrick, D., Blackmore, D., Kelman, J., Muller, M., Sadoff, C., 2013. Water security in one blue planet: twenty-first century policy challenges for science. Philos. Trans. R. Soc. 371, 10.

Habib, H., Anceno, A.J., Fiddes, J., Beekma, J., Ilyushenko, M., Nitivattananon, V., Shipin, O.V., 2013. Jumpstarting post-conflict strategic water resources protection from a changing global perspective: gaps and prospects in Afghanistan. J. Environ. Manage. 129, 244–259.

Hall, J.W., Grey, D., Garrick, D., Fung, F., Brown, C., Dadson, S.J., Sadoff, C.W., 2014. Coping with the curse of freshwater availability. Science 346 (6208), 429–430.

Hamby, A.L., 1985. Truman quoted in, Liberalism and its challengers: FDR to Reagan: New York, 416 pp.

Harvey, I., 2009. Dahla Dam restoration the centerpiece of Afghanistan reconstruction: Daily Commercial News. http://dailycommercialnews.com/Infrastructure/News/2009/11/Dahla-Dam-restoration-the-centrepiece-of-Afghanistan-reconstruction-DCN036271W/ (accessed August, 2015).

He, X.Y., Wang, Z.Y., Huang, J.C., 2008. Temporal and spatial distribution of dam failure events in China. Int. J. Sed. Res. 23, 398–405.

Hewitson, B., 2015. To build capacity, build confidence. Nat. Geosci. 8, 497–499.

Hirsch, A.M., 1956. Utilization of international rivers in the Middle East. Am. J. Int. Law 50 (1), 85.

ICOLD (International Commission on Large Dams), 2015. Failure of a dam: dictionary entry. http://www.icold-cigb.org/GB/Dictionary/dictionary.asp (accessed July, 2015).

IEP (Institute for Economics and Peace), 2015. Global peace index. http://www.visionofhumanity.org/#/page/indexes/global-peace-index (accessed July, 2015).

IRIN, 2005. Afghanistan: dam burst causes flooding in Ghazni: IRIN Humanitarian News and Analysis, Ankara, Turkey. http://www.irinnews.org/report/28378/afghanistan-dam-burst-causes-flooding-in-ghazni (accessed August, 2015).

IROA (Islamic Republic of Afghanistan), 2014. Realizing self-reliance: commitments to reforms and renewed partnerships. London Conference on Afghanistan. http://www.afghanistan-un.org/wp-content/uploads/2014/12/REALIZING-FINAL-SELF-RELIANCE-25-November-2014.pdf (accessed May, 2015).

Kaika, M., 2006. Dams as symbols of modernization: the urbanization of nature between geographical imagination and materiality. Ann. Am. Assoc. Geograph. 96 (2), 276–301.

Koh-e-Albroz, 2012. Completed projects, Band-e-sultan/DAM: AlbruzGroup, 1 pp.

Kohi, Q., 2015. Construction work on Almar Dam resumes: Pajhwok Afghan News. http://archive.pajhwok.com/en/2015/03/15/construction-work-almar-dam-resumes (accessed August, 2015).

Korman, R., 2014. Why we need large dams. Eng. News-Record 274 (4), 126.

Kumar, R., Prasad, A.N.N., Mahure, N.V., Sharma, P., Gupta, S.L., Ratnam, M., 2014. Concreting at low ambient temperatures at Salma Dam project Afghanistan—a case study. Int. J. Emerging Technol. Adv. Eng. 4 (2), 916–920.

Lane, N., 2008. Aging infrastructure: dam safety: Congressional Research Service Report for Congress, RL33108, 16 pp.

Lashkaripour, G.R., Hussaini, S.A., 2008. Water resource management in Kabul River basin, eastern Afghanistan. Environmentalist 28, 253–260.

Linsley, R.K., Franzini, J.B., Freyberg, D.L., Techobanoglous, G., 1991. Water-Resources Engineering, 4th ed. Tata McGraw Hill Publishers, NJ, USA841.

Maierhofer, C.R., 1961. Drainage and relate problems of the Helmand Valley Development Project, Afghanistan. US Department of the Interior, Bureau of Reclamation, Denver, CO, 54 pp.

McDonald, K., Bosshard, P., Brewer, N., 2009. Exporting dams: China's hydropower industry goes global. J. Environ. Manage. 90, S294–S302.

Meinzen-Dick, R., 2007. Beyond panaceas in water institutions. Proc. Natl. Acad. Sci. 104 (39), 15200–15205.

MEW (Ministry of Energy and Water), 2006. Feasibility Study of the Baghdara Hydropower Project: Report on Hydrological Study. Fichtner Consulting Engineers, Germany, 203 pp.

MEW (Ministry of Energy and Water), 2007. Feasibility Study of the Baghdara Hydropower Project, Volumes 1 and 2: Preliminary Environmental and Social Impact Report. Fichtner Consulting Engineers, Germany, 156 pp.

Michel, A.A., 1959. The Kabul, Kunduz, and Helmand Valleys and the National Economy of Afghanistan. National Academy of Sciences, National Research Council, Washington, DC, 441 pp.

MK (Morrison-Knudsen Company), 1951a. Kajaki Dam: Keystone to Modern Afghanistan. Em-Kayan 10 (2), 1–2.

MK (Morrison-Knudsen Company), 1951b. Arghandab Dam Rises in Afghanistan. Em-Kayan 10 (3), 1.

Mogelson, L., 2012. The Scariest Little Corner of the World: New York Times. http://www.nytimes.com/2012/10/21/magazine/the-corner-where-afghanistan-iran-and-pakistan-meet.html (accessed August, 2015).

Murray, J., 1886. Ghazni and Ghilzi Country: Supplementary Papers, v. 1. Royal Geographical Society, p. 347.

Mustafa, K., 2011. India to help Afghanistan build 12 dams on Kabul River: The News International. http://www.thenews.com.pk/Todays-News-13-5933-India-to-help-Afghanistan-build-12-dams-on-Kabul-River (accessed August, 2015).

NGIA (National Geospatial-Intelligence Agency), 2001a. Kandahar, Afghanistan: sheet 2180, series U611 edition 7-NGA, scale 1:100,000, 1 sheet.

NGIA (National Geospatial-Intelligence Agency), 2001b. Ghazni, Afghanistan: sheet 2684, series U611 edition 6-NGA, scale 1:100,000, 1 sheet.

NGIA (National Geospatial-Intelligence Agency), 2001c. Sorubay, Afghanistan: sheet 2986, series U611 edition 8-NGA, scale 1:100,000, 1 sheet.

NGIA (National Geospatial-Intelligence Agency), 2001d. Dizak, Afghanistan: sheet 1582, series U611 edition 5-NGA, scale 1:100,000, 1 sheet.

NGIA (National Geospatial-Intelligence Agency), 2002. Kajaki Olya, Afghanistan: sheet 2081, series U611 edition 6-NGA, scale 1:100,000, 1 sheet.
NYT (New York Times), 1950. Texts of Truman orders to implement Point IV Plan.
Opsingh, 2011. Almar, Afghanistan, photograph. http://www.panoramio.com/photo/58109014 (accessed May, 2015).
Owens, G.P., 1970. Farm Economic Survey: Helmand-Arghandab Valley of Afghanistan. U.S. Agency for International Development, Kabul. December 1971, 106 pp.
PAN (Pajhwok Afghan News), 2010. Planning to build four mega water dams underway. http://e-ariana.com/ariana/eariana.nsf/allDocs/c8bc5ca4657551908725771900590d49!OpenDocument&Click= (accessed August, 2015).
PAN (Pajhwok Afghan News), 2011. Initial work on Kamal Khan Dam begins. http://www.pajhwok.com/en/2011/12/18/initial-work-kamal-khan-dam-begins (accessed August, 2015).
PAN (Pajhwok Afghan News), 2013. Plot to blow up dam hatched in Quetta: NDS. http://www.pajhwok.com/en/2013/04/01/plot-blow-dam-hatched-quetta-nds (accessed August, 2015).
PAS (Public Administration Service), 1959. A proposed management improvement program for the Helmand Valley Authority Afghanistan: Public Administration Service, USOM/Afghanistan, 80 pp.
Perkins, D.C., Culbertson, J.K., 1968. Hydrographic and sedimentation survey of Kajaki reservoir, Afghanistan. In: United States Geological Survey Water-Supply Paper 1608-m, United States Government Printing Office, Washington, DC, 49 pp.
Peterson, S., 2013. Why a dam in Afghanistan might set back peace: Christian Science Monitor. http://www.csmonitor.com/World/Asia-South-Central/2013/0730/Why-a-dam-in-Afghanistan-might-set-back-peace (accessed August, 2015).
Pisaniello, J.D., Burritt, R.L., Tingey-Holyoak, J., 2011. Dam safety management for sustainable farming businesses and catchments. Agric. Water Manage. 98, 507–516.
PISU (Program Implementation Support Unit), 2005. River basins, hydrometric network and proposed projects: Planning Department, Department of Hydrology, map, 1 sheet.
Potter, M., 2010. Security standoff stalls Canadian dam project in Kandahar: The Toronto Star. http://www.thestar.com/news/canada/2010/06/09/security_standoff_stalls_canadian_dam_project_in_kandahar.html (accessed August, 2015).
Poullada, L.B., Poullada, L.D.J., 1995. The Kingdom of Afghanistan and the United States: 1928–1973. The Center for Afghanistan Studies at the University of Nebraska at Omaha, NE, USA, 96 pp.
Qureshi, A.S., 2002. Water Resources Management in Afghanistan: the issues and options. In: Pakistan Country Series No. 14, Working Paper 49, International Water Management Institute, 24 pp.
RGA (Royal Government of Afghanistan), 1956. Final design report on the Kajaki, Arghandab Dams and Boghra Canal projects. International Engineering Company for Morrison-Knudsen Afghanistan.
Roberts, J.J., 1992. Afghanistan and Western policy: 1929–1956. Ph.D. Dissertation, Submitted to Ohio State University in 1990, Printed by UMI Dissertation Information Service, USA.
Rohde, D., 2006. Afghan symbol for change becomes a symbol of failure: New York Times. http://www.nytimes.com/2006/09/05/world/asia/05afghan.html?pagewanted=all&_r=0 (accessed August, 2015).
Roser, M., 2014. Population density in 2014/2100. http://ourworldindata.org/data/population-growth-vital-statistics/world-population-growth/ (accessed August, 2015).
RRERS (Regional Rural Economic Regeneration Strategies), 2013. Provincial Profile for Nimruz Province, 6 pp.
Sadat, S.H., 2012. Large Infrastructure Management: Analysis of Dam Failure in Italy and Afghanistan. Lap Lambert Academic Publishing, Netherlands, 141 pp.
Saffi, M.H., Kohistani, A.J., 2013. Water Resources Potential, Quality Problems, Challenges and Solutions in Afghanistan: DACAAR Scientific Investigation Study in Afghanistan, 274 pp.
Sayed, M., 2014. Abdullah campaigns in Ghazni: ToloNews. http://www.tolonews.com/en/afghanistan/15115-abdullah-campaigns-in-ghazn (accessed August, 2015).
Scudder, T., 2005. The Future of Large Dams: Dealing With Social, Environmental, Institutional and Political Costs. Earthscan, London, 389 pp.
Shroder, J.F., 2007. Afghanistan's development and functionality: renewing a collapsed state. GeoJournal 70, 91–107.
Shroder, J.F., 2014. Natural Resources in Afghanistan: Geographic and Geologic Perspectives on Centuries of Conflict. Elsevier, Amsterdam, Netherlands, 572 pp.
SHVA (Scott's Helmand Valley Archives), 2015. A personal collection of official and unofficial documents relating to the U.S. involvement in the Helmand Valley, Afghanistan between 1950 and the present: Richard B. Scott, USAID (retired), Helmand analyst. http://scottshelmandvalleyarchives.org (accessed May, 2015).

Stewart, A.K., 2009. Dam Assessment Report, Band-e Soltan masonry dam. Department of the Army Technical Report, 10 pp.

Sulaiman, M.S., 1957. The Irano-Afghan dispute over the Helmand River: a study of its legal aspects and implications. Unpublished Master's Thesis, George Washington University, 195 pp.

Tabish, J., 2011. Iran Faces Renewed Afghan Dam Sabotage Claims: Institute for Water & Peace Reporting. https://iwpr.net/global-voices/iran-faces-renewed-afghan-dam-sabotage-claims (accessed August, 2014).

TEC (Tudor Engineering Company), 1956. Report on Development of Helmand Valley, Afghanistan: Prepared for International Co-operation Administration. Tudor Engineering Company, Washington, DC, 203 pp.

Terraserver, 2015. Satellite and aerial imagery. http://www.terraserver.com (accessed August, 2015).

Tilt, B., Braun, Y., He, D., 2009. Social impacts of large dam projects: a comparison of international case studies and implications for best practice. J. Environ. Manage. 90, S249–S257.

Torell, G.L., Ward, F.A., 2010. Improved water institutions for food security and rural livelihoods in Afghanistan's Balkh River Basin. Int. J. Water Resour. Dev. 26 (4), 613–637.

Tortajada, C., 2014. Water infrastructure as an essential element for human development. Int. J. Water Resour. Dev. 30 (1), 8–19.

Tortajada, C., Biswas, A.K., 2014. Infrastructure and development, an editorial. Int. J. Water Resour. Dev. 30 (1), 3–7.

Troell, J., Weinthal, E., 2013. Shoring up peace: water and post-conflict peacebuilding. In: Weinthal, E., Troell, J., Nakayama, M. (Eds.), Water and post-conflict peacebuilding. Earthscan, London, pp. 1–24.

Tullos, D., Tilt, B., Liermann, C.R., 2009. Introduction to the special issue: understanding and linking the biophysical socioeconomic and geopolitical effects of dams. J. Environ. Manage. 90, S203–S207.

UN (United Nations), 2007. Dam safety in Central Asia: capacity-building and regional cooperation. Economic Commission for Europe, Water Series No. 5, 83 pp.

UNEP (United Nations Environment Program), 2013. Natural resource management and peace building in Afghanistan: United Nations Country Team Brief. http://www.unep.org/disastersandconflicts/portals/155/countries/Afghanistan/pdf/UNEP_Afghanistan_NRM.pdf, 85 pp.

UNEP (United Nations Environment Program), 2015. Afghanistan. http://www.unep.org/disastersandconflicts/CountryOperations/afghanistan/tabid/79589/Default.aspx (accessed May, 2015).

Urban, M., 2011. What went wrong with Afghanistan Kajaki power project?: BBC News. http://www.bbc.com/news/13925886 (accessed August, 2015).

USACE (United States Army Corps of Engineers), 2004. Project Management Plan, Helmand Valley Water Management Plan, 23 pp.

USACE (United States Army Corps of Engineers), 2007. Water balance and regulation alternative analysis for Kajakai reservoir using HEC-ResSim: Phase I and II Final Report PR-63, Institute for Water Resources, Hydraulic Engineering Center, Davis, CA, 64. http://www.hec.usace.army.mil/publications/ProjectReports/PR-63.pdf (accessed August, 2015).

USACE (United States Army Corps of Engineers), 2009. Southeast Afghanistan water resources assessment: Project Report number 148252. http://www2.shovic.com:8080/henryshovic/AssetsGeography/Documents/Afghanistan/SE-AFG%20Water%20Resources%20Assessment%20Final%2031Oct09ReducedSize.pdf, 148 pp. (accessed May, 2015).

USACE (United States Army Corps of Engineers), 2012. Dahla Dam from the air, photograph by Ray, M. https://www.flickr.com/photos/usace-tas/7371244172/in/set-72157630125504148 (accessed May, 2015).

USAID (United States Agency for International Development), 1967. Afghanistan-Kajaki hydroelectric power plant: capital assistance paper: AID-DLC/P-546, 30 pp.

USAID (United States Agency for International Development), 1983. The Helmand Valley in Afghanistan, A.I.D. Evaluation Special Study No. 18, USAID/Washington, 69 pp.

USAID, 1973, USAID-Assisted development in the Helmand-Arghandab Valley: Notes on Program Successes and Problems, USAID/Afghanistan, Kabul, December 12, 1973, 17p, http://www.scottshelmandvalleyarchives.org/docs/evl-73-01.pdf

USAID, 2014, USAID Afghanistan and United States Forces—Afghanistan (USFOR-A) Response to Special Inspector General for Afghanistan Reconstruction (SIGAR) 14-68-SP-IL Inquiry Letter—Kandahar Bridging Solution, dated 10 June 2014, 5 p., in SIGAR 14-87-SP, https://www.sigar.mil/pdf/Special%20Projects/SIGAR-14-87-SP.pdf

USAID (United States Agency for International Development), 2015. Infrastructure: Afghanistan. http://www.usaid.gov/afghanistan/infrastructure (accessed May, 2015).

USCB (United States Census Bureau), 2015. International Programs, Country Rank database. http://www.census.gov/population/international/data/countryrank/rank.php (accessed August, 2015).

USMC (United States Marine Corps), 2011. Photograph of Kajaki Dam, 110508-M-EF299-185 by Richardson, J.R. http://cryptome.org/eyeball/kajaki-dam/pict37.jpg (accessed May, 2015).

Vigne, G.T., 1840. A personal narrative of a visit to Ghuzni, Kabul, and Afghanistan, and of a residence at the court of Dost Mohamed, with notices of Runjit Sing, Khiva, and the Russian expedition (1840): Books in English, Paper 9. http://digitalcommons.unomaha.edu/afghanuno/9 (accessed May, 2015).

Vining, K.C., Vecchia, A.V., 2007. Water-Balance Simulations of Runoff and Reservoir Storage for the Upper Helmand Watershed and Kajaki Reservoir, Central Afghanistan: USGS Scientific Investigations Report 2007-5148, 24pp.

Vining, K.C., Vecchia, A.V., 2007. Water-Balance Simulations of Runoff and Reservoir Storage for the Upper Helmand Watershed and Kajaki Reservoir, Central Afghanistan: United States Geological Survey Scientific Investigations Report 2007-5148, Washington, DC, 17pp.

Wadsam, 2012. Why are major Afghan dam construction projects running so slowly?: Wadsam Afghan Business News Portal. http://wadsam.com/afghan-business-news/why-are-major-dam-construction-projects-running-so-slowly-in-afghanistan-211/ (accessed August, 2015).

Ward, F.A., 2010. Financing irrigation water management and infrastructure: a review. Water Resour. Dev. 26 (3), 321–349.

Watson, P., 2012. Canada's Afghan legacy: failure at Dahla Dam: the Toronto Star. http://www.thestar.com/news/world/2012/07/14/canadas_afghan_legacy_failure_at_dahla_dam.html (accessed August, 2015).

WBG (World Bank Group), 2012. OP 7.50—Projects on International Waterways. http://web.worldbank.org/WBSITE/EXTERNAL/PROJECTS/EXTPOLICIES/EXTOPMANUAL/0,,contentMDK:20064667~menuPK:64701637~pagePK:64709096~piPK:64709108~theSitePK:502184,00.html (accessed August, 2015).

WBG (World Bank Group), 2015. Public data, World Development Indicators. http://data.worldbank.org/data-catalog/world-development-indicators?cid=GPD_WDI (accessed August, 2015).

WCD (World Commission on Dams), 2000. Dams and Development: A New Framework for Decision-Making. Earthscan Publications, London, 404pp.

Wegerich, K., 200. Water strategy meets local reality: Afghanistan Research and Evaluation Unit, Issues Paper Series, 84pp.

Whitney, J.W., 2008. Geology, Water, and Wind in the Lower Helmand Basin, Southern Afghanistan: USGS Scientific Investigations Report 2006-5182, 50pp.

WWF (World Wildlife Fund), 2003. DamRight!: an investor's guide to dams: WWF's Dams Initiative, 35pp.

Yuldashev, A., 2011. The Tajik company has won the tender on dam construction in Afghanistan: ASIA-Plus News. http://news.tj/en/news/tajik-company-has-won-tender-dam-construction-afghanistan (accessed August, 2015).

Yusufi, M.Q., 1988. Effects of the war on agriculture. In: Huldt, B., Jansson, E. (Eds.), The Tragedy of Afghanistan. Croom Helm, London, pp. 197–216.

Zafarnejad, F., 2009. The contribution of dams to Iran's desertification. Int. J. Environ. Stud. 66 (3), 327–341.

Zaland, S., 2015. The construction work on Salma Dam delayed due to security concerns: Afghan Zariza. http://www.afghanzariza.com/2015/01/10/the-construction-work-on-salma-dam-delayed-due-to-security-concerns#sthash.5KIvBijK.dpuf (accessed August, 2015).

CHAPTER

# 10

# Water Quality Matters

P. McNamara

University of Nebraska at Omaha, Omaha, NE, United States

OUTLINE



## Abstract

The focus of most discussions on water-resources management is on the quantity of water available for various purposes, the infrastructure that carries that water, and the allocation of that water among users. This chapter focuses on the other important element in water resources, the quality of that water. In Central and South Asia generally, and Afghanistan specifically, the quality of water varies greatly, although in general water quality in the region is generally very poor. This chapter explores some of the larger themes of water quality, next examines the data available on water quality in Afghanistan and its neighbors. It ends with some policy recommendations on improving water quality.

**Keywords:** Water Quality, Water and Health, Water Conflict, Water Policy



http://dx.doi.org/10.1016/B978-0-12-801886-6.00010-0

## 10.1 INTRODUCTION

A glass of clear, fresh water is much more than just a quenching drink. Beliefs about water ownership, cultural, and religious ideologies, water-management policies, economics of water use, engineering of delivery systems, legal debates, and local, national and international politics are behind each sip. Around the world, these elements have led to conflicts and cooperation, victories for some and challenges for many. In Central and South Asia, the transboundary waters shared by Afghanistan and its neighbors are a primary example of the complexities of water.

Iyer (2007) argued that water is a simple and familiar feature of everyone's daily life, but is in fact a highly complex phenomenon with multiple aspects. How we talk about water frames our perceptions of it; whether we name it a common-pool resource, a mineral to be extracted, a commodity, or a gift from God (Sementelli, 2008). Because water in its many aspects is essential for life, and because the demands outpace the supply of water in many areas, some predictions are that water wars are imminent (Chellaney, 2013; Pierce, 2006; Shiva, 2002; Wolf, 1998). Although I disagree with the more dire predictions, I do think that water stress will increase in the decades ahead and that we will need better jointly constructed, international science projects and regional policy-making approaches to deal with this.

The sciences that inform policies and practices related to water are hotly debated topics in the literature. These range from debates about the effects of global climate change (Dinar et al., 2012; Pollack, 2010; Ringler et al., 2010), to transboundary-water conflicts and cooperation (Giordano et al. 2013; Subramanian et al., 2012; Khalid, 2011; Solomon, 2010; Pierce, 2006), from security (Floyd and Richard, 2013; Kaplan, 2011), and to the sacred symbolism of water in religious traditions (Olivelle, 2009; McNamara, 2014; Jhunjhunwala, 2014; Wolf, 2012). South Asia is of particular interest to scholars of water conflict (Shroder, 2014; Iyer, 2009; Joy et al., 2008; Chellaney, 2011; Asia Society, 2009; Economist, 2011).

Two overarching issues occur in water management: quantity and quality. Most of the attention on water is focused on larger scale issues of insufficient quantities of water. This chapter deals mainly with the latter, the quality of the water. Although it is fairly easy to say that a reservoir contains 3 million cubic meters of water, it is much more difficult to say that this reservoir contains safe water. Because of the hundreds of potential contaminants that water might pick up on its journey from clouds to Earth to reservoir to final user, the tests of water-quality indices are complex and generally only trained scientists or hydro-engineers can read the results.

This chapter begins with an exploration of what is meant by "water quality." It then explores the problems of water quality in the region around Afghanistan. Particular attention focuses on the crucial role that clean water plays in health. The chapter ends with recommendations regarding policies and practices to ensure better water quality.

## 10.2 DEFINITIONS OF WATER QUALITY

Water is the most precious of all resources because without it life cannot survive. All communities must have safe water in order to thrive. From the Roman Empire grew some of the earliest systems for delivery of clean water and disposal of waste water in urban areas.

And Edwin Chadwick's groundbreaking 1842 report on the relationship between clean water and disease may be the birth of the science of public health (Tebbutt, 1998). John Snow's study linking the London cholera outbreak of the mid-1850s to water from a hand-pump on Broad Street was another milestone in the field of public health (Hempel, 2006).

The quality of water depends upon its intended use. Standards for drinking water are different than for water used in fighting fires or irrigating fields. Water for domestic use needs to be cleaner than water for washing farm machinery. Humans have health concerns that farm machines do not. Taking water into our bodies necessitates a purity in the water that use by non-living things does not. Furthermore, water used in some laboratories of the pharmaceutical industry needs to be even freer of contaminants.

Various reasons for ground-water contamination exist. Some are human factors that can be corrected more easily; some are geological factors that are less easily addressed. Table 10.1 summarizes some of the reasons that ground water can be unsafe for use.

Nothing is simple about providing clean, safe, potable water to people in South and Central Asia, with a population of 1.845 billion (Population Reference Bureau, 2013). Yet, this simply must be done if health conditions are to be improved, economies are to grow, and societies are to thrive. The framework in this section to explore why some countries are thriving and others have mounting challenges uses the broad terms "bad water" and "good water."

## 10.2.1 Bad Water

When it comes to bad water, two broad reasons exist: human-caused pollutants and natural-caused contaminants. The sources of water pollutants obviously differ from community to community. Two main sources of human-caused pollutants are agricultural and industrial-urban. In developed countries, agricultural pollutants are overtaking industrial sources as the greatest cause of bad-water quality (Jones, 2010, p. 50).

The Green Revolution of the 1970s produced higher crop yields, but the tradeoff was increasing use of nitrates, pesticides, and herbicides causing agricultural pollution in streams, rivers, and ground-water sources. This is further complicated by genetically modified seeds being used in recent decades that are bred to withstand higher doses of pesticide and herbicide

TABLE 10.1 The Main Factors of Ground-Water Contamination

a. Poor efforts for water-quality monitoring, management, and water-protection systems.
b. Poor land use, which results in contamination of both surface and ground water.
c. Poor legislation and regulation of ground-water quantity and quality protection.
d. Insufficient waste treatment or disposal management.
e. Permeable characteristics of overlying cover layers of aquifers.
f. Poor sanitation and hygiene practices.
g. Lowering of ground-water table levels, which has caused deterioration of ground-water quality.
h. Cross contamination occurs in wells due to poor well-site selection and construction.
i. Lack of awareness about importance of water quality.
j. Strong urbanization without sewerage system and poor solid waste management.

*From Saffi, M.H., Kohistani, A.J., 2013. Water Resources Potential, Quality Problems, Challenges and Solutions in Afghanistan: An Evaluation of the Past and Present Situation of Water Resources and Consequences on the Basis of Previous Water Resource Investigation and Present National Groundwater Monitoring Wells Network Data Management, Mapping, Visualization and Assessment. DACAAR. http://www.dacaar.org/Publications/6%20Research%20and%20Studies/WASH/Report%20on%20Water%20Resource%20Porential_opt.pdf (accessed 09.11.15.).*

applications, which in turn wash off into waterbodies and seep into soil, eventually reaching ground-water sources (Jones, 2010, p. 53). Evidence also exists that the Green Revolution coincided with a resurgence of malaria in South Asia (Verghese, 2007, p. 240).

The amounts of agricultural pollutants in the waters of Afghanistan and its neighbors are not well documented. The pollutants, however, are certainly not at the same level as occurs the high-yield agricultural production in many developed countries. So, the pollutants in the waters of Afghanistan from nitrates, pesticides, and herbicides are not as high.

Infectious diseases transmitted through water are another major concern. The World Bank estimates that 90% of the 2 million people who die per year from sanitation-related disease are children (Jones, 2010, p. 81). Grouped together these have been coined "no-plumbing disease" (Sauer, 2009).

In Afghanistan and its neighbors, the systems of plumbing—both pipes carrying good water to people and pipes carrying bad water from people—are not well developed. In a study of Badakhshan Province in Afghanistan (Kanji et al., 2012) only 32% of remote villagers had access to improved-water sources whereas 75% of non-remote villagers had access; only 31% of remote villagers had access to improved toilets versus 65% of non-remote villagers. Other countries in South and Central Asia fare better than Afghanistan, which has water quality rated among the lowest in the world. But in any case, the region as a whole is not rated well in terms of water quality. According to the World Health Organization (2000), the following data are available for only some countries (Table 10.2).

Countrywide information for Tajikistan was not included in the WHO (2000) report, but a case study mentioned some specific regions of the country (p. 49). Historically, the agricultural people in the Khatlon Province of southern Tajikistan settled near the Vakhsh or Pyandj (Panj) rivers that have their origin in the Pamir glaciers. The drinking-water system developed during Soviet-times involved piping water from wells, but in the 1980s these sources were found to be contaminated by the Vakhsh Chemical Company. By the mid-1990s, "less than 10% of the rural population of Khatlon Province had access to safe drinking water and less than 5% to sewerage systems" (WHO, 2000, p. 49).

TABLE 10.2 Water Quality Statistics for Afghanistan and Neighboring Countries

| Country | Urban Population With Water Supply (%) | Rural Population With Water Supply (%) | Urban Population With Improved Sanitation (%) | Rural Population With Improved Sanitation (%) |
|---|---|---|---|---|
| Afghanistan | 19 | 11 | 25 | 8 |
| India | 92 | 86 | 73 | 14 |
| Iran | 99 | 89 | 86 | 74 |
| Kazakhstan | 98 | 92 | 100 | 98 |
| Kyrgyzstan | 98 | 66 | 100 | 100 |
| Pakistan | 96 | 84 | 94 | 42 |
| Uzbekistan | 96 | 78 | 100 | 100 |

The data clearly show that Afghanistan is far below the other countries in all categories of water quality.
*Summarized from World Health Organization, 2000. Global Water Sanitation and Health Assessment. WHO, Geneva. http://www.who.int/water_sanitation_health/monitoring/Glassessment7.pdf?ua=1 (accessed 01.26.15.).*

Three health concerns related to waste water are the transmission of germs from human feces to food sources; the contamination of surface water through lack of proper sewage drains; and, not washing hands after defecating or urinating. As examples of typical South Asian conditions, no city exists in India with a complete effluent treatment system, the result being open sewage seeping into streams, rivers, and seas (New York Times, Jul. 15, 2014). A recent report has noted that 70% of India's water supply is polluted with sewage effluents, and water quality in India ranks 120th among the 122 nations (UNICEF, 2013). In a different ranking from Yale University, India was 124th out of 178 nations, whereas Tajikistan was 105th, Pakistan was 123rd, and Afghanistan was 164th (Environmental Performance Index, 2014).

The Afghanistan water and health situation is made worse by lack of practices for hygiene and sanitation. A Sustainable Water Supply and Sanitation Project of the US Agency for International Development has helped select provinces including Baghlan, Takhar, Badakhshan, Laghman, Nangarhar, and Lowgar (USAID Afghanistan, 2012). These areas in northeastern Afghanistan are highlighted in the map of Fig. 10.1.

With these USAID-funded efforts, non-governmental organizations (NGOs) greatly improved the sanitation and health of the water supply in those provinces. Over half of the households now treat their water by boiling and another 36% by chlorine tablets. Others use sunlight or filter systems to treat water. Hand washing has increased, including use of soap, as part of these project efforts. Open defecation, which is a problem in many parts of Afghanistan, has decreased because of new or improved latrines in these select provinces. More such efforts are needed, but with security concerns on the rise in many parts of Afghanistan, this is difficult. The human behavior factors can be more easily changed through interventions such as public education campaigns to promote handwashing with soap. The infrastructure factors, such as building latrines and water-treatment plants, are more difficult to impact (Fig. 10.2).

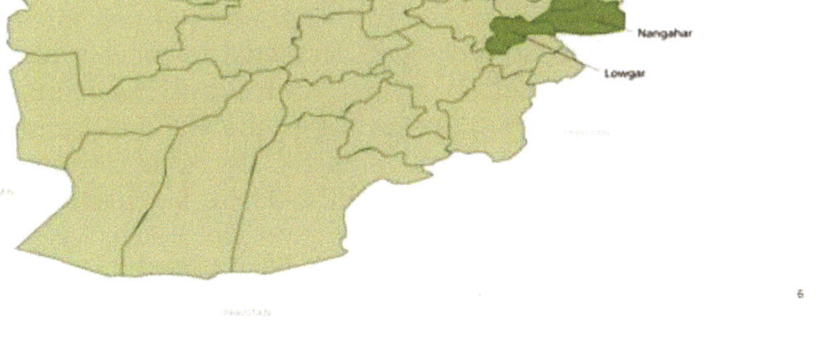


FIG. 10.1 Afghanistan sustainable water supply and sanitation project. *From USAID Afghanistan, 2012. Afghan sustainable water supply and sanitation (SWSS) project: sustainable health outcomes unit project final report. https://sanitationupdates.files.wordpress.com/2012/12/afghanistan-swss2012.pdf (accessed 11.01.15.).*

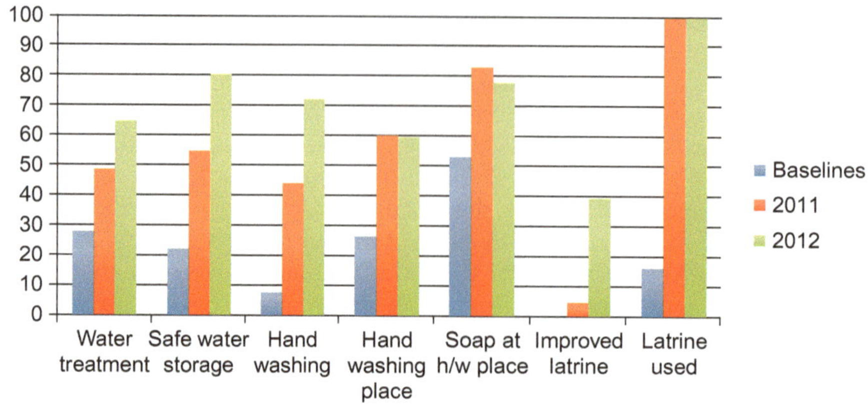


FIG. 10.2 Project-weighted average rates for USAID-required sanitation and hygiene indicators. *From USAID Afghanistan, 2012. Afghan sustainable water supply and sanitation (SWSS) project: sustainable health outcomes unit project final report. https://sanitationupdates.files.wordpress.com/2012/12/afghanistan-swss2012.pdf (accessed 11.01.15.).*

In a study of waste-water treatment that the author did in Andhra Pradesh, a state in southern India, a sad state of environmental affairs was observed. Media coverage of a public lecture the author gave reported that, "Explaining the functioning of the sewage treatment plant near Appu Ghar here in Visakhapatnam, [the speaker] said that only one turbine of the six at the sewage treatment plant was in working condition. Broken pump equipment was not replaced for the last few years, and except for a security guard, there was no staff to handle the plant. The waste was being released into the sea without any treatment" (India Express, Jul. 31, 2014).

Natural-caused contaminants are more difficult to prevent than those that are human-caused; they can be just as deadly, however. Two of these natural causes for bad water are fluoride and arsenic. Fluoride is a naturally occurring chemical in some water that has shown positive benefits for preventing tooth decay, especially in children. Some municipal water supplies are laced with low-levels of fluoride for the dental benefits. But in larger doses it can stain teeth, cause bone decay and possible increased risk of cancer (Tebbutt, 1998).

Arsenic pollution is prevalent in some of the water supplies of Central and South Asia. Jones (2010, p. 78) has noted that hydrothermal waters permeating through fractured granites in mountain ranges such as the Himalaya, and the weathering of volcanic rocks can release arsenic, which may then be carried by rivers and deposited in concentrated form in flood-plain sediments. Arsenic can also come from human sources such as paints and dyes, detergents, agricultural chemicals, and mining. According to the U.S. Environmental Protection Agency, drinking water with dangerous levels of arsenic can cause dermatological damage, circulatory problems, and increased risk of cancer (EPA, 2015). Arsenic in Afghanistan drinking water is discussed in detail in Chapter 3 herein.

Although a complex relationship exists between certain water-borne diseases and water management practices, natural causes exist for many of the most prevalent diseases. For instance, when flooding occurs, a higher probability of standing water results that is breeding grounds for diseases such as cholera, malaria, and dengue. Floods also increase other disease risks such as sewers overflowing, houses being destroyed causing residents to sleep outside exposed to the severe weather and insect bites, and washing out the infrastructure of roads and bridges for transportation of food and other necessities.

Earthquakes are common in the region around and especially inside Afghanistan, which has the world's largest number of large magnitude earthquakes (>7–8 magnitude) recorded than anywhere else in the world (Shroder, 2014, p. 243). These earthquakes often cause other disasters such as landslides or floods, negatively impacting water access and quality.

## 10.2.2 Good Water

The flipside of bad water is, of course, good water. And the goal of every community is to have it. Drinking water is the most precious resource because life cannot survive without it. Drinking-water quality is measured in terms of its physical characteristics including color, taste, odor, salinity, acidity, and the presence of many potentially harmful chemicals. These physio-chemical characteristics make up water-quality indices of which there are hundreds. The first water-quality index (WQI) was introduced in Germany in 1848 to indicate the fitness of a water source based on the absence or presence of certain organisms (Abbasi and Abbasi, 2012, p. 5). Later, Horton (1965) would develop the first modern era WQI using 10 commonly measured variables including dissolved oxygen, coliform bacteria, pH, alkalinity, specific conductance, and chloride (Abbasi and Abbasi, 2012, p. 5). More recently, the UN Environment Program Global Environment Monitoring System (2007) has come up with recommended WQIs. Different factors are included in various WQI calculations, but the basic formula is the same. The factors are calculated such as in the formula in Fig. 10.3 and a resulting single number is determined that is used to rate the water as excellent, good, fair, marginal, or poor (UN Environment Program, 2015).

The benefits of using a universally agreed-upon WQI is that by defining what is good water then appropriate resource allocations can be made, along with environmental regulations and enforcement, public education, and scientific research. The WQI selected depends on the

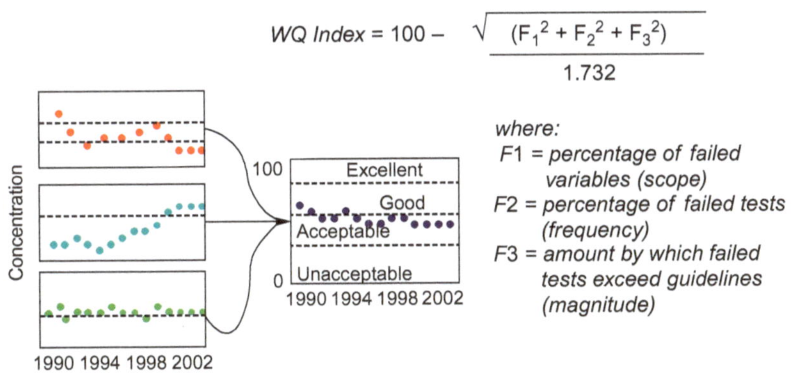


FIG. 10.3 Index calculation of water quality. Important WQ variables compared to appropriate guidelines; results combined to produce a single number categorizing WQ as excellent, good, fair, marginal, and poor. *From UN Environment Program, 2015. Water Quality Indices. UNEP, Nairobi http://www.unep.org/gemswater/Default.aspx?tabid=101094 (accessed 11.01.15.).*

intended use of the water, the environment from which the water comes, and the measurement tools available to those testing the water.

One of the United Nation's Millennial Development Goals (2015) (MDG) is to "halve, by 2015, the proportion of the population without sustainable access to safe drinking water and basic sanitation." According to an UNICEF (2014) report, Afghanistan met the MDG goal with 49% of its 2012 population gaining access to safe drinking water and basic sanitation since 2000; Tajikistan met the target with 25% of its population gaining access; as of 2014, Pakistan was not on track to meet the MDG goal with 18% gaining access, and India was not on track to meet the MDG goal with only 14% gaining access since 2000.

A major factor determining whether water is good enough for its intended use is the infrastructure. In urban areas this characteristically comprises a distribution system that includes reservoirs, pumping stations, filtration, pipes, and ultimately sinks, toilets, and showers. In rural areas, especially in developing countries, this system may be a watering hole or stream that is believed to be relatively clean, a well that might have a pump attached, or a more elaborate piping system that carries water from a distant source to a village for use.

As an example, in rural Maharashtra, India, the author visited a village that had invested, with the help of an NGO, in a system to pipe water from the hills a few kilometers away to the village where some homes actually had running taps. But a developer purchased the highlands to build a hotel. Tube-wells were drilled by the developer, reducing the ground water that had flowed down to the village. Within months, the flow of water to the village was one-third of what it had been. The villagers had no legal rights to the water under the land purchased by the developer. They went back to using the brackish water in local wetlands to supplement the small amount that now trickled out of the pipe from highlands. A difficult lesson was learned by the villagers and the NGO leaders. Such types of occurrence are all too common in much of Asia, including Afghanistan.

Another major factor in access to quality of water is good governance. The infrastructure must be built and maintained, the disputes between users must be resolved, and the policies that govern water must be debated, approved, and implemented. A case study of Tajikistan's water (NPR, 2013) revealed a crumbling infrastructure and lack of good governance in the post-USSR period. This led to a decrease in the access of Tajiks to good water.

The dispute-resolution mechanisms that exist around water clearly differ based on local customs, laws, cultures, and relationships. One innovative mechanism for conflict resolution grows out of the work of the NGO Tarun Bharat Sangh that was founded by Rajendra Singh (2005). Called "The Waterman of India," Singh won the Man Say Prize (known as the Nobel Prize for Asia) for his work in rural villages in Gujarat State. The *Jal Sansad* or Water Parliaments of the Arvari River (Sisodia, 2014; Sinha et al., 2013; Glendenning, 2009) were described as an informal decision-making and conflict-resolution body that was based on traditional customs of the small Arvari River in Rajasthan, India (Wolf et al., 2005; as noted in Gehrig and Rogers, 2009, p. 83). Pushing for decentralized decisions and tapping into local wisdom seems to have worked in this case.

The processes for conflict resolution, however, are more complex when dealing with transboundary water disputes. In groundbreaking work by a team at Oregon State University led by Aaron Wolf, the Transboundary Freshwater Dispute Database (TFDD) was used to create a water event intensity scale (WEIS) to rate the conflicts in over 1800 cases (Wolf et al., 2003). The summary analysis of the TFDD has impressively dispelled

TABLE 10.3 The BAR (Basins at Risk) Scale, Which Is Used to Determine the Severity of the Water Event

| BAR Scale | Event Description |
|---|---|
| −7 | Formal declaration of war |
| −6 | Extensive War Acts causing deaths, dislocation, or high strategic cost |
| −5 | Small scale military acts |
| −4 | Political-military hostile actions |
| −3 | Diplomatic-economic hostile actions |
| −2 | Strong verbal expressions displaying hostility in interaction |
| −1 | Mild verbal expressions displaying discord in interaction |
| 0 | Neutral or non-significant acts for the inter-nation situation |
| 1 | Minor official exchanges, talks or policy expressions; mild verbal support |
| 2 | Official verbal support of goals, values, or regime |
| 3 | Cultural or scientific agreement or support (non-strategic) |
| 4 | Non-military economic, technological or industrial agreement |
| 5 | Military economic or strategic support |
| 6 | International Freshwater Treaty; Major strategic alliance (regional or international) |
| 7 | Voluntary unification into one nation |

*From Oregon State University, 2015. Transboundary waters. http://www.transboundarywaters.orst.edu/database/event_bar_scale.html (accessed 11.01.15.).*

the myth of water wars, with more than two-thirds of over 1800 water-related "events" having fallen on the cooperative scale of the WEIS, whereas less than one-third occur on the conflict scale (Zeitoun and Mirumachi, 2008, p. 300). The scale of basins at risk (BAR) determines the severity of the water event, as shown in Table 10.3, summarizing the WEIS online database.

Although water can be a source of conflict, the vital substance can also be a means to cooperation. The positive numbers on the BAR scale demonstrate this. When viewed not as a scarce resource to be fought over, but as a precious resource to share between different households, communities, and nations, water brings people together. The author has seen places—such as communities in rural Rajasthan, India—where peace through water is made real with People's Parliament of Water (Sisodia, 2014).

## 10.3 WHY WATER QUALITY MATTERS

The first reason water quality matters is for public health. Evidence has grown for the linkage between clean water and health (Prakash and Chourey, 2011). Many diseases are carried through water. In India, almost 70% of hospitalizations are caused by some form of water-borne disease and annually over 100,000 people die of water-borne diseases (UNICEF, 2013). Cases of malaria or dengue fever get headlines, but the much more frequent hospital visits from diarrhea, vomiting, and resultant dehydration caused by bad water get less attention. Tuberculosis is also caused by poor sanitation. These diseases and many others are carried in unclean waters. It is astounding to think that over two-thirds of the hospital visits and 100,000 deaths could be prevented if the water in India were clean. Similar statistics apply in Afghanistan.

A second reason is environmental. Although we think of the immediate consequences of bad water on humans, the long-term sustainability of the ecosystem is in doubt as we pollute the waters of the Earth. Beyond the consequences for humans, disastrous impacts occur on all life. A report from a recent conference hosted by the Daugherty Water for Food Institute at the University of Nebraska (2015, p. 44) makes this link: "Though the idea that human health is affected by animal and environmental health—the One Health concept—isn't new, recent global changes have led to renewed focus on these interconnections."

Scientists are envisioning a future in which climate change exacerbates the challenges identified in the One Health approach. "Climate change increases the risk of diarrheal disease because pathogens proliferate faster in warmer temperatures and wash into waterways during heavy rains, while droughts encourage people to use pools of standing water where pathogens or disease vectors reside. In addition, climate change will reduce food production's potential gains, thereby increasing the number of undernourished and stunted children by 2050" (Daugherty Water for Food Institute at the University of Nebraska, 2015, p. 47).

How this will play out in different countries is not yet known. Those with better infrastructure, conflict-resolution mechanisms, and good governance will fare better in the future. The governance issues in the region are a continuing concern, but Afghanistan has unique problems in this regard. Since Sept. 11, 2001, Afghanistan has been on the radar of the international community. After a war that overthrew the Taliban government, the level of international donor coordination has been unprecedented (Qassem, 2009). Afghanistan still remains a troubled state 15 years later. Now we look at four countries in the neighborhood of south-central Asia and explore in further detail the water quality there.

## 10.4 STATE OF WATER QUALITY IN CENTRAL AND SOUTH ASIA

### 10.4.1 Afghanistan

Around the Central and South Asia region, access to clean water varies widely. Afghanistan's rural population, which makes up 79% of the country's 29 million people, has among the lowest percentage in the world of people who have access to clean water at 20%. This is closely related to the fact that only 5% of the country has access to improved sanitation facilities with 20% practicing open defecation (HydrateLife, 2015). The Danish Committee for Aid to Afghan Refugees (DACAAR) did a comprehensive study of Afghanistan's water. Table 10.4 summarizes some of the key findings of that study.

Treaties between Afghanistan and its neighbors do have some frameworks for building agreements on issues of water quality, but do not have specific targets with enforcement

TABLE 10.4 Main Water Supply and Sanitation Management Indicators in Afghanistan

a. National overage access to safe drinking water in Afghanistan is only 27%, in which only ~20% of the rural population has access, which are among the world's lowest values.
b. Estimates are that 191 out of 1000 children die before their 5th birthday, out of which 44 die because of unsafe water and poor sanitation.
c. No clear and scientific image of ground water potential (availability) exists to assess supply and demand.
d. No water safety plan exists.
e. A high density of wells per unit area exists, without regard to spacing norms or number of households constructed; a standard well has mains that supply safe drinking for 25 households with 8 people per household.
f. Only limited technical options exist for water use, development, or storage.
g. No practical alternatives exist for investigation and development of water resources.
h. Cross contamination of wells occurs because of poor selection and construction of well sites.
i. A lack of technical capacity occurs at the national and local levels in terms of effective operation and maintenance of the water supply systems.
j. The capacity to access spare parts is also lacking due to distance to markets where such products are available.
k. According to safe drinking water standards, such is a challenge for millions of rural inhabitants around the country.
l. Poor access to safe drinking water, inadequate sanitation, and poor hygiene cause water-related diseases, which potentially affect the overall health of people in the country.
m. The water private sector is poorly equipped and has poor skills.

*From Saffi, M.H., Kohistani, A.J., 2013. Water Resources Potential, Quality Problems, Challenges and Solutions in Afghanistan: An Evaluation of the Past and Present Situation of Water Resources and Consequences on the Basis of Previous Water Resource Investigation and Present National Groundwater Monitoring Wells Network Data Management, Mapping, Visualization and Assessment. DACAAR. http://www.dacaar.org/Publications/6%20Research%20and%20Studies/WASH/Report%20on%20Water%20Resource%20Porential_opt.pdf (accessed 09.11.15.).*

mechanisms. The 1958 Treaty between the Afghanistan and Tajikistan, which latter country was then part of the Union of Socialist Soviet Republics (USSR), provided for cooperation on preventing water pollution (Fipps, 2007, p. 3). A number of agencies (US Geological Survey or USGS, Danish Committee for Aid to Afghan Refugees or DACAAR, and others) have investigated ground-water quality in many parts of Afghanistan that could be sampled and tested without too many security concerns, but particularly in the capital city of Kabul. The results (Saffi and Kohistani, 2013) have shown that the chief contaminants of concern in Afghanistan are: salinity, nitrate, nitrite, sodium, chloride, boron, bromide, fluoride, arsenic, chromium, sulfate, hardness, and *Escherichia coli* bacteria from fecal waste. For example, a large number of wells in the Kabul River Basin are contaminated, with 108 wells tested between Jul. and Nov. 2004 that showed the following numbers of wells with certain harmful contaminants that exceeded USEPA drinking-water standards (Thomas and Eqrar, 2011):

- dissolved solids—54 wells
- nitrate—13 wells
- sulfate—14 wells
- chloride—11 wells
- boron—44 wells
- fluoride—0 wells
- bromide—22 wells
- fecal coliform bacteria—73 wells.

It is clear from these and other studies that the shallow ground water in wells throughout the country is quite contaminated, with high levels also in other monitored wells in 16 provinces, many of which are blamed for outbreaks of intestinal disease, especially those that are quite harmful to children. In a number of cases with high salinity and boron, the water is not even useful for irrigation, because the salt is harmful to plants, and the boron causes liver, stomach, and kidney problems. Currently, Afghanistan lacks the technical or laboratory capacity to properly assess or regularly monitor the quality and surface variations of ground-water resources because of a lack of trained personnel and testing equipment (Saffi and Kohistani, 2013).

Evidence from U.S. Geological Survey (Mack, 2010) indicates that the ground water near urban areas of Afghanistan that is less than 100 m down is of poor quality because of the quite inadequate sanitation and virtual absence of any sort of waste-water treatment. However, the quality of the deeper ground water in aquifers occurring in the Kabul Basin that likely date back many thousand years is unknown. Although these deep aquifers may sustain the projected needs of municipal users, it is unlikely that the much greater needs of agricultural users can be sustained (Mack, 2010).

Access to clean, safe domestic water has been shown in one study to be an important factor in the risk for pregnancy-related mortality in Afghanistan (Gon et al., 2014). In Kabul, the quality of water is a concern because several surveys have indicated a significant degree of contamination in the upper layers of the aquifer, which is most probably linked with the poor waste-disposal practices of the inhabitants (Geneva Foundation, 2002, p. 1).

In surveys of 148 wells in the Kabul River Basin, the ground-water quality differed greatly with some being suitable for human and irrigation uses, whereas other ground water containing dangerously high concentrations of nitrate, boron, dissolved solids, and fecal contamination. Over 10% of the wells sampled had water that was unsuitable for public use (Broshears et al., 2005).

The treatment of this water is important for health and development in Afghanistan. A public education program also needs to ensure that people know what they are consuming. One project that the author is currently working on takes a "citizen scientist" approach with field studies in Nebraska, India, and Chile. In this project, high school students in these three countries are taught to test the local water supply with simple dip-stick technology, and then analyze if that water is safe for drinking and other purposes. A similar project could be started in Afghanistan, training youth to test water supplies and report unsafe conditions. This might be particularly successful in remote areas where security concerns prevent visits and training of citizen scientists could be done via videoconference or cell phone (Tables 10.5–10.7).

TABLE 10.5 Challenges to the Delivery of Clean Water and Improved Sanitation in Afghanistan

1. Lack of institutional capacity.
2. Lack of sufficient material and spare parts for water-system infrastructure.
3. Lack of water management information systems.
4. Lack of coordination among aid partners.
5. Lack of monitoring and evaluation.
6. Low demand for sanitation.
7. Poverty barriers to access to improved water and sanitation.
8. Gender barriers to access to improved water and sanitation.
9. Inefficient community mobilization and participation.

*From Alim, A.K., Ahmadzai, A., Rizk, J., 2011. Chapter 3: water for human consumption and water for sanitation. In: Afghanistan Human Development Report 2011. Centre for Policy and Human Development, Kabul.*

TABLE 10.6 Requirements in Afghanistan for Water and Sanitation to Meet the Principles of Human Rights as Expressed by the United Nations Committee on Economic, Social, and Cultural Rights in 2002

*Water*

1. Affordable for all (money and status no barrier).
2. Accessible to all, physically and legally (no barriers because of class, status, politics, gender, or other factors).
3. Adequate (clean, acceptable, safe).
4. Sufficient (at least 20 l/person/day).
5. Sustainable (for families and communities).

*Sanitation*

6. Safe (excreta away from human contact and disposed of).
7. Physically accessible (children, women, elderly, disabled).
8. Affordable (construction and maintenance).
9. Culturally sensitive (dignity, privacy, safety).

*From Alim, A.K., Ahmadzai, A., Rizk, J., 2011. Chapter 3: water for human consumption and water for sanitation. In: Afghanistan Human Development Report 2011. Centre for Policy and Human Development, Kabul.*

TABLE 10.7 Delivering Safe Drinking Water and Proper Sanitation in Afghanistan Will Take Strong Attention to the Details Outlined Below

1. Important political leadership.
2. Recognition of the right to good water and proper sanitation.
3. Reducing inequality and extending coverage to the poor.
4. Role of the private sector could be greatly increased to help.
5. New independent service authority and social policy strategies can help.
6. Developing, testing, and applying subsidies and pro-poor pricing should be tried.
7. Community participation is essential.
8. Monitoring the quality of drinking water must be done.
9. Using household filters should be introduced.
10. Focus must be upon long-term operation and maintenance.
11. Hygiene must be promoted on a large scale.
12. Traditional latrines must be upgraded.
13. Low-cost disposal systems should be constructed.
14. Waste can be converted into a resource.
15. Technological fixes to different latrines can be implemented.

*From Alim, A.K., Ahmadzai, A., Rizk, J., 2011. Chapter 3: water for human consumption and water for sanitation. In: Afghanistan Human Development Report 2011. Centre for Policy and Human Development, Kabul.*

## 10.4.2 Tajikistan

Davide Costa of Oxfam in Tajikistan (NPR, 2013) has noted that during the Soviet times functioning water systems occurred throughout the country, but because the infrastructure has not been maintained during post-Soviet independence, only 5% of rural Tajiks now have access to safe water or adequate sewage facilities, leading to outbreaks of typhoid and other water-borne diseases.

From 2010 to 2013, USAID invested more than $1.7 million in improving 57 drinking water systems that provide access to clean drinking water to 150,000 people in Tajikistan (Blau, 2013). Yet, it remains a problem for many. The United Nations' Environmental Performance Review of Tajikistan (2012) found that approximately one third of the 7.2 million population has access to chlorinated piped water, whereas another third rely on spring water and the remaining approximate third must rely only upon river and ditch-water sources.

In Jun. 2015, Dushanbe hosted the High Level International Conference on the Implementation of the International Decade for Action Water for Life 2005–2015. This conference brought together top water experts from around the world. In the statement that came out of the conference, progress was acknowledged on water cooperation within and between countries, yet gaps remain between rich and poor, males and females, affordability and accessibility in water resources (UN Water for Life, 2015).

A parallel track during this UN Water for Life Conference focused on water in mountain countries. Some of the key messages coming out of this dialog between mountain states included: (1) water is a source of life upstream and downstream, as well as a source of conflict; (2) water can bring together countries in cooperative arrangements over shared-water resources; and (3) poor countries and peoples are especially hard hit by the condition of the environment, socioeconomic factors, and distribution of resources (Mountain Partnership, 2015).

One suggestion for addressing some of these inequities in key message number three above would be for downstream riparian states to compensate those countries higher on the watershed if they maintain clean water and released it to flow downstream (discussed in Chapter 6 as payment for ecosystem services). It benefits the downstream countries to have their upstream neighbors take good care of that shared water resource. Surely a monetary value can be assigned to that clean water. And this would benefit the upstream states by providing much needed income, as well as the downstream states by providing much needed safe water.

### 10.4.3 Pakistan

A case study (Bokhari, 2006) of a village near Abbottabad, Pakistan, illustrates the power of women to overcome a dire state of water quality. After forming a community-based women's organization in their village, they were able to drill a well by using donations from some of the local families. This has transformed the health, hygiene, and sanitation in the village, including increased frequency and privacy of bathing, especially for girls and women; increased security regarding cleanliness of water source; decreased contamination of water from animal waste; and increased understanding of sanitation (Bokhari, 2006, p. 18).

The wider issue of gender and water is of particular significance in South and Central Asia. Through many years of experience living and traveling in the region, the author has observed that women and girls are often the family members that carry water from source to home, that they are most vulnerable to insecurity based on their going out of their homes to urinate and defecate, and that they do not get an equitable share of water for agricultural use. The relationship between gender and access to water needs further exploration.

The National Sanitary Policy of 2006 set up for smaller communities establishes goals for public toilets, sanitation standards, waste management, water efficiency, and other issues (Government of the Islamic Republic of Pakistan, 2006). Whether those goals are met or not is hard to determine. According to Wateraid (2015), the situation remains dire in much of Pakistan, with 93 million or over half the population without adequate sanitation, 15.3 million without access to safe drinking water, and 40,000 children a year dying from diarrhea.

### 10.4.4 India

Another example from the region is India, which has 18% of the world's population, but only 4% of the world's fresh water (Levacher, 2014). A leader in water management in South Asia once told the author, "It is not that we have too little water. It is unequally distributed. We have too much during two months of the year and not enough during the other ten." Anyone who has experienced the beautiful and sometimes destructive monsoon season in that part of the world knows this to be correct. If only that rain could be effectively harvested and stored for later use—then transported from damp areas of the country to dry areas—this would solve many of the water-quantity issues and could positively impact issues of water quality as well.

The region's rain might be harvested in rain barrels for storage or with check-dams to recharge the ground water that is fast depleting, especially in urban settings of India where the combination of few rainwater-harvesting units and many illegal boreholes is dropping the water table by one meter per year (Times of India, Jul. 15, 2014b). Projects for rainwater harvesting have been successfully implemented on a small scale. But these need to be scaled up in order to meet the growing demand for fresh water. Currently, 65% of all rainwater in India drains as runoff into the streams, rivers, and seas (Times of India, Mar. 22, 2014a).

Afghanistan receives a lot less rainfall per year than India; only an average of about 327 mm versus 1083 mm for India, according to the World Bank's (2014) data. If only a portion of that rainfall was harvested it could be a literal lifesaver because if stored properly this is the cleanest drinking water available.

At a conference center in Panchgani, Maharashtra in India, the NGO Initiatives of Change has built a campus that harvests rainwater during the monsoons and stores it in huge tanks under the buildings that hold enough for year-round drinking usage. And some houses like the eco-friendly one the author visited in Bangalore are designed or retrofitted with rooftop rain barrels. These examples are indicative of what could be scaled up to provide safer drinking water for people in the region.

Characteristically it is surface water that is most dangerous to drink. The untreated sewage, combined with the toxic waste from small-, medium-, and large-scale industries, makes the surface water deadly. When villages have wells that draw up ground water, they usually have positive health benefits. However, ground water in one-third of India's 600 districts is not fit for drinking, because the concentration of fluoride, iron, salinity, and arsenic exceeds safe levels (UNICEF, 2013).

These micro-solutions, however, are not being scaled up fast enough to meet the great need in South and Central Asia. A recent UN Report predicts that by 2050 water demand will exceed water supply in India (Levacher, 2014). It will take focus and effort from government, business, and NGO sectors in order to meet this gap. It is fairly certain that if this condition prevails in India all the neighboring countries of South and Central Asia are in similar straights. Although water quantity is a major part of the problem, the main issue for public health is water quality.

## 10.5 RECOMMENDED POLICIES AND PRACTICES TO IMPROVE WATER QUALITY

This final section concludes with some recommendations for policy and practice. The world's water supply is depleting. Inasmuch as the total amount of water contained in the Earth's ecosystem is finite—in other words, water is a closed system—the easily accessible surface water and shallow ground water are increasingly scarce. This is driving users to look at more difficult sources. Priscoli and Wolf (2009, p. 18) have suggested that in the near future we will have to further develop different water sources such as the deep fossil-water aquifers, waste-water reclamation sources, and by interbasin water transfers that are not restricted to the confines of watershed boundaries.

One framework we can use draws from the recycling revolution in the U.S. and applies the lessons to water. *Reduce, reuse, recycle* is the mantra. First, we can reduce our use of water by more efficient infrastructure and retention practices so that the water we have is not wasted. Second, we can reuse greywater, which is the waste water from all but toilets. This greywater can provide safe fertilizer and irrigation for topsoil (Let's Go Green, 2015). Third, recycling water (like reusing water) means taking water that has been used for something else and repurposing it for new uses such as to flush toilets, cool industrial plants, or recharge ground water. This three-pronged approach to water use could be effective if scaled up for use in this South Central Asia region.

Another framework we can use with regards to water recommendations is broken down by before and after; said a different way, prevention and treatment. Policies and practices that fall on the prevention side include the following:

- Increase public-education campaigns to prevent open defecation and encourage hand washing.
- Harvest and store rainwater through retrofit of existing structures and mandatory rain barrels for all new homes, underground rain collection tanks for all new public buildings or multi-family units, and focusing projects on simple check-dams construction, and other ground-water recharge infrastructure.
- Introduce appropriate pricing structure—targeting water use by wealthy households, agribusiness, and industrial users—to allow market-mechanisms to work on preventing water waste.

And on the treatment side, policies and practices include the following:

- Drink ground water that is tested for safety rather than unsafe surface water.
- Reuse gray water for appropriate purposes, such as industrial uses, some domestic uses, and irrigation when safe.
- Provide, at the local level, techniques for conflict resolution of water-user disputes.
- Set up transparent and consistent legal frameworks at state and central government, as well as international levels.
- And potentially most impactful would be to ensure enforcement of environmental regulations, especially those mandating sewage and industrial waste water is treated and meets standards before being discharged into streams, rivers, or seas.

Additionally, a strong need exists in all of South and Central Asia to measure and track progress on water quality. The Millennial Development Goals began to do some of that and the next version of those, called UN Global Goals and announced in 2015, will continue that effort. Baselines of current behaviors that pollute the waters should accompany those MDGs, then testing of particular interventions for changing those behaviors must be done over time. Some issues of interest for this longitudinal research with regard to public-health concerns might include reducing open defecation, increasing hand washing, and greater use of verified safe ground water for drinking and cleaning (World Health Organization, 2012).

Other recommendations that parallel some of those discussed above, include four policy strategies to address water-quality problems: prevent pollution; treat water once polluted; dispose of wastewater safely; and restore and protect vulnerable ecosystems (UN Water Policy Brief, 2012, p. 4). In the end, not a single answer exists to the issues of effective water management. As Verghese (2007) has written, no reason exists to adopt a single pattern everywhere because adapting traditional systems to modern needs would be the best.

Although some of the ongoing research called for here should be micro-level, many of these must be transboundary and comparative studies because river basins generally lie in multiple jurisdictions. For instance, the Ganga River Basin includes China, Nepal, Bhutan, Burma, Bangladesh, and India (Verghese, 2007). Or the Chitral-Kunar-Kabul-Indus River system has tributaries that originate in northern Pakistan, flow into Afghanistan, then back into central Pakistan, and south to the Arabian Sea. And the Amu Darya system forms the border between Afghanistan and Tajikistan, runs through Turkmenistan and Uzbekistan, and finally drains into the dwindling Aral Sea.

## 10.6 CONCLUSION

Afghanistan is an exemplar of regionally deficient water quality that pitifully deficient education, lack of information, dramatic overpopulation, and failure of the state have brought access to clean water to a generally fairly wretched status. All of the countries in Central and South Asia have deficient water supplies, with Afghanistan being one of the countries with the worst access to improved and healthy water supplies. Water-borne pathogens, among the world's highest child mortality rates, and primitive understandings, or religiously biased views of water cleanliness contribute to ever-present problems. Concerted efforts by the United Nations, international donors, and many voluntary aid agencies have helped in some places, but the general lack of good and safe water supplies is a major retardant to modern development in a country bedeviled by a host of other problems as well.

## References

Abbasi, T., Abbasi, S.A., 2012. Water Quality Indices. Elsevier, Oxford.

Asia Society, 2009. Asia's Next Challenge: Securing the Region's Water Future. Asia Society, New York. http://asiasociety.org/asias-next-challenge-securing-regions-water-future (accessed 03.08.2014.).

Blau, M., 2013. In Tajikistan, "Little drops make the river". Frontlines, March/April. http://www.usaid.gov/news-information/frontlines/water-neglected-tropical-diseases/Tajikistan-little-drops-make-river (accessed 05.11.15.).

Bokhari, J., 2006. Chapter 3: Pakistan: initiative of one, relief for all – women's leadership in the Banda Golra water supply scheme. In: Gender, Water and Sanitation Case Studies on Best Practices. United Nations, New York.

Broshears, R.E., Akbari, M.A., Chornack, M.P., Mueller, D.K., Ruddy, B.C., 2005. Inventory of Ground-Water Resources in the Kabul Basin, Afghanistan. US Department of Interior, US Geological Survey, Washington, DC. http://pubs.usgs.gov/sir/2005/5090/sir20055090.pdf (accessed 11.21.15.).

Chellaney, B., 2011. Water: Asia's New Battleground. HarperCollins, New Delhi.

Chellaney, B., 2013. Water, Peace and War: Confronting the Global Water Crisis. Rowman & Littlefield, Lanham, MD.

Daugherty Water for Food Institute at the University of Nebraska, 2015. In: Proceedings of the 2014 Water for Food Global Conference, Seattle.

Dinar, S., De Stefano, L., Duncan, J., Stahl, K., Strzepek, K.M., Wolf, A.T., 2012. No wars for water: why climate change has not led to conflict. Foreign Affairs, 18 October.

Economist, 2011. Unquenchable thirst: a growing rivalry between India, Pakistan, and China over the region's great rivers may be threatening South Asia's peace. The Economist, 19 November, 27–29.

Environmental Performance Index, 2014. Yale University. http://epi.yale.edu/epi/issue-ranking/water-and-sanitation (accessed 05.14.15.)

EPA, 2015. http://water.epa.gov/drink/contaminants/basicinformation/arsenic.cfm (accessed 05.14.15.)

Fipps, G., 2007. Appendix to the Transboundary Water Policy of Afghanistan: Transboundary Water Issues, Draft. GIRoA, Kabul. 10 pp, http://gfipps.tamu.edu/Afghanistan/Transboundary%20Water%20Issues26-04-07.pdf.

Floyd, R., Richard, M. (Eds.), 2013. Environmental Security: Approaches and Issues. Routledge, London.

Gehrig, J., Rogers, M.M., 2009. Water and Conflict: Incorporating Peace Building Into Water Development. Catholic Relief Services, Baltimore, MD. http://www.crsprogramquality.org/storage/pubs/peacebuilding/waterconflict.pdf (accessed 08.01.14.).

Geneva Foundation, 2002. Thirsty Cities in War. Geneva Foundation, Tacoma, WA. http://www.thirstycitiesinwar.com/wp-content/uploads/2013/03/kabuloccasionalpaper7.pdf (accessed 01.26.15.).

Giordano, M., Drieschova, A., Duncan, J.A., Sayama, Y., De Stefano, L., Wolf, A.T., 2013. A review of the evolution and state of transboundary freshwater treaties. Int. Environ. Agreem. Polit. Law. Econ. 14, 245–264.

Glendenning, C.J., 2009. Evaluating the Impacts of Rainwater Harvesting (RWH) in a Case Study Catchments: The Arvari River, Rajasthan, India. Ph.D. Dissertation, University of Sydney, Australia.

Gon, G., Monzon-Llamas, L., Benova, L., Willey, B., Campbell, O.M.R., 2014. The contribution of unimproved water and toilet facilities to pregnancy-related mortality in Afghanistan: analysis of the Afghan Mortality Survey. Tropical Med. Int. Health 19 (12), 1488–1499.

Government of the Islamic Republic of Pakistan, 2006. National Sanitary Policy. Ministry of Environment, Government of Pakistan, Islamabad. http://www.environment.gov.pk/NEP/SanitationPolicy.pdf (accessed 05.15.15.).

Hempel, S., 2006. The Medical Detective: John Snow, Cholera, and the Mystery of the Broad Street Pump. Granta Books, London.

Horton, R.K., 1965. An index-number system for rating water quality. J. Water Pollut. Control Fed. 37 (3), 300–305.

HydrateLife, 2015. Informing the world about our current water and sanitation crisis. http://www.hydratelife.org/?p=105 (accessed 01.26.15.).

India Express, 2014. Waste being released into sea without proper treatment: expert. India Express, 31 July 1.

Iyer, R.R., 2007. Towards Water Wisdom: Limits, Justice, Harmony. Sage, New Delhi.

Iyer, R.R. (Ed.), 2009. Water and the Laws in India. Sage, New Delhi.

Jhunjhunwala, B. (Ed.), 2014. Water: Impact of Dams on Its Qualities. Kalpaz, Delhi.

Jones, J.A.A. (Ed.), 2010. Sustaining Groundwater Resources: A Critical Element in the Global Water Crisis. Springer Science + Business Media, New York.

Joy, K.J., Gujja, B., Paranjape, S., Goud, V., Vispute, S. (Eds.), 2008. Water Conflicts in India: A Million Revolts in the Making. Routledge, New Delhi.

Kanji, N., Sherbut, G., Fararoon, R., Hatcher, J., 2012. Improving quality of life in remote mountain communities: looking beyond market-led approaches in Badakhshan province, Afghanistan. Mt. Res. Dev. 32 (3), 353–363. http://dx.doi.org/10.1659/MRD-JOURNAL-D-12-00012.1.

Kaplan, R.D., 2011. Monsoon: The Indian Ocean and the Future of American Power. Random House, New York.

Khalid, I., 2011. Trans-boundary water sharing issues: a case of South Asia. J. Polit. Stud. 1 (2), 79–96.

Let's Go Green, 2015. http://www.letsgogreen.com/greywater-recycling.html (accessed 05.15.15.)

Levacher, C., 2014. UN Report Stresses Upcoming Water Crisis in India. In: Global Food and Water Crisis Research Programme, April 16. www.futuredirections.org.au (accessed 03.08.2014.).

Mack, T.J., 2010. Conceptual Model of Water Resources in Kabul Basin, Afghanistan. U.S. Department of the Interior, U.S. Geological Survey, Reston, VA.
McNamara, P., 2014. Rivers of peace: the role of water in world religions. In: Presentation to Kashi Kumbh conference on Ganga Action Plan, Varanasi, India.
Millennial Development Goals, 2015. United Nations. http://www.un.org/millenniumgoals/environ.shtml (accessed 05.13.15.)
Mountain Partnership, 2015. Agriculture and Food Security Under Changing Climate in Central Asian Mountains: Integrated Watershed Management as a Viable Solution for Coping with and Adapting to Changes. Mountain Partnership Central Asia Hub, Bishkek. http://www.mountainpartnership.org/fileadmin/user_upload/mountain_partnership/docs/Dushanbe%20Mountain%20Forum2015.pdf (accessed 11.21.15.).
New York Times, 2014. Malnutrition in well-fed children is linked to poor sanitation. New York Times, 15 July, A1.
NPR, 2013. Rich with water but little to drink in Tajikistan. http://www.npr.org/blogs/health/2013/07/10/198103517/TAJIK-WATER (accessed 05.11.2015.).
Olivelle (Trans.), P. (Ed.), 2009. The Law Code of Manu. Oxford University Press, Oxford.
Pierce, F., 2006. When the Rivers Run Dry: Water—The Defining Crisis of the Twenty-First Century. Beacon Press, Boston.
Pollack, H., 2010. A World Without Ice. Avery Trade, New York.
Population Reference Bureau, 2013. http://www.prb.org/Publications/Datasheets/2013/2013-world-population-data-sheet/world-map.aspx#map/world/population/2013 (accessed 01.26.15.).
Prakash, A., Chourey, J., 2011. Interlacing Water and Human Health: Case Studies from South Asia. SAGE India, New Delhi.
Priscoli, J.D., Wolf, A.T., 2009. Managing and Transforming Water Conflicts. Cambridge University Press, Cambridge.
Qassem, A.S., 2009. Afghanistan's Political Stability: A Dream Unrealized. Ashgate, Surrey.
Ringler, C., Biswas, A.S., Cline, S.A. (Eds.), 2010. Global Change: Impacts on Water and Food Security. Springer, Heidelberg.
Saffi, M.H., Kohistani, A.J., 2013. Water Resources Potential, Quality Problems, Challenges and Solutions in Afghanistan: An Evaluation of the Past and Present Situation of Water Resources and Consequences on the Basis of Previous Water Resource Investigation and Present National Groundwater Monitoring Wells Network Data Management, Mapping. Visualization and Assessment, DACAAR. http://www.dacaar.org/Publications/6%20Research%20and%20Studies/WASH/Report%20on%20Water%20Resource%20Porential_opt.pdf (accessed 09.11.15.).
Sauer, J., 2009. No-plumbing disease. On Tap 9 (2), 38 (accessed 04.09.2016).
Sementelli, A.J., 2008. Naming water: understanding how nomenclature influences rights and policy choices. Public Works Manag. Policy 13 (1), 4–11.
Shiva, V., 2002. Water Wars: Privatization, Pollution and Profit. South End Press, Canada.
Shroder, J., 2014. Natural Resources in Afghanistan: Geographic and Geologic Perspectives on Centuries of Conflict. Elsevier, Waltham, MA.
Singh, R., 2005. Community driven de-centralized management. In: Issue-based Workshop on Gaining Public Acceptance, Nairobi, Kenya.
Sinha, J., Sinha, M.K., Adapa, U.R., 2013. Flow—River Rejuvenation in India: Impact of Tarun Bharat Sangh's Work. Swedish International Development Agency, Stockholm.
Sisodia, R., 2014. Democratising Rivers: A Comparative Study of Local River Conflicts in Rajasthan. Master's Thesis, Tata Institute of Social Sciences, Mumbai.
Solomon, S., 2010. Water: The Epic Struggle for Wealth, Power and Civilization. Harper Perennial, New York.
Subramanian, A., Brown, B., Wolf, A.T., 2012. Reaching Across the Waters: Facing the Risks of Cooperation in International Waters. The World Bank Press, Washington, DC.
Tebbutt, T.H.Y., 1998. Principles of Water Quality Control. ButterWorth-Heinemann, Oxford.
Thomas, V., Eqrar, N., 2011. Chapter 2: managing water resources, scarcity, and climate shocks. In: Afghanistan Human Development Report 2011. Centre for Policy and Human Development, Kabul.
Times of India, 2014a. World water day, UN report predicts grim scenario for India. Times of India, 22 March. http://timesofindia.indiatimes.com/home/environment/the-good-earth/World-Water-Day-UN-report-predicts-grim-scenario-for-India-experts-pitch-for-making-water-conservation-a-national-obsession/articleshow/32507693.cms (accessed 03.08.2014.).
Times of India, 2014b. Water table dipping 1 metre every year in Gugaon. Times of India, 15 July. 3.
UN Environment Program, 2015. Water Quality Indices. UNEP, Nairobi. http://www.unep.org/gemswater/Default.aspx?tabid=101094 (accessed 11.01.15.).

UN Environment Program Global Environment Monitoring System, 2007. Global drinking water quality index development and sensitivity analysis report. http://www.un.org/waterforlifedecade/pdf/global_drinking_water_quality_index.pdf (accessed 05.14.15.).

UN Water for Life, 2015. High Level International Conference on the Implementation of the International Decade for Action "Water for Life", 2005–2015. http://waterforlifeconf2015.org/eng/wp-content/uploads/2015/01/Itogovy-j-dokument-konferentsii-final-ny-j_20.08.2015.pdf (accessed 11.01.15.).

UN Water Policy Brief, 2012. Water Quality. http://www.unwater.org/downloads/waterquality_policybrief.pdf (accessed 01.26.15.).

UNICEF, 2013. Water in India: Situation and Prospects. UNICEF, New Delhi.

UNICEF, 2014. Progress on drinking water and sanitation: 2014 update. http://www.unicef.org/gambia/Progress_on_drinking_water_and_sanitation_2014_update.pdf (accessed 05.14.15.).

United Nations, 2012. Environmental Performance Review of Tajikistan. http://www.un.org/apps/news/story.asp?NewsID=43774#.VVEEvPlVhBc (accessed 05.11.15.).

USAID Afghanistan, 2012. Afghan sustainable water supply and sanitation (SWSS) project: sustainable health outcomes unit project final report. https://sanitationupdates.files.wordpress.com/2012/12/afghanistan-swss2012.pdf (accessed 11.01.15.).

Verghese, B.G., 2007. Waters of Hope: Facing New Challenges in Himalaya-Ganga Cooperation. India Research Press, New Delhi.

Wateraid, 2015. http://www.wateraid.org/uk/where-we-work/page/pakistan (accessed 05.15.15.).

Wolf, A.T., 1998. Conflict and cooperation along international waterways. Water Policy 1, 251–265.

Wolf, A.T., 2012. Spiritual understandings of conflict and transformation and their contribution to water dialogue. Water Policy 14, 73–88.

Wolf, A.T., Yoffe, S., Giordano, M., 2003. International waters: identifying basins at risk. Water Policy 5, 29–60.

World Bank, 2014. World development indicators 2014. http://wdi.worldbank.org/table/3.2 (accessed 01.26.15.).

World Health Organization, 2000. Global Water Sanitation and Health Assessment. WHO, Geneva. http://www.who.int/water_sanitation_health/monitoring/Glassessment7.pdf?ua=1 (accessed 01.26.15.).

World Health Organization, 2012. All for Equity: World Conference on Social Determinants of Health Summary Report. WHO, Geneva.

Zeitoun, M., Mirumachi, N., 2008. Transboundary water interaction: reconsidering conflict and cooperation. Int. Environ. Agreem. Polit. Law Econ. 8 (4), 297–316.

CHAPTER

# 11

# Issues of Hydrologic Data Collection by Remote Sensing in Afghanistan and Surrounding Countries

*J. Shroder*

University of Nebraska at Omaha, Omaha, NE, United States

OUTLINE



## Abstract

The paucity of data about the many subsystem parts of the water cycle in Afghanistan leads to new analytical measures through the use of satellite sensors that operate in space and that download their collected data electronically. Numerous satellites (Landsat, SPOT, Terra, ASTER, TRMM, Aqua, GRACE, Global Precipitation Measurement Core Observatory, etc.) have been launched in the past half century by different nations to assess water in all its forms and locations; the imagery and data from these can be used to considerable advantage to assist Afghanistan and neighboring countries.



http://dx.doi.org/10.1016/B978-0-12-801886-6.00011-2

**Keywords:** Water data collection, Earth Observatory (EO) Portal Directory, Landsat satellite, SPOT satellite, Terra satellite, ASTER satellite, TRMM satellite, MODIS satellite imagery, Aqua satellite, GRACE satellite, Global Precipitation Measurement (GPM) Core Observatory

## 11.1 INTRODUCTION

The US National Research Council (NRC) has noted that observing and accurately predicting Earth's environment is critical for health, safety, and prosperity (National Research Council, 2003). Countries such as Afghanistan, Tajikistan, and Pakistan that cannot do this consistently are doomed to many kinds of failure (Shroder, 2014) unless they become more capable of working with advanced systems of collecting information on their resource base. In recent years, it has become obvious to many scientists that major new opportunities exist for advances in prediction and in other weather and climate-information products, which are of course, critical portions of the hydrological cycle in any country. The NRC has further said that realizing the potential benefits of the investments in satellites requires rapid, efficient transitions of measurement and modeling capabilities in the research communities, to the observing and prediction systems of the operational agencies. Not only do these recommendations apply inside the United States, but it is not too much to suggest that similar benefits could also accrue to other countries in Central and Southwest Asia that also seek to utilize at least some of these advanced methodologies to assess their own water resources.

In countries where information on water availability can be oxymoronically relegated into the realm of the state secret, any technology that can overcome the dearth of accurate information has much to recommend it. In the past half century, great strides have been made in the use of satellite technology and remote sensing to discover essential attributes of phenomena having to do with almost all aspects of water. The result is that the development of many new ways of detecting aqueous phenomena reveals wide varieties of essential aspects of the $H_2O$ system that need further explanation.

The global water cycle encompasses not only the land surface, but also the atmospheric transport of all the phases of $H_2O$ (solid snow and glacier ice, liquid water, water vapor) via frozen precipitation, melting to liquid, infiltration into the soil, recharge of ground water, surface runoff, stream and river flow, and evaporation and transpiration (evapotranspiration) of water vapor from exposed surface water and from plants. The monitoring and measuring of all these many vital resources has long been undertaken in the developed world in recognition of their importance. In recent years, however, due to budgeting and programmatic constraints, a number of these measurement programs have been cut back in many parts of the world. In Afghanistan, however, the original rudimentary data-collection systems for the weather and for river discharge have been curtailed since the late 1970s with the outbreak of the several forms of war (invasion, civil war, insurgency, etc.) in the country since then. The absence or loss of data about all forms of water resources, for whatever reason, thus helped lead to the many new methods of remote sensing to overcome the information deficit.

With the advent of the Earth Observing System (EOS) in the 1990s, the American National Aeronautics and Space Agency (NASA), the European Space Agency (ESA), and the Japanese Aerospace Exploration Agency (JAXA), these organizations all launched a number of spaceborne satellite sensors to observe and record the various components of the terrestrial water cycle. For example, soil moisture was detected by the Advanced Microwave Scanning Radiometer (AMSR) and the Soil Moisture and Ocean Salinity systems; precipitation by the Tropical Rainfall Measuring Mission (TRMM); vegetation by the Moderate Resolution Imaging Spectroradiometer (MODIS); ocean-surface water level by JASON—1 and JASON—2 and TOPEX—POSEIDON; and underground water by the Gravity Recovery and Climatic Experiment (GRACE). These instruments, numerous varieties of them, and many others as well, were launched on a fleet of different satellites put into various orbital positions by different rocket launches (Fig. 11.1). In addition, a whole new suite of sensors have been developed to study the water cycle and are scheduled to be launched within the next decade as well. These many older and newer sensor systems led to a number of new methods for estimating and monitoring the most important components of the hydrologic cycle, such as: (1) precipitation; (2) evapotranspiration; (3) surface water; (4) snow; (5) glacier ice; (6) soil moisture; (7) ground water; and (8) water modeling and data assimilation.

This diversity of sensors and methodologies will enable those with the skill to assess such data to begin anew or to continue to collect plentiful new information about the hydrologic cycle across Afghanistan. Coupled with newly collected data from ground monitoring stations to add greater verisimilitude to the sometimes rather esoteric data will allow far greater accurate assessment of the water situation in Afghanistan. In all probability, at least for the

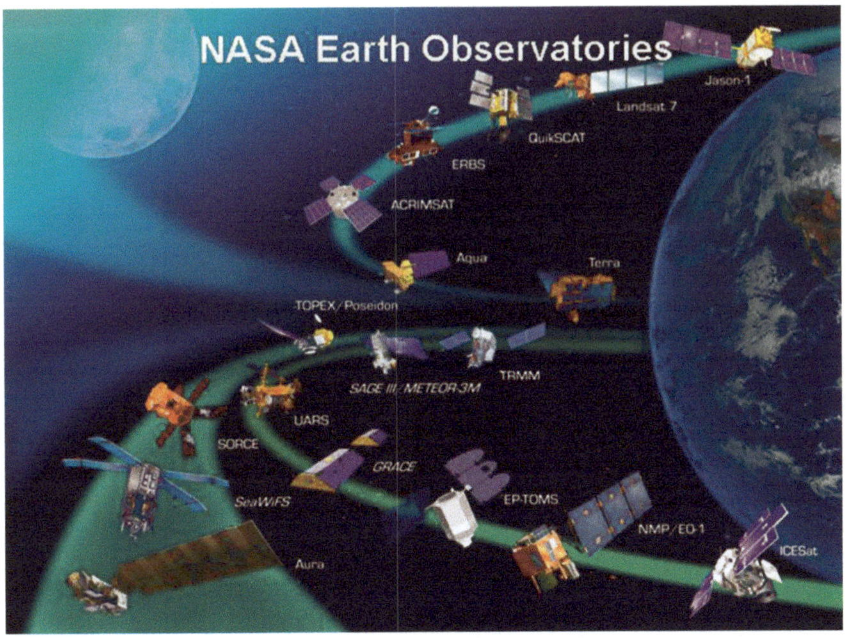


FIG. 11.1 Eighteen of the Earth observatory satellites launched by rockets from the US National Aeronautical and Space Administration (NASA). Many of these satellites collect essential information on water resources. Additional useful satellites have already been launched, or are planned for future years.

foreseeable future for the next decade or so, it will be necessary for most Afghan researchers to associate with foreign scientists to avail themselves of the capabilities of remote-sensing assessment. This is because the quantitative assessment of such profuse data, along with the considerable expense of the resulting satellite imagery, is complex, esoteric, and expensive. With newly gained skill, equipment, and new data collections, however, no reason exists that Afghanistan cannot develop its own capabilities for comprehensive remote sensing of its water cycle. All that will be required will be for Afghan professors and their students to learn as much as they can about availing themselves and their universities of the expertise and equipment needed to do this kind of work. Most likely, the best way for this to happen will be to develop a series of small projects to investigate certain aspects of the hydrologic cycle, or other resources that Afghanistan needs to develop, so that familiarity with the techniques and equipment can be properly assimilated over time.

## 11.2 EARTH OBSERVATORY (EO) PORTAL DIRECTORY

The number of satellites and sensor systems that have been launched on rockets into space to orbit around the Earth in the past half century or so is completely bewildering to the average geoscientist who is not a specialist but who may wish to learn to work with the valuable satellite imagery. Because of the many hundreds of different satellites and sensor systems from many different countries, not all of which are available to people of differing nationalities, some means of accessing the diverse information is quite valuable. Fortunately, the EO Portal Directory for Earth Observation Resources is available online to sort out the huge quantity of necessary information (EO, 2015).

## 11.3 $H_2O$ SYSTEM SENSING

This key component of the climate and the hydrologic cycle in any environment exists in a number of liquid and solid forms (rain, snow, sleet, etc.), of course, which can be detected and measured in a variety of fashions. The electromagnetic spectrum (Fig. 11.2A) sensed by the satellites is divided into different wavelengths that provide various kinds of useful information: (1) visible light (wavelengths 0.4–0.7 μm); (2) near and shortwave infra-red (wavelengths 0.7–3.0 μm); (3) thermal infra-red (TIR) (wavelengths 8.0–15.0 μm); and (4) microwaves (MW) (1.0–300 mm). Some of these wavelengths coming from the Sun do not pass through the atmosphere, and others such as most visible light, for example, pass through the atmosphere to reflect off features on the ground back up through the atmosphere to be detected by satellites passing overhead (Fig. 11.2B).

None of these wavelengths can directly measure rainfall, but several clever techniques can be used to get close to useful results of measuring precipitation quantities. For example, some sensors can make indirect estimates of rainfall by measuring such things as the thickness of clouds, or the temperature of the cloud tops (Ceccato and Dinku, 2010).

Techniques of remote sensing include two general types; the active types and the passive remote sensing. In the active types, a pulse of electromagnetic radiation is bounced off the target such as raindrops, as is done in weather radar, for example. Passive remote

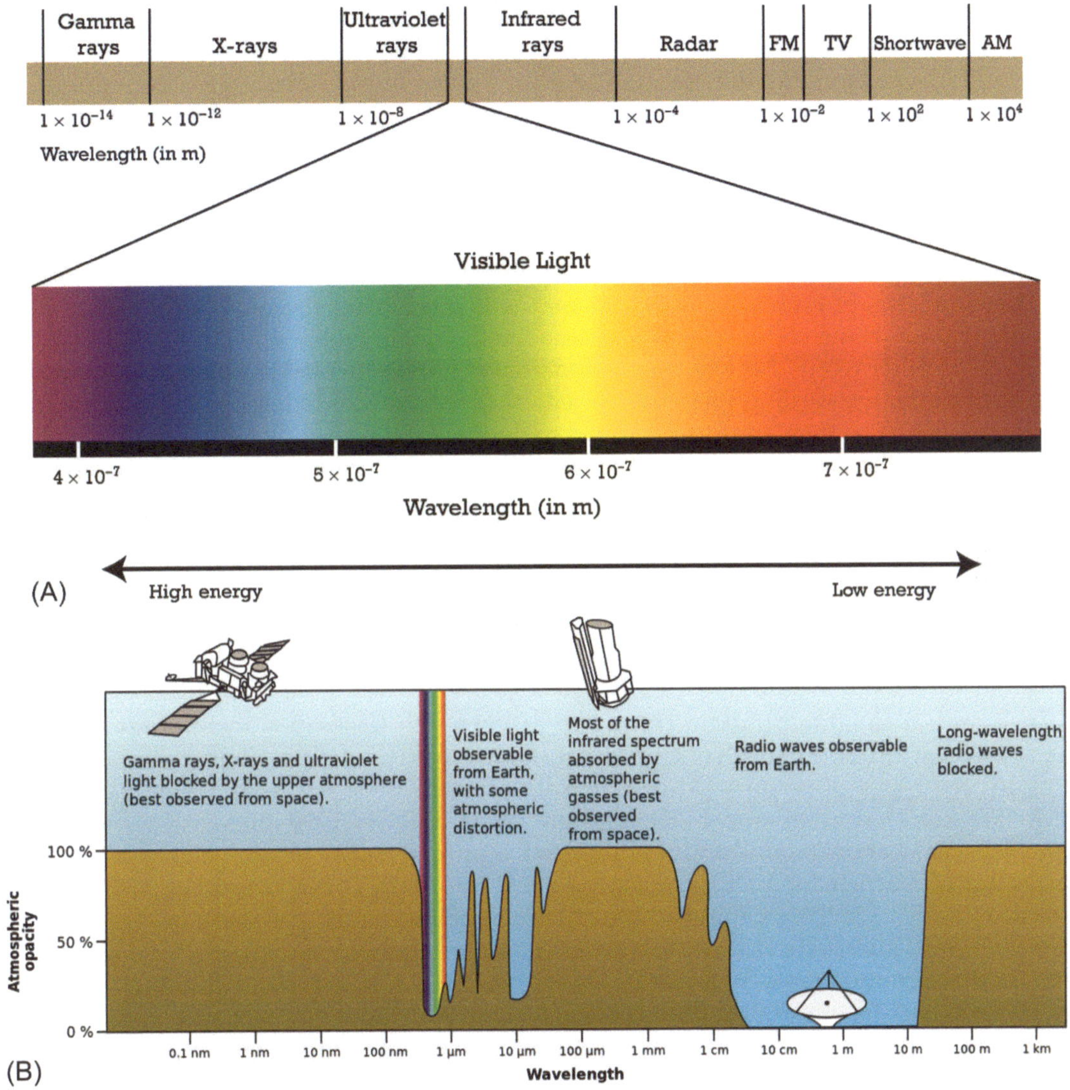


FIG. 11.2 (A) The electromagnetic spectrum showing the range of different forms of radiation that can be sensed by different satellites to detect various aspects of the hydrologic cycle. The range of visible light that the human eye can detect is shown in expanded form below and (B) is a plot cartoon of the Earth's transmittance or opacity to various wavelengths of electromagnetic radiation through the atmosphere. As can be seen, some wavelengths pass easily through the atmosphere and others do not. The high energy, high frequency wavelengths, such as gamma and X-rays do not easily pass through the atmosphere, which is fortunate for life forms because such energy is damaging to living cells.

sensing, on the other hand, simply records the naturally incoming radiation, such as visible light, much as a photograph is taken of a scene, but the process can also record microwave and infrared wavelengths that cannot be seen by the human eye unless converted into some sort of false-color presentation. Microwave radiation can "see" deep inside clouds, whereas TIR infrared can record only cloud tops and therefore can be used only to determine raining from non-raining clouds. Commonly the MW data can be combined with the TIR, and even ground measurements, to produce reasonable estimates of precipitation rates and amounts.

One of the biggest problems for new researchers with remote sensing is learning the physical characteristics of each sensor system that will most affect the quality and quantity of the data that are being collected. A major issue, in fact, will be the scale or resolution of the imagery—in other words, how much ground area might be covered in any single scene. This may be important in dealing with phenomena that require close viewing at high resolution and large scale of certain landform features to determine advance or retreat of glaciers with Advanced Spaceborne Thermal Emission and Reflection Radiometer (ASTER) imagery, for example, or the lower resolution of broad areas to map regional snow cover, as with MODIS imagery. Many other sorts of variations of the imagery have to be carefully selected for the appropriate study at hand so as not to assess the incorrect imagery, and to get useful and reliable results to enable accurate predictions.

## 11.3.1 Landsat

The Landsat satellite program (formerly called Earth Resources Technology Satellite), has a >40 year duration and has employed the use of six satellites (Landsats 1, 2, 3, 4, 5, 7) and three types of sensors [Multispectral Scanner (MSS; 1972–83), Thematic Mapper (TM; 1984–2011), and enhanced Thematic Mapper (ETM+; 1999–present)] (USGS, 2014) (Fig. 11.3). The Landsat data archive provides the unique ability to analyze particular areas on the Earth's surface for a >40 year period (1972–present). This spatio-temporal historical extension is ripe for the gleaning of information such as landscape and land-use changes to be detected, interpreted, and explained. Pitfalls of analysis using Landsat data include the fact that ground resolutions are 80–30 m$^2$ (>0.5 ha), and that the satellite revisit times are 16–18 days, depending on the sensor. These ground resolutions cause the minimum mappable unit to be somewhat large as compared to much finer resolution sensors such as

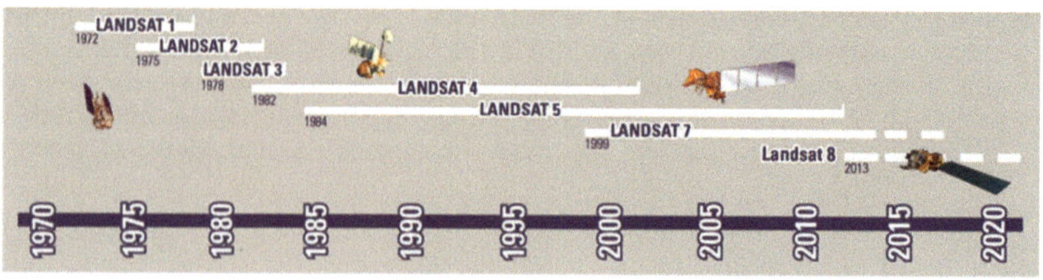


FIG. 11.3 Landsat timeline with different satellites launched at different times that worked effectively for a number of years until they ceased operation. Landsat 7 is still aloft, although it has some problems. LDCM is the Landsat Data Continuity Mission, now renamed as Landsat 8.

Système Pour l'Observation de la Terre (SPOT) (10 $m^2$ pixel resolution); however, such sensors do not afford the same spatio-temporal breadth as do the Landsat sensors. The revisit intervals of 16–18 days provides roughly two scenes per month, and therefore surpasses the temporal resolutions of much data acquired by means of aircraft sensors of high spatial resolution.

The most important band of Landsat TM and ETM for detecting water is band 5 (1150–1250 nm), and for MSS, band 4 (800–1100 nm), which can be used to differentiate between vegetation and soil-moisture levels (Jensen, 2007; Ozesmi and Bauer, 2002). Because deep, pure water absorbs nearly all incident near- and middle-infrared radiation, with little to no scattering, the wavelengths between 740 and 2500 nm are the most appropriate for discriminating land from water (Jensen, 2007). Additionally, Ozesmi and Bauer (2002) have noted that "middle infrared Landsat TM bands provide for much of the separability between wetland types" (Ozesmi and Bauer, 2002).

### 11.3.2 SPOT Satellite

*Satellite Pour l'Observation de la Terre*, or satellite for the observation of the Earth, is a series of sensors put aloft by the French Space Agency since 1986. SPOT satellites are numbered 1–7. SPOT 5 was launched with 2.5, 5 m, and 10 m resolution capabilities. SPOT 6 and 7 were launched by India's Polar Satellite Launch Vehicle in 2012 and 2014, with 1.5 m resolution capabilities in both natural color and panchromatic (black and white). Each scene has a 60 km × 60 km footprint on the ground. These satellites are expected to be productive through to the year 2024. The imagery derived will be useful in a range of applications such as study of natural and man-made disasters, exploration for natural resources, agriculture, environmental protection, and geographic information systems.

### 11.3.3 Terra Satellite

In 1999, the Terra device was launched as the first EOS (EOS AM-1) and became the flagship of the NASA effort to attain global data on the state of the atmosphere, land, and oceans, as well as their interactions with solar radiation and between one another (Fig. 11.4). These are complex connections between the Earth's water sources, water storages, and land upon and within which the water is stored, and the energy balances. These factors are being studied, of course, in order to understand the Earth's climate and climate change, so as to be better able to hypothesize the most likely water futures of the planet. These analyzes will be augmented progressively by assessments at increasingly larger scales, down to regional levels that will be able to help countries such as Tajikistan, Afghanistan, and Pakistan. This system begins to enable the mapping the impact of human activity and natural disasters on communities and the ecosystems around them.

The Terra flagship system has five sensors onboard, which are: (1) ASTER; (2) CERES, or Clouds and Earth's Radiant Energy System; (3) Multi-angle Imaging SpectroRadiometer; (4) MODIS; and (5) Measurements of Pollution in the Troposphere. Each of these sensor systems collects a variety of data, some of which we can use to assess the parts of the hydrologic cycle in which we are most interested or expert.

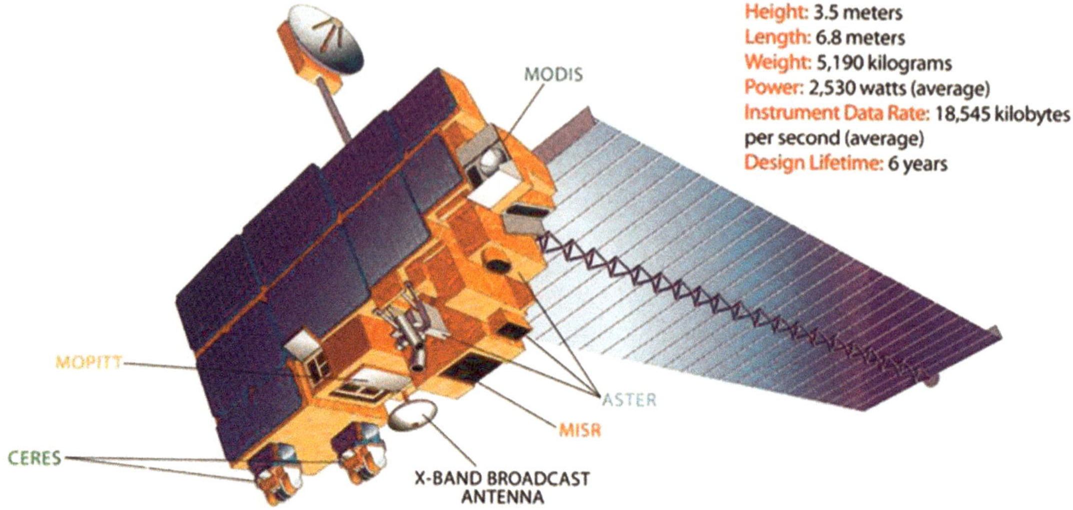


FIG. 11.4 The Terra flagship mission with a number of different sensor systems on board. One of the systems of most use for sensing weather systems is MODIS, but ASTER has provided huge quantities of useful imagery for mapping glaciers, for example.

## 11.3.4 ASTER Imagery

ASTER satellite image data contribute to a wide array of global change-related application areas, including the dynamics of vegetation and ecosystems, monitoring of hazards, geology and soils, land-surface climatology, hydrology, glacier changes, land-cover change, and the generation of digital elevation models (DEMs). The ASTER high-resolution satellite is capable of producing stereo imagery for creating detailed digital terrain models, which are also known as DEMs (Fig. 11.5). ASTER is unique in having a visible near infra-red telescope on board with a high-resolution (15 m ground resolution), along-track stereoscopic observation capability. In addition the system also has a short-wave infra-red capability at 30 m resolution, and a TIR facility at 90 m resolution. These sensor-detection capabilities can be used in a wide variety of investigatory activities associated with the hydrologic cycle in countries worldwide, and an example of direct relevance here for Afghanistan and Pakistan is the project to assess glacial ice.

The Global Land Ice Measurements from Space (GLIMS) Project, for example, for over a decade has enabled a close look at the ice of the world. GLIMS is a project designed to monitor the world's glaciers primarily using data from optical satellite instruments, such as ASTER, although Landsat and other imagery have been used in a few areas. GLIMS began as an ASTER Science Team project. Through this connection, the many GLIMS personnel have guided the ASTER instrument to acquire imagery of Earth's glaciers that was optimized (best season and instrument gain settings) for glacier monitoring. The GLIMS head, Jeffrey S. Kargel of the University of Arizona Department of Hydrology and Water Resources, also put together a network of international collaborators who analyze imagery of glaciers in their regions of expertise. Analysis results include digital glacier outlines and related metadata, and can also include snow lines, center flow lines, hypsometry data, surface-velocity fields,

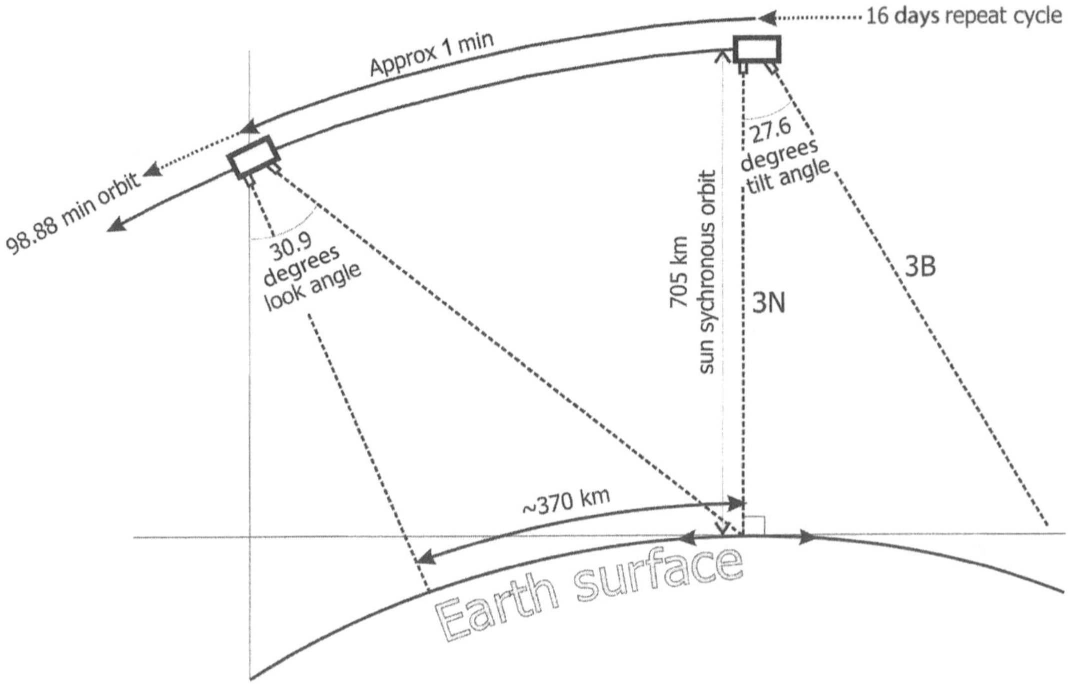


FIG. 11.5 The imaging geometry of the ASTER sensor, which is on board the Terra satellite, 705 km above the Earth in a Sun-synchronous orbit that revisits each area every 16 days. The downward-looking, nadir (3N), and backward-looking (3B) views produce a stereoscopic scene when viewed together, which yields a three-dimensional representation of the Earth's topography referred to as a digital elevation model, or digital terrain model, which can be used to effectively model glaciers and river channels. *From Hirano, A., Welch, R., Lang, H., 2003. Mapping from ASTER stero image data: DEM validation and accuracy assessment. ISPRS J. Photogrammerty Remote Sens. 57 (5/6), 356–370; Quincey, D.J., Bishop, M.P., Kääb, A., Berthier, E., Flach, B., Bolch, T., Buchroithner, M., Kamp, U., Khalsa, S.J.S., Toutin, T., Hatitashya, U.K., Racoviteanu, A., Shroder, J.F., Raup, B.R., 2014. Digital terrain modeling and glacier topographic characterization. In: Kargel, J.S., Leonard, G.J., Bishop, M.P., Kääb, A., Raup, B. (Eds.), Global Land Ice Measurements from Space (GLIMS). Springer, New York, 113–143.*

and literature references. Results from analysis done by the Regional Centers are sent for archive to the National Snow and Ice Data Center (NSIDC) in Boulder, Colorado, United States, where the data are stored. The GLIMS group has continued to develop tools to aid in glacier mapping and for transfer of analysis results to NSIDC. These include GLIMSView, documented procedures for GLIMS analysis, and web-based tools for data formatting and quality control. Over 60 institutions across the globe have been involved in GLIMS up to 2015.

In the mid-1990s, in the absence of responses from the Taliban government in Afghanistan, or the more skillful government of Pakistan to the GLIMS leadership, the GLIMS Regional Center for Southwest Asia (Afghanistan and Pakistan) (GRCSWA) was established at the University of Nebraska at Omaha (UNOmaha) in 1998. The UNOmaha GLIMS group was funded by the NASA, the U.S. Geological Survey, and the U.S. National Science Foundation through 2011. GRCSWA was established for the express purpose of assessing the glacier and snow resources of Southwest Asia using ASTER and other satellite imagery. UNOmaha has had an historical interest in education and scientific research in Southwest Asia for over four decades and has developed close working relationships with a number of government agencies and universities throughout the region in order to facilitate these efforts.

A number of stewards facilitated further working assessments of snow and ice resources between the GRCSWA and other individuals or agencies who may have wished to access ASTER satellite imagery at no cost to themselves. Such imagery is available on a cost-free basis as long as the information on snow and ice resources obtained from it is made available to GRCSWA and is not classified or made secret. ASTER imagery may not be distributed to other individuals or agencies not working in direct support of the GLIMS goals. Comparative assessments of snow, ice, glacier lake hazards, climate, and hydrologic data have been shared between the interested parties in order to facilitate the mutual research efforts.

The results of the prior collateral research on glaciers for the U.S. Geological Survey by one of us (JS) in the 1970s to the 1990s, and the Nanga Parbat, K2, and GLIMS projects from the mid-1990s through the first decade of the 2000s, resulted in a wide variety of publications about glaciers in the Hindu Kush, Afghanistan Pamir, Hindu Raj, and Western Himalaya (Shroder, 1980, 1989, 2011a,b; Shroder and Bishop, 2010a,b; Shroder et al., 1998, 2007; Haritashya et al., 2009; Bishop et al., 1995, 1998, 2000, 2001; 2004, 2007, 2010, 2014a,b; Bulley et al., 2013; Bush et al., 2004; Copland et al., 2009, 2012; Kargel et al., 2005; Kieffer et al., 2000; Phillips et al., 2000; Shroder et al., 2000; Quincey et al., 2014; Sarikaya et al., 2012, Sarıkaya et al., 2013; Seong et al., 2008). This work was able to establish some fundamental understandings about the: (1) theoretical foundations of remote sensing for glacier assessment and mapping; (2) digital terrain modeling and glacier topographic characterization; and (3) glacier activity in Afghanistan and Pakistan.

The theoretical foundations included discussions of the radiation transfer cascade, which so affects glacier ice through solar irradiance, and surface irradiance and reflectance. The surface-energy interactions of snow, glacier ice, and water must be understood from the points of view of information extraction about snow cover, ice sheets, alpine glacier mapping, the snow lines and equilibrium line altitudes, as well as ice-flow velocity determinations. Various forms of numerical modeling of climate, energy balances, and glacier mass balances are essential (Bishop et al., 2014a).

Digital terrain modeling with ASTER imagery to produce DEMs has been helped greatly by assessing glacier topographic characterization through satellite-imagery stereoscopy (Fig. 11.5). In this process, it is necessary to understand the DEM-error uncertainty in the assessment of geomorphometric variables in the mapping of glaciers (Quincey et al., 2014).

We used field work and large numbers of ASTER images to assess almost all glaciers across Afghanistan and Pakistan, with special attention to the significant roles of glaciers in local and regional water resources, in natural hazards (Shroder and Bishop, 1998; Shroder and Weihs, 2010; Shroder et al., 2011a,b), and in geopolitical stability. Many glaciers in both countries have lost significant ice mass and have evolved into numerous smaller individual glacier masses. In Pakistan, however, we especially noted that because of global warming and likely increased marine evaporation, any resulting summer precipitation increase above 5000 m would augment the glaciers and cause advance. Coupled with increased melting of lower-altitude glacier ice because of the augmented melting during global warming has apparently increased glacier surge activity in the Karakoram Himalaya of Pakistan, and perhaps of breakout water floods as well (Copland et al. 2009, 2012).

## 11.3.5 MODIS Imagery

The Moderate Resolution Imaging Spectroradiometer is a key sensor aboard the Terra (EOS AM) and Aqua (EOS PM) satellites (Fig. 11.6). These two in the morning and afternoon have been viewing the entire Earth's surface in 1–2 days, acquiring data in 36 spectral bands, or groups of wavelengths. Ground resolution is rather coarse (250 m, 500 m, and 1 km), which provides capabilities for broad weather patterns, cloud characteristics, water vapor, and so on.

## 11.3.6 Tropical Rainfall Measuring Mission Satellite

The TRMM is a joint space mission between NASA and JAXA from 1998 to about 2015 that was designed to measure a variety of precipitation characteristics. TRMM is an Earth-observation satellite (Fig. 11.7) that has been investigating conditions on Earth by measuring precipitation in tropical and subtropical areas from 35 degrees both north and south of the Equator, with both active and passive means. The satellite uses five different sensors: (1) the TMI Microwave Imager; (2) Visible Infrared Scanner; (3) Precipitation Radar; (4) Lightening Imaging Sensor; and (5) CERES (Fig. 11.8).

FIG. 11.6 MODIS true-color image of the low pressure zone of the powerful monsoon storm covering the India-Pakistan border region of Kashmir on Sept. 5, 2014. The field of view shows the Indian Ocean obscured by dust in the lower left, the green lower Indus valley in the left middle, and Afghanistan from Kabul to the Amu Darya in the upper left. The irrigated green in the upper center is the Yarkand region of southwestern China, and the lakes of Tibet occur in the center right, with the India subcontinent just south of the central Himalaya in the lower right. *Image credits NASA/MODIS. From Monsoon floods kill 420 in India and Pakistan. Jeff Masters' Wunderblog. http://www.wunderground.com/blog/JeffMasters/comment.html?entrynum=2791 (accessed 03.09.2015.).*

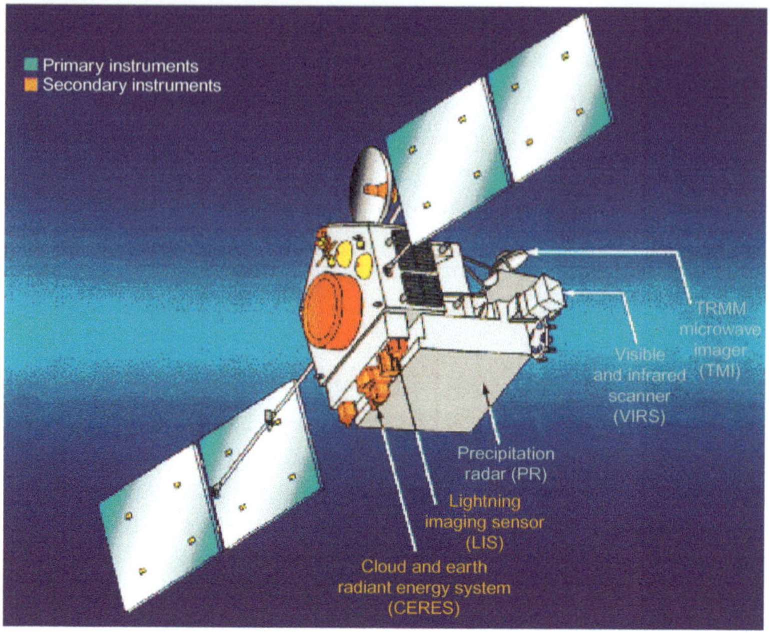


FIG. 11.7 TRMM satellite showing active and passive sensor systems that detect rainfall and weather characteristics.

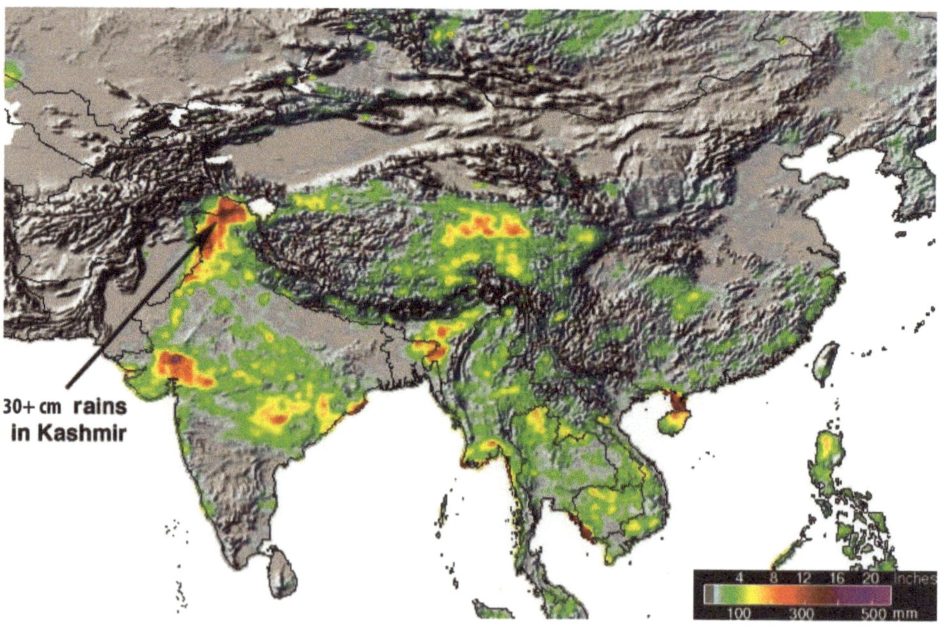


FIG. 11.8 TRMM data showing precipitation totals for the seven-day period ending at 18 UTC Sept. 8, 2014 show that up to 356 mm fell along the India-Pakistan border in Kashmir. This rainfall was the result of the arrival of a strong monsoon low pressure over the region that is illustrated in Fig. 11.6. *Image credits NASA/TRMM. From Monsoon floods kill 420 in India and Pakistan. Jeff Masters' Wunderblog. http://www.wunderground.com/blog/JeffMasters/comment.html?entrynum=2791 (accessed 03.09.2015.).*

Obviously rainfall is related to the daily weather. Precipitation provides water, which is essential to the survival of all living things, but it can also commonly be the cause of natural disasters such as floods and sediment-deposition disasters. Recently, attention in many countries such as largely arid Afghanistan has focused on the importance of maintaining water resources. To that end, the observation of rainfall is extremely important so that TRMM can make global measurements of rainfall.

When moisture in the air changes into rain, the latent heat of condensation is emitted into the atmosphere. Because this heat is the driving force for the circulation of the atmosphere, data and information on global precipitation are essential to explain and forecast that circulation. Since tropical and subtropical rainfall comprises more than two-thirds of global rainfall, it has a significant influence on the circulation of the atmosphere. A serious lack of data on precipitation occurs in tropical and subtropical regions, because not enough land-based observation points exist in those regions. To get such information, we have to make our observations from space with instruments such as the TRMM satellite, or a new replacement satellite system launched as the old one degrades in the next few years.

The TRMM satellite data can also be used to diagnose conditions likely to lead to a variety of natural hazards that are related to plentiful precipitation. Flooding is a scourge in Central and Southwest Asia, mainly because rapid runoff from deforested and overgrazed watersheds lack slow infiltration into the soil and instead the rainfall and snowmelt runs off the land rapidly and inundates the lowlands repeatedly. The Global Flood Monitoring System (GFMS) is a "real-time" (imagery available within hours), quasi-global (50° N–50° S), hydrological run-off and routing model that runs on a 1/8th degree latitude/longitude grid, at a relatively high spatial (~12 km) and temporal (3-hourly) resolution (NASA, 2015a,b; GFMS, 2015). Wu et al. (2014) have developed techniques to estimate flood routing and inundations using this satellite imagery with a coupled land surface model upon which to plot the floods (Fig. 11.9).

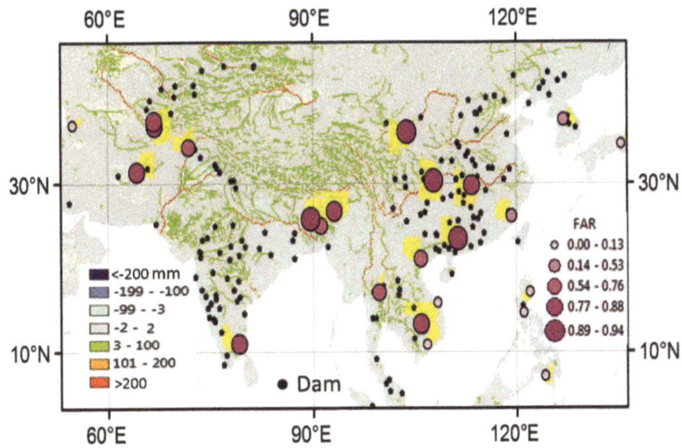


FIG. 11.9 Example of floods in Asia where yellow indicates well-reported areas and the other colors of rivers *(red, green, blue)* are the levels of flooding in mm. Dams are in black as pentagram figures, many of which cause errors (false alarm ratios) illustrated in purple circles. *Real-time global flood estimation using satellite-based precipitation and a coupled land surface and routing model. Water Resour. Res. 50 (3), 2693–2717.*

In addition to flooding, many parts of the world are subject to catastrophic mass movements such as rapid, wet debris flows and other landslides generated by rapid high precipitation (Petley, 2008). The real-time TRMM data have been used to assess where in Asia heavy or intense rains would set up conditions for possibly fatal landslides, and for associated flooding. The idea was to use a satellite-based system for issuing warnings against such events.

An additional element of investigation has been the development of new assessment methods of snowpack and runoff generation by combining TRMM precipitation estimates together with information from other satellites (Vuyovich and Jacobs, 2011). This involved deriving the snow-water equivalent (SWE) from the AMSR—EOS (AMSR-E) and the Special Sensor Microwave/Imager, which are mounted in the Aqua satellite. The passive microwave SWE was found to provide valuable water-resource information and to have potential value for forecasting reservoir infilling with spring snowmelt, which was an improvement over other models to estimate spring runoff prediction using only the simple temperature-index, snow model alone.

## 11.3.7 Aqua Satellite

This device carries six state-of-the-art instruments in a low-Earth orbit that is near-polar in orientation. The six instruments are the: (1) Atmospheric Infrared Sounder; (2) Advanced Microwave Sounding Unit; (3) Humidity Sounder for Brazil; (4) AMSR-E; (5) MODIS; and (6) CERES (Fig. 11.10). The Aqua (Latin for water) satellite system (NASA, 2003) is part of the so-called afternoon or A-train of other satellites (Cloudsat, CALIPSO, PARASOL, Aura, and

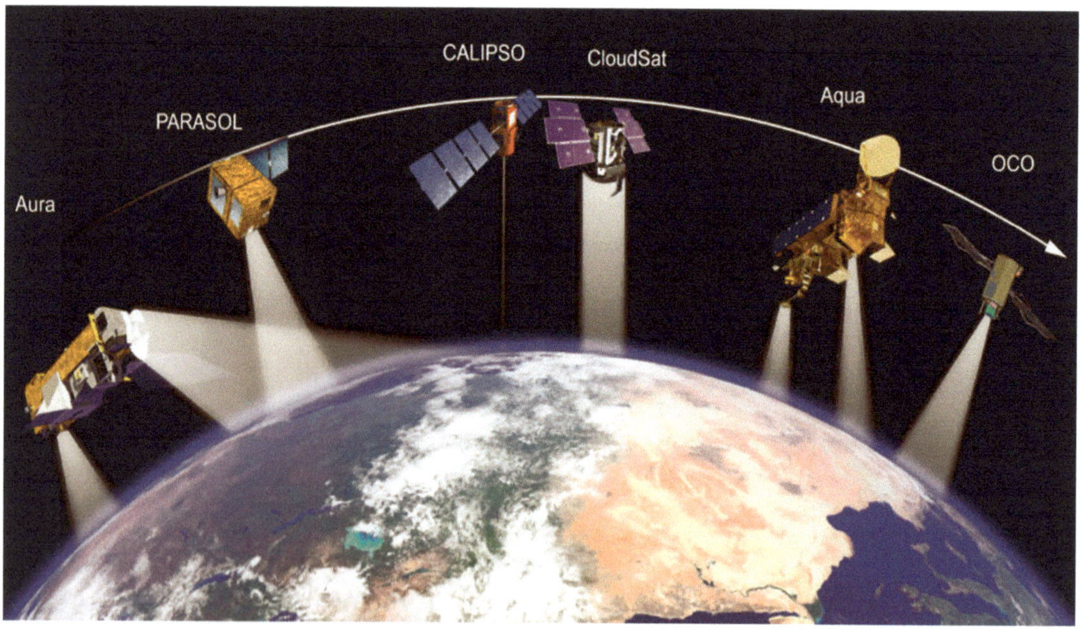


FIG. 11.10 The afternoon or "A-train" satellite constellation of six satellites (Aqua, Cloudsat, CALIPSO, PARASOL, Aura, and OCO) all cross the equator in the early afternoon. Collectively, these many satellites are collecting a variety of data about all the phases of water.

OCO) that focus upon the multi-disciplinary study of the planet's interrelated processes that involve the atmosphere, oceans, ice, and the land surface, and their relationships to changes in the Earth processes systems. The large amount of information that the mission is collecting about Earth's water cycle includes evaporation from the oceans and water vapor in the atmosphere, clouds, precipitation, soil moisture, sea and land ice, and snow cover on the land and ice. Additional data being collected also include radiative energy fluxes, aerosols, vegetation cover on land, phytoplankton and dissolved organic matter in the oceans, and temperatures of air, land, and water. Aqua was launched in 2002, and was supposed to provide data for only 6 more years, but it is still working quite well and is expected to keep producing data into the early 2020s.

### 11.3.8 GRACE Satellites

The two Gravity Recovery and Climate Experiment satellites (Ward, 2014) were put into orbit in 2002 to detect variations in Earth's time-variable gravity field (Fig. 11.11). This is being done by very precisely measuring the varying distance between the two satellites due to changing amounts of water or ice mass below them. This water and ice mass on or within the Earth, of course, occurs as glaciers, permafrost ground ice within the Earth at high altitude and latitude, and as underground water, all of which vary through time. A great many countries in the arid and semiarid parts of the world are known to have been "mining," or extracting more water from their underground aquifers than has been accumulating from natural replenishment sources. Many of these aquifers were largely filled in previous climatic periods some tens to hundreds of thousands of years ago during which recharge occurred through infiltration in the unconfined parts of the aquifer (Sultan et al., 2015). Present-day water production from underground aquifers may be unsustainable over long time periods but must be assessed to ascertain more ideal and reduced future rates of withdrawal. The GRACE

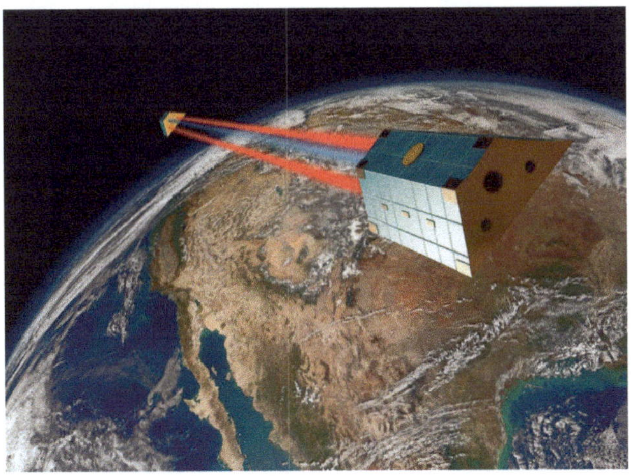

FIG. 11.11 Gravity Recovery and Climate Experiment (GRACE) satellites that use precise laser interferometry to measure the slightest deviation in distance between them caused by small variations in the gravitational attraction of the Earth below.

satellite capability offers a superb opportunity to detect wide-spread (km-scale) changes in glacier ice mass or in underground water storage, both of which could be vital to understanding water shortages through ground-water drawdown or oversupply through overpumping, or because of water logging and flooding in Tajikistan, Afghanistan, and Pakistan.

## 11.4 GLOBAL PRECIPITATION MEASUREMENT

Advances made in remote-sensing detection of precipitation in the past few decades have led to a new constellation of satellites known as the Global Precipitation Measurement Mission (GPM) (Fig. 11.12). The mission is primarily an international partnership between NASA and JAXA, but it also includes a number of other satellites, including several from France, India, and Europe. This network of satellites is providing the next generation of global observations of precipitation from space. The GPM Core Observatory launched in 2014 is set up to carry an advanced system of radar and radiometer devices that will serve as a reference standard to

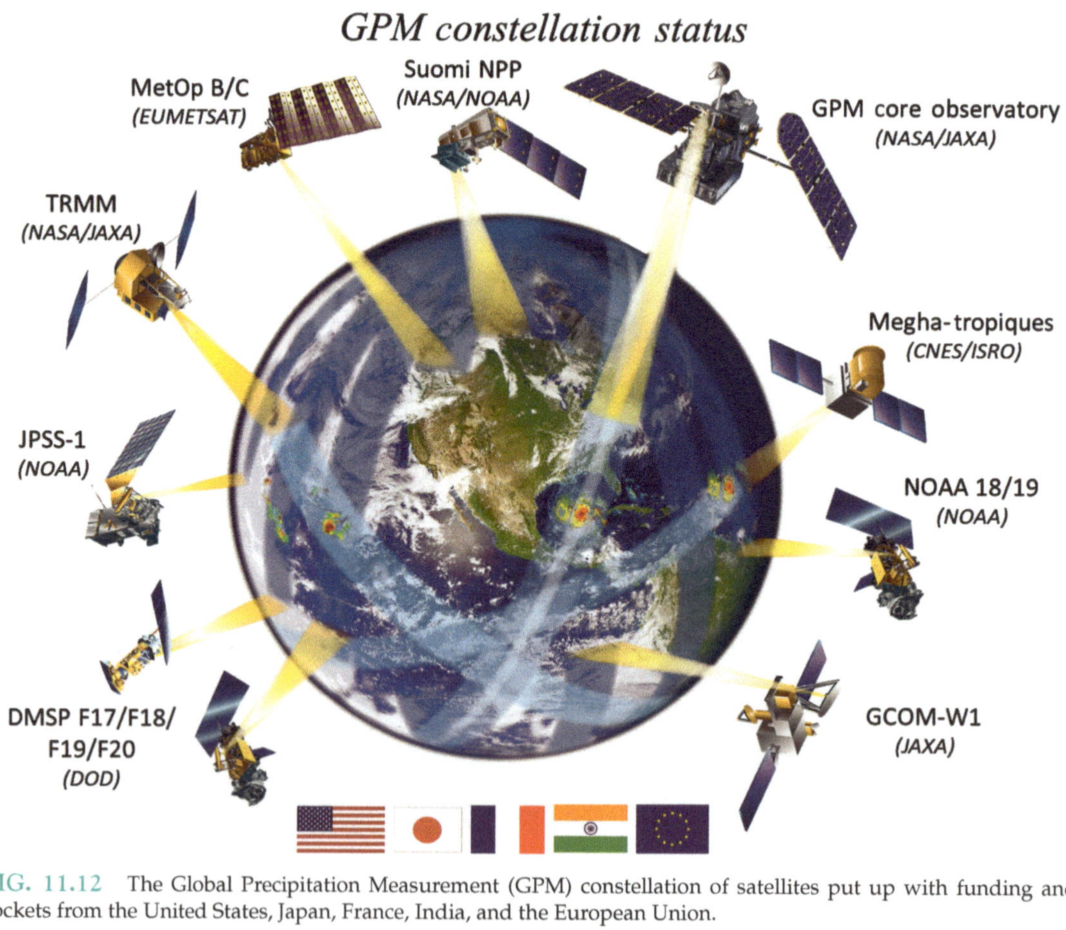


FIG. 11.12 The Global Precipitation Measurement (GPM) constellation of satellites put up with funding and rockets from the United States, Japan, France, India, and the European Union.

unify the measurements of precipitation from all the other satellites that are flying within the constellation. This new mission is a direct result of the success of the TRMM satellite, but the new GPM Core Observatory has extended the coverage from 35° N and S latitude to 65° N and S, or the Arctic and Antarctic Circles. This means that the latitudes from 41° N in northern Tajikistan to 23° N in southern Pakistan are now well covered for all future assessments of precipitation. In fact the GPM constellation now provides measurements on the:

- intensity and variability of precipitation;
- 3D structure of cloud and storm systems;
- microphysics of the ice and liquid particles within clouds; and
- amount of water falling to Earth's surface.

In addition to the basic GPM data obtained, in combination with basic terrain data on any region, the constellation information can be used for producing:

- weather-forecast models;
- climate models;
- integrated hydrologic models of watersheds; and
- forecasts of floods, droughts, and landslides.

The GPM Core Observatory carries the GPM Microwave Imager (GMI) and Dual Frequency Precipitation Radar (DPR). These devices will collect more advanced information by observing inside clouds. With these advanced capabilities, scientists will be able to study the internal structure of moisture-bearing storms throughout their life cycles to see how the storms change over time. Both of these instruments, together with data from the other satellites in the GPM will provide a database of measurements that can be meaningfully compared and combined to make uniform datasets of global precipitation. In addition, measurements from the GMI have been designed as a reference standard for cross-calibration of other satellites in the GPM group of satellites by imaging the same storms and making cross comparisons.

The GMI is a multichannel microwave radiometer that was designed to sense all of the precipitation within cloud layers, including such illusive targets as light rain and snowfall. This is done by measuring the intensity of lower and higher frequency, microwave energy that is constantly being emitted or scattered by all of the precipitation elements. The contrast between the signals received by the different GMI channels enable distinguishing between rain-drop sizes, snow fall, calculation of precipitation rates, and quantification of intensity of precipitation.

The DPR is making three-dimensional measurements of precipitation structures and rates across a great deal of the Earth's surface. Different radar bands detect drop-size distributions, or how many raindrops of different sizes are in the various cloud layers and how they are spread throughout any given storm system. The improved data collection of bulk precipitation properties such as precipitation intensity, water fluxes, and amount of water content, as well as information on distribution and sizes of precipitation particles, together with the radiometer measures, will improve the accuracy of rain and snow fall estimates.

The GPM systems are a real change for assessing precipitation over Afghanistan, Tajikistan, and Pakistan, which will provide unprecedented detail for the critical "water towers" of the Pamir, Hindu Kush, and Western Himalaya. By providing measurements of the microphysics of these regions, GPM will greatly advance Earth-system analysis and modeling in this region to enable description of how conditions that exist today might change in future. The greatly improved forecasting capabilities for floods, droughts, and landslides will meet the demand for better understanding and predicting these natural hazards that are so prevalent in the region, and will greatly facilitate the more timely staging of search and rescue in these remote and transport-poor regions. Satellite data from the GPM mission will provide essential precipitation estimates over land that can be used to improve forecast models of agricultural forecast models. Crop production will thereby be more easily assessed to help avoid famine in drought years. Water- resource managers must rely upon accurate precipitation measures to monitor the freshwater resources that are necessary for critical human activities, such as public consumption, irrigation, livestock, sanitation, mining, and powering industries. Overall the GPM constellation of satellites will allow the scientific community to find out how any changes in the freshwater supply will occur, and will thereby enable far better forecasting of any changes in a region that far too often has been hovering at the edge of criticality.

## 11.5 CONCLUSION

Collection of information about water resources is regarded as an essential aspect of any nation state in order to be able to plan better for domestic uses and agricultural activities, as well as for generation of hydroelectricity. In remote and poorly developed countries such as Afghanistan, Tajikistan, and Pakistan, knowledge about water is essential, yet too little of such data are readily available. These areas of lower technology can be improved a great deal by attention to satellite-sensing capabilities that can collect important data on water. The expense and sophistication of the high technologies involved, however, are such that much of the capability has not been used a great deal in Afghanistan. In future, however, attention to the details of these sensor systems will bring greater development potential to the region, providing that sufficient attention to the satellite media and education about the technology can be brought about.

### References


Bishop, M.P., Shroder Jr., J.F., Ward, J.L., 1995. SPOT multispectral analysis for producing supraglacial debris-load estimates for Batura Glacier. Geocarto Int. 10, 81–90.

Bishop, M.P., Shroder Jr., J.F., Hickman, B.L., Copland, L., 1998. Scale-dependent analysis of satellite imagery for characterization of glacier surfaces in the Karakoram Himalaya. Special volume on "remote sensing in geomorphology". Geomorphology 21, 217–232. S. Walsh and D. Butler (editors).

Bishop, M.P., Kargel, J.S., Kieffer, H.H., MacKinnon, D.J., Raup, B.H., Shroder Jr., J.F., 2000. Remote-sensing science and technology for studying glacier processes in high Asia. Ann. Glaciol. 31, 164–170.

Bishop, M.P., Bonk, R., Kamp Jr., U., Shroder Jr., J.F., 2001. Terrain analysis and data modeling for alpine glacier mapping. Polar Geogr. 24 (4), 257–276.

Bishop, M.P., Barry, R.G., Bush, A.B.G., Copland, L., Dwyer, J.L., Fountain, A.G., Haeberli, W., Hall, D.K., Kääb, A., Kargel, J.S., Molnia, B.F., Olsenholler, J.A., Paul, F., Raup, B.H., Shroder, J.F., Trabant, D.C., Wessels, R., 2004. Global land-ice measurements from space (GLIMS): remote sensing and GIS investigations of the Earth's cryosphere. Geocarto Int. 19 (2), 57–84.

Bishop, M.P., Shroder Jr., J.F., Haritashya, U.K., Bulley, H.N.N., 2007. Remote sensing and GIS for alpine glacier change detection. In: Baudo, R., Tartari, G., Vuillermoz, E. (Eds.), In: Shroder, J.F. Jr. (Ed.), Mountains, Witnesses of Global Changes: Research in the Himlaya and Karakoram. Under the Auspices of SHARE-Asia. Stations at High Altitude for Research on the Environment in Asia. Developments in Earth Surface Processes, 10. Elsevier, Amsterdam.

Bishop, M.P., Bush, A., Copland, L., Kamp, U., Owen, L.A., Seong, Y.B., Shroder, J.F., 2010. Climate change and mountain topographic evolution in the Central Karakoram, Pakistan, special issue on climate change. Ann. Assoc. Am. Geogr. 4, 772–793.

Bishop, M.P., Bush, A.B.G., Furfaro, R., Gillespie, A.R., Hall, D.K., Haritashya, U.K., Shroder Jr., J.F., 2014a. Theoretical foundations of remote sensing for glacier assessment and mapping. In: Kargel, J.S., Leonard, G.J., Bishop, M.P., Kääb, A., Raup, B. (Eds.), Global Land Ice Measurements From Space. Springer, Heidelberg. (Chapter 2), ISBN-10: 354079817X | ISBN-13: 978–3540798170.

Bishop, M.P., Shroder Jr., J.F., Ali, G., Bush, A.B.G., Haritashya, U., Roohi, R., Akif Sarikaya, M., Weihs, B.J., 2014b. Remote sensing of glaciers in Afghanistan and Pakistan. In: Kargel, J.S., Leonard, G.J., Bishop, M.P., Kääb, A., Raup, B. (Eds.), Global Land Ice Measurements From Space. Springer, Heidelberg (Chapter 23), ISBN-10: 354079817X | ISBN-13: 978–3540798170.

Bulley, H.N.N., Bishop, M.P., Shroder, J.F., Haritashya, U.K., 2013. Integration of classification tree analyses and spatial metrics to assess changes in supraglacial lakes in the Karakoram Himalaya. Int. J. Remote Sens. 34 (2), 387–411. http://dx.doi.org/10.1080/01431161.2012.705915.

Bush, A.B.G., Prentice, M.L., Bishop, M.P., Shroder Jr., J.F., 2004. Modeling global and regional climate systems: climate forcing and topography. In: Bishop, M.P., Shroder, J.F. Jr. (Ed.), Geographic Information Science and Mountain Geomorphology. Springer, Chichester.

Ceccato, P., Dinku, T., 2010. Introduction to Remote Sensing for Monitoring Rainfall, Temperature, Vegetation and Water Bodies. International Research Institute for Clinate and Society, Palisades, NY. IRI Technical Report 10-04; 15 pp. http://iri.columbia.edu/docs/publications/Introduction%20to%20remote-sensing-environmental%20 monitoring.pdf (accessed 02.28.2015.).

Copland, L., Pope, S., Bishop, M., Shroder, J., Clendon, P., Bush, A., Kamp, U., 2009. Glacier velocities across the central Karakoram. Ann. Glaciol. 50 (52), 41–49.

Copland, L., Sylvestre, T., Bishop, M.P., Shroder, J.F., Seong, Y.B., Owen, L.A., Bush, A., Kamp, U., 2012. Expanded and recently increased glacier surging in the Karakoram. Arct. Antarct. Alp. Res. 43 (4), 503–516.

EO, 2015. Sharing earth observation resources: EO portal directory. https://directory.eoportal.org/web/eoportal/satellite-missions/a (accessed 03.01.2015.).

GFMS, 2015. Global Flood Monitoring System. http://flood.umd.edu/ (accessed 03.01.2015.).

Haritashya, U.K., Bishop, M.P., Shroder, J.F., Bush, A.B.G., Bulley, H.N.N., 2009. Space-based assessment of glacier fluctuations in the Wakhan Pamir, Afghanistan. Clim. Chang. 94, 5–18.

Jensen, J.R., 2007. Remote Sensing of the Environment: An Earth Resource Perspective, second ed. Prentice Hall, Upper Saddle River, NJ. 592 pp.

Kargel, J.S., Abrams, M.J., Bishop, M.P., Bush, A., Hamilton, G., Jiskoot, H., Kääb, K., Lee, H.H., Paul, E.M., Rau, F., Raup, F., Shroder, B., Soltesz, J.F., Stainforth, D., Stearns, L., Wessels, R., 2005. Multispectral imaging contributions to global land ice measurements from space. Remote Sens. Environ. (Terra/ASTER Sci.) 99, 187–219.

Kieffer, H., Kargel, J.S., Barry, R., Bindschadler, R., Bishop, M., Mackinnon, D., Ohmura, A., Raup, B., Antoninetti, M., Bamber, J., Braun, M., Brown, I., Cohen, D., Copland, L., Duehagen, J., Engeset, R.V., Fitzharris, B., Fujita, K., Haeberli, W., Ove Hagen, J., Hall, D., Hoelzle, M., Johansson, M., Kaabe, A., Koenig, M., Maisch, M., Paul, F., Rau, F., Reeh, N., Rignot, E., Rivera, A., de Ruyter de Wildt, M., Schaper, J., Scharfen, G., Shroder, J., Solomina, O., Thompson, D., van der Veen, K., Wohlleben, T., Young, N., 2000. Satellite measurements of glaciers and ice sheets. EOS Trans. (Am. Geophys. Union) 81 (24), 265–271.

NASA, 2003. Formation flying: the afternoon "A-Train" satellite constellation. NASA Facts FS-2003-1-053-GSFC; 6 pp. http://aqua.nasa.gov/doc/pubs/A-Train_Fact_sheet.pdf (accessed 03.09.2015.).

NASA, 2015a. TRMM-based climatology. http://pmm.nasa.gov/TRMM/TRMM-based-climatology (accessed 03.01.2015.).

NASA, 2015b. Global flood and landslide monitoring. Current heavy rain, flood, and landslide estimates. http://pmm.nasa.gov/node/187 (accessed 03.01.2015.).

National Research Council, 2003. Satellite Observations of the Earth's Environment: Accelerating the Transition of Research to operations. The National Academies Press, Washington, DC. 163 pp.

Ozesmi, S.L., Bauer, M.E., 2002. Satellite remote sensing of wetlands. Wetl. Ecol. Manag. 10 (5), 381–402.

Petley, D., 2008. TRMM and landslide monitoring. The Landslide Blog—AGU Blogosphere; 2 pp. http://blogs.agu.org/landslideblog/2008/06/27/trmm-and-landslide-monitoring/ (accessed 02.28.2015.).

Phillips, W.M., Sloan, V.F., Shroder Jr., J.F., Sharma, P., Clarke, M.L., Rendell, H.M., 2000. Asynchronous glaciation at Nanga Parbat, northwestern Himalaya Mountains, Pakistan. Geology 28, 431–434.

Quincey, D.J., Bishop, M.P., Kääb, A., Berthier, E., Flach, B., Bolch, T., Buchroithner, M., Kamp, U., Khalsa, S.J.S., Toutin, T., Hatitashya, U.K., Racoviteanu, A., Shroder, J.F., Raup, B.R., 2014. Digital terrain modeling and glacier topographic characterization. In: Kargel, J.S., Leonard, G.J., Bishop, M.P., Kääb, A., Raup, B. (Eds.), Global Land Ice Measurements From Space (GLIMS). Springer, Heidelberg, pp. 113–143.

Sarikaya, M.A., Bishop, M.P., Shroder, J.F., Olsenholler, J.A., 2012. Space-based observations of eastern Hindu Kush glaciers between 1976 and 2007, Afghanistan and Pakistan. Remote Sens. Lett. 3 (1), 77–84. http://dx.doi.org/10.1080/01431161.2010.536181.

Sarıkaya, M.A., Bishop, M.P., Shroder, J.F., Ali, G., 2013. Remote-sensing assessment of glacier fluctuations in the Hindu Raj, Pakistan. Int. J. Remote Sens. 34 (11), 3968–3985. http://dx.doi.org/10.1080/01431161.2013.770580.

Seong, Y.B., Owen, L.A., Bishop, M.P., Bush, A.B.G., Clendon, P., Copland, L., Finkel, R., Kamp, U., Shroder Jr., J.F., 2008. Quaternary glacial history of the Central Karakoram. Quat. Sci. Rev. 26, 3384–3405.

Shroder Jr., J.F., 1980. Special problems of glacial inventory in Afghanistan. In: World Glacier Inventory Proceedings Reideralp Workshop, September 1978 (IAHS-AISH) Publication No. 126. Hydrological Sciences Bulletin, London, pp. 142–147.

Shroder Jr., J.F., 1989. Glacierized areas of Afghanistan. In: Haeberli, W., Bosch, H., Scherler, K., Ostrem, G., Wallen, C.C. (Eds.), World Glacier Inventory, Status 1988. IAHS (ICSI)-UNEP-UNESCO, Teufen, pp. C39–C40. C346–C353.

Shroder Jr., J.F., 2011a. Hindu Kush. In: Singh, V.P., Singh, P., Haritashya, U.K. (Eds.), Encyclopedia of Snow, Ice and Glaciers. Springer, Dordrecht, pp. 523–525.

Shroder Jr., J.F., 2011b. Himalaya. In: Singh, V.P., Singh, P., Haritashya, U.K. (Eds.), Encyclopedia of Snow, Ice and Glaciers. Springer, Dordrecht, pp. 510–520.

Shroder Jr., J.F., Bishop, M.P., 1998. Mass movement in the Himalaya: new insights and research directions. Special volume on "mass movement in the Himalaya". Geomorphology 26, 13–35. J.F. Shroder Jr. (editor).

Shroder Jr., J.F., Bishop, M.P., 2010a. Glaciers of Afghanistan. In: Williams, R.S. Jr., Ferrigno, J.G. (Eds.), Satellite Image Atlas of Glaciers of the World—Asia. U.S. Geological Survey, Denver, CO, pp. 167–199. U.S. Geological Survey Professional Paper 1386-F3.

Shroder Jr., J.F., Bishop, M.P., 2010b. Glaciers of Pakistan. In: Williams, R.S. Jr., Ferrigno, J.G. (Eds.), Satellite Image Atlas of Glaciers of the World—Asia. U.S. Geological Survey, Denver, CO, pp. 201–257. U.S. Geological Survey Professional Paper 1386-F4.

Shroder Jr., J.F., Weihs, B.J., 2010. Geomorphology of the Lake Shewa landslide dam, Badakshan, Afghanistan, using remote sensing data. Geogr. Ann. 92A (4), 471–486.

Shroder Jr., J.F., Bishop, M.P., Scheppy, R., 1998. Catastrophic flood flushing of sediment, western Himalaya, Pakistan. In: Kalvoda, J., Rosenfeld, C.L. (Eds.), Geomorphological Hazards in High Mountains Areas. Kluwer, Dordrecht, pp. 27–48.

Shroder Jr., J.F., Bishop, M.P., Sloan, V., Copland, L., 2000. Debris-covered glaciers and rock glaciers in the Nanga Parbat Himalaya, Pakistan. Geogr. Ann. 82A, 17–31.

Shroder Jr., J.F., Bishop, M.P., Bulley, H.N.N., Haritashya, U.K., Olsenholler, J., 2007. Global land ice monitoring from space (GLIMS) project regional center for Southwest Asia (Afghanistan and Pakistan). In: Baudo, R., Tartari, G., Vuillermoz, E. (Eds.), Mountains, Witnesses of Global Changes: Research in the Himalaya and Karakoram. Under the auspices of SHARE-Asia. Stations at High Altitude for Research on the Environment in Asia. Developments in Earth Surface Processes. In: Shroder, J.F. Jr. (Eds.), 10. Elsevier, Amsterdam.

Shroder Jr., J.F., Weihs, B., Schettler, M., 2011a. Mass movement in northeast Afghanistan. J. Phys. Chem. Earth 36, 1267–1286. http://dx.doi.org/10.1016/j.pce.2011.03.003.

Shroder, J.F., Schettler, M.J., Weihs, B.J., 2011b. Loess failure in northeast Afghanistan. J. Phys. Chem. Earth 36, 1287–1293. http://dx.doi.org/10.1016/j.pce.2011.03.001.

Shroder, J., 2014. Natural Resources in Afghanistan: Geographic and Geologic Perspectives on Centuries of Conflict. Elsevier, Amsterdam.

Sultan, M., Ahmed, M., Wahr, J., Yan, E., Emil, M.K., 2015. Monitoring aquifer depletion from space: case studies from the Saharan and Arabian aquifers. In: Lakshmi, V., Alsdorf, D., Anderson, M., Biancamaria, S., Cosh, M.H., Entin, J., Entin, J., Huffman, G.J., Kustas, W., Oevelen, P.v., Painter, T.H., Parajka, J., Rodell, M., Rüdiger, C. (Eds.), Remote Sensing of the Terrestrial Water Cycle. Wiley, Hoboken, NJ, pp. 349–366. Geophysical Monograph 206, American Geophysical Union.

USGS, 2014. Landsat missions timeline. http://landsat.usgs.gov/about_mission_history.php (accessed 03.04.2015.).

Vuyovich, C., Jacobs, J.M., 2011. Snowpack and runoff generation using AMSR-E passive microwave observations in the Upper Helmand Watershed, Afghanistan. Remote Sens. Environ. 115, 3313–3321.

Ward, A., 2014. Gravity Recovery and Climate Experiment (GRACE). http://earthobservatory.nasa.gov/Features/GRACE/ (accessed 03.07.2015.).

Wu, H., Adler, R.F., Tian, Y., Huffman, G.J., Li, H., Wang, J.J., 2014. Real-time global flood estimation using satellite-based precipitation and a coupled land surface and routing model. Water Resour. Res. 50 (3), 2693–2717.

CHAPTER

# 12

# Afghanistan Water Treaties and Relevant International Water Law

*J. Shroder*

**University of Nebraska at Omaha, Omaha, NE, United States**

OUTLINE



## Abstract

Sustainable water use in the world requires rules about shared water usage with neighboring states that overcome *absolute territorial sovereignty* where an upstream state can do whatever it likes to the shared water resources, in contrast to the downstream state where *absolute territorial integrity* allows that state to an uninterrupted flow of usable water. Limited territorial integrity is a solution, along with precise definitions of international water courses. New water law in Afghanistan has been developed even while public discussion has been limited by constraints from past presidential decree. New treaty negotiations are required but are yet to get underway because of a history of prior failures and water constraints, coupled with fears of water avarice from neighboring states.

**Keywords:** Absolute territorial sovereignty, Absolute territorial integrity, Water diplomacy, Water treaties, Sustainable water use, International water course, Transboundary river, Riparian state, Water Sector Strategy (WSS), Water justice, River basin management (RBM)



http://dx.doi.org/10.1016/B978-0-12-801886-6.00012-4

## 12.1 INTRODUCTION

As freshwater in general throughout the world moves into a crisis quantity state, and in Afghanistan in particular, assessing paths to sustainable water use is paramount. The history of international law and practice about water is replete with different theories that govern the use of international rivers (Le Marquand, 1978; Hussein and Al-Jayyousi, 2001) (Table 12.1). In the realm of international water law, such as it is, no universally accepted definition of equity exists in the diversion of waters between users (Moore, 1992). This is because the characteristics of each international water resource are specific in diverse hydrologic, institutional, and legal aspects, so that any rules and regulations of universal applicability are not realistic unless they can be kept as broad and flexible as possible. Multiple conflicts have arisen in past years over water resources shared between different states because of different approaches to the issues of sovereignty. Such conflicts have occurred repeatedly between upstream riparian countries and downstream users.

Gleick (1998) has noted a wide range of ecological and human crises that result from inadequate access to, and the inappropriate management of, freshwater resources. He took particular note of seven "sustainability criteria" that should be sought as part of an effort to reshape long-term water planning and management, which we have applied to Afghanistan and its neighboring countries downstream (Table 12.2). In addition to the criteria designed to ensure long-term sustainability, Gleick (1998) also offered a working definition of sustainable water use as "the use of water that supports the ability of human society to endure and flourish into the indefinite future without undermining the integrity of the hydrological cycle or the ecological systems that depend on it." This is all a major order for a country such as Afghanistan so we must regard it at the present time as aspirational goals. It is indeed part of the goals of this book to try to move things concerning water in Afghanistan in these directions toward sustainability, but it will be a large struggle, given population growth, climate change, and opposition to many of these endeavors by surrounding nation-states downstream who will fear losing water to which they feel so entitled.

TABLE 12.1 Five Theories Governing Use of International Rivers

1. Absolute territorial sovereignty (Harmon doctrine), in which the upper riparian state has absolute sovereignty over any river flowing through its territory and can therefore do as it wishes to the water.
2. Absolute territorial integrity, which guarantees to the lower riparian states the use of the rivers in an unaltered condition.
3. Limited territorial sovereignty or equitable utilization theory, which permits the use of the rivers, as long as no harm is done to the other riparian states.
4. Limited territorial integrity, which recognizes the existence of a community of interests among the riparian states that in turn gives rise to a series of reciprocal obligations and rights.
5. Drainage basin development, or the community of interests theory, which stresses the common development of rivers in which all riparian states in a drainage basin sit down regularly to apprise each other of all water uses and water-development works within the drainage basin.

In general the list ranges from least favored and older practices to the most favored situations in the modern viewpoint.
*Utton, A.E., Teclaf, L., 1978. Water in a Developing World: The Management of a Critical Resource. Western Special Studies in Natural Resources and Energy Management. United Nations Development Programme, New York; From Hussein, I., Al-Jayyousi, O., 2001. Management of shared waters: a comparison of international and Islamic law. In: Faruqui, N.I., Biswas, A.K., Bino, M.J. (Eds.), Water Management in Islam. United Nations University, Tokyo, 95–100.*

TABLE 12.2 Sustainability Criteria for Water Planning in Afghanistan and Surrounding Countries Downstream

1. A basic water requirement is to guarantee to all humans enough to maintain human health.
2. A basic water requirement will be to guarantee to restore and maintain the health of ecosystems.
3. Water quality should be maintained at certain levels to meet certain minimum standards, depending upon location and how the water is to be used.
4. Human actions should not impair the long-term renewability of freshwater stocks and flows.
5. Data on water-sources availability, use, and quality should be collected and made accessible to all parties.
6. Institutional mechanisms should be set up to prevent and resolve conflicts over water.
7. Water planning and decision making should be democratic, ensuring representation of all affected parties and fostering representation of all affected parties and fostering direct participation of affected interests.

This is an aspirational list, because not all of these criteria are being met at the present time.
*From Gleick, 1998.*

Some of the earliest controversies about international water courses have involved differences between upstream and downstream states (Dinar et al., 2013). Early extreme positions naturally involved ideas of *absolute territorial sovereignty*, in that the rights of an upstream state were to do as it wished with the waters of its territory, regardless of the adverse effects on any downstream state. This was countered with ideas of *absolute territorial integrity*, or the right of a downstream state to an uninterrupted flow of a fixed quantity of usable water from the upstream state. These ideas have had to be modified over time due to changing necessities and development of new understandings.

International law is rather different than the internal laws of any particular nation-state in that international law is a decentralized system that generally lacks any particular rules of enforcement or even enforcement mechanisms, either by an international police force, or with courts having a compulsory jurisdiction. For these reasons, questions are sometimes raised about whether it can even be classified as "law" (Dinar et al., 2013). For backward or largely illiterate nations that are only weakly developed or in a status of failed state, international law may seem rather irrelevant to them, or they may claim this status. Nonetheless, the two main sources of international law are treaties and customary international law.

Treaties can be referred to in many ways: treaty, convention, agreement, protocol, accord, covenant, exchange of notes, charter, or memorandum of understanding. By whatever name, the instrument of agreement is negotiated, signed, and ratified by the respective governments. On the other hand, the rules of customary international law develop slowly over time as a basis of state conduct. The rules are observed generally with expectations of reciprocal treatment from the other states. Thus the Law of International Watercourses, like the Law of the Sea, developed first out of necessities in terms of navigation. Later in terms of international watercourses came the ideas of actual quantitative allocations of water amounts delivered across international boundaries. This was followed by ideas on maintaining the quality of those international water courses to avoid unwarranted pollution, as well as the need for upstream states to notify downstream states about any significant changes underway that could affect the downstream nation. International law for water underground is very much in its infancy because of an out-of-sight, out-of-mind treatment of the resource, but scientifically the two are the same, even if they must be treated differently because of their locations and treatments. Legally they may be divorced; scientifically they are integral, even if they must be treated differently in practicality.

## 12.2 DEFINITIONS OF AN INTERNATIONAL WATER COURSE

The idea of exactly what constitutes an international watercourse means different things to different people, especially when all the different parts of an international drainage basin can be considered in terms of the many different locations and phases of the $H_2O$ systems that can physically exist there. For example, an international watercourse could include the glacier ice on the high peaks, the meltwaters that derive from the glaciers, the river downstream that comes from the meltwaters, and the ground-water base flow that seeps into the *gaining* river in the uplands, as well as in the lowlands where the river might reverse flow outward through its channel to soak into the sediments and rock fractures where the river might then become a *losing* stream, and one that also has numerous canal outtakes on the surface for use in the agricultural fields. Thus an international drainage basin is more than just a transboundary river; it is the entire geographic area of a watershed because water is not only spread throughout it above ground as solid glacier ice, and liquid water in rivers and lakes, but below ground as well in different phases in many places and open spaces (soil moisture, rock and sediment porosity, rock fractures, and so forth). In recognition of this integral scientific complexity, in 1997 a United Nations convention defined an international water course to include all interconnected waters, including rivers, tributaries, lakes, glaciers, underground aquifers, reservoirs, and canals. The legal and physical scope of watercourse system components and water use are complex but can be resolved into different water types that are covered by the UN treaty, uses of the water in all its forms, and elements of the hydrologic cycle that are not covered in the treaty (Fig. 12.1).

Because, according to the UN convention treaty, an international watercourse exists where any portion, whether a tributary, lake, or other surface waterbody receives water from, or contributes water to another nation-state, mountain glaciers in an upper riparian nation would require certain rights and duties to apply to the use of that glacier and its meltwater. Many lower riparian nations have been pleased with this understanding, and as a result have ratified the UN treaty. In part, however, because of this apparent possible long-distance application of the concept of such extensive international waterbodies, upper riparian states have commonly failed to ratify the treaty. Only South Africa and Uzbekistan (who originally abstained in the vote), as upper riparian states, have ratified the treaty, and of course, Uzbekistan has considerable downstream riparian interests to protect as well. Both India and Pakistan, as well as many others, abstained from voting on the Convention during the General Assembly debate, because they evinced suspicion and general disregard of the treaty. In 1997, the government of Afghanistan was dysfunctionally in the Taliban's hands and did not contribute to the discussion. Iran and the Russian Federation voted in favor. China has roundly refused to ratify any water agreements, whether bilateral, regional, or international, and along with only Burundi and Turkey voted directly against the UN Convention. This vote stands in stark contrast to President Xi's assertion that the country opposes all forms of hegemony (Wouters, 2014), perhaps not quite understanding the concepts or implications of the hydro-hegemony that China has been exerting against its neighbors so effectively.

The result of the complete lack of consensus on international water in the nation-states of the Hindu Kush and Himalaya (HKH) is that some of the countries are most likely to aggressively exploit all their glacier meltwaters, and arguably do not feel subject to a duty to share or cooperate in the exploitation of these resources (Uprety, 2014). Of all the nations served by waters arising in the HKH, only Uzbekistan is signatory to the UN Watercourse Convention.

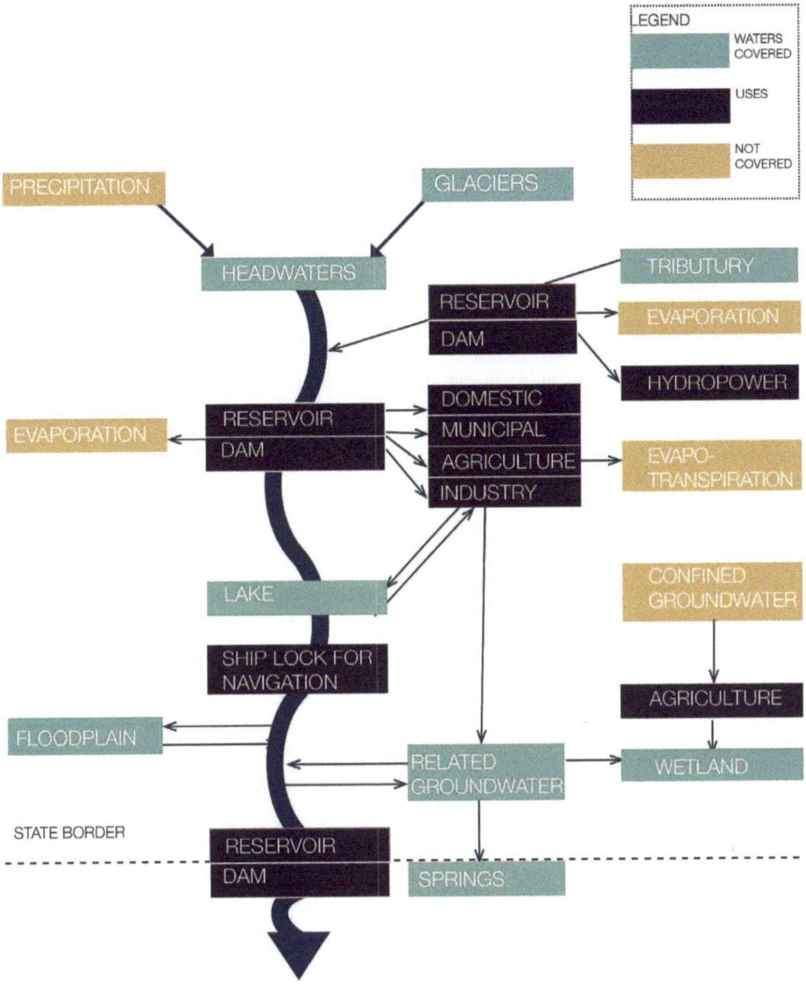


FIG. 12.1 Legal and physical scope of the UN convention on water courses that defines the legal coverage of a treaty about waters and water use. Diagram of water types (solid ice, liquid water, gaseous vapor) that are both covered or not covered under the UN Watercourses Convention of 1997. The diagram is not intended as a comprehensive assessment of all possibilities but it does show how core uses in one nation-state can have impact on transboundary waters of another nation-state, and whether or not these uses will be covered under the scope of the agreements. Under some special cases of the liquid locations of water only the confined (artesian) water is not covered. *From Rieu-Clarke, A., Moynihan, R., Magsig, B.-O., 2012. UN Watercourses Convention: User's Guide. IHP-HELP Centre for Water Law, Policy and Science, Dundee; 271 pp. http://www.gwp.org/Global/Our%20Approach/Strategic%20Allies/User's%20Guide%20to%20the%20UN%20Watercourses%20Convention%20(2012).pdf (accessed 03.09.2014.). By permission, Rieu-Clarke, email Sept. 4, 2014.*

Nonetheless, as of May 19, 2014, Vietnam became the 35th party to ratify the terms of the 1997 Convention, which triggered the coming into force of the UN document on Aug. 17, 2014, some 17 years after its original proposal (Table 12.3) (UN, 1997)—better late than never, it would seem. Accordingly by this act, the concept of *limited territorial integrity* has come to the fore over the old doctrines of absolute territorial sovereignty and integrity (Rieu-Clarke et al., 2012).

TABLE 12.3 Scope and Main Provisions of the UN Watercourses Convention That Make Requirements of the Signatory Nation-States

- To cooperate in adopting agreements that implement and/or adjust the Convention to their specific circumstances and needs.
- To participate actively and equitably in development and protection of international watercourses with regular relevant information exchange (water not as a state secret).
- To avoid causing significant harm to watercourses, or eliminate or mitigate such harm in good faith with fair and balanced manners.
- To follow procedures of notification, consultation, negotiation, and data exchange.
- To protect ecosystem of international water courses.
- To prevent, reduce, and control pollution.
- To prevent introduction of exotic species into international watercourses.
- To take all appropriate measures to prevent and mitigate harmful conditions.
- To consult over establishment of joint management mechanisms, such as basin organizations, transboundary management plans, and water-quality standards.
- To seek peaceful settlements of disputes.

*Loures, F., Rieu-Clarke, A., Vercambre, M.L., Witmer, L., 2014. Everything You Need to Know About the UN Watercourses Convention. World Wildlife Fund, Gland; 36pp. http://wwf.panda.org/what_we_do/how_we_work/policy/conventions/water_conventions/un_watercourses_convention (accessed 23.08.2014.).*

In this new thinking, the doctrine stipulates that in all watercourses, states shall enjoy an equal right to the utilization of the shared resource, and each watercourse nation must respect the sovereignty and reciprocal rights of the other watercourse nation-states. A strong advantage of this principle of limited territorial integrity is that it simultaneously recognizes the rights of both the upstream and downstream nations without sacrificing the principle of sovereignty.

Uprety (2014) has stressed that water disputes in the HKH exacerbate already strained bilateral relations, with conspiracy theories and blame games prevailing throughout South Asia. This "blame culture" reflects the absence of trust that plagues intraregional relationships, and makes multi-country, river-sharing arrangements particularly difficult to materialize. The countries of South Asia apparently find the principles of water use emanating from the Convention to be intuitively appealing, but too difficult to operationalize and implement in the particularly heated geopolitical environment that prevails in the region. Many of the principles contained in the Convention that were designed to ensure cooperation in the abstract are considered too vague, if not actually contradictory in reality. Furthermore, they are deemed unsuitable for resolving South Asian-specific issues, where the prevailing concepts are to treat water as a security priority, wherein water scarcity and riparian rights to water allocations are part and parcel of state security. Such nationalistic thinking produces a reluctance to engage in multilateral agreements over water because these agreements inherently seem to require the secession of some portion of their national sovereignty. In addition, some historical materialists may endorse multilateral water-sharing cooperation, while at the same time believing that such cooperation can only be an unfair allocation of water skewed toward the more developed states, the bigger states, or the former colonial masters. In fact, almost all of such states occur in downstream positions, leaving the headwater or upper riparian states such as Afghanistan, in an apparently vulnerable position, should they sign up to the UN Convention.

King and Sturtewagen (2010) have pointed out that policy makers in Afghanistan in 2007 originally drafted a Water Sector Strategy (WSS) where they noted a potentially negative impact that Afghanistan water development could have on downstream riparian nations. The policy makers proposed development of a clearly drafted water strategy with significantly improved hydrotechnical capacity and knowledge, apparently in order to engage their neighbors on transboundary water issues on more of an equal footing. They saw that without the ability to measure water resources and their use, Afghanistan might lose out in any of the regional political processes and economic advantages accruing as maximum benefits of cooperation on shared waterways. This wise position in 2007, however, did not appear in the WSS of the following year, and instead the need for enhanced development of the dam and irrigation infrastructure supplanted the regional approach, as any mention of transboundary water issues or regional cooperation were omitted.

Much treaty-making activity in South and Central Asia appears rather unique, in that prevailing mutual distrust, inability to delink hydropower generation from water resources, and strong reservations against generalizing policy to avoid establishing precedent, leads to engagement in water-related initiatives only at a superficial level. Practical advancement in water negotiations seems to only occur if it can serve some specific interests, but never just for the common good. Under these constraints, if a specific situation is thought to be too complex because of both upstream and downstream considerations that would require a differentiated strategy, any uniform formula such as the UN Watercourse Convention for water management is therefore deemed to be impractical (Uprety, 2014).

Water justice in the HKH region thus requires new ways of thinking about shared water resources. Any sort of true water accord in the region is quite far from reality. A new approach is required that is centered on the idea of water as a common good and a human right that is universally held and in need of acknowledgment. It seems that the only solution to the dilemma to help achieve this might be strong scientific discourse to investigate water issues and raise the level of education about them. Sensible scientific discussions between upstream and downstream riparian nations would seem the only recourse available at present. Nonetheless, under President Karzai, the Government of Afghanistan (GoA) has taken the unusual position that even international scientific discussions on water should not be held in its territory because of the inherent danger to their water sovereignty, as if they were worried about getting talked into making water deals that they do not understand. This is an unfortunate situation of apparently willful ignorance that can only be solved by detailed water-cognizant, capacity-building in the country, which is partly what this volume is about. But in fact the only way that a country like Afghanistan can possibly know how best to think about the water-resource problem of which they are the headwater custodians, is for them to get well educated in the hydrological and collateral sciences of related fields. An understanding of the UN Watercourses Convention of 1997 is thus required to be promulgated inside the country in order to achieve some sort of better understandings that would enable proper treatment of their treasured resource (Table 12.3), rather than its present (and diminishing) condition (Shroder, 2014).

## 12.3 WATER LAW IN AFGHANISTAN

Following the removal of the dysfunctional Taliban government in late 2001, the re-establishment of the rule of law and a succession of new governments in Kabul led to the passing of many new laws as the country tried to recover from its long nightmare of civil war and destruction of normal life. Early on it was recognized that new water law needed to be established and by 2004, policy makers and international donors introduced and promoted concepts of "good" water governance to reform the water sector in the country (Thomas, 2013). Accordingly, the Water Sector Strategy of Feb. 2008 led to the adoption of the Afghanistan Water Law of Apr. 2009, published in the Ministry of Justice Official Gazette No. (980), as a rather aspirational and somewhat unrealistic document. Nonetheless, the new law was regarded as some sort of "holy trinity" of integrated water-resource management (IWRM), river-basin management (RBM), and participation by multi-stakeholder platforms (MSPs). This was done deliberately to move toward greater devolution of decision-making power, both away from the central government toward subbasin levels of water organizations, and away from the government, directly to the water users.

The new law was quite comprehensive; including some 40 separate articles that specify such things as the relevant water duties of the Ministry of Energy and Water, as well as of Agriculture, Irrigation, and Livestock, along with the establishment of river-basin agencies and councils, correct uses of water, protecting water resources from pollution, and establishing such things as dam maintenance, deep well drilling, enforcement of the new water law, and a host of other water-relevant actions and activities. Establishment of sewage treatment systems in the villages were even discussed as the province of several ministries, which will be news to almost all of the villages in a country where not even the capital city of Kabul enjoys very much in the way of any sort of sewage control. The absence of virtually any prior studies of what was most needed in water-law construction in the country led to overly broad and certainly rather vague generalizations that tended to be based on what were thought to be universal models, but that were based on external models from outside the country, rather than conforming to local cultural values and traditions (Thomas, 2013).

As usual all too much, when external power brokers impress their own external views of water realties down upon an Afghan reality, some unequal, local-level, power structures can trump liberal principles of equality and participation as promoted by advisors from the European Union (Warner, 2007). Significant obstacles to rethinking of reforms to Afghanistan water law emerged as the EU hegemonic norms of IWRM, RBM, and MSPs were realized as being barriers to the emergence of alternative models. In large part this was because any alteration would seem to partly invalidate or constitute an indirect acknowledgment of the limits of its own model.

Thus yet again as failures emerged in the new water law, water-governance reform emerged as a necessary concept in the minds of the largely European advisors. On the other hand, the Afghan government appeared rather uninterested in such actions, preferring instead to assert its pre-war focus on water-resource development through large-scale infrastructure projects such as dams and irrigation projects (Kakar and Thomas, 2012; Thomas, 2013).

## 12.4 WATER TREATIES

Almost all of the rivers of Afghanistan, as well as almost all of the supply of water for irrigation, drinking, and maintenance of wetland ecosystems throughout the country, derived from precipitation that falls within the borders of the country. All of the large and small rivers, except for the Pakistani headwaters of the Kunar (Konar) River, originate in the central Hindu Kush or the high Pamir and Hindu Raj mountains of the far northeast. Thus Afghanistan, at the top of the multiple watersheds, shares its waters outwards to the many surrounding countries; the Amu Darya from the confluence of the Panj on the border and the Vakhsh from Tajikistan downstream to Uzbekistan, Turkmenistan, and the Aral Sea; the Kabul River confluent with the Kunar and downstream to Pakistan and the Indus River; and the Helmand, Hari Rud-Tenjen and Murghab downstream to Iran and Turkmenistan (FAO/UNEP and OSU, 2003). These transboundary water resources have flowed relatively unimpeded for millennia into the surrounding lowlands, where the surrounding nation-states have grown quite accustomed to what always appeared to be an expectable and unexhausted supply. Indeed, the unwelcome floods came down from the Afghanistan highlands too, but that particular hazard tended to be overlooked as long as the rivers provided their essential services as vital water sources or approximate border markers. Establishment of treaties between the upstream riparian Afghanistan and the surrounding downstream riparians has long been generally agreed upon as an essential precondition to guarantee no loss of these riverine capabilities.

Thus for more than a century Afghanistan has concluded agreements with Russia, the USSR, and the Central Asian states relating to the Amu Darya that serves as a rough border marker, except where the river channel has changed its location laterally as a result of avulsion and strong bank erosion. Agreements in 1873, 1946, and 1958 primarily focused on the river as an international boundary (Horsman, 2008). Nonetheless, the 1946 Agreement established a joint commission that was supposed to meet regularly, and the 1958 Agreement agreed to joint measures to prevent and correct course changes in the frontier rivers, share costs equally, prevent water pollution, exchange regular information and level and volume of water, meteorological data, and set up a flood-warning system, and establish joint, integrated utilization of the frontier sections of the Amu Darya. Needless to say, hardly any of this was ever done, with major changes occurring, mostly to the detriment of Afghanistan, as the wealthier more northern Tajikistan hardened its banks with riprap to prevent erosion and the river advanced differentially southward 5–10 km, since the border was configured over a century ago along the thalweg of the river as it existed in the late 19th century (Fig. 12.2).

The Helmand River that rises in the Paghman Mountains to the northwest of Kabul, flows southwest until it crosses the border into Iran, where it empties into a series of large ephemeral lake basins known as *hamouns*. Originally set up by the British Seistan Commission under Goldsmid in 1872 (Shroder, 2014), the border jogs eastward to the disadvantage of Afghanistan, which was deliberate on the part of the British who were about to fight a second Anglo-Afghan War in the country. In 1950, a Helmand River Delta Commission was created to measure and divide the river flows between the two signatory countries (Favre and Kamal, 2004). Iranian-Afghan negotiations in 1972–73 resulted in the Helmand River Water

FIG. 12.2 Northern border of Afghanistan with Tajikistan where the original river thalweg (channel) meandered across the valley bottom from where the international border *(yellow line)* was set in 1873. The river has since migrated over 5 km to the south *(bottom)*. The straight yellow line on the lower right is 10 km in length.

Treaty that stipulated Iranian water consumption to 26 $m^3/s$ (0.82 $km^3$) of unpolluted water in a normal year, with lesser amounts in droughts (Samii, 2005; Hanasz, 2012). The treaty was not ratified by the Afghanistan government until 1977. After that, increased political and social instability in Afghanistan sent the treaty into dormancy, followed by years of recurrent drought so that in recent years incursions by Iranian border police into Afghanistan territory to divert water from Afghan irrigation canals have occurred repeatedly and resulted in armed skirmishes (Palau, 2013).

The Kabul River flows out of the Hindu Kush and into Pakistan some 27 km north of the Khyber Pass. Following cessation of the third Anglo-Afghan War in 1919, in 1921 a treaty on its waters was established between Afghanistan and His Britannic Majesty's Government in 1921. Water rights granted to British India included use of the Kabul River for navigation, as well as maintenance of existing rights of irrigation (Favre and Kamal, 2004). No subsequent water agreements have ever been negotiated with Pakistan over transboundary water flow into their territory; indeed, Afghanistan has long refused to recognize the creation of Pakistan from the partition of British India, so apparently to the GoA there would be little point in such a negotiation.

## 12.5 IMPLICATIONS FOR REASONABLE MANAGEMENT OF TRANSBOUNDARY WATER RESOURCES

Clearly, if one examines the dates of the old water treaties pertaining to the rivers that emanate from Afghanistan, such treaties were all signed a long time ago with governments that no longer exist. Some Afghanistan government people believe that this fact entitles their new government to do fairly much as it pleases with "its" own water, and thus does not need to renegotiate with any riparian nations downstream. This viewpoint is anathema to European and American advisors, the World Bank, and other such agencies concerned with maintaining peaceful relations across the borders as much as possible, and who recognize the long-term continuity of old water treaties, no matter what government is in authority at the moment. Nonetheless, Afghanistan's recent intransigence under the Karzai government on these issues is a combination of a number of important factors that must be considered, including the following:

- Thirty-six years of war have reduced the educational system of Afghanistan to a shambles, which means that it takes time to rehabilitate and reeducate local people and professionals to adequate understandings.
- Lack of adequately skilled human resources with solid knowledge of international water law, and ability to negotiate in international forums (King and Sturtewagen, 2010).
- Limited hydro-meteorological capacity (King and Sturtewagen, 2010).
- Scientific capacity in understanding water must be built in Afghanistan first, to better enable negotiation at a later time from a position of strength through this increased scientific awareness.
- Loosely constrained estimates are that this water-cognizant capacity-building will take 1–3 years.
- Afghans greatly fear the designs of neighboring countries who have delivered war across their borders for decades, and who are known to want the water (Mashal, 2012).
- Foreigner-driven advice and scientific water conferences are seen by Afghans as ill-informed, favoring technical or engineering answers, rather than understanding local arrangements, or political or economic solutions (Kakar and Thomas, 2012; Thomas, 2013).
- The short-term interests of Afghanistan apparently were seen by the government as best served by hindering progress on water negotiations, especially many of those being pressed by foreigners.
- Dominant thinking in the government of Afghanistan is that once dams upstream are funded and built, the downstream riparians will not have much to say.
- Accordingly, no one in the Afghanistan government below presidential cabinet level in the Karzai years (2004–14) was allowed to discuss water issues, although these rules were relaxed under President Ghani in 2015.

Thus, although most government officials recognize the inherent difficulties and even economic or militaristic dangers that this outlook is likely to produce, nothing could be done until the new post-Karzai government consolidated its rule. Then the new policies of the next president—diplomatic, technical, and academic—will determine new approaches to water. It was anticipated that new president, Ashraf Ghani, a former technocratic employee of the World Bank, would undertake new water-directed negotiations, as appropriate.

The global water-security implications to these developments in Afghanistan are judged as fairly serious, inasmuch as significant parts of at least two (Amu Darya, Indus) of the seven international rivers of the world were selected by the U.S. Intelligence Committee (2012) as most deserving of their attention in assessing global water security. This leads directly to assessments of the coming water situation in Afghanistan as a possible driver for peace if there were any hydro-statesmen available to help with better guidance, but which will most likely lead to the use of water not only as leverage, but even as weapon, with terrorism in

TABLE 12.4 Strategies for River-Basin Cooperation for Afghanistan With Neighboring Nation States

| Strategic Goal | Specific Items |
|---|---|
| Strengthen policy and regulation, both domestic and international. | (1) Building and strengthening local capacity to understand all relevant actions on water issues.<br>(2) Strengthening and improving data and information collection and analysis systems, and sharing of data and information.<br>(3) Increasing awareness and involving stakeholders in decision-making processes, both domestic and international.<br>(4) Building academic water-management curricula and programs at Kabul University and Kabul Polytechnic University. |
| Support long-term national and regional programs. | (1) Adopting basin-wide approaches for Amu Darya, Kabul River, and Helmand-Hari Rud rivers.<br>(2) Promoting consensus on importance and urgency of international water problems.<br>(3) Promoting harmonization of water regulation among the basin countries.<br>(4) Encouraging participation of all sectors, including colleges. |
| Evaluate basin-wide development alternatives | (1) Identifying, modifying, and promoting successful regionally shared water agreements.<br>(2) Facilitating exchange between riparians, including educational efforts.<br>(3) Hydrological data services—data and information sharing set up between agencies and universities. |
| Expand areas of interaction and joint interest. | (1) Promoting joint water-related and non-water related projects benefiting all countries in basin.<br>(2) Establishing dispute-resolution mechanisms.<br>(3) Including out-of-basin expertise, both university and agency. |

*From Dinar, A., Dinar, S., McCaffrey, S., McKinney, D., 2013. Bridges Over Water: Understanding Transboundary Water Conflict, Negotiation and Cooperation. World Scientific Series on Energy and Resource Economics, second ed., vol. 11. World Scientific Publishing, Singapore.*

the region mixed in by various players. This can be easily deduced with reference to the published documents of the U.S. Intelligence Committee (2012), such that regional instability with water as a focal point will be most difficult to curtail in this region of Southwest Asia, if indeed that will even be possible. Portents on the water horizon do not appear to be especially promising, given the many reasons for pessimism (Shroder, 2014). Nonetheless, some opportunities do exist for river-basin cooperation with neighboring states (Table 12.4), with the result that 14 specific items have been proposed to assist in this endeavor.

## 12.6 CONCLUSION

International water law has been developed to help resolve transboundary water issues between states but its application in Afghanistan has yet to be resolved to any local diplomatic satisfaction. Access to water is regarded as a major security issue in a fractious neighborhood. Various strategies have been developed to enable greater river-basin cooperation that may eventually lead to new diplomatic solutions and treaties.

## References


Dinar, A., Dinar, S., McCaffrey, S., McKinney, D., 2013. Bridges over water: understanding transboundary water conflict, negotiation and cooperation, second ed. World Scientific Series on Energy and Resource Economics, vol. 11. World Scientific Publishing, Singapore.

FAO/UNEP and OSU, 2003. The Atlas of International Freshwater Agreements. FAO/UNEP and OSU, Nairobi. http://www.transboundarywaters.orst.edu/publications/atlas/atlas_pdf/4_Treaties_asia.pdf (accessed 24.08.2014.).

Favre, R., Kamal, G.M., 2004. Watershed Atlas of Afghanistan: II Note on Transboundary Riparian Issues. Afghanistan Information Management Service (AIMS), Kabul. pp. 71–72.

Gleick, P.H., 1998. The World's Water 1998–1999: The Biennial Report on Freshwater Resources. Island Press, Washington, D.C.

Hanasz, P., 2012. The Politics of Water Security Between Afghanistan and Iran. Strategic Analysis Paper. Future Directions International, Independent Strategic Analysis of Australia's Global Interests, Dalkeith, WA. 6 pp. http://reliefweb.int/sites/reliefweb.int/files/resources/The%20Politics%20of%20Water%20Security%20between%20Afghanistan%20and%20Iran%20-%20March%201%202012.pdf (accessed 08.25.2014.).

Horsman, S., 2008. Afghanistan and transboundary management on the Amu Darya: a political history. In: Rahaman, M.M., Varis, O. (Eds.), Central Asian Waters: Social, Economic, Environmental and Governance Puzzle. Water & Development Research Group, Helsinki University of Technology, Espoo.

Hussein, I., Al-Jayyousi, O., 2001. Management of shared waters: a comparison of international and Islamic law. In: Faruqui, N.I., Biswas, A.K., Bino, M.J. (Eds.), Water Management in Islam. United Nations University, Tokyo, pp. 95–100.

Intelligence Committee, 2012. Global water security. Intelligence Community Assessment, ICA 2012-08, 02.02.2012. Defense Intelligence Agency (DIA), NGA, CIA, State/INR, and DOE. http://fas.org/irp/nic/water.pdf (accessed 05.09.2014.)

Kakar, K., Thomas, V., 2012. Water governance reform in Afghanistan: early lessons for a water-secure future. In: Bigas, H., Morris, T., Sandford, B., Adeel, Z. (Eds.), The Global Water Crisis: Addressing an Urgent Security Issue. United Nations University, Institute for Water, Environment and Health, Hamilton, ON, pp. 110–119.

King, M., Sturtewagen, B., 2010. Making the most of Afghanistan's river basins: Opportunities for regional cooperation. East-West Institute, New York, Gerda Henkel Stiftung. http://www.eastwest.ngo/sites/default/files/ideas-files/Afghanistan_Water.pdf (accessed 01.04.16.).

Le Marquand, D., 1978. Water in a developing world: the management of a critical resource. In: Utton, A.E., Teclaf, L. (Eds.), Western Special Studies in Natural Resources and Energy Management. United Nations Development Programme, New York.

Mashal, M., 2012. What Iran and Pakistan want from the Afghans: water. Time. 3 pp. http://world.time.com/2012/12/02/what-iran-and-pakistan-want-from-the-afghans-water/ (accessed 25.08.2014.).

Moore, J., 1992. Water Sharing Regimes in Israel and the Occupied Territories—A Technical Analysis. Project Report 609. Operational Research and Analysis Establishment. Department of National Defense, Ottawa.

Palau, R.G., 2013. Water Security: Afghanistan Transboundary Water Resources in Regional Context, Transboundary Issues. Civil-Military Fusion Centre (CFC), Norfolk. 15 pp. http://www.scribd.com/doc/230677281/Water-Security-Afghanistan-Transboundary-Water-Resources-in-Regional-Context (accessed 25.08.2014.).

Rieu-Clarke, A., Moynihan, R., Magsig, B.-O., 2012. UN Watercourses Convention: User's Guide. IHP-HELP Centre for Water Law, Policy and Science, Dundee. 271 pp. http://www.gwp.org/Global/Our%20Approach/Strategic%20Allies/User's%20Guide%20to%20the%20UN%20Watercourses%20Convention%20(2012).pdf (accessed 03.09.2014.).

Samii, B., 2005. Iran/Afghanistan: still no resolution for century-old water dispute. Radio Free Europe, September 7. 3 pp. http://www.rferl.org/content/article/1061209.html.

Shroder, J.F., 2014. Natural Resources in Afghanistan: Geographic and Geologic Perspectives on Centuries of Conflict. Elsevier, Amsterdam. 572 pp.

Thomas, V., 2013. "Good" Water Governance Models in Afghanistan: Gaps and Opportunities. Policy Note Series. Afghanistan Research and Evaluation Unit, Kabul. 4 pp. http://www.areu.org.af/EditionDetails.aspx?EditionId=626&ParentId=7&ContentId=7&Lang=en-US (accessed 24.08.2014.).

United Nations (UN), 1997. UN convention on the law of the non-navigational uses of international watercourses. 9 pp. http://legal.un.org/ilc/texts/instruments/english/conventions/8_3_1997.pdf (accessed 23.08.2014.).

Uprety, K., 2014. A South Asian perspective on the UN watercourses convention. International Water Law Project Blog. 3 pp. http://www.internationalwaterlaw.org/blog/2014/07/14/dr-kishor-uprety-a-south-asian-perspective-on-the-un-watercourses-convention (accessed 23.08.2014.).

Warner, J. (Ed.), 2007. Multi-stakeholder Platforms for Integrated Water Management. Ashgate, Farnham.

Wouters, P., 2014. Considering China's approach to the UN watercourses convention—time to revisit? International Water Law Project Blog. 4 pp. http://www.internationalwaterlaw.org/blog/2014/07/28/professor-patricia-wouters-considering-chinas-approach-to-the-un-watercourses-convention-time-to-revisit/ (accessed 23.08.2014.).

CHAPTER

# 13

# Hydro-Cognizant Capacity-Building in Afghanistan

*J. Shroder*

**University of Nebraska at Omaha, Omaha, NE, United States**

OUTLINE



## Abstract

Building awareness about water in the government and population of Afghanistan is complex capacity-building to train people so that eventually intelligent negotiations can be practiced with representatives from the neighboring countries. Capacity types include not only knowing about the actions of technical instruments in job functions, but also ideas about ownership, physical characteristics of water resources, adaptation, and maintaining focus. Capacity-building in the ministries involves ten major activities that have led to capacity-building assessments by various agencies of the US government, who have discovered that a number of problems exist with attempts to understand and deal with water issues in Afghanistan that must be addressed. The uses of the many principles of water science specific to the situations that exist in Afghanistan must be used to raise the levels of understanding about water in the country.

**Keywords:** Capacity-building, Water cognizance, Library database, Water-quality laboratory, Project assessment, Ownership capacity, Adaptation capacity, Stream gaging, Focus maintenance capacity



*Transboundary Water Resources in Afghanistan*
http://dx.doi.org/10.1016/B978-0-12-801886-6.00013-6

## 13.1 INTRODUCTION

The problems of water in Afghanistan are myriad; from the actual and depleted sources of the water, to pollution, sanitation, assessment, monitoring, hazards, irrigation, energy, etc., and all the other problems of access to the vital resource. This is a fundamental existential threat to Afghanistan that must be dealt with effectively and studied assiduously to solve major problems. Water services in the nation are limited and access to an improved water infrastructure is among the lowest in the world. In fact, all across rural Afghanistan, and even in most cities, some 89% of people defecate in the open, 75% drink contaminated water, and every one in ten children die directly from preventable, water-borne diseases before their tenth birthday. Common, water-borne diseases such as amoebic dysentery, giardia, hepatitis A, polio, salmonella, and others are endemic throughout the country. In fact, an overwhelming majority of the Afghan population lacks an adequate, safe supply of water because of contamination by a wide variety of sources, a serious lack of water-resource regulation and management, and a lack of basic infrastructure, compounded by increasingly closely spaced drought cycles and seasonal flooding.

In the post-9/11 environment of Afghanistan, in order to begin to create a well-trained, self-sufficient, and sustainable water-resources sector in the country, water-resource activities were carried out by various agencies such as the U.S. Geological Survey (USGS), the U.S. Agency for International Development (USAID), and others, in close cooperation with Afghan water officials in efforts to address immediate needs that would be well accepted. The general approach was to:

- build the capacity of local officials, technicians, and scientists so that they could monitor, assess, and manage the water resources of the country;
- provide training, equipment, and supplies;
- strengthen existing institutions;
- establish a national water-quality laboratory; and
- develop a national water-resources database.

In addition to these factors in the past decade and a half, a USGS multidisciplinary water team, for example, began several subtasks that were planned to include evaluation of the ground water throughout the country; characterization of the 1991 flood in the Helmand River Basin; recovery of the historic surface-water quantity data for the Helmand River; do a hydrologic, time-series analysis for the Helmand River; install and demonstrate a streamflow gaging station; finish a Helmand River Basin ground-water assessment; and assess the potential effects of the change of glaciers on water resources and hazards, among other things.

The primary objectives of the water-resources portion of hydro-development by the USGS in Afghanistan were to accomplish a number of capacity-building tasks. First was to work with Afghan officials to assess and understand the basic hydrology of the country on issues related to the interactions between ground and surface water, ground-water sustainability, contamination of drinking water, flood prediction and prevention, and other such water-relevant factors. Next was to help Afghans officials re-install the critical network of streamflow gages needed to allow early flood warning, as well as to more effectively manage and distribute the surface waters of the country. In addition it was obvious that Afghan officials needed help to begin a water-quality monitoring program to identify and protect good drinking water. They especially needed to learn how to estimate the amount of safe water available for use in key population centers of Kabul, Mazar-i-Sharif, Herat, and Kandahar so that new and sustainable water sources could be identified and documented.

The USGS geologists worked with scientists from the Afghanistan Geological Survey (AGS) of the Ministry of Mines and Petroleum (MMP), as well from the Afghanistan Ministry of Energy and Water (MEW), and the Ministry of Agriculture, Irrigation, and Livestock (MAIL), along with some personnel from non-government organizations such as the Danish Committee for Aid to Afghan Refugees (DACAAR). The quest to rebuild the pre-war data sets was partially successful, as well as many of the original goals and much of the capacity-building. Up to the present time (mid-2015), more than 40 reports on maps and ground water, water quality, and surface water have been produced by the collaboration between USGS and AGS, including 24 geohydrologic assessments of important mineral prospects throughout the country. The ground water of the Kabul Basin has received special attention over much of the past decade, with some 150 wells throughout the 100 km-long valley being regularly monitored. Water levels (Fig. 13.1) and samples were collected and analyzed for physical, chemical, and microbiological contents, which formed the basis of the first collaboration between USGS and AGS (Broshears et al., 2005; Akbari et al., 2007; Mack et al., 2010a,b, 2013; Taher et al., 2014). USGS and DACAAR scientists and developed a national ground-water monitoring well network and database that now contains >150 water-level monitoring wells across Afghanistan that constitute the beginnings of a national database (DACAAR, 2011).

Starting in the early 1950s, the USGS and a predecessor to USAID had begun a stream-gaging program to measure water-discharge characteristics in the Helmand River Basin (Mack et al., 2014). By the late 1970s, however, when everything collapsed after the communist takeover, some 169 stream gages had been distributed all across Afghanistan to collect hydrologic data. In recent years, with the assistance of the Ministry of Energy and Water, some of the original data were recovered by the USGS, and about 127 historical stream gages were reestablished to assist in the recovery of hydrologic capabilities in Afghanistan (Figure 2.2 in Chapter 2).

The building of a knowledge base about water in a country such as Afghanistan is a matter of finding out as much as one can about the water that is the treasured resource dealt with by all people, and seeking ways to improve their skills and capabilities with it through schooling, training, management, and increased understanding of all its vagaries. Capacity-building in Afghanistan of whatever type can be difficult and intimidating, however, especially in complex organizations such as the Ministry of Education (MoE), the numerous other ministries (MMP, MEW, MAIL) that look after all its elements, or the various universities where the best thinking and best teaching may occur. Basically if it is to be successful, capacity-building about water in Afghanistan has to be a complicated, long-term process because it must involve not only changes in the knowledge and abilities of different individuals, but also changes in how large organizations may operate (Holland, 2010). These organizational changes are not just technical ones, but also they are inherently political, mainly because they can so alter access to authority, resources, and opportunities, perhaps even cutting off access to the easier money of the endemic corruption of the place.

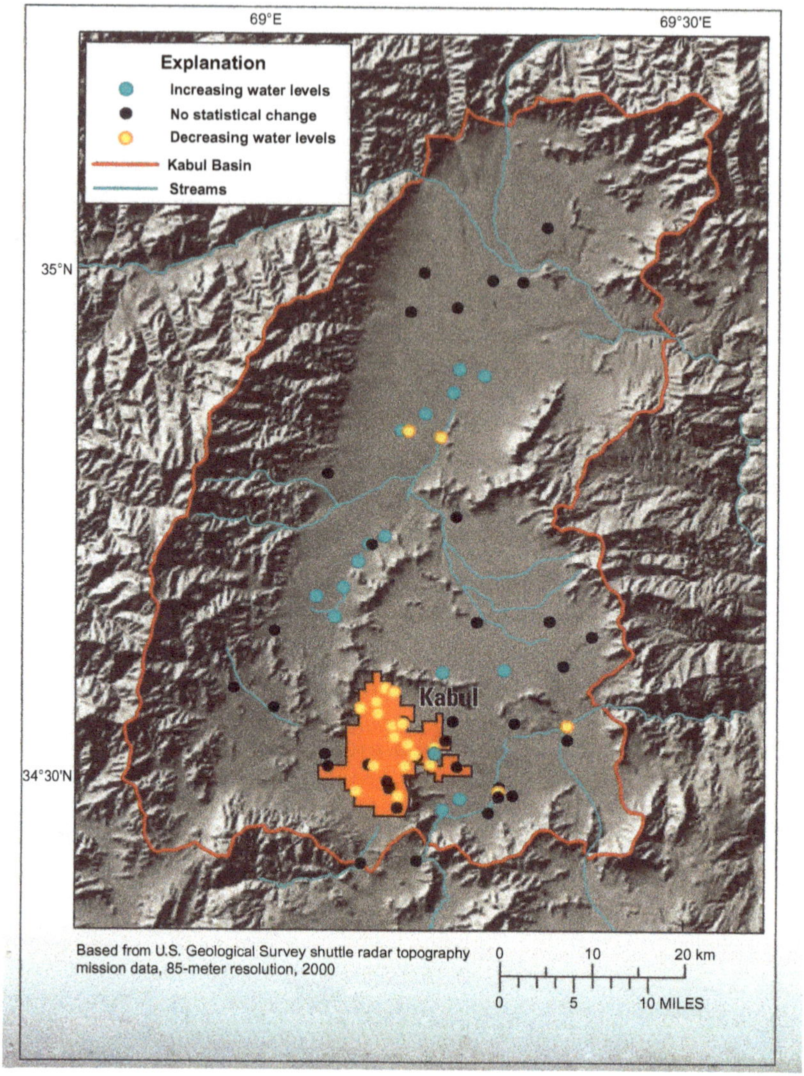


FIG. 13.1 Relief map of Kabul Basin, Afghanistan, showing water wells used for monitoring relative water levels over about the past decade a half. Note the large number of wells being drawn down in the more heavily pumped areas of Kabul. Information sources such as this, as well as profuse other graphical and written text throughout this volume are essential to help increase water knowledge in the country of Afghanistan. *From Mack, T.J., Chornak, M.P., Vining, K.C., Amer, S.A., Zaheer, M.F., Medlin, J.H., 2014. Water resources activities of the U.S. Geological Survey in Afghanistan from 2004 through 2014. Fact Sheet 2014-3068; USGS Afghanistan Project Product No. 265, August 19. http://pubs.usgs.gov/fs/2014/3068/ (accessed 10.07.2014.).*

## 13.2 CAPACITY TYPES

Capacity thus exists at both individual and collective organizational levels, is not just simple education or technical training, but instead includes five kinds of capabilities:

- Instrumental action, which is the capability of carrying out job functions and tasks properly, and is the most common way of thinking about capacity in any country.
- Ownership, in which people commit to, and engage in certain activities that relate to the importance of motivation and taking a deep interest in their chosen fields.
- Resource attraction, which is the capability of being able to relate to other people and simultaneously attract resources and support to sustain existing sources and produce new sources of funding, staffing, and learning.
- Adaptation, which is the capability to adapt and self-renew. This can be especially influential where a high level of dependence upon external grants requires resilience when the funding streams dry up.
- Maintain focus, which helps balance coherence and diversity. In large and complex organizations, this means preventing fragmentation and loss of attention to things, while encouraging innovation.

In these fashions then, capacity-building about water in Afghanistan is a great deal more than just simple education and training about any given topic. Instead it is a fairly all-encompassing effort to improve the capabilities of individuals and organizations.

The larger a government organization is, the greater and more complex are the interactions and activities within it. This means that it can be quite problematic to assume that "the government," "the ministry," or "the department" shares the same ideas, outlooks, or priorities that will carry it forward effectively. As an example, the MoE of Afghanistan is a large and complex organization that has needed changes in both individuals and portions of its internal organization. In addition, the National Education Strategic Plan that was developed in 2006 (Holland, 2010) involved the technical assistance of a number of non-Afghan international specialists. Some have argued that non-Afghan expertise in any given governmental organization cannot be used in consideration or assessment of its capacity capability. Instead it may be argued that the expertise in any Afghan organization should be based upon local capabilities. In any case, a major part of any capacity-building process has to be to find out what works for them and what does not, because some suggestions by donor agencies may not work especially well in that particular milieu. In spite of all, at the end of the day we do know that education and education revision are the future for Afghanistan if it is to have any hope of better lives for millions of people, so many different means must be tried and only the best whittled out for further support.

Recommendations for overall capacity-building through policy making recommended by the Afghanistan Research and Evaluation Unit (Holland, 2010), include six major points (Table 13.1) that should be adhered to in order to try to guarantee some success in the sorely beset country. For example, because capacity-building includes both instruction of individuals, as well as of organizations as a whole, care must be taken to ensure that both are kept in mind, and in the realization that considerable attention must be paid to the fact that individual and organizational capacity are distinct, yet not quite interrelated phenomena. This means designing projects to enhance hydro-cognizance requires thinking about the multiple ministries that

TABLE 13.1 Recommendations Concerning Capacity-Building and Policy Making in Afghanistan

1. Explicitly define what is meant by capacity.
2. Plan and include capacity-building within programs.
3. Look beyond instrumental views of capacity.
4. Expand effective capacity-building models.
5. Distribute ownership evenly in a complex organization.
6. Monitor how capacity elements interact.

*From Holland, D., 2010. Capacity-Building Through Policymaking. Afghanistan Research and Evaluation Unit. Briefing Paper Series, AREU Publication Code 1033E, October; 20 pp. http://www.iiep.unesco.org/fileadmin/user_upload/Cap_Dev_Technical_Assistance/pdf/2010/Afghanistan_National_Ed_Strat_Plan2010.pdf (accessed 10.05.2014.).*

handle the various aspects of water. Perhaps the fragmentation of water aspects between the tasks of the different ministries (agricultural irrigation, energy production, domestic supply, wastewater, environment, etc.) will lead to certain governmental reorganization, but this would be quite disruptive in times that are uncertain enough, so it would not be recommended, although formation of a supra-water commission of some sort might be brought into existence to help out with the lack of interministerial communication and coordination, as long as the commonly associated corruption could be either rooted out or controlled with some vigor.

Capacity-building itself, as a principle of planning and activity in its own right, needs to be monitored and evaluated to maintain true quality, rather than contributing more to bureaucratic obfuscation and lethargy. Sufficient time for individuals to develop skills and knowledge has to be allowed for, even as the organizations can develop the novel operational systems necessary to make effective use of their new skills. Project assessments, questionnaires, and evaluations; all need obvious and transparent timetables and commitments to monitoring and evaluation. In the course of planning and inserting capacity-building into programs, however, the assessment beyond instrumental views of capacity effects needs further work as well. This would include development of capacity relating to ownership of projects, attraction of diverse resources to help programs, adaptive changes made by the organization, and maintenance of focus upon the most important priorities.

Expansion of known effective capacity-building models such as "peering" can be effectively used as a model for other efforts in capacity-building. Technically speaking, peering is the arrangement of traffic exchange between internet service providers. But properly speaking, peering is thus simply the agreement to interconnect and exchange routing information, although this can be fraught in some situations in Afghanistan where internet traffic can be blocked for a number of reasons (filters for pornography, blockage of radical propaganda, etc.). Many different models of information exchange exist to enable special programs of hydro-capacity-building that can be developed to accelerate the processes of education and training about water.

Even in complex organizations such as the multiple ministries who are responsible for handling all the different aspects of water in Afghanistan, a sense of ownership of this vital resource is not felt throughout. Instead efforts must be made to work toward a sense of greater ownership of all the different aspects necessary to enhance, protect, predict, augment, and all other aspects of using water in Afghanistan. Finally, the monitoring of how all the capacity elements interact is a factor that needs considerable development and assessment to improve the process. Research on the processes of capacity-building in Afghanistan to examine the relevance of the elements outlined above is also necessary to avoid neglect or compromise in the aid-dependent environment to ensure that certain aspects of all the forms of handling water are not harmed in any way.

## 13.3 CAPACITY-BUILDING IN THE MINISTRIES

The Ministry of Education in Afghanistan, for example, is not responsible for water per se, of course, but it has a strong role in upgrading educational and capacity-building activities about water in the country. It is obviously so essential and important to learn things about water necessary for helping the country, that the processes to develop good capacity-building about water in that ministry are essential, so they can be adapted to educate others. Thus Sigsgaard's (2011) detailing of a number of lesson recommendations for the MoE can be used effectively across ministries and universities to assist in making useful changes to capacity in dealing with water nationwide:

1. Building trusting partnerships takes time and is required for the needed high-level political backing. Decade-long engagements in Afghanistan by some agencies have allowed gains of credibility and development of trusting partnerships.
2. People come and go, but systems remain. Delivery of services and implementation of policy hinges on systems, which enable planning for water issues based on facts, and can reduce corruption and too much reliance on certain, less-desirable individuals, because of their incompetence, corruption, or lethargy.
3. Put processes before products. United Nations and other agency collaboration in the production of policy documents can give impetus to capacity development and enable donor coordination, which in the process allows development of self-confidence, a prerequisite for the ability to commit and engage in water issues.
4. Donor flexibility and long-term commitment are helpful. Donor support for capacity-development activities should adhere to prior principles on aid effectiveness by engaging over several years, showing flexibility and taking responsible risks by permitting sudden process changes and accepting participatory design, instead of fixed long-term plans for water development that can be ineffective.
5. Donor coordination is needed for salary harmonization. Coordination mechanisms such as the Human Resources Development Boards or other relevant bodies can be instrumental in solving major staffing challenges, some of which are the severe salary disparities between government civil servants and national technical assistants coming from outside that are funded by donors.
6. Choose pragmatic and basic solutions. Capacity-development partnerships can begin with simple basic infrastructure work such as supplying office space and the pragmatic teaching of such skills as the English language or computer literacy.
7. Gender is also a human-resource issue. Only about a quarter of the workforce in Afghanistan is female, which occurs because of reprehensible stereotypes of female inferiority in the country, which wastes much talent that could be tapped for improvement in water issues.
8. Nation-building should be based on decent, non-ideological education. Use of equitable, non-ideological provision of decentralized education on water issues or enhanced water cognizance might one day produce higher relevance to the citizenry and reduce need for foreign aid on that topic.

9. Plans are statements of will and self-confidence. In Afghanistan's existing political process, ambitious national plans signal a will for drastic change and may create hope and self-confidence, which are invaluable resources when all things are priorities and everything a challenge, so that criticisms of projects as being unrealistic may not be valid.
10. Sustained financial support is a must for achieving national development objectives. On-going nation building and socio-economic growth absolutely depend on uninterrupted funding to keep things going until the rich mineral-resource base can come on stream to market to begin to pay more of the GoA's bills (Shroder, 2014).

## 13.4 CAPACITY-BUILDING ASSESSMENTS

The results of prior projects on water in Afghanistan that were funded by USAID have been investigated by the Special Inspector General for Afghanistan Reconstruction (SIGAR, 2014), under the auspices of the U.S. Department of Defense. The SIGAR audit report on Afghanistan's water sector has found that USAID has spent about $518 million to support, among a variety of other things, rural and urban water supplies, and water sanitation, as well as agricultural productivity. Since fiscal year 2010 alone, USAID has funded nine other water projects in Afghanistan. In any case SIGAR assessed four of these USAID water projects that dealt with: (1) community and farm-level irrigation; (2) water supply and sanitation in Gardez, Ghazni, Jalalabad, and Mazar-i-Sharif; (3) general rural sustainable water supply and sanitation; and (4) siting of small irrigation dams and micro-hydropower projects across Afghanistan. The objectives of the audit report were to see if the project goals and objectives for these projects were met, as well as to see the extent to which they were dealt with in the USAID-U.S. Government Inter-Agency Water Strategy for Afghanistan in 2010.

SIGAR (2014) found that USAID did not meet three key objectives in the Inter-Agency Water Strategy for Afghanistan because it did not get enough of the budget necessary to do the work, among other problems. In addition, perhaps once it realized things were not going so well with some of its projects, USAID wisely changed its approach from its typical large infrastructure projects—such as dams and commercial water and sewer systems—toward building capacity at the Afghanistan ministries to manage these projects themselves. Even though their priorities were changed, midstream as it were, their strategies were not updated, nor were the appropriate political and security transition changes made when most of the Western Coalition military forces withdrew from the country in late 2014. Without updating the strategy to reflect more current priorities, USAID risked planning and implementing water projects that were not aligned with its goals for the development of Afghanistan's water sector. In addition, part of USAID's problems were because they could not measure some performance goals because of budget cuts, unrealistic performance targets, and higher than estimated costs. As a result of this rather negative assessment by SIGAR (2014), they recommended that USAID needed to develop a new water-sector strategy for Afghanistan with updated short-, medium-, and long-term goals and objectives that reflect USAID's current water-sector priorities, as well as developing and implementing a performance-measuring plan upon completion of a new strategy that could be used to evaluate USAID's performance in meeting the new strategies, goals, and objectives, while ensuring that the new strategy included clear lines of responsibility and accountability of who within USAID would implement the strategy. USAID agreed with, and accepted the audit.

The assessment of recent prior work by the World Bank on technical assistance for capacity-building in the water sector in Afghanistan (Anon., 2013) has been similarly frank to that of USAID's assessment by SIGAR (2014). This World Bank project's objectives were:

- staffing of the Water Resources Planning Unit and Project Preparation Unit included adequately trained staff and consultants;
- developing a knowledge base and analytical tools for Kabul Basin and other basins;
- identification of investment priorities in Kabul Basin and other basins;
- complete a number of preparation studies for priority investment options in water (at least to pre-feasibility levels); and
- increase the number of staff trained on key water-resource issues (including integrated water-resources management and transboundary water management).

These key indicators are quite central to the core wishes of the World Bank to effectively develop water resources in different countries so that they will have sufficient hydroelectric power generation, irrigation waters, and sanitary water supplies, as well as proper waste disposal. Unfortunately the performance ratings by the World Bank's Implementation Completion and Results division for outcomes, Bank performance, and grantee performance were all judged as moderately unsatisfactory, whereas the only high rating was the likelihood of high risk to development outcomes. This quite disappointing audit result is apparently at least partly a reflection of the stonewalling by the Karzai government in 2014 against discussing water, but actually reflects a general lack of appropriate local knowledge about water that needs a great deal of new capacity-building attention, which is, of course, the main topic of this chapter.

## 13.5 USES OF WATER SCIENCE FOR CAPACITY-BUILDING

The uses of the hydrologic sciences are obviously profound and essential to the well-being of the world's people. Furthermore, the findings of water science clearly play a role in the politics of allocation and management (Jarvis and Wolf, 2010). Nonetheless, in the real world any water science will generally be subordinate to the politics. It has been pointed out that water commonly flows uphill to money and power. Perhaps the reason for this is that the scientific enterprise in its purest form is commonly incompatible with the practical demands of decision making for the greatest public good. Unfortunately, these political decision-making processes may be caught up in altogether too much short-term thinking, with little regard for the longer-term consequences. Defending the status quo, or the way things have always been done, generally takes precedence because it seems such common sense, even in the face of dire predictions of future shortages. It is the business of this discussion to present the issues as best we understand them, in hopes that rational scientific discourse will help alleviate at least some of the forthcoming difficulties emergent in the capacity-building development of better hydro-cognizance and awareness in Afghanistan.

Water-cognizant capacity-building must be established in any place in full recognition of all the disparate elements of water knowledge. This book is an attempt to put most of the essential elements of hydro-cognizance capacity-building in one place for the use of the government of Afghanistan. This effort would obviously begin with introductory discussions of the basic physics of liquid water ($H_2O$), solid ice, and gaseous water vapor, as well as the basic physical principles in the hydrologic cycle (Figure 1.1A–E from Chapter 1). This includes the physical geographic processes of moving different $H_2O$ phase types between their different locations, in addition to the energy sources that run the hydrologic cycle (solar energy, Earth rotation, winds, latent heats of evaporation, condensation, melting, freezing, etc.). The specific details of the hydrologic cycle in Southwest and Central Asia are also essential aspects of hydro-cognizance, and would include the nine main factors of wind types and water sources, snow and glacier ice in the mountains, snow and ice melt to river water, rain and snowmelt on hillsides, runoff water flow, soil erosion, drainage systems, and climate controls. The main rivers and tributaries of the region need to be discussed in some detail; the Kabul, Helmand, Hari Rud, Murghab, and Amu Darya, as well as the other rivers that drain internally entirely within Afghanistan. The lakes of Afghanistan are quite important as well, particularly because they are so diverse: glacier lakes, landslide-dammed lakes, carbonate-precipitate lakes, diastrophic (tectonic and volcanic) lakes, mixed origin types, and the numerous wind-blown deflation types, many of which are dry, salt pans for much of the year.

Underground water in Afghanistan is another essential part of the hydrologic cycle in the country that must be understood in order to avoid future problems, with rampant overpumping, contamination, and water-table decline being critical issues. Possible solutions involve rainwater harvesting and delivery of extra winter precipitation runoff into specially constructed ground-water recharge basins that are used seasonally to bring water tables back up in clever ground-water recharge schemes (Reddy, 2008), as well as long-distance aqueducts to do the same (Chapter 4 herein). Desalinated ground water and clean-up of polluted supplies need to be understood as well so that in future, corrective measures can be taken.

Environmental issues in Afghanistan that negatively impact water supplies are another essential issue that needs to be well understood so that problems can be anticipated and dealt with more effectively (Shroder, 2012, 2014). Centuries of environmental rapacity in the country have eroded away the formerly rich soils through rampant deforestation, dearbification, over-grazing, over-irrigation (water logging), soil salinization, and wind-blown deflation, so much so that precipitation water with inadequate soil infiltration runs off far too rapidly, which leads to massive downslope movement of masses of mud and rocks in periodic, rapid, wet debris-flow slurries (*sailab, duab*) that constitute a major hazard to life and property throughout the region (Chapter 5 herein). The fact that the water that produces this common problem does not have time to infiltrate the soil and contribute to annual recharge of ground water is a constant problem that can be partially alleviated by construction of terraces and check-dams on slopes and hillside depressions.

The Hindu Kush and associated ranges of Afghanistan, as with the greater Himalaya and other mountains, constitute the so-called mountain water towers of Asia upon which millions of people must rely for their very existence. The implications of observed climate change in this region are looming hugely (Shroder, 2014; Chapter 6 herein) and must be planned for with intelligent measures that will help Afghanistan face coming hardships beyond what they must endure already. Inasmuch as the nation is at the top of the watersheds that emanate from the high ground in the center and northeast of the country, and all the rivers flow out more-or-less radially into the surrounding countries, it is quite clear that great attention must be paid to understanding all the aspects necessary for effective transboundary-water management of these vital resources. Within this context, new thinking and new research are clearly required to solve delicate transboundary water issues so that maximum positive outcomes can result. This requires careful diplomacy by people knowledgeable about water, which is why this idea of hydro-cognizance is so important to Afghanistan if it is to avoid being outclassed at the negotiating table by neighboring countries who have long traditions of dealing with water issues, with water engineering, and with respectable intellectual traditions of considerable long-term continuity, which Afghanistan regrettably does not have.

A number of other important measures for dealing with water in Afghanistan are also necessary, especially the considerable problem of flooding in the country. Normally, in any developed nation, long decades of keeping records of past flood locations and flood heights enable a statistically robust method of establishing mathematical probabilities of how often through time one can normally expect floods of certain depths to occur. In Afghanistan, however, the early robust record keeping of floods that was started originally by the U.S. Geological Survey became destroyed in the past three decades of war so that new methods of flood prediction had to be developed. This was accomplished by developing a special new methodology utilizing a computer-driven, cellular automata system wherein some control was provided by some satellite-image records of flooding, and then by applying successively larger amounts of virtual water in the computer to high-resolution digital terrain elevation data (DTED) provided by the US Department of Defense (Hagen et al., 2010). This procedure enabled the production of a flood prediction map of the whole of Afghanistan that has been used to help the Western Coalition of military forces to avoid problems with their siting of various installations and other uses, but unfortunately because of classified restrictions for the use of such DTED maps where the Taliban could unjustifiably profit, the maps are not available to most people or to the GoA. This is an unfortunate result of the ongoing war with the Taliban insurgency but may be changed in coming years as a result of special needs and procedures with the new government of President Ashraf Ghani.

In any case, the problem of flooding coming from Afghanistan is also something that the Government of Pakistan would especially like to avoid, mainly because of recurrent flooding as a result of climate change that has been repeatedly devastating the lowlands alongside the Indus River in that country (Shroder, 2014), including floods once again in the summer of 2014 rather much like those that have occurred every year for at least the past four. A whole dialog needs to occur in the region about construction of run-of-the river, low dams for hydroelectric power generation and some irrigation to stave off the floods, while at the same time providing vital energy that both Afghanistan and Pakistan need desperately. Afghanistan should be a net exporter of electricity rather than sending the destructive floods downstream, instead of importing electricity to Kabul all the way from Tajikistan far in the north as it has to do now. This means that capacity-building also needs to focus in on training a cadre of specialists in water engineering so that types and distinctions between large dams, weirs, barrages, headworks, and run-of-the river dams can be all be assessed with intelligence in order the meet the needs of the maximum number of people among the riparian users all along the river banks in the maximum number of places. Then a series of carefully managed, full-scale discussions need to be had about hydropower and irrigation planning and structures on the Kabul and its tributaries, including the Chitral-Kunar River coming into Afghanistan from Pakistan. In the course of this activity, a full discussion of the hydro-hegemony of Afghanistan also needs to be introduced as a part of the water capacity-building of the country, in the manner expressed in Chapter 14 herein.

Capacity-building for better water cognizance in Afghanistan requires a certain amount of understanding of international water law, which Chapter 12 herein endeavors to explain as a first-order attempt at introducing the topic for people unfamiliar with the ramifications. International and national water laws are actively changing in many parts of the world in partial or full recognition that not to do so is to invite the many different sorts of disasters associated with water all over the world of transboundary rivers. Although international water law is not binding on any country that does not agree voluntarily to be bound by it, the fact is that the law is based upon wise principles of water management that can overcome recurrent transboundary water issues with diplomatic finesse when necessary. Outright water wars resulting from failures of diplomacy are unusual; less easy to understand is the potential for water problems to reverberate alongside, or from within other violence-inducing, natural, or human-caused problems that can be an increasingly unpleasant development that needs to be avoided. Reference to Chapter 14 herein on hydro-hegemony will give sufficient understanding to many Afghan civil servants to enable adequate information for achieving knowledgeable negotiation.

In sum, the hydro-cognizant capacity-building in Afghanistan requires a variety of scientific, applied engineering, and clever anthropology—sociology, coupled with astute political and legal understandings to accomplish. It is probably safe to assume that no matter how much capacity capabilities can be raised in Afghanistan, more will always be needed because of the deep levels of new scientific finding and advanced thinking that are constantly leading to disruptions in complacency and forcing reactions. But by facing up to the levels of information that are needed, the GoA can begin to think of solutions to the problems before they get any further out of control than they are already. Insofar as many of these hydro-cognizance issues develop into serious future problems, early warning by newly trained scientists and engineers in Afghanistan will be a most necessary planning tool to assist in government competence, which Afghanistan so sorely needs.

## 13.6 CONCLUSION

Increasing the awareness of all the aspects of water in Afghanistan for the benefit of the government, the academic communities, and ultimately for the generally water-short people of Afghanistan, is the purpose of this book and the idea behind a series of workshops we have been holding in Afghanistan, Tajikistan, and Pakistan. In spite of these efforts, and because of the scientific, sociological, political, legal, and other complexities that occur around water, considerable future effort and funding will have to be expended in Afghanistan in the foreseeable future to achieve the goal of hydro-cognizance or *water-knowing* in the country.

## References

Akbari, M.A., Tahir, M., Litke, D.W., Litke, M.P., 2007. Ground-water levels in the Kabul Basin, Afghanistan, 2004-07. U.S. Geological Survey Open-File Report 2007-1294; 45 pp. http://pubs.usgs.gov/of/2007/1294/pdf/ofr2007-1294web.pdf (accessed 10.08.2014.).

Anon., 2013. Implementation completion and results report (TF-93637) on a grant in the amount of US$5.5 Million to the Government of Afghanistan for a technical assistance for water-sector building project. Document of the World Bank Report No: ICR00002747. Environment and Water Resources Unit, Afghanistan Country Department, South Asia Region. http://www-wds.worldbank.org/external/default/WDSContentServer/WDSP/IB/2013/10/17/000356161_20131017113516/Rendered/PDF/ICR27470ICR0Af0t0PUBLIC00Box379845B.pdf (accessed 10.06.2014.).

Broshears, R.E., Akbari, M.A., Chornak, M.P., Mueller, D.K., Ruddy, B.C., 2005. Inventory of ground-water resources in the Kabul Basin, Afghanistan. U.S. Geological Survey Scientific Investigations Report 2005-5090; 34 pp. http://pubs.usgs.gov/sir/2005/5090/ (accessed 10.06.2014.).

DACAAR (Danish Committee for Aid to Afghan Refugees), 2011. National Ground-Water Monitoring Wells Network Activities in Afghanistan From July 2007 to December 2010, Kabul, Afghanistan. Danish Committee for Aid to Afghan Refugees, Danish. 23 pp. http://dacaar.org/programmes/water-sanitation-and-hygiene/ (accessed 10.08.2014.).

Hagen, E., Shroder Jr., J.F., Lu, X.X., Teufert, J.F., 2010. Reverse engineered flood hazard mapping in Afghanistan: a parsimonious flood map model for developing countries. Quat. Int. 226 (1–2), 82–91.

Holland, D., 2010. Capacity-building through policymaking, briefing paper series. Afghanistan Research and Evaluation Unit. AREU Publication Code 1033E, October; 20 pp. http://www.iiep.unesco.org/fileadmin/user_upload/Cap_Dev_Technical_Assistance/pdf/2010/Afghanistan_National_Ed_Strat_Plan2010.pdf (accessed 10.05.2014.).

Jarvis, T., Wolf, A., 2010. Managing water negotiations and conflicts in concept and in practice. In: Earle, A., Jägerskog, A., Öjendal, J. (Eds.), Transboundary Water Management: Principles and Practice. Earthscan, Stockholm International Water Institute, Stockholm, pp. 126–141.

Mack, T.J., Akbari, M.A., Ashoor, M.H., Chornak, M.P., Coplen, T.B., Emerson, D.G., Hubbard, B.E., Litke, D.W., Michel, R.L., Plummer, L.N., Rezai, M.T., Senai, G.B., Verdin, J.P., Verstraeten, I.M. 2010a. Conceptual model of water resources in the Kabul Basin, Afghanistan. U.S. Geological Survey Scientific Investigations Report 2009-5262; 239 pp. http://pubs.usgs.gov/sir/2009/5262/ (accessed 10.08.2014.).

Mack, T.J., Chornak, M.P., Coplen, T.B., Plummer, L.N., Rezai, M.T., Verstraeten, I.M., 2010b. Availability of water in the Kabul Basin, Afghanistan. U.S. Geological Survey Fact Sheet 2010-3037; 4 pp. http://pubs.usgs.gov/fs/2010/3037/pdf/fs2010-3037.pdf (accessed 10.08.2014.).

Mack, T.J., Chornak, M.P., Taher, M.R., 2013. Groundwater-level trends and implications for sustainable water use in the Kabul Basin, Afghanistan. Environ. Syst. Decis. 33 (3), 457–467. 11 pp. http://link.springer.com/article/10.1007%2Fs10669-013-9455-4#page-1 (accessed 10.08.2014.).

Mack, T.J., Chornak, M.P., Vining, K.C., Amer, S.A., Zaheer, M.F., Medlin, J.H., 2014. Water resources activities of the U.S. Geological Survey in Afghanistan from 2004 through 2014. Fact Sheet 2014-3068; USGS Afghanistan Project Product No. 265, August 19. http://pubs.usgs.gov/fs/2014/3068/ (accessed 10.07.2014.).

Reddy, K.R., 2008. Enhanced aquifer recharge. In: Darnault, C.J.G. (Ed.), Overexploitation and Contamination of Shared Groundwater Resources. Springer, New York, pp. 275–288.

Shroder Jr., J.F., 2012. Afghanistan: rich resource base and existing environmental despoliation. Environ. Earth Sci. 67 (7), 1971–1986. http://dx.doi.org/10.1007/s12665-012-1638-7.

Shroder, J.F., 2014. Natural Resources of Afghanistan: Geographic & Geologic Perspectives on Centuries of Conflict. Elsevier, San Diego, CA.

SIGAR, 2014. Afghanistan's water sector: USAID's strategy needs to be updated to ensure appropriate oversight and accountability. SIGAR 14-52 audit report, April. http://www.sigar.mil/pdf/audits/SIGAR-14-52-AR.pdf (accessed 10.06.2014.).

Sigsgaard, M., 2011. One the road to resilience: capacity development with the Ministry of Education in Afghanistan. International Institute for Educational Planning, Education in Emergencies and Reconstruction, United Nations Educational, Scientific and Cultural Organization. http://www.iiep.unesco.org/fileadmin/user_upload/Info_Services_Publications/pdf/2011/Afghanistan_Resilience.pdf (accessed 10.06.2014.).

Taher, M.R., Chornak, M.P., Mack, T.J., 2014. Groundwater levels in the Kabul Basin, Afghanistan. U.S. Geological Survey Open-File Report 2013-1296; 51 pp, http://pubs.er.usgs.gov/publication/ofr20131296 (accessed 10.06.2014.).

CHAPTER

# 14

# Hydro-Hegemony in Afghanistan and Surrounding Countries

*J. Shroder*

University of Nebraska at Omaha, Omaha, NE, United States

OUTLINE



## Abstract

Domination over water issues in any group of countries requires four major pillars: geographic position in the watershed, material power (population, literacy, military strength, infrastructure, etc.), bargaining power (knowing rules of water and diplomatic negotiations), and ideational power, or power of ideas. Afghanistan has five major river systems (Kabul, Helmand, Hari Rud-Murghab, Amu Darya) that it shares to greater or lesser extents with Pakistan, Iran, Turkmenistan, Uzbekistan, and Tajikistan. Assessment of levels of hydro-hegemony in this region reveals necessary maneuvers being made to maximize control of the resource by all neighboring nations. Afghanistan positioned at the top of most of the resource watershed is in an enviable position if its water-cognizant capacity can be raised sufficiently to allow eventual negotiations for new treaties with all neighbors.



http://dx.doi.org/10.1016/B978-0-12-801886-6.00014-8

**Keywords:** Watershed top, Hegemonic pillars, Geographic position, Material power, Literacy, Military strength, Bargaining power, Water diplomacy, Kabul River, Helmand River, Hari Rud-Murghab river system, Amu Darya, Transboundary water sharing, Water hegemon, Riparian state, Pakistan, Iran, Turkmenistan, Uzbekistan, Tajikistan, Water-cognizant capacity (WCC), Ideational power

# 14.1 INTRODUCTION

Leadership in water issues and power over that resource in the river basins of Afghanistan (Fig. 14.1) are a function of a number of competing or antithetical interests that are only slightly below the major front-line problems created by the insurgent Taliban. Many other difficulties with water also crisscross the old British-built Durand Line that constitutes the modern border of Afghanistan with Pakistan, as well as with the other borders: the Ridgeway line on the north with the Central Asian "Stans" republics, and the Goldsmid Line to the west with Iran. But a strong undercurrent question abroad in Afghanistan in knowledgeable quarters is exactly what the state or condition of available water actually consists of in the country, because most people, educated or not, are well aware of many

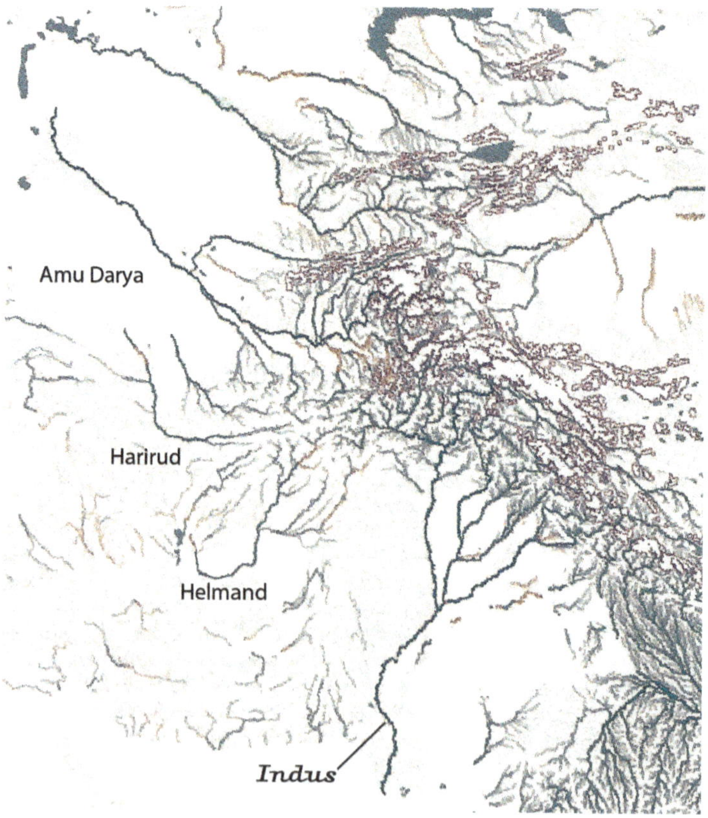


FIG. 14.1 Index map of the rivers of Afghanistan and their extents into neighboring Pakistan, Iran, Turkmenistan, Uzbekistan, and Tajikistan.

different sorts of water problems now, and they worry for their future. The present state of affairs appears to be that of a state of active aspirant, hydro-hegemony maneuvering that exists in South Asia, in which many different people and national states are seeking to gain control of the remaining hydrologic resources in order to more reasonably assure a future for themselves. As has been noted, "upstreamers use water to get more power, and downstreamers use power to get more water" (Warner, 2004; Zeitoun and Warner, 2006). Afghanistan has arrived at a state of interesting dilemmas and possibilities as various ideas about how best to deal with its combined problematic, and perhaps promising, water situations have developed. As an aspiring transboundary water hegemon, Afghanistan at the present time is attempting to gain more water-cognizant capacity (WCC) to gain more control of water within its own borders, which by its very nature, will alter the water availability downstream. This needs to be done by Afghanistan in as clever and non-threatening a way as possible, and it can be done with a number of new cross-border development linkages and plentiful scientific discourse so that no one is too surprised or too unhappy. This chapter explains the details of how that might be done.

Hydro-hegemony in any river basin is the ideas and actions concerning the four pillars of power and leadership (Cascão and Zeitoun, 2010) that exist with any transboundary water resource: (1) *geographical power* is associated with position in the drainage basin as the upstream and downstream riparian landowners; (2) *material power* is that which is the most obvious in terms of the relative amounts of educational background, economic power, military might, technological prowess, and international political and financial support that is available; (3) *bargaining power* exists in knowing the rules of the water and diplomatic game, the setting of agendas, the ability to work the systems in terms of negotiations and agreements, the ability to provide, and/or influence incentives that may encourage weaker parties to comply, or to apply leverage by linking non-water issues to such things as regional security or political alliances; and finally, (4) *ideational power* is the "power over ideas" wherein a riparian player is able to impose and legitimize particular ideas and narratives. These four factors then allow the somewhat subjective graphing of the levels of power into the four pillars, which can be a useful tool in discussing what needs to be done in trans-boundary water sharing and who needs to be doing whatever, whenever, however, and wherever. The result of the processes of assessing all the factors can then allow the riparian negotiators and their facilitators to better organize the scientific discussions and negotiations that can be set up later as issues for discussion framed within this useful four-pillar graphing procedure.

Transboundary water management is a political process subject to the fundamentally political processes of water control, utilization, and allocation, just as it is subject to the laws of nature, operating under the physical processes of the hydrologic cycle (Cascão and Zeitoun, 2010). It is perhaps best understood with reference to the complexities of the modern and quantitative systems analysis and mathematical programming in environmental engineering and decision making (Soncini-Sessa et al., 2007a,b). Various riparian states have highly asymmetric capacities to use both overt and covert forms of power that determine to a significant extent the fundamentally political distributional issue of who gets however much water,when, where, and why. Power, and power asymmetry are, however, constantly being contested and challenged by nonhegemonic ("weaker") riparians, who can, in theory at least, "level the playing field," influence the regional agendas and the negotiations, and possibly contribute to change the hegemonic configuration of the river basin.

Control over transboundary waters is an open competition that can occur, of course, under hostile circumstances, or it can be quite cooperative (Cascão and Zeitoun, 2010; Wegerich, 2008). The four forms or "fields" of hydro-hegemonic power (geographic, material, bargaining, and ideational) are utilized concurrently and generally combine to determine the ultimate hydro-hegemonic configurations in a river basin. The graphic configurations of these four pillars of hydro-hegemony, as representations of reality, are not actually static forms, and because of the considerable subjectivity or arguable uncertainties, can be evaluated in different fashions and the graphs changed over time, or can be interpreted differently by different analysts from the several countries involved.

Determining relative degrees of importance or levels of hegemony to assign to each of the pillars can be an exercise, with a number of diverse considerations, and comparing them all may become difficult, for subjective or nonquantitative reasons. Many different concepts of ideational hydro-hegemony can be applied, for example. An example of a hegemonic use of ideational power would be a refusal to share data on the actual water quantities available or the usages of the water. Arguments against any sort of equitable sharing of water can be based on issues of state security and are quite common. Government thus can transform water issues into matters of top national security. This process of securitization of water issues is then used interchangeably to either silence debate or then overplay certain issues, all of which can seriously politicized in the interests of promoting ideational hydro-hegemony in one nation-state or another. Already in the background of ongoing negotiations on transboundary water resources underway in Southwest Asia, evidence for water data restriction by several governments is known by the editors of this volume who are associated with the Center for Afghanistan Studies at the University of Nebraska at Omaha.

A further issue can occur in the contrast of the conceptions of transboundary water resources versus "shared water resources"; the questions are always at what levels, and when and how the water sharing occurs. Sharing can occur in many different ways, for example, with equitable sharing being an ideal condition, but periodic sharing, historical sharing, or theoretical sharing also needing to be considered in some cases. Another issue is that the normally paramount upstream geographic position is actually only relevant if it is also combined with other material, financial, and geopolitical power. Otherwise it may not count for all that much, unless the upstream riparian elects to go it alone and erect a series of its own dams to control the resource for its own benefit. This lack of local cooperation can backfire badly, however, with all sorts of diplomatic and insurgent reactions, the more violent actions of which can be supported by the downstream riparian against the headwater country. This, of course, is largely the situation in Afghanistan, with some of the Taliban being paid by outsiders to disrupt access to the water.

## 14.2 CONTROL STRATEGIES TO DOMINATE WATER RESOURCES

A resource-capture strategy may involve the actions of a riparian, in the absence of formal understandings or treaties, to move ahead with projects that affect the flow and quality of the water resource (Zeitoun and Warner, 2006). This can be carried out by creating "facts on the ground" such as dams or irrigation works that enable control over access to the resource.

This could also be where transboundary ground water is captured by higher-capacity and deeper wells and pumps on one side of a border, which can draw down the water table out of reach of the other side's ability to plan, construct, and operate large hydrologic infrastructure projects on the other side of a border that would capture the resource effectively. This is the case with Iran on the border of Afghanistan in the lower Helmand. A number of other tactics exist in water-resource control strategies to accomplish the goals of consolidated water control by the aspirant hegemon (Fig. 14.2).

Riparian position does allow an aspirant hegemon such as Afghanistan to exploit water as a coercive resource outside of the struggle for a better position in the balance of hydro-hegemonic power between nation-states. The power ratio between competing riparian states is based upon a state's: (1) overall political/economic/military power; (2) overall technological power; and (3) that extremely important *geographical position* in the watershed. According to Zeitoun and Warner (2006), the levels of power of the first dimension are military and economic, those of the second dimension are active stalling incentives, and the third-level dimension are securitization and sanctioned discourse (Table 14.1).

The hydro-hegemonic position for any country in a region is a function of a wide variety of physical and social attributes that may be determined by the characteristics of river basin (Table 14.2) or national state (Table 14.3). These factors include who the neighbors are, what the size of the river basin is, the population and its density in the basin, its literacy, technical or engineering capacities, gross domestic product per capita, and military troop strength, either domestic or borrowed from elsewhere. Less tangible factors may include infrastructure levels, clandestine insurgency strengths advised and/or controlled by another state, environmental qualities and controls, transport types and levels, political stability, education quality, and understandings of hydro-hegemonical factors (Dinar et al., 2013).

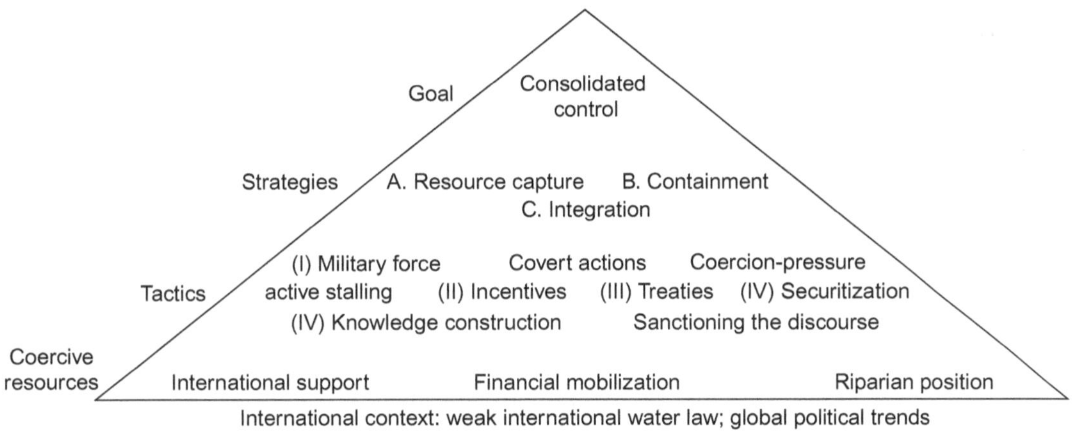


FIG. 14.2 Water resource control strategies and tactics. The Roman numerals inside the triangle shape relate to Lustick's (2002) classification of increasingly efficient compliance-producing mechanisms in seeking hydro-hegemony for any nation state. Mechanisms: (I) coercion; (II) utilitarian exchange; (III) instigating normative agreement; (IV) inducing ideologically hegemonic beliefs. The list is judged as nonexhaustive by Zeitoun and Warner (2006) from whom this figure was obtained.

TABLE 14.1 Hydro-Hegemonic Compliance-Producing Mechanisms and Concepts Concerning Transboundary Water That Are Used in Strategies and Tactics for Water-Resource Control

| | |
|---|---|
| *Active stalling* | Apparent compliance by a nonhegemon during transboundary water negotiations |
| *Coercion pressure* | In water-resource control strategy, threats (military, economic, political) by an aspirant or actual hydro-hegemon are applied to a neighboring state to make it drop certain claims about the resource |
| *Coercive resources* | The more powerful state in any hydro-hegemonic competition gains its political edge through international support, financial mobilization, and riparian position |
| *Containment* | A bilateral or multi-lateral strategy that requires engagement with competitors to integrate them or contain them in as asymmetric a position as possible through the use of coercive, utilitarian, hegemonic, and other compliance-producing mechanisms. Drafting and signing a treaty favoring the aspirant hydro-hegemon is the preferred tactic when pursuing a containment strategy |
| *Financial mobilization* | Ability to rapidly or easily secure funding for water infrastructure policy and project development is a luxury afforded to relatively rich states, access to wealthy allied states outside the common Western donors (World Bank, etc.), or those willing to bleed their coffers dry |
| *Hydro-hegemony* | Dominance at the river basin level, achieved through water-resource control strategies such as resource capture, integration, and containment |
| *Ideologically hegemonic beliefs* | Strong political position taken on water issues essential for very survival of nation |
| *Incentives* | An aspirant or actual hydro-hegemon can utilize various trade incentives, diplomatic recognitions, military linkages, etc. to form compliance-encouraging "carrots" instead of coercive pressure "sticks" |
| *Integration* | A strategy that seeks to encourage compliance with agreements through incentives in a utilitarian fashion in which the riparians cooperate and share control in some fashion, such as by forming joint governing boards of hydrologic basins |
| *International support* | A form of coercive resource in which having powerful friends lends efficiency to securing access to funds for water infrastructure or having a favorable global political position |
| *Knowledge construction* | Competition over water may be guided by popular belief—accurate or inaccurate—rather than science-based knowledge so that quite different perspectives ("spin") can be promoted as one view to the local people, another to the international donors, and yet others to riparian competitors or friends |
| *Normative agreement* | Population agrees that political position on water issues is right thing to do |
| *Resource capture* | A riparian state, in the absence of formal understandings (water treaties), goes ahead with hydro-projects (dams, irrigation canals, etc.) that affect the quality and quantity of flow of the resource |
| *Riparian position* | The all-important geographical location of the aspirant hydro-hegemon riparian state in the upstream, midstream, or downstream position on the watershed |
| *Sanctioned discourse* | Definitions of water reality viewpoints endorsed by the more powerful (hegemonic) side |
| *Securitization* | Hegemonic thought-control mechanism wherein water projects, information, and data are considered national security issues and therefore are placed outside or above any possible discussion or criticism |
| *Utilitarian exchange* | Action on water issues correct if it tends to promote the most happiness |

*From Zeitoun, M., Warner, J., 2006. Hydro-hegemony—a framework for analysis of trans-boundary water conflicts. Water Policy 8, 435–460.*

**TABLE 14.2** River Basin Statistics for Afghanistan and Its Neighboring Countries

| | |
|---|---|
| *Afghanistan total water* | |
| ~73 $km^3$ annual renewable total countrywide | |
| ~57 $km^3$ surface water | |
| ~18 $km^3$ ground water | |
| ~20 $km^3$ annual irrigation water used | |
| *Kabul River Basin* | |
| Riparian states: Afghanistan, Pakistan | |
| Basin area: 76,908 $km^2$ (smallest river basin area in Afghanistan) | |
| Annual discharge flow: ~24 $km^3$ (highest flow in Afghanistan) | |
| Irrigated area: ~ <0.5 M ha | |
| Population in basin: ~7.2 million | |
| Population density: 93 persons $km^{-2}$ | |
| *Helmand River Basin* | |
| Riparian states: Afghanistan, Iran, Pakistan | |
| Basin area: 353,400 $km^2$ (largest river basin area in Afghanistan) | |
| Annual discharge flow: ~0.5 $km^3$ (least flow in Afghanistan) | |
| Irrigated area: ~1.1 M ha | |
| Population in basin: ~7.8 million | |
| Population density: 22 persons $km^{-2}$ | |
| *Hari Rud-Murghab River Basin* | |
| Riparian states: Afghanistan, Turkmenistan (Murghab and Hari Rud), Iran (Hari Rud only) | |
| Basin area: 61,000 $km^2$ (Murghab); 92,600 $km^2$ (Hari Rud) | |
| Annual discharge flow: ~3 $km^3$ | |
| Irrigated area: ~0.2 M ha | |
| Population in basin: ~1.06 million in Murghab; ~5.02 million in Hari Rud | |
| Population density: 17 persons $km^{-2}$ in Murghab; 54 persons $km^{-2}$ in Hari Rud | |
| *Amu Darya River Basin* | |
| Riparian states: Afghanistan, Tajikistan, Uzbekistan, Turkmenistan, Pakistan | |
| Basin area: 309,400 $km^2$ | |
| Annual discharge flow: ~17 $km^3$ | |
| Irrigated area: ~1.16 M ha | |
| Population in basin: ~15.6 million | |
| Population density: 50 persons $km^{-2}$ | |

*From Palau, R.G., 2011. Afghanistan's Transboundary Water Resources: Regional Dimensions. Civil-Military Fusion Center; 8 p. https://www.cimicweb.org/Documents/CFC%20AFG%20Infrastructure%20Archive/CFC_Afg_Monthly_Transboundary_Water_Resources_July2011.doc.pdf; Cookson, F., Thirkill, D., Firogh, A.A., Girma, M., Azoy, G.W., 1992. Afghanistan water constraints overview analysis. Nathan Berger Afghanistan Studies Project Final report; 121 p. http://pdf.usaid.gov/pdf_docs/PNABL957.pdf.*

TABLE 14.3 Background Data to Some Levels of Hydro-Hegemony Material and Bargaining Power in Afghanistan and the Surrounding Countries

| | |
|---|---|
| *Afghanistan* | *Turkmenistan* |
| Population: ~27 million | Population: ~5 million |
| Literacy: 32% | Literacy: 98% |
| Technical capacity: Very low | Technical capacity: Moderate |
| GDP/capita: $800 | GDP/capita: $4300 |
| Security forces: ~380,000 | Troops: ~18,000 |
| *Tajikistan* | *Iran* |
| Population: ~7 million | Population: ~66.1 million |
| Literacy: 98% | Literacy: 72% |
| Technical capacity: Moderate | Technical capacity: Moderate |
| GDP/capita: $1140 | GDP/capita: $6300 |
| Troops: ~7000 | Troops: ~543,000 |
| Russian troops: 25,000 in 2001 | *Pakistan* |
| *Uzbekistan* | Population: ~145 million |
| Population: ~26 million | Literacy: 43% |
| Literacy: 99% | Technical capacity: Good |
| Technical capacity: Moderate | GDP/capita: $2000 |
| GDP/capita: $2400 | Troops: ~587,000 |
| Troops: ~74,000 | |

*Data derived from National Geographic, 2001; Mashal, M., 2014. From Karzai. The Atlantic, July/August, 50–58.*

## 14.3 HYDRO-HEGEMONY IN THE KABUL RIVER BASIN

Afghanistan's upstream position on the Kabul River drainage basin (Fig. 14.1), for example, is not necessarily a great advantage over Pakistan, unless many other factors could be brought to bear, which are not likely to happen as long as Pakistan can control the Taliban's depredations so effectively. Thus in the case of the Kabul River Basin in detail, we can assess the following hydro-hegemonic factors that apply to the two countries of Afghanistan and Pakistan, as follows:

### 14.3.1 Afghanistan

*Geographic power*: Afghanistan is at the top of the Kabul River watershed, which includes two of its most important main tributaries, the Panjshir and Ghorband rivers, with their many mountain glacier and snowbank water sources in the Hindu Kush to the north of Kabul City; and the Logar River, with its sources in the Paghman Mountains and Kohi Baba Ranges to the west of Kabul. Afghanistan, however, does not have the headwaters of the important Chitral-Kunar (Konar) River system, for example. That belongs to Pakistan. The important Swat River tributary to the Kabul River in Pakistan is also not a part of Afghanistan's territory (Fig. 14.1).

*Material power*: Afghanistan brings hardly any material power into any hydro-hegemonic understandings in the region because it has been undergoing over three decades of war and near-constant insurgency, which has destroyed much of its capability and capacity in so many things. The most that can be said for its material power is that it does enjoy the support of the United States and most of the NATO (ISAF) countries, which does give it some backing from the United Nations, the World Bank, and other important agencies. Afghanistan's low literacy, population >5 times smaller than Pakistan's, very low technical capacity, one-third the number of troops, and low GDP per capita (Table 14.3) all indicate a low material power.

*Bargaining power*: Other than its upstream geographical position, Afghanistan lacks almost all bargaining power in transboundary water issues, which is one of the reasons that it has such minimal water structures in its own drainages; other countries such as Pakistan, which have long wanted plentiful water delivered to them from the Kabul River drainage, have certainly expressed much dismay in recent years that Afghanistan would even be contemplating any potential diminution of water supplies downstream because of dam construction upstream. Of course, plentiful war, issues of backward and backwater development, or bureaucratic incompetence and malfeasance in Kabul also may be reasons for the lack of water-project development on the Kabul River drainage system in Afghanistan, rather than any Pakistani manipulation.

*Ideational power*: In any sort of "power over ideas" wherein a largely upper riparian nation such Afghanistan would be able to impose and legitimize particular ideas and narratives about water, only one concept about water exists that could possibly be considered, in even a convoluted fashion. For example, some might hypothesize that in the process of sovereign reality of administration of the Durand Line by Afghanistan on one side in its own self-constructed "Pashtunistan," and Pakistan and its pervasive military Interservice Intelligence (ISI) operating malignantly on the other side (Gall, 2014), somehow Afghanistan as a result had been cheated out of being able to dam its own waterways and reap any of the benefits. Otherwise, no big ideas about water seem to be afoot in Afghanistan, other than the rather paranoid factor that Pakistan might have been able in some fashion to foreclose on Afghanistan being able to develop its water resources. This is a thin supposition unsupported by any credible documentation, however. Pakistan's desires to have strategic depth to the west away from India and into the Islamic Republic of Afghanistan may be part of the equation, of course, that could be considered. In that case, however, if Pakistan wanted greater credibility with its ideas of strategic depth rather than being seen as being driven by simple greed, it could take a different tack. For example, by offering to step up to the aid plate with some of its own technical assistance and actually helping the Afghans to gain more education about scientific water issues to do the damming and hydroelectric power production; that would alter the ideational power structure significantly. In order to raise its own ideational power value in that manner, of course, Afghanistan would have to buy into, and promote this notion of technical help from Pakistan.

## 14.3.2 Pakistan

*Geographic power*: Although Afghanistan is the nation with much of the discharge of the Kabul River within its borders, Pakistan is the upstream hegemon of the waters of the Chitral River, which is one of the main sources of the Kunar River tributary to the Kabul River within Afghanistan. The Chitral River then, is also a main water contributor to the lower Kabul River through the Kunar tributary, which gives Pakistan some important geographic power contribution to the Kabul River system as well. In addition to the Chitral-Kunar tributary, Pakistan also has the entire drainage system of the Swat tributary to the Kabul River drainage basin, which adds additional water into the system and more positional geographic power to Pakistan.

*Material power*: Pakistan has much of the true educational background about water, as well as the economic power, military might, technological prowess, and international political and financial support that are available in Southwest Asia, so its score in this hydro-hegemonic accounting system must be regarded as fairly high in comparison to Afghanistan.

*Bargaining power*: Pakistan is very advanced in knowing the rules of the water game, especially because of its long bargaining with India concerning the Indus River and its many tributaries over the past half century since the 1947 partition. Clearly the setting of negotiation agendas, the ability to work the systems in terms of negotiations and agreements, the ability to provide, and/or influence incentives that may encourage weaker parties to comply, and to apply leverage by linking nonwater issues to such things as regional security or political alliances are things with which Pakistan has considerable familiarity. This gives it by far the greater power in this hydro-hegemonic pillar.

*Ideational power*: At the present time in Pakistan, many water statistics are low-level state secrets, which is a negative form of ideational power. Pakistan has the potential to gain control of the ideational power structure about the Kabul River Basin in a positive way, however, by offering to help Afghanistan construct a high number of dam- and power-development projects. Perhaps they can do this with help from China (Vick, 2014; Tiezzi, 2015). That this would also require Pakistan to put on the pressure to control the Taliban insurgency might be a great deal more than the existing government in Islamabad could guarantee, but it is an idea large enough to have them think about it seriously, one would think. After the dreadful murder of so many Pakistani school children by the Taliban, however, the Government of Pakistan actually has begun to put pressure on those insurgents in 2015. It is also clear, however, that the ISI in Pakistan could help to control the Taliban, but whether it ever would in reality, or even could force that result effectively into the tribal areas, is altogether another story. In any case, if Pakistan would look northwest toward Afghanistan in less of an apparently avaricious or colonialist mode and rather more of a benign and developmentally helpful fashion to augment hydroelectric energy sources, flood control, and augmented irrigation upstream in Afghanistan, greater strategic depth of considerable legitimacy could be obtained. This, however, has not yet happened seriously; with the result that Pakistan has a rather diminished ideational power in the hydro-hegemonic accounting system under discussion here.

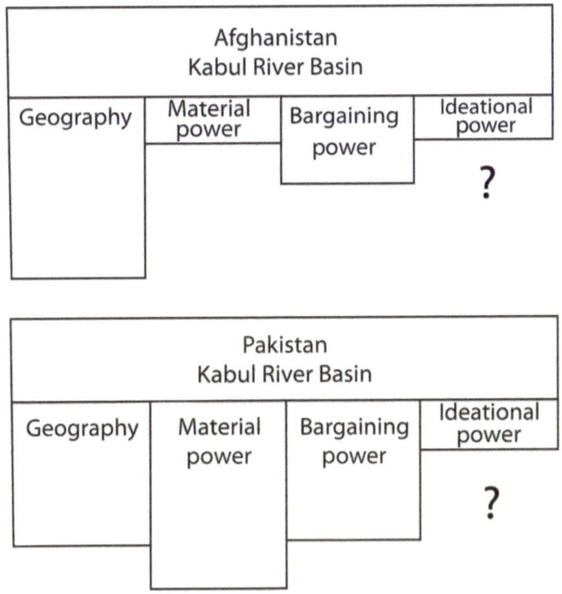


FIG. 14.3 Hydro-hegemonic assessment of the Kabul River Basin between Afghanistan and Pakistan. Collectively because of its overt and strengths in material and bargaining power, coupled with considerable covert strength through the Taliban, Pakistan outranks Afghanistan hydro-hegemonically, but this may not always remain true.

A glance at the graph (Fig. 14.3) on hydro-hegemony in the Kabul River Basin shows the somewhat subjective ranking of each of the four pillars for the two countries, as well as considerable uncertainty about ideational power in both nations. Illustrated with paired query symbols, the ideational power relations are the regions where considerable intellectual gains could be made. What would be the new ideas that could produce advancing or changing levels of hydro-hegemony between the two nations? At the moment, Pakistan seems to have some hydro-hegemonic advantage over Afghanistan in both material and bargaining powers, with these power factors perhaps being slightly balanced or offset by Afghanistan's somewhat higher geographic advantage. It will be the task of proposed transboundary water-sharing workshops in Kabul and Islamabad to contribute to the development of the ideational power structure for both countries, if possible.

## 14.4 HYDRO-HEGEMONY IN THE HELMAND RIVER BASIN

The Helmand (Hilmand or Hirmand to some) River starts in the Paghman Mountains 80 km west of Kabul, and flows >1150 km southwest through Afghanistan to the Seistan Depression on the Iranian border where it empties into a series of four or more *hamoun* lake basins that occur there before overflowing back inside Afghanistan into the Gaudi Zirreh (Shroder, 2014). The size of the Helmand drainage basin measures some 353,400 $km^2$ (Table 14.2), with a highly variable discharge with a median of ~2.2 $km^3$. A glance at a map of Afghanistan (Fig. 14.1) reveals a pronounced eastward bend of the border around the end of the Helmand River in the otherwise predominantly north-south-oriented border with Iran. This bend in the border was made by the British Goldsmid or Seistan Arbitration Commission in 1872–73 to the benefit of the Persians, in part because the British had fought and lost one war with the Afghans in the 19th century, were about to fight another with them, and were not particularly disposed kindly to the Afghans at the time (Shroder, 2014). As it was, the British had to pay a heavy indemnity or bribe to the Afghan ruler, Sher Ali, at the time to get his very reluctant consent to the new border being constructed to the detriment of Afghanistan.

The Helmand River basin, viewed from the four pillar points of the hydro-hegemony application (Cascão and Zeitoun, 2010), of geographical position, material power, bargaining power, and ideational power, shows interesting relations between Afghanistan and Iran (Fig. 12.4). For example, geographically Afghanistan clearly has an overabundance of territory of the Helmand Basin headwaters and the entire trunk of the main river, except for a few thin distributaries of the Helmand delta that exit into Iran. This obviously gives Afghanistan quite strong geographical positioning as the upstream riparian landowner. And because in recent years the lower Helmand has dried up entirely in the strong drought times (Shroder, 2014), the *hamoun* basins have themselves been entirely devoid of water too, which has driven tens of thousands of inhabitants away as refugees on both sides of the border. This has resulted in a severe water dispute and much diplomatic discussion between Tehran and Kabul (Samii, 2014; Hanasz, 2012; Diba, 2011), where Afghanistan has been cast as the malefactor by Iran. In fact, however, the years of recurrent drought may be more related to climate change that is adversely impacting South Asia, and not to any bad or unethical behavior by Afghanistan in holding back water deliberately.

The levels of material power of both Afghanistan and Iran are unbalanced, with Iran clearly having a more modern economy deriving from a formerly well-fueled hydrocarbon production industry, considerable infrastructure, good roads, a regionally high literacy rate of 82% (97% literacy among the youth), and over a half million men at arms. In comparison, Afghanistan has a very low literacy rate, half as many in the military, widespread poverty, and largely destroyed industry, poor roads, and a moribund economy that has been relying on foreign aid and the Western Coalition troops for development. Thus the material power of Afghanistan may be judged to be only half that of Iran. Basically, Afghanistan's only material strength in this area is its extensive hydraulic development in the Helmand Valley project with the Kajaki Dam and the widespread irrigation canal system that was built largely in the mid-20th century with plentiful foreign aid from the United States as a result of the Cold-War competition with the Soviet Union.

Comparisons of bargaining-power strengths between Iran and Afghanistan show that the latter country, of course, has considerable strength simply from its position astride over 90% of the watershed, as well as the extensive Helmand Valley irrigation network. Afghanistan does enjoy a certain moral high ground as a result of the punishment it received from the British in the border delineation of the 19th century, and the overpumping in the present day of ground water by the Iranians in the lower Helmand border areas, which has drawn down the water table across the border in Afghanistan. In addition, in 1969, Kabul was noted by the Iranian government as agreeing to ensure water flow to Iran in exchange for credit facilities, improved access to Iranian port facilities, as well as development assistance. Subsequent political upheavals on both sides of the border, coupled with natural precipitation variations, have, however, negated any of this from happening. For its part, Iran has periodically been coming across the poorly defended border into Afghanistan to bulldoze its own distributary canals to get the water it wants, although this has resulted in multiple skirmishes with border police. Iran is also supposed to "buy" some of the water it wants, but has not been paying for it (Diba, 2011). No active water agreement exists between the two countries (Hanasz, 2012), the Iranians do not compensate the Afghans for any water received as they once agreed to do, nor do they consult with the Afghan government on any other downstream water activities that they undertake, as they once assented to as well. In addition, although Iran's official policy is to reach formal agreements and pursue the benefits of cooperation in an Iran-Afghanistan commission to negotiate the discharge flow of the Helmand, reports and hard evidence have also surfaced that Iran uses the Taliban as proxies to disrupt Afghanistan's water projects (Houk, 2011). It thus would appear that it might be in the best short-term interests of Afghanistan to stall new negotiations on transboundary water rights because once they get more upstream dams in place, the downstream riparian will not have much to say, outside of more violence. Given this complex back-and-forth, the bargaining strength between the two countries thus might be judged as about equal and more-or-less stalemated between the two powers.

Neither Afghanistan nor Iran has produced any particularly new thinking about water cooperation or sharing in the region so the ideational power between the two is distinctly limited. The result in Fig. 14.4 is that both countries are judged about equally deficient in ideational power at the present time, although once the Salma Dam is finished, that may not

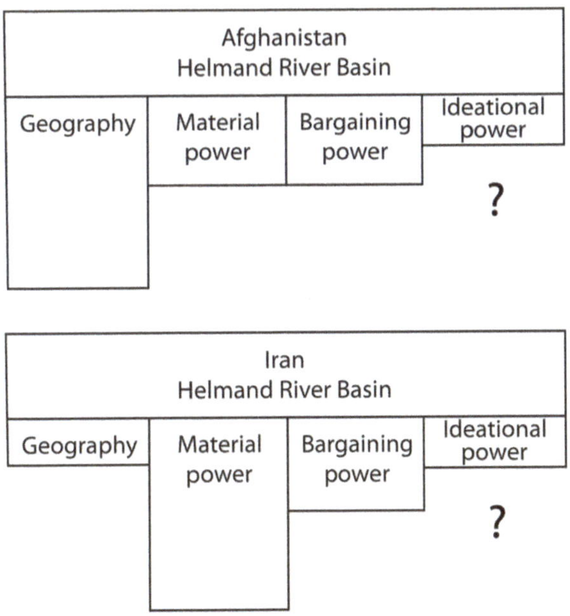


FIG. 14.4 Hydro-hegemonic assessment of the Helmand River Basin between Afghanistan and Iran. Collectively, perhaps because of overt strengths in material power Iran and possible covert strengths with Taliban, Iran might out-balance hydro-hegemonic strengths with Afghanistan, but this is an unstable situation with considerable geographic strengths in upstream Afghanistan.

continue to be true unless Iran chooses to pay a modicum for water that Afghanistan delivers downstream to them.

## 14.5 HYDRO-HEGEMONY IN THE HARI RUD-MURGHAB RIVER BASIN

The Hari Rud-Murghab River Basin is actually two basins; the one large Hari Rud Basin that begins in the center of Afghanistan and extends as a narrow strip to the northwest of the nation, and the other smaller Murgab Basin just in the northwest adjacent to the Hari Rud (Fig. 14.1; Table 14.2). Because the two basins are among the smaller basins under consideration in this chapter and they are directly adjacent in their lower reaches, they are considered here together for the terms of this analysis.

The Hari Rud fluvial system represents about 12% of the water resources of Afghanistan and is dominant in the heavily irrigated regions around Herat (Fig. 14.1). The waters of the Hari Rud are mainly sourced in central Afghanistan and flow in a nearly straight westerly direction along a narrow, fault-guided valley trace toward the Iranian border, before looping north to form Afghanistan's far northwest border with Iran, and there between Iran and Turkmenistan, before the last remnants soak into the sands of the Karakum desert. All the lower reaches of the river have dried up entirely at least once in recent years between 1998 and 2002. However, in the mid-20th century, as much as 1.3 $km^3$ annual flow could occur.

Directly downstream from Afghanistan, the several million people in the rapidly expanding city of Mashad, Iran, rely on the Hari Rud for their drinking water that is pumped 182 km from the Doosti Dam impoundment to supplement their declining ground-water levels. Rapid population growth in the lower Hari Rud basin is expected to add pressure in both Afghanistan and Iran for better water resources (Palau, 2011; Houk, 2011; Fipps, 2007; Favre and Kamal, 2004) The Murghab River in this basin also flows out of Afghanistan to soak into the Karakum desert as well, at first flowing in Afghanistan through a strong canyon only ~25 km upstream from the border where the narrow valley width does not allow much room for irrigation along its banks but with adequate engineering might serve as a reasonable dam site. Downstream in the city of Mary (Merv) in Turkmenistan, the Murghab River and the Karakorum canal from the Amu Darya bring additional water to the region.

Application of the hydro-hegemonic notions to the Hari Rud-Murghab River basin shows us (Fig. 14.5) that upstream riparian Afghanistan controls the extensive headwaters, and in fact, with Indian assistance, is building the Salma Dam across the Hari Rud. This greatly worries the Iranians downstream, but neither they nor the Turkmenistanis bothered to consult with Afghanistan when they jointly built the Doosti Dam across the lower reaches of the river between their two countries. In terms of geographical position therefore, with Afghanistan having so much territory in the headwaters, obviously produces a full complement of hegemony in both basins, with Iran perhaps only half as much and Turkmenistan at only about a quarter. In material power (Tables 14.2 and 14.3), Iran clearly enjoys greater strength with its large population, superior military authority, and higher literacy. Turkmenistan also has high literacy but its small population and diminutive troop strength relegate its material power to the lower levels. Afghanistan, in contrast, has a greater material power here than Turkmenistan merely because it does not have to project this power very far to control the hydrologic resource. Bargaining power is also thus highest with the Afghans, a bit less with the Iranians, and very

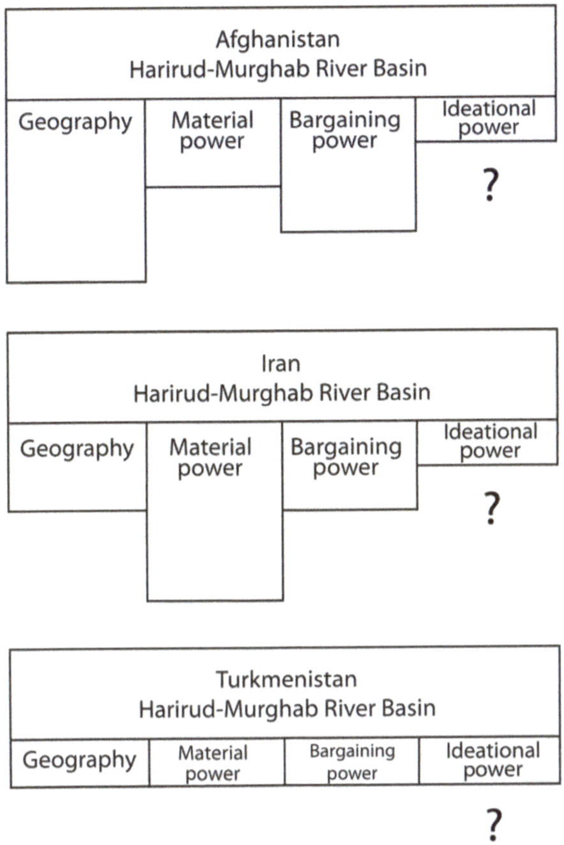


FIG. 14.5 Hydro-hegemonic assessment of the Hari Rud-Murgab River Basin between Afghanistan and Iran and Turkmenistan. Collectively, perhaps because of overt strengths in material power Iran and possible covert strengths with Taliban, Iran might out-balance hydro-hegemonic strengths with Afghanistan, but this is an unstable situation with considerable geographic strengths in upstream Afghanistan. Turkmenistan is outbalanced by both Iran and Afghanistan.

small with the Turkmenistanis. As with all the other neighbors with Afghanistan, at the present time, no one seems to have many good new ideas on what to do about the declining water situation in the country (Hutchinson, 2013), with the result that all three countries have only minimal ideational power in the hydro-hegemony calculation (Fig. 14.5).

## 14.6 HYDRO-HEGEMONY IN THE AMU DARYA RIVER BASIN

The Amu Darya River exists in the Aral Sea Basin with its headwaters in the Pamir Mountains of Afghanistan and Tajikistan, and with flow through Uzbekistan and Turkmenistan downstream to end in the drastically declining Aral Sea that overlaps into in part of Kazakhstan as well. The average annual flow on the Amu Darya is about 74–79 $km^3$, which is quite comparable to the annual discharge of the well-known Nile River in Africa, which averages about

$82\,km^3$. The flow of the Amu Darya and its tributaries is dominantly formed in the territory of Tajikistan (72.8%), 14.6% in Afghanistan, and 8.5% in Uzbekistan. Although the Panj River that forms the northeastern border of Afghanistan with Tajikistan is a prominent tributary of the Amu Darya system, a number of other tributaries add considerable discharge; the Vakhsh, Kafirnigan, Surkhan Darya, and Sherabad on the north (right) bank from Tajikistan, and the Kokcha and Kunduz from the south (left) bank in Afghanistan. At the present time almost none of the flow of the main river is used in Afghanistan, but the greater part of all of the flow is diverted off for irrigation in Turkmenistan and Uzbekistan downstream so that the Aral Sea, once the fourth largest lake in the world, has now drastically declined and more than three-quarters dried up as a result of overuse. The original 1946 treaty by Afghanistan with the Soviet Union was to be able to claim and use $9\,km^3\,yr^{-1}$ or a third of the flow of the Amu Darya, so that possibility looms in the future for the Central Asian countries to face up to (ICG, 2002; Horsman, 2008). In 1977, when Afghanistan officials arrived in Tashkent to work out a water-sharing agreement for their claim of that $9\,km^3\,yr^{-1}$, the Soviets would only agree to $6\,km^3\,yr^{-1}$, so no new agreement was reaffirmed (Rycroft and Wegerich, 2009). Then in 1987, close to the end of their disastrous war in Afghanistan, and without consulting the Afghans, the Soviets decided to formally divide up the Amu river waters among all the Central Asian republics, with an assigned water usage for Afghanistan of only $2.1\,km^3\,yr^{-1}$, whereas already in 1965, Afghanistan was even then known to be using some $3.85\,km^3\,yr^{-1}$ (Rycroft and Wegerich, 2009). After the collapse of the USSR in 1991, the following year the Almaty Agreement set water quotas close to the originals for the wealthier downstream agricultural regions of Uzbekistan, Kazakhstan, and Turkmenistan, whereas the upstream Tajikistan and Kyrgyzstan were given much smaller allotments because they were not thought to need so much water, but this is highly contentious (ICG, 2002). In any case, the water allotments have been routinely ignored and different countries have taken as much as they could get away with.

The problems of water use in the Aral Sea Basin region are manifold and serious. To start with, the highland areas that trap so much of the winter moisture also need to make plentiful hydroelectricity during the cold winter season to use as a heating source, when the downstream riparian users would prefer that the moisture be stored for later use during the summer growing season. Under the old Soviet regime even with mismanagement of the levels of the Aral Sea, at least the region once had a basin-wide water management approach. No longer; now the region has an uncoordinated competition between upstream and downstream users that is growing ever more serious (Siegfried, 2009).

In determining the hydro-hegemony levels of the Amu Darya River Basin, geographical power, material power, bargaining power, and ideational power for the riparian landowners of Afghanistan, and the bordering Tajikistan, Uzbekistan, and Turkmenistan, we can see that only Tajikistan compares reasonably with Afghanistan (Tables 14.2 and 14.3). This is because both countries are at the top of the watershed, and although both countries can be treated as roughly geographically equivalent, perhaps because Tajikistan's tributary drainages are thought to contribute significantly more of the total discharge, it has been given more geographical power over Afghanistan in Fig. 14.6. The other two countries that border Afghanistan along the Amu Darya, Uzbekistan and Turkmenistan, have significantly less geographic power because of their far shorter mutual riparian lengths, their higher material and bargaining powers are because of larger populations, high literacy rates, higher GDP

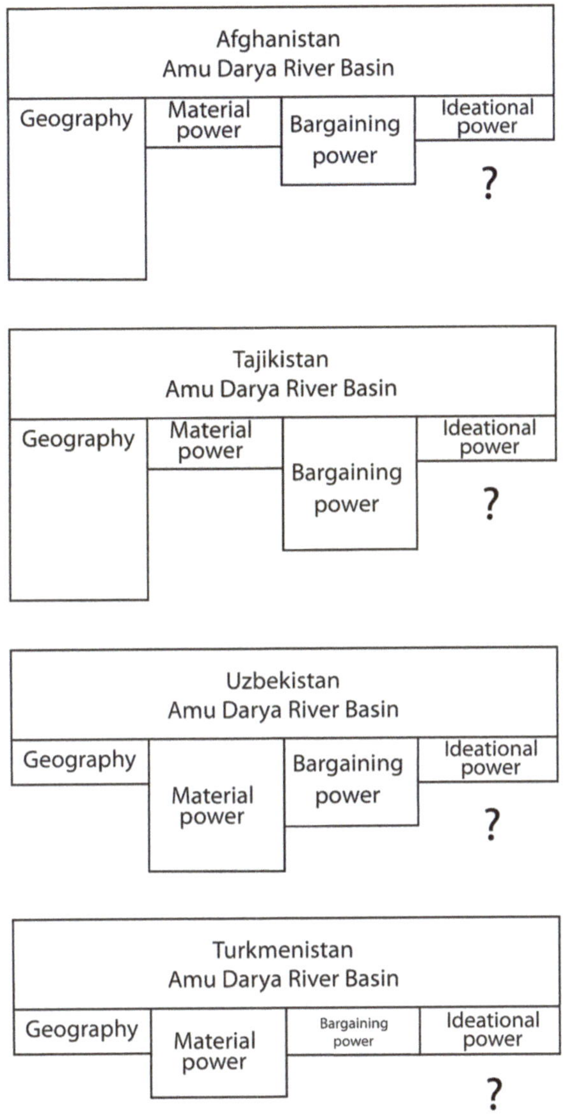


FIG. 14.6 Hydro-hegemonic assessment of the Amu Darya River Basin between Afghanistan and Tajikistan, Uzbekistan, and Turkmenistan. Collectively, Tajikistan outbalances all the other countries, and Afghanistan outbalances or approximately equals Uzbekistan and Turkmenistan.

per capita, and larger number of military troops. In detail, Tajikistan perhaps has the most bargaining power of the three former Soviet republics because of the presence of a significant Russian troop presence all along the border with Afghanistan, ostensibly to protect against any cross-border incursions or smuggling from unstable Afghanistan. Tajikistan's material power, however, is considered fairly low because of its small population and lowest per-capita income compared to Uzbekistan and Turkmenistan (Table 14.3). On the other hand, Tajikistan supplies much of the hydroelectric power to Kabul, which links the two high-watershed nations and adds to their mutual hydro-hegemony. Uzbekistan and Turkmenistan have significantly less geographical power because of their short (<110 km) river borders with Afghanistan. In contrast, however, Uzbekistan's large population, which is similar in size to Afghanistan's, coupled with a three-times higher income, high literacy, and the highest troop strength among the Central Asian nations bordering Afghanistan, raise its material and bargaining-power strengths. All four countries are given equivalent ideational power ratings with attached queries because if any advances are going to be made in eventual water negotiations, new ideas from any side will be most welcome advances, and any side can offer new thinking and new solutions.

## 14.7 SUMMARY AND CONCLUSION

In any sort of crude summation of the hegemonic strengths of Afghanistan over its neighbors, clearly the geographic strength at the top of the watersheds does mean something in that generally suspicious, even paranoid (for good reason), and deceitful region where any small advantage is sought over the neighbors as a matter of course. Poverty-stricken people do not make particularly good neighbors, with the result that cooperation among them is a rare entity. Sharing of transboundary water resources is done only reluctantly even through the waters may have been used upstream already for little more than a trash conveyance.

The 37 years of war in Afghanistan, in part already the result of diminished available resources, have greater reduced the technical capacity of the Afghan people. The result is that water issues in Afghanistan have resulted in a strong lack of trust by Afghans against the neighboring countries, who past or present have delivered so much violence into Afghanistan. The most fear is of Pakistani colonialist machinations through the Taliban, followed by fear of Iranian manipulations, and some anxiety about the ex-Soviet "Stans" countries to the north (Table 14.4).

A variety of foreigner-driven conferences have been proposed to resolve some of these water issues but have been seen in Afghanistan and some neighboring countries (ICG, 2002) as ill-informed, favoring technical answers, rather than understanding local water arrangements, or political or economic solutions. In the course of our trying to set up such a water conference, a high government official in Afghanistan has indicated that "might makes right" regarding bilateral water relations, so that scientific capacity must be built first, to better negotiate from a position of intellectual strength at a later time. Afghans seem not quite ready to negotiate yet, even between scientists, although indications are that this will change soon. They greatly fear having something put over on them in negotiations where they don't understand issues or the translation languages are inadequate. Thus it has been perceived that it is in the short-term interests of Afghanistan to hinder progress on water negotiations, because as the upstream aspirant hydro-hegemon, once upstream dams are funded and built, the downstream riparians will not have much to say (Table 14.4).

In summer 2014, when this chapter was being written, the Afghanistan government would not allow water discussions by anyone below Presidential Cabinet level. Government ministers had been forbidden to discuss water issues yet. In mid-2015, as we prepared for our water

TABLE 14.4 Water Issues in Afghanistan Concerning Capacity Needs and Fears

- Educational system drastically reduced by 36 years of war
- Strong lack of trust against neighboring countries, who deliver violence to Afghanistan; most fear of Pakistani colonialist machinations through Taliban, followed by fear of Iranian manipulations, and anxiety about the ex-Soviet "Stans" countries to the north
- Foreigner-driven conferences on water issues seen as ill-informed, favoring technical answers, rather than understanding local arrangements, or political or economic solutions
- Government official indicates that might makes right regarding bilateral water relations
- Scientific capacity in water must be built first, to better negotiate from a position of strength at a later time
- Afghans are not ready to negotiate yet, even between scientists
- It is in the short-term interests of Afghanistan to hinder progress on water negotiations
- Once upstream dams are funded and built, downstream riparians will not have much to say
- Long-term interests are put off for short-term reward
- Long-term good relations and shared futures are not seen or understood yet
- The US Embassy, World Bank and friendly countries are pushing for water negotiations
- The World Bank will withhold all development money unless negotiations are held
- The Afghan government under President Karzai would not allow water discussions by anyone below Presidential Cabinet level
- Ministers are forbidden to discuss water issues yet
- The policy of the next president—diplomatic, technical, academic—will determine new approaches to water
- Must wait for the new post–Karzai government of Ashraf Ghani to consolidate its rule
- Might be ready to discuss water issues after 1–3-year capacity-building phase
- Cooperation is only long-term option for human survival and prosperity
- Cooperation in human history increases range and variety of payoffs
- Short-sighted Afghan bureaucracy bases noncooperation in water issues on old and outmoded behaviors
- Far more to be gained in regional cooperation in resource corridor development than insular grasping at ever more limited water supplies

workshops in South Asia, we were not pushing the GoA yet for permissions, but that would happen later in the year. The policies and ruling styles of the new president, Ashraf Ghani—diplomatic, technical, and academic—presumably would determine any new approaches to water issues. Furthermore, everyone had to wait for the new post–Karzai government to consolidate its rule so that they might be ready to discuss water issues after as long as a 1–3-year capacity-building phase. Long-term interests have thereby been put off for short-term reward, with the result that long-term good relations and shared futures are apparently not seen or much understood, unless a too-clever scheme is afoot that has not come to the fore as yet. On the other hand, the US Embassy, World Bank and several friendly countries have been pushing for water negotiations by Afghanistan, and the World Bank has threatened to withhold all development money unless negotiations on transboundary water resources are held. If we take the Afghans at their word that they really do need to be trained to all the subtleties of hydro-hegemonic negotiations, then our proposed capacity-building effort in hydro-science and sensibilities could bear important fruit for the Afghans in future. Increasing the WCC of the general population and government bureaucracy is seen as an essential requirement for the educational system in Afghanistan to ensure wise decision making and eventual water-treaty negotiations with all neighboring states. Increased WCC can be done through concerted secondary school and university education and training in a national effort for survival.

Cooperation is thought by many as the only long-term option for human survival and prosperity in these difficult times of catastrophically expanding populations in the midst of changing climates and environments. In fact cooperation in human history has increased the range and variety of payoffs for those involved. It would seem that the rather short-sighted Afghan bureaucracy bases its noncooperation in water issues on old and outmoded behaviors that might have seemed to confer some advantage in times past. In fact far more is to be gained in regional cooperation in development of resource corridors than insular grasping at ever more limited water supplies. For example, if development of the rich mineral-resource base could get off the ground (Shroder, 2014, 2015), the increase in technology and in government profits could provide the financing to desalinate water, among other advances that would greatly profit Afghanistan's water-delivery systems. Many development options with water exist that are feasible for Afghans, if other, less pleasant options do not develop further (Mashal, 2014).

## References

Cascão, A.E., Zeitoun, M., 2010. Power hegemony and critical hydropolitics. In: Earle, A., Jägerskog, A., Öjendal, J. (Eds.), Transboundary Water Management: Principles and Practice. Earthscan, Stockholm International Water Institute, Earthscan, London & Washington, DC, pp. 27–42.

Diba, B.A., 2011. Iran-Afghanistan: water dispute over Hirmand or Helmand River. Iranian.com Archives; 3 p. http://iranian.com/main/blog/diba/iran-afghanistan-water-dispute-over-hirmand-or-helmand-river.html.

Dinar, A., Dinar, S., McCaffrey, S., McKinney, D., 2013. Bridges Over Water: Understanding Transboundary Water Conflict, Negotiation and Cooperation, second ed. World Scientific Series on Environmental and Energy Economics and Policy, vol. 11. Hackensack, NJ, USA.

Favre, R., Kamal, G.M., 2004. Watershed Atlas of Afghanistan: Working Document for Planners, first ed. FAO and Afghanistan Information Management Agency (AIMS), Kabul.

Fipps, G., 2007. Appendix to the transboundary water policy of Afghanistan: transboundary water issues. Draft; 10 p. http://gfipps.tamu.edu/Afghanistan/Transboundary%20Water%20Issues26-04-07.pdf.

Gall, C., 2014. What Pakistan knew about Bin Laden. The New York Times Magazine 45 (March 23), 30–35, 45.

Hanasz, P., 2012. The politics of water security between Afghanistan and Iran. Future Directions International; Independent Strategic Analysis of Australia's Global Interests; 6 p. http://www.futuredirections.org.au/publications/food-and-water-crises/416-the-politics-of-water-security-between-afghanistan-and-iran.html.

Horsman, S., 2008. Afghanistan and transboundary water management on the Amu Darya: a political history. In: Rahaman, M.M., Varis, O. (Eds.), Central Asian Waters: Social, Economic and Governance Puzzle. Helsinki University of Technology, Water & Development Research Group, Helsinki, Finland, pp. 63–74.

Houk, A., 2011. Transboundary Water Sharing: Iran and Afghanistan. The Stimson Center, Washington, DC, 3 p. http://www.stimson.org/spotlight/transboundary-water-sharing-iran-and-afghanistan/.

Hutchinson, K., 2013. The other Afghan conflict. Geographical Magazine. 6 p. http://www.geographical.co.uk/Magazine/Afghanistan_water_-_Jan_13.html.

International Crisis Group (ICG), 2002. Central Asia: water and conflict. ICG Asia Report No. 34. Osh, Brussels; 38 p. http://www.crisisgroup.org/~/media/Files/asia/central-asia/Central%20Asia%20Water%20and%20Conflict.pdf

Lustick, I.S., 2002. Hegemony and the riddle of nationalism: the dialectics of nationalism and religion in the Middle East. Logos 1 (3, Summer), 18–44.

Mashal, M., 2014. After Karzai. The Atlantic July/August, 50–58.

National Geographic, 2001. Afghanistan: Land in Crisis and Afghanistan and Pakistan (Maps, Topographic, Scale 1:3,363.300).

Palau, R.G., 2011. Afghanistan's Transboundary Water Resources: Regional Dimensions. Civil-Military Fusion Center, Norfolk, Virginia, USA, 8 p. https://www.cimicweb.org/Documents/CFC%20AFG%20Infrastructure%20Archive/CFC_Afg_Monthly_Transboundary_Water_Resources_July2011.doc.pdf.

Rycroft, D.W., Wegerich, K., 2009. The three blind spots of Afghanistan: water flow, irrigation development, and the impact of climate change. Central Asia—Caucasus Institute & Silk Road Studies Program. China and Eurasia Quarterly 7 (4), 115–133.

Samii, B., 2014. Iran/Afghanistan: Still No Resolution for Century-Old Water Dispute. Radio Free Europe, Radio Liberty, Washington, DC and Prague, Czech Republic, 3 p. http://www.rferl.org/content/article/1061209.html.

Shroder Jr., J.F., 2014. Natural Resources in Afghanistan: Geographic and Geologic Perspectives on Centuries of Conflict. Elsevier, Amsterdam.

Shroder, J.F., 2015. Progress with Afghanistan extractive industries: will the country know resource success or failure evermore? Extract. Indust. Soc. 2, 264–275.

Siegfried, T., 2009. Water and energy conflict in Central Asia. State of the Planet; Water Matters. News from the Columbia Water Center. http://blogs.ei.columbia.edu/2009/08/18/water-and-energy-conflict-in-central-asia/.

Soncini-Sessa, R., Castelletti, A., Weber, E., 2007a. Integrated and participatory water resources management. In: Jakeman, J. (Ed.), Developments in Integrated Environmental Assessment; Theory Volume 1A. Volume 1B, Elsevier, Amsterdam.

Soncini-Sessa, R., Cellina, F., Pianosi, F., Weber, E., 2007b. Integrated and participatory water resources management. In: Jakeman, J. (Ed.), Developments in Integrated Environmental Assessment; Practice 1B. Volume 1B, Elsevier, Amsterdam.

Tiezzi, S., 2015. The China-Pakistan alliance: the key to Afghan stability. The Diplomat. 11 February. http://thediplomat.com/2015/02/the-china-pakistan-alliance-the-key-to-afghan-stability/ (accessed 27.02.15).

Vick, M., 2014. Steps towards an Afghanistan-Pakistan water-sharing agreement. Water Resour. Dev. 30 (2), 224–229.

Warner, J., 2004. Mind the Gap—Working with Buzan: The Illisu Dam as a Security Issue. SOAS Water Issues Study Group, School of Oriental and African Studies/King's College, London (Occasional Paper 67).

Wegerich, K., 2008. Hydro-hegemony in the Amu Darya Basin. Water Policy 10 (Supplement 2), 71–88.

Zeitoun, M., Warner, J., 2006. Hydro-hegemony—a framework for analysis of trans-boundary water conflicts. Water Policy 8, 435–460.

CHAPTER

# 15

# Water in Islam

*J. Shroder*

**University of Nebraska at Omaha, Omaha, NE, United States**

OUTLINE



## Abstract

The position of water in Islam is foremost, in large part because of its scarcity in so many Muslim countries. Sharia law actually began in pre-Islamic time in relation to water. Issues of Islamic law and equity include rights of thirst and of irrigation, as well as protection of the environment, and water rights under Islam. Water conservation is seen as essential and educating about water through the mosques has been a new effort in some places in Afghanistan and elsewhere. Water-resource management has included a new subdivision into certain geographic/geological zonal characteristics that enable better understandings, among which the *water-stressed zone of physical scarcity* applies best to Afghanistan, but also to a certain extent to Pakistan and Tajikistan. Other water-related subdivision zones that also apply are the *tension-driven zonation* and *water disaster-prone zonation*, which also occur predominantly in these countries. The impact of Islam on problematic water issues of the present day has been somewhat limited and new research and other scholarly efforts are indicated as advisable to take maximum advantage of useful religious viewpoints to improve problematic societal actions. New linkages established between the discoveries of modern science, and the very wise environmental strictures of ancient Islam would be a real boon.

**Keywords:** Islamic law, Sharia, Koran, Equity, Environment, Water conservation, Water-resource management, Water-use rules, Water pricing, Water education, Water rights, Water ownership, Waste-water use, Causal Layered Analysis, Water-stressed zone, Physical scarcity, Tension-driven zone, Disaster zone, Transboundary water management


http://dx.doi.org/10.1016/B978-0-12-801886-6.00015-X

## 15.1 INTRODUCTION

Water in Islam occupies an essential position, just as it does in most other cultures, but in Islam it is marked by an overwhelming scarcity in most regions so that its importance is perhaps somewhat more magnified (Fig. 15.1). Many scholarly sources discussed below repeatedly emphasize the number of times (63) that water is mentioned in the Koran, as well as all the other references to its importance and central nature in ritual cleanliness and other religious rituals associated with the faith (Fig. 15.2). Given its secular and religious importance, the fact is surprising that some of water's associations to Islam have been recommended recently as needing considerable additional intellectual development in order to further the important issues.

Ideas of environmental protection in Islam have been developed directly in recent years from the numerous admonitions and critical pronouncements about the natural world in the Koran, and from the ancient religious authorities such as the Prophet Mohammad (pbuh) and the Caliphs or other thinkers (Table 15.1). This positive attitude about the environment of the world, and by extension, the water within this environment, involves taking measures to improve all aspects of life on Earth (Islamreligion.com, 2013). In this fashion, the

(A) (B)

FIG. 15.1 The importance of water in Islam is exemplified by spectacularly beautiful fountains and pools in mosques: (A) Fountain in the Court of the Lions in the Alhambra in Spain, built by the Spanish Moors; (B) Fountain and pool in a mosque.

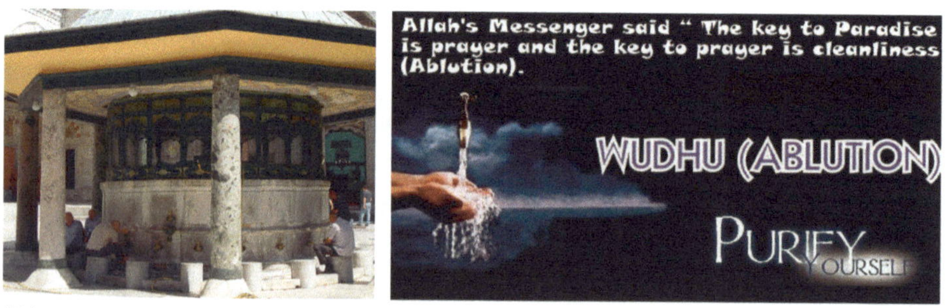


(A) (B)

FIG. 15.2 (A) An Islamic washing station outside a mosque in Turkey. (B) Admonitions to purify oneself in Islam.

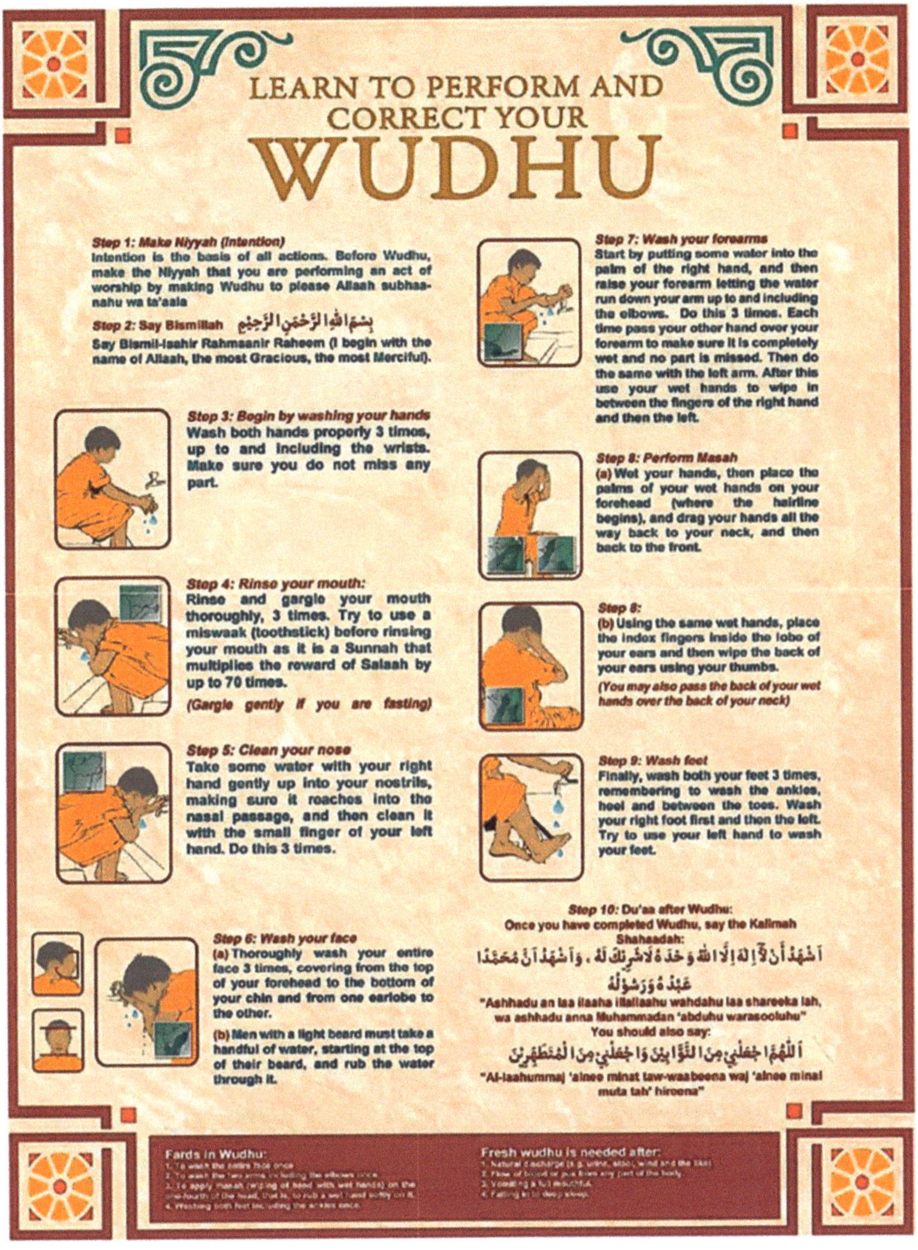


(C)

FIG. 15.2, CONT'D (C) Instructions for ritual cleansing by washing, or ablution (wudu) in Islam before prayers.

TABLE 15.1 Verses Used to Advocate Environmental Protection in Islam

1. "Verily, all things have We created by measure." (Koran 54:49)
2. "Everything to Him is measured." (Koran 13:8)
3. "And We have produced therein everything in balance." (Koran 55:7)
4. "Have you not seen that God is glorified by all in the heavens and on the Earth – such as the birds with wings outspread? Each knows its worship and glorification, and God is aware of what they do." (Koran 24:41)
5. "He Who has spread out the Earth for you and threaded roads for you therein and has sent down water from the sky: With it We have brought forth diverse kinds of vegetation. Eat and pasture your cattle; verily in this are the signs for men endowed with understanding." Koran (20:53–54)
6. "We have not created the heavens and the Earth and all that is in between carelessly. We have not created them but for truth." Koran (44:38–39).
7. "God it is that splits the seed and the date stone, brings the living from the dead and the dead from the living: That is your God – how are you turned away?" (Koran (6:95)
8. "Blessed is He in Whose Hand is dominion, and He has power over every thing; He who has created death and life to try you, which of you work the most good." (Koran 67: 1–2)
9. "There is a reward in doing good to every living thing." (Saheeh Al-Bukhari)
10. "The world is beautiful and verdant, and verily Allah, be He exalted, has made you His stewards in it, and He sees how you acquit yourselves." (Saheeh Muslim)
11. "And he has set within it mountains standing firm, and blessed it, and ordained in it its diverse sustenance in four days, alike for all that seek." (Koran 41:10)
12. "It is He Who has produced you from the Earth and settled you therein." (Koran 11:61)
13. "If any Muslim plants a tree or sows a field, and a human, bird, or animal eats from it, it shall be reckoned as a charity from him." (Saheeh Al-Bukhari, Saheeh Muslim)
14. "If anyone plants a tree, neither human being nor any of God's creatures will eat from it without its being reckoned as charity from him." (Ahmad, Tabarani)
15. "If the day of resurrection comes upon anyone of you while he has a seedling in hand, let him plant it." (Ahmad Bukhari, Abu Dawud)
16. "Partake of it gladly, so long as you are a benefactor, not a despoiler; a cultivator, not a destroyer." (Ali ibn Abi-Talib, the fourth Caliph)

*Notes:* For people who desire to use the writings and interpretations of ancient religious authorities (the Prophet Mohammad [pbuh]; Caliphs, etc.), these statements lend particular credence to interpretations for the protection of the world's environment. In other words, these statements provide the religious justification and authority necessary to accomplish particular environmental objectives.

*From Islamreligion.com, 2013. Environmental protection in Islam. http://islam.ru/en/content/story/environmental-protection-islam (accessed 04.06.15).*

health, nutrition, and the psychological and spiritual dimensions for mankind's benefit and the maintenance of human welfare are well taken care of. In addition, the betterment of life for all future generations, obviously requires attention to the water and the overall environment of the planet to ensure that harm is not done. As is shown in the Prophetic declarations of Table 15.1, the aim of both the conservation and development of the world's environment in Islam is to try to ensure universal good for all created things.

This chapter explores some Islamic perspectives on water, some of which are controversial and in need of further assessment, in terms of Islamic law and equity, water conservation, use of wastewater, water tariffs, water markets, community-based, water-management systems, and so forth. The basic idea of this chapter on Islam and water, is that by developing necessary concepts on water from within an Islamic context, much greater success is likely to be obtained within Muslim countries in developing important protections of the resource, especially in these times of climate change and overpopulation growth everywhere that so threaten everyone's future.

Khalid (2002, 2010) has made a learned effort to show essential elements of Islam and the environment wherein he has also shown some of the contrasting viewpoints between Islamic and other religions and the Western and scientific viewpoints. He tellingly points out what he sees as the collision course between the values expressed between the ideas of the Enlightenment in the West and Islam, such that numerous Western models of living in today's world (socialism, capitalism, etc.) that have produced a current world order that is untenable in terms of keeping the planet in good repair. He has noted that an awakening has been occurring among some Muslims that the realities of environmental change have led to efforts in Muslim countries to deal with the problem by secular agencies rather than by their religious ones, which has weakened the influence of the Islamic authorities on environmental issues. This has produced a paradox to the effect that an off chance exists that Muslims may discover that they are on the "wrong side of the tracks" in this regard (Khalid, 2002, 2010). Getting the religious authorities to act more on these issues may be a better solution with more attention to good results, especially with their exhortative preaching and potential for wise environmental education from an Islamic point of view.

## 15.2 ISLAMIC LAW AND EQUITY

As might be expected of a religion born in the hyperarid Arabian Peninsula, water is a central or core issue at the heart of Islam. In fact the very concept of Islamic law, or *sharia*, originally meant "the law of water." Before the establishment of Islam in Arabia, the *shari'ah* was actually a series of rules about water use; with the *shuraat al-maa* being the rights to drinking water. Later the terms evolved to include the body of other laws and rules given to Mohammad by Allah (De Chatel, 2002). Thus the present day Sharia (Islamic Canonic law) itself, is intimately bound to water because the term means something like "the watering source" or "the path that leads to a source of water" (Gilli, n.d.). This is quite significant in terms of water in that in the first interpretation, the metaphor is that of divine law quenching the thirst of knowledge, whereas the second metaphor is that of a path leading to the source of truth, which obviously is a target to be sought by every pious Muslim (Mallat, 1995).

Two fundamental precepts guide the rights to water in Sharia law: *shafa*, the right of thirst, which establishes the universal right of humans to quench their thirst, as well as that of their animals; and *shirb*, the right of irrigation, wherein all people may water their crops (De Chatel, 2002). But both of these rules can be interpreted in different ways by the various schools of Islam, with actual implementation varying from region to region, and from village to village. Each community may apply the law differently to suit different geographic, social, and ethnic circumstances. Domestic water use has precedence over agricultural or industrial use.

One of the issues of water in Islam is the fact that water belongs to the communities of believers and no one is supposed to own it or to have individual water rights over it. This is quite unlike most Western concepts of water law, which are themselves subject to considerable controversy, especially in times of water shortage. On the other hand in Islam, if certain people have made a special effort to carry the water through to recipients, or to distribute it in various ways at some expense to themselves, this would give them the capability to have a right of ownership over the water (Mallat, 1995), or perhaps in a better way of consideration, the value added to the water by their labor, rather than the actual water itself, enables pricing and trade for delivery of the water. These rules of water delivery in Islam were codified over the centuries, and shaped by regional traditions and behaviors. For example, the Majallah (Mejelle), the civil code of the Ottoman Empire, presented 92 articles on the subject of water management, and some of these rules still survive to the present day, especially in the Arab Levant (Mallat, 1995). Other countries have solved recovering costs for providing water in different ways, and, for example, in Iran irrigation water must be sold at the average cost price with both operation and maintenance costs factored in, along with capital depreciation costs included as well (Faruqui, 2001; Wickström, 2010).

In the case of naturally occurring water sources, in Islam the riparian landowners generally have senior usufruct rights (user rights without damage of the resource that belongs to others) of the water resource that is not sufficiently plentiful to allow unrestricted use. In Islam all communities along a river have a right to benefit from the water, with the most upstream riparian users taking amounts allotted to their domestic and irrigation needs, and then the next users, and so on downstream. If a new farm is established someplace upstream in the midst of the other earlier users, that farm is supposed to receive its water share only after the prior appropriations are taken. This resembles the rights of prior appropriation in Western water law in the United States, for example. In Islam, however, this does not mean that upstream riparian states can entirely ignore downstream states (Wickström, 2010). In addition in terms of well water, the owner of a well has senior usufruct rights similar to riparian landowners. The owner, for example may not withhold surplus water beyond his needs, nor may he pollute the well, or even deplete or draw down the well to the point that it will cause previously established wells nearby to fail, but whether or not these strictures are held to and followed in the different Islamic countries is altogether another story.

Thus in Islam, water is generally recognized as an absolutely critical and vital resource, to which everyone and everything has the right to a fair share (Table 15.2). Furthermore, beyond humans and their animals and crops, the environment also has clear and unmistakable rights to water as well. Humankind is thus responsible as a steward to ensure that human-environmental interactions are guided by the notion of humans as *khulafa*, or viceregents or stewards of the Earth, who are not lords and masters but its friends and guardians (Halal Alternative Solutions, 2013). The environment of Earth is supposed to be protected from humans by specific injunctions against upsetting the natural order of things by producing pollution and the making of such mischief (*fassad*) upon the planet (Table 15.3). The meaning of *fassad* can be interpreted directly as the despoliation of the natural function of the world or the degrading or spoiling of natural resources (Halal Alternative Solutions, 2013). Water is to be absolutely conserved, because the supply of water is fixed and it must not be

TABLE 15.2 Water as a Social Good in Islamic Practice

| | |
|---|---|
| 1. | Water is first and foremost a social good in Islam—a gift from God and a part of, and necessity for the sustaining of all life |
| 2. | Water belongs to the community as a whole—no individual can literally own water |
| 3. | The first priority of water use is access to drinking water of acceptable quality and quantity to sustain life |
| 4. | Every human has a right to this basic water requirement |
| 5. | The second and third requirements are water for domestic animals and for irrigation |
| 6. | Humankind is the steward of water on and in the Earth |
| 7. | The flora and fauna of the environment have very strong and legitimate rights to water, so the water must be protected from pollution |
| 8. | Individual people, organizations and nation states are liable for the harm that they cause the environment, or the environmental rights of others, including water-use rights |
| 9. | Water resources must be managed and used in sustainable ways |
| 10. | Sustainable and equitable water management ultimately depends upon following universal values such as fairness, equity, and concern for others |

*Notes:* This is a strong mandate for environmental protection within Islam, with religious authority behind it in the form of Koranic injunctions and the *hadith* rules [the collection of traditions containing sayings of the prophet Muhammad that, with accounts of his daily practice (the *Sunna*), that apart from the Koran, constitute the major source of guidance for Muslims].
*From Faruqui, N.I., 2001. Islam and water management: overview and principles. In: Faruqui, N.I., Biswas, A.K., Bino, M.J. (Eds.), Water Management in Islam. United Nations University Press, Tokyo, pp. 1–32; Halal Alternative Solutions, 2013. Water and equity. http://www.halalsolutions.org/water-management-and-conservation-in-islam (accessed 26.05.15).*

TABLE 15.3 Demand-Management of Water in Islam

| | |
|---|---|
| 1. | Water conservation is central to Islam, which means that mosques, religious institutes, and religious schools should be used to disseminate this principle in order to complement other secular and religious efforts |
| 2. | Wastewater reuse is permissible in Islam as long as the recycled water meets the required level of treatment to ensure purity and health for its intended purpose |
| 3. | Full-cost recovery for water use is permissible, which means that the full cost of supplying, treating, storing, and distributing the water is allowed to be charged for, as well as the cost of waste-water collection, treatment, and disposal, as long as the pricing is equitable and efficient |
| 4. | Privatization of water-service delivery is permissible in Islam, but the government has a duty to ensure equity in pricing and service |

*From Faruqui, N.I., 2001. Islam and water management: overview and principles. In: Faruqui, N.I., Biswas, A.K., Bino, M.J. (Eds.), Water Management in Islam. United Nations University Press, Tokyo, pp. 1–32; Halal Alternative Solutions, 2013. Water and equity. http://www.halalsolutions.org/water-management-and-conservation-in-islam (accessed 26.05.15).*

wasted. Water obviously must be carefully managed because supplies cannot be infinitely increased (Table 15.4), and already they are sorely stressed and environmentally contaminated in extensive part. One of the great problems for future Islamic jurisprudence will be how to determine what is a fair method to establish how to share transboundary water fairly with other nation states.

TABLE 15.4 Integration of Water-Resources Management in Islam

| | |
|---|---|
| 1. | Proper water management requires *shura* (consultation) with all the stakeholders |
| 2. | All community members, including both men and women, can play effective roles in water management and should be encouraged in this endeavor |
| 3. | Communities must be proactive or act in anticipation of future problems, needs, or changes in order to ensure equitable access to water resources |
| 4. | All nation states have an obligation to share water fairly with other nation states |
| 5. | Integrated water management is a necessary tool in order to balance equity across all sectors and regions |

*From Faruqui, N.I., 2001. Islam and water management: overview and principles. In: Faruqui, N.I., Biswas, A.K., Bino, M.J. (Eds.), Water Management in Islam. United Nations University Press, Tokyo, pp. 1–32; Halal Alternative Solutions, 2013. Water and equity. http://www.halalsolutions.org/water-management-and-conservation-in-islam (accessed 26.05.15).*

## 15.3 ISLAM AND THE ENVIRONMENT

Ever since the emergence of the well-known global environmental issues on the world stage at such a high level over a half century ago, some Muslim scholars have been engaging in the global debate on green issues. The environmentalists' call for a holistic effort on behalf of world environment has been seen as a practical platform and a quite comfortable and compatible position to help echo Islam's holistic view of the world as well (Bagader et al., 1994; Dien, 2000). Numerous verses in the Koran have been cited to quite clearly show that the holy book is well suited to justify strong environmental controls on all the main issues of concern (Islamreligion.com, 2013).

Khalid and O'Brien (1992) have detailed natural resources in Islam with water resources classified in Islamic law into three main categories: rivers, wells, and springs. In turn, rivers were subdivided into three categories:

- Large, naturally occurring rivers with permanent water that may be freely used by anyone who wishes to dig channel outtakes to irrigate land, water livestock, and other purposes.
- Small, naturally occurring rivers of two types:
  - those with enough water for the needs of local people so that additional channels may be dug; and
  - those with insufficient water for local needs unless they are dammed and distributed downstream carefully according to needs and other local circumstances.
- Karez come under the category of rivers, except that they flow underground.

Artificial canals and irrigation channels treated as joint property, but subject to control by those who dug them. Once the priorities of rights of usage are established, days or hours may be used to apportion the water.

Wells are divided into three categories that have various categories of water availability:

- general public use;
- temporary use; and
- personal use.

Springs are divided according to whether they are naturally occurring, exploited by individual endeavor on public land, or if they are on private land, so apportionment of water is based on whether the waters are public or private, and whether or not the water is needed in an emergency.

Bagader et al. (1994) have usefully detailed Islam's attitude toward the universe, natural resources, and the relationship between man and nature. They also assessed the protection of man and the environment from the harmful impacts of products and processes generated by man. Also assessed were the legislative principles, policies, and institutions of Islamic law that govern the procedures as measures for the protection and conservation of the environment. Their focus on this protection and conservation of the basic natural resources of the Earth includes plants and animals, the land and soil, air, and first and foremost, the water of the planet.

Conservation of the vital water element was seen (Bagader et al., 1994) as absolutely fundamental to the preservation and continuation of life in its various forms: plant, animal, and human. In Islamic law, whatever is indispensable to fulfill the imperative obligation of preserving life, is itself obligatory. Any actions that impair or obstruct the biological and social functions of this vital compound, water, whether by polluting or destroying it so as to impair its function as the basis of life, are prohibited in Islam. Thus such procedures as the dumping of compounds and chemicals into water as a convenient waste-disposal system, which is so common in so many parts of the world, Islamic and nonIslamic alike, is quite improper. In addition, the long experience of Muslim jurists in allocation of water rights in their arid lands has given rise to prominent and sustainable uses of the scarce resource, which are increasingly relevant in a world where such abundant resources are also increasingly scarce.

## 15.4 ISLAM AND WATER CONSERVATION

The use of Islamic concepts to promote well-being is a common procedure throughout the Middle East and South Asia. Water conservation is generally seen as a basic component and common good for water-resources management, so that public awareness and education are the basic tools needed to help guarantee the participation and involvement in water conservation (Atallah et al., 2001). Such conservation is seen as one of the most reliable and cost-effective solutions to water shortages in the region but has been too much focused on only the producers of the water (local and government water agencies, as well as decision-makers), and has not been as much stressed to the public consumers of the water. The poor consumer awareness and lack of public participation seem to be the main reasons behind this essential focus. Accordingly, in recent years a push has been on to get out the word for the effectiveness of using Islamic concepts to raise public awareness, and a number of such projects have been put forth throughout the Islamic world.

In the Islamic world in general, and in Afghanistan in particular, for example, the World Health Organization (WHO, 1992, 1995, 1996a,b,c, 1997) launched a campaign of "Health Education and Awareness through Mosques" in 1997 that promoted good health practices and raised awareness about water conservation and the importance of safe water and adequate sanitation, as well as hygiene and disease prevention. The imams from the mosques were trained by religious experts and water specialists as an essential part of the course work. Clear messages were developed from the available literature for the imams, who after their training developed special Friday sermons that were delivered twice on the holy day (WHO, 1997). Some initial evaluation showed that the messages were well received (Atallah et al., 1999).

## 15.5 WATER-RESOURCE MANAGEMENT (WRM) IN THE MUSLIM WORLD

Insight into the availability and usage of water in the Muslim world should be based upon the global data available or reasonably interpreted with respect to specific geographical regions and an extensive literature review. This enables the categorization of distinct regions with similar hydrogeological and socio-economic make-ups into six distinct water zones, half of which are directly relevant to our countries of most interest: Afghanistan, Pakistan, and Tajikistan (Absar, 2013). The futures assessment technique of Causal Layered Analysis (CLA) enabled examination of the manner in which water has been valued and managed, as well as the systems and worldviews that have influenced the current human-to-water relationships in the Muslim world. The six distinct socio-economic and hydrogeological zones are a clumping together to form characteristic zones, based on whether or not they have similar natural water distribution systems and strategies for the use of water. These systems and strategies have been best defined by Absar (2013) as: (1) oil barons in the desert; (2) water-stressed; (3) Mediterranean to tropical; (4) global-charity; (5) tension-driven; and (6) disaster-prone. Many countries with these overlapping zonal attributes can be assigned or classified into more than one of these divisions of the Muslim world.

Afghanistan fits squarely into the *water-stressed zone of physical scarcity* where the CLA level of societal approach or thinking has four main characteristics: (1) the litany of what they say; (2) the social causes of what they do; (3) the world view of how they think; and (4) the myth and metaphor about whom they really are. In terms of the litany for this particular zone, damming the rivers and pumping the ground water so far provides the necessary water for competing uses, whereas agriculture takes up the largest share of usage of fresh water. In the social causes of what they do, the building of dams, reservoirs, and tube wells is thought to be generally sufficient enough to ensure a supply of water to supply the drinking consumption, food requirements, and energy needs of the population. Water is seen as an absolute right for domestic and agricultural use, whereas disposal is seen as the government's problem. The world view of how the Afghans think is that the downstream impacts of dams, or the diversion of water from rivers for agriculture, are ignored or denied in order to preserve self-benefit and maximum potential upstream. Agricultural productivity is regarded as the only way to prosperity, and traditional irrigation methods are felt to be time-tested, no matter how wasteful that they might actually be. In terms of the myth and metaphor of any country, the idea is that power comes from dams, so that water abundance and control of that water indicates wealth and security, which is a right assigned by Allah, so that such people clearly have the right to exploit the water.

Pakistan and Tajikistan also fit into this category of water-stressed zones as well, although they may be more beneficially or accurately thought of as existing in other zones as well, especially because both have many internal and external troubles caused by critical neighbors over their plans for dams. In addition, Pakistan and Tajikistan fit well into the *tension-driven zone* and the *disaster-prone zone*, which are discussed in the next section.

In the *tension-driven zone* of some Islamic countries such as Pakistan, and, to a certain extent, also Afghanistan and Tajikistan, the CLA level of societal approach or thinking has the same four main characteristics of a litany of what they say, social causes of what they do, world view of how they think, and myth and metaphor of who they are. The result in this instance is that if a river passes over or by their land, or if ground water is available, then they have a right to the water in all cases, even while increasing population growth requires them to make the most of all available water resources even if it results in water logging from over irrigation or salinization buildup. Most multilateral treaties fail to reach any far-reaching agreement or framework in the countries in these zones, especially where contentious issues exist. River basins obviously transcend national boundaries but exclusive ownership of water is preferred. Unilateral agreements that do exist are not holistic or inclusive, especially those with shared ground-water aquifers. Upper riparian nations have the rights to the water, and the country with the most power to exert has the right over the water in any given basin, such that might is thought to make right, which is a rather dangerous position to take against the neighbors. Power is thought to come from the individual ownership of a portion of nature, so that the proverbial idea of the sovereign and individual state as a single entity leads to an inequitable sharing due to the political strength one nation exerts over another.

In the modern *disaster-prone zone* of Afghanistan, Pakistan, and Tajikistan, floods and droughts are increasingly common, along with a lack of effective use of water. Population increases and competing uses are putting great stress on water resources. Because watersheds are not managed properly, they are prone to recurrent disasters. Each sector of society has distinct beliefs about water, in many cases in conflict with each other, so that these ideas go unquestioned. Water that gets flushed away is thus out of sight and out of mind in this thinking. The societies are confounded by how or when to tackle any problem, and industry in denial looks for the most cost-effective methods, even while someone else is invariably seen as responsible for any environmental degradation. The mistakes made by other nations are repeated endlessly by these countries because they have been making use of their rivers in these fashions for centuries, and they feel that they can continue to do the same as long as they might wish. The myth and metaphor in all of this is that the wrath of "Mother Nature" (or Allah better, perhaps) is viewed as the punishment they deserve for their sins, even while they are in denial of their dying rivers, inefficient water usage, and melting glaciers. On the other hand, any success that they might have is viewed mainly as because of plenty of irrigation water and a bountiful crop, so prior problems are easily forgotten and nothing more is done for the future.

According to this futures analysis of Absar (2013), the alternative CLA approaches to existing water management that incorporates Islamic ideology can alter these three different analysis zones of the Muslim world that are so characteristic of Afghanistan, Pakistan, and Tajikistan. Absar (2013) feels that instead of the status quo, that Muslim religious philosophy and the importance of water needs to be revisited as a potential linchpin that could influence future policies and begin to answer some of the water-demand challenges that these nations face in the 21st century.

Accordingly, in the water-stressed group, Absar (2013) has noted additionally that environmental flows in rivers can allow for healthy watersheds, even as treated effluent and clean water preserves rivers and biodiversity. Social causes can be changed by exploring alternative methods of augmenting water supply in times of water shortages, such as rain-water harvesting, water treatment, and putting available resources to their most efficient uses. The world view is that water is finite and an increase in demand does not always come with an increase in supply. Conservation of the available water resources and maintenance of the health of watersheds are required. The myth and metaphor idea associated with this kind of thinking is that among the most beloved of deeds to Allah is the one that is continuous, even if it is only a little bit. Water is thus seen as a blessing from Allah that must be preserved in a manner that will allow future generations to benefit from it.

Similarly, in the *tension-driven group* of Muslim nations, the countries were thought as necessary to unite to create neutral water-management bodies for integrated transboundary, water-resource management, because equitable distribution of water resources can prevent water wars. The joint management of the watersheds can provide synergistic solutions, and the decrease of political and social tensions supports domestic harmony. Human beings are seen as the stewards of Allah on Earth and they must share in the abundance or scarcity of all the blessings of Allah because they are quite finite and humans as the inheritors are accountable to Allah for their actions on Earth (Absar, 2013).

For the *disaster-prone* Muslim nations, Absar (2013) also thought that integrated watershed management, with treatment of drinking water and sewage, and better planning of urban settlements all help to reduce the impacts of natural disasters. The protection against all the stressors on water systems is the key to protecting the water resources. Access to water is the fundamental right of every human being and this blessing should be sanctified and equitably distributed. Finally, Absar (2013) has noted that a disowned future shows no mercy. Instead the Muslim countries need to address the issues in sustainable manners to mitigate the impacts of natural disasters on their populations and future generations to create better futures.

The Muslim world has been seen as not being able to waste a single drop of water so that governments must urgently implement sustainable water-management policies that rationalize demand to ensure more efficient use. Abdurrahman (2000) and Absar (2013) have both noted that this can be achieved by attaching an economic value to water, shifting from flood irrigation to more efficient irrigation such as drip types, and introduction of new crop types that are salinity- and aridity-resistant. Recycling, treatment, and reuse of wastewater are required, as is development of affordable water desalination. The need for an integrated management of water resources has been derived from a new understanding that the water crises in so many places in the Islamic world cannot be thought of as a hydrological crisis alone. Instead a crisis of governance has been seen to be primary, because the institutional and management scarcity are hindering sustainable, equitable, and efficient uses of water resources (Hübschen, 2011). Islamically inspired water-management principles have been seen as in need of elaboration, because Amery (2001) has noted that many of the hydrological dimensions of Islam have been largely underdeveloped up to now.

Insofar as Saudi Arabia is considered as a wellspring of high Islamic principles, Abdurrahman's (2000) exposition has shown that modern thinking is quite possible in Muslim countries and the older or more poorly understood basic Islamic principles can be rethought and used successfully to bring Islamic water-management principles to the fore. For example, new measures introduced and legislation passed in accordance with Islamic laws in Saudi Arabia helped to accomplish a more effective management of water demand for domestic, agricultural, and industrial purposes. New specialized agencies were founded for water production and distribution and were able to enhance protection of the water interests of the communities, while reducing national demand for water and augmenting available water resources. Implementation of new water-pricing policies, improved leakage detection and flow control, and the promotion of awareness for water conservation have resulted in a further diminution in water requirements, which is a smart advance. By reducing government support for wheat production, water demand for that crop was reduced by 25%, for example. Because of considerable ignorance about, and resistance against, treated wastewater, a religious *Fatwah* had to be issued at a national level to enhance use of wastewater effluent. Encouragement of farmers to use modern irrigation systems, and adoption of irrigation scheduling and water-metering approaches further reduced water demand. Thus the successful experience of Saudi Arabia shows that Islamic laws can be used in a dynamic fashion and are sufficiently reasonable to respond to changing conditions and the challenges of the Muslim nations on such important issues as water in all its manifestations (Abdurrahman, 2000). A possible caveat to some of this is that Saudi Arabia is a far wealthier place than the other Islamic countries of South and Central Asia, but nonetheless some clear guidelines for improvement of water management are obvious in spite of the financial disparity between that Muslim country and so many of the others.

Identifying Islamic approaches to good water management has now been well underway for at least the past decade and a half (Faruqui, 2001; Scott, 2001). Some of the rather controversial topics addressed in the work on water management in Islam edited by Faruqui (2001) include water equity, allocation, reuse, environmental quality, pollution, water-demand management that includes family planning and water pricing, as well as water rights, water ownership, water markets with fair trading and pricing, and transboundary water management. A conclusion is that Islam can be quite resilient in facing the water problems of today. The analysis is an innovative blend of critical analysis of pressing issues in water management, with a religious Islamic perspective on the acceptability of current approaches to these issues. Wickström (2010) has echoed these findings about Islam as a social and political power in these matters, as well as its possible role in helping to resolve transboundary water issues and an increase in the awareness for critical water conservation. She notes critically, however, that the extent to which Islam actually influences practices, including those practices about water, in Muslim societies is a largely uncharted area, and to date the research on the impact of Islam on environmental issues has been quite limited. Insofar as the considerations of Khalid (2002, 2010) are relevant in terms of possible confrontational relationships between Western-dominated science and Islam, new linkages established between the discoveries of modern science, and the very wise, but generalized environmental strictures of Islam would be a real boon.

## 15.6 CONCLUSION

To anyone who has spent much time observing the rampant environmental despoliation and generally polluted and even disgusting water in almost all urban areas, and even in many rural ones as well, in such places as Afghanistan and Pakistan (Shroder, 2012, 2014), these are astute observations in need of greater discussion and wider emphasis. Nonetheless,

Wickström (2010) has noted that in order to conceptualize the contemporary environmental situations in Muslim societies in this regard, especially about water, further research and fieldwork is of immediate interest and must be pursued objectively.

## References

Abdurrahman, W.A., 2000. Water demand management and Islamic management principles. Water Resour. Dev. 16 (4), 465–473.

Absar, S.M., 2013. The future of water resource management in the Muslim world. J. Future Stud. 17 (3), 1–20. http://www.jfs.tku.edu.tw/17-3/A01.pdf (accessed 26.05.15).

Amery, H.A., 2001. Islamic water management. Water Int. 26 (4), 481–489.

Atallah, S., Ali Khan, M.Z., Malkawi, M., 1999. Water conservation through Islamic public awareness in the eastern Mediterranean region. East Mediterr. Health J. 5 (4), 785–797.

Atallah, S., Ali Khan, M.Z., Malkawi, M., 2001. Water conservation through public awareness based on Islamic treachings in the eastern Mediterranean region. In: Faruqui, N.I., Biswas, A.K., Bino, M.J. (Eds.), Water Management in Islam. United Nations University Press, Tokyo, pp. 46–54.

Bagader, A.A., El-Sabbagh, A.T.E.-C., Al-Glayand, M.A., Samarrai, M.Y.I.-D., 1994. Environmental Protection in Islam. IUCN—The World Conservation Union. IUCN Environmental Policy and Law Paper No. 20 Review; http://cmsdata.iucn.org/downloads/eplp_020reven.pdf (accessed 04.06.15).

De Chatel, F., 2002. Drops of faith: water in Islam. http://www.onislam.net/english/shariah/contemporary-issues/human-conditions-and-social-context/418062-water-islam-environment-green-christianity-ablutio.html?Social_Context= (accessed 26.05.15).

Dien, M.I., 2000. The Environmental Dimensions of Islam. Lutterworth Press, Cambridge.

Faruqui, N.I., 2001. Islam and water management: overview and principles. In: Faruqui, N.I., Biswas, A.K., Bino, M.J. (Eds.), Water Management in Islam. United Nations University Press, Tokyo, pp. 1–32.

Gilli, F., n.d. Islam, water conservation and public awareness campaigns. http://www.greenfaith.org/files/Islam-%20Water%20Conservation%20and%20Public%20Awareness%20Campaigns.pdf (accessed 26.05.15).

Halal Alternative Solutions, 2013. Water and equity. http://www.halalsolutions.org/water-management-and-conservation-in-islam (accessed 26.05.15).

Hübschen, K., 2011. Integrated Water Resources Management as a Governance Challenge for Countries of the Middle East with Special Focus on Yemen, Jordan and Syria. Logos Verlag, Berlin 280 pages.

Islamreligion.com, 2013. Environmental protection in Islam. http://islam.ru/en/content/story/environmental-protection-islam (accessed 04.06.15).

Khalid, F.M., 2002. Islam and the environment. In: Timmerman, P., Munn, T. (Eds.), Encyclopedia of Global Environmental Change. Social and Economic Dimensions of Global Environmental Change, vol. 5. John Wiley & Sons, Ltd, Chichester, pp. 332–339. http://www.ifees.org/islamEnviroment.pdf (accessed 04.06.15).

Khalid, F., 2010. Islam and the environment ethics and practice. In: The Environment in Islam. The 15th General Conference, 27–29 September. The Royal Aal al-Bayt Institute for Islamic Thought, Amman, The Hashemite Kingdom of Jordan.

Khalid, F., O'Brien, J., 1992. Islam and Ecology. Cassell Publishers, Ltd., London.

Mallat, C., 1995. The quest for water use principles: reflections on Shari'a and custom in the Middle East. In: Allan, J.A., Mallat, C. (Eds.), Water in the Middle East: Legal, Political and Commercial Implications. Tauris Publishers, London, UK, pp. 127–137.

Scott, C., 2001. Book review: water management in Islam. Water Resour. Dev. 17 (4), 647–650.

Shroder Jr., J.F., 2012. Afghanistan: rich resource base and existing environmental despoliation. Environ. Earth Sci. 67 (7), 1971–1986. http://dx.doi.org/10.1007/s12665-012-1638-7.

Shroder, J., 2014. Natural Resources in Afghanistan: Geographic and Geologic Perspectives on Centuries of Conflict. Elsevier, San Diego, CA.

WHO (World Health Organization), 1992. Regional Strategy for Health and Environment. WHO/EMRO/CEHA, Amman, Jordan.

WHO (World Health Organization), 1995. Beirut Declaration on Action for a Healthy Environment. WHO/EMRO/CEHA, Amman, Jordan.

WHO (World Health Organization), 1996a. Health Promotion Through Islamic Lifestyles: The Amman Declaration: The Hidden Path to Health: Health Education Through Religion, 5. WHO/EMRO, Alexandria, Egypt.
WHO (World Health Organization), 1996b. Water and Sanitation in Islam The Right Path to Health: Health Education Through Religion, 2. WHO/EMRO, Alexandria, Egypt.
WHO (World Health Organization), 1996c. Environmental Health in the Islamic Perspective (in Arabic) The Right Path to Health: Health Education Through Religion 7. WHO/EMRO, Alexandria, Egypt.
WHO (World Health Organization), 1997. Centre for Environmental Health Activities (CEHA) Newsletter, No. 23. WHO/EMRO, Amman, Jordan.
Wickström, L., 2010. Islam and water: Islamic guiding principles on water management: FIIA Report (Finnish Institute of International Affairs, Helsinki, Finland). ISN, The International Relations and Security Network, ETH, Zurich, Switzerland. http://www.isn.ethz.ch/Digital-Library/Publications/Detail/?id=124606 (accessed 26.05.15).

CHAPTER

# 16

# Water and the Opium Economy in Afghanistan

*J. Shroder*

University of Nebraska at Omaha, Omaha, NE, United States

OUTLINE



## Abstract

Illicit cultivation of opium poppies in Afghanistan has been increasing at high rates in recent years, especially in the Southwest in Hilmand (Helmand) Province, and a number of other provinces in the West of the country. As the climate of Afghanistan changes and droughts increase, the drought-tolerant poppy is an attractive crop to extremely poor people with little other recourse. Various ideas have been put forth to curtail the illegal activity, including ground eradication, aerial spraying, interdiction of precursor chemicals, stopping drugs at border crossings, arrest of drug lords, and other mechanisms, but all to little avail, except for some minor advances made from time to time. Close to US $7 billion has been outlaid to crush the trade, but it continues to do better than ever, with a record 6400 tons produced in 2014. New ideas of legalizing opium production to make inexpensive pharmaceuticals for poor developing countries have been shown to be economically not



*Transboundary Water Resources in Afghanistan*
http://dx.doi.org/10.1016/B978-0-12-801886-6.00016-1

viable, especially in the face of resistance from Turkey, India, and Australia, and a few other countries that are licensed already to grow licit opium. Given on-going climate change and most probable ever-drier conditions in Afghanistan, the likelihood for continued increase in opium production is expected unless a most unlikely, complete reorganization of the government of Afghanistan and concomitant drastic diminution of corruption occurs under new President Ashraf Ghani.

**Keywords:** Illicit drug trade, Climate change, Drought-tolerant poppy, Opium poppy, Heroin, Morphine, Warlords, Druglords, Criminal gangs, Alternate crops, Pharmaceuticals, Eradication

## 16.1 INTRODUCTION

The opium poppy (*Papaver somniferum*) can be thought of as a flower of war (Parenti, 2015), and as such is emblematic of Afghanistan, where it certainly accounts for >80% of the world's illicit opium production (Vicens, 2014), and perhaps as high as 90% at certain times (Rehabs.com, 2014) (Fig. 16.1). The narcotic attributes of this particular plant, and the unusual history of the persistence of cultivation of the opium poppy in Afghanistan after many decades of attempts at eradication, provide an argument that a new linking paradigm of illicit drug production as an answer to climate change appears have arisen in Afghanistan. The reason that the two concepts are linked in this fashion is that opium poppies are a very drought-tolerant crop, with the result that Afghan farmers may have little choice in the matter, even if they know that it is an evil crop. Not much of a contest exists if it is a matter of feeding one's children or not. In fact, opium poppies require only one-fifth to one-sixth of the water requirements needed by traditional crops such as wheat, for example (Parenti, 2015).

FIG. 16.1 Farmers harvesting opium poppies in Afghanistan. Note the dark opium tar on the wide, flat knife collector in the hand of the man in the foreground. The author has seen piles of such opium tar for sale in the bazaars of Afghanistan and Pakistan many times in the past. *Photograph by Associated Press, by permission.*

Thus the huge literature available concerning opium growing in Afghanistan (Peters, 2009, and many others) would seem to be caught up in issues of climate change and water availability, so much so that some of the issues related to its control would seem germane to any discussion of transboundary water resources.

## 16.2 AFGHANISTAN CONFLICT SITUATIONS AND OPIUM-HEROIN SUPPLY

Since the early 1970s when I first entered Afghanistan, opium-poppy production has been an issue that was under near-constant attention by the US Embassy—Kabul and the US Agency for International Development (USAID), as well as presumably, by relevant agencies in the Government of Afghanistan (GoA). In those earlier days in Afghanistan, the issue was mainly to ensure that opium-poppy agriculture was not being aided in any way by irrigation from the US-sponsored Helmand Valley Project through water impounded behind the dams on the Helmand and Arghandab rivers and delivered through any of the associated irrigation-feeder canals.

In fact poppy is known to have been grown in Badakhshan Province at least for a century or more (Pain, 2009, 2010, 2012), and Lacouture (2014) noted that a relationship between Afghanistan and the opium poppy had actually existed for thousands of years prior (Alexander the Great is reputed to have brought the first poppies to Afghanistan, c.~2300 years ago). Thus it is no surprise that Richard Scott (email communication to JS, Jun. 25, 2015) recalls that in the early 1970s in Afghanistan, as a USAID employee, he was primarily involved with monitoring poppy in Helmand, but that he and another USAID employee, Robert Riley Reilly, did go up into the Jurm Valley in Badakhshan, which was a large opium-producing area where "lots of people were addicted." By the late 1970s, poppy was cultivated in half the provinces of the country (Goodhand, 2005). Although an extensive drought did occur in Afghanistan in the early 1970s, no one would be likely to imply a causal relationship between drought tolerance and opium production in those earlier years; unlike the situation of the present day.

Both the Daoud government in Kabul that came to power in the 1973 coup and the People's Democratic Party of Afghanistan that took power in the 1978 coup had strict anti-narcotic regimes in place, but then in 1979, following the Soviet invasion of Afghanistan, the US Central Intelligence Agency (CIA) developed a covert war against the communists that included overlooking opium production (Prados, 2002; Parenti, 2015), or using it against the Soviet soldiers. The Mujahedeen rebels against the communist regime were well financed through drug trafficking, which was ignored by the United States even though it was quite inimical to US interests on the domestic front. Thus during the CIA's secret war, Afghanistan became one of the world's leading producers of heroin from the raw opium. This fallout or blowback from the CIA's effort seemingly was to sacrifice the Reagan administration's war on drugs to fight the Cold War. C.G. Cogan, one of the former CIA directors, admitted that in the desire to damage the Soviet effort they quite ignored the drug trade because it helped fund their overall effort to keep the Mujahedeen active (McCoy, 1997; Cogan, 1993).

During the post–Soviet civil war that ensued in Afghanistan as the various factions fought each other, and the Taliban rose to power from the ashes of the Mujahedeen, Prados (2002) somehow thought incorrectly that opium production had been largely eliminated. Actually at about the same time the United Nations (UNODC, 2003) had also noted that under the Taliban from 1996 to 1999, opium production had actually doubled and peaked at over 4600 tons (Fig. 16.2). In 2000 the Taliban banned cultivation of opium poppies but not the trade in the actual opium that had been produced in prior years. This move by the Taliban was condemned by some as little but a cynical ploy to drive up prices (Rehabs.com, 2014; Farrell and Thorne, 2005), because the prices went up by a factor of 10 thereafter. The opium trade was thus de facto legal in Afghanistan, both before and throughout the Taliban period, but in Jan. 2002, the new Karzai regime banned it (UNODC, 2003), as it has remained since.

By 2002 (UNODC, 2003), in spite of the ban by the Karzai government, when the Afghanistan's opium production resumed after the brief Taliban ban (3400 tons in 2002), it had actually increased overall more than 15-fold since 1979, the year that the Soviets invaded Afghanistan. Also in 2002, opium was being cultivated by several different ethnic groups in the south (Helmand), east (Nangrahar), and north (Badakhshan). Opium markets in southern Afghanistan were fragmented and competitive, whereas in the east and north they were oligopolistic. Price levels and structures varied accordingly at that time, but they were converging (UNODC, 2003). In 2002, gross income from poppy rose to $1.2 billion, or $6500 per family, which was extraordinary revenue in a country where average wage did not exceed $2 per day.

In fact, since the 9/11 attacks in the United States and the Western Coalition invasion of Afghanistan in 2001, opium cultivation has increased precipitously to ever-rising record levels (Bove, 2011; Lacouture, 2014). In fact, the steadily deteriorating security situation in Afghanistan following the arrival of the US and its NATO allies and the formation of

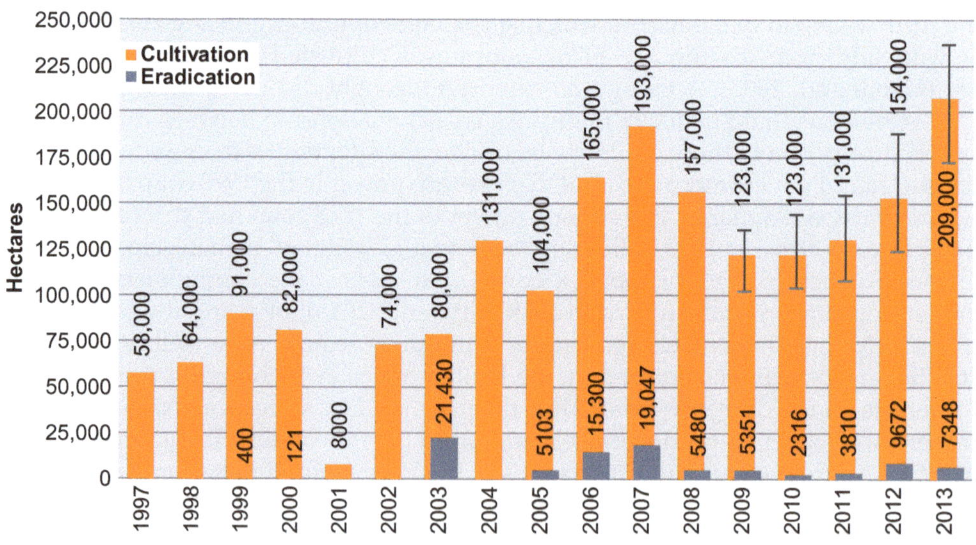


FIG. 16.2 Opium poppy cultivation in Afghanistan, 1997–2013. Data on verifiable eradication efforts reflect the highly variable political nature of the controversial process and the volatility or changing of advisability about alienating farmers. *From 1997–2002 UNODC; From UNODC, 2003. The Opium Economy of Afghanistan: An International Problem. United Nations Office of Drugs and Crime, Vienna, 222 pages; National Illicit Crop Monitoring System supported by UNODC.*

International Security Assistance Force (ISAF), propelled the Taliban into the vanguard of the insurgency, and completely revived opium production and its vital economy (Lacouture, 2014). As a result, a marriage of convenience has been created between the drug lords and the leaders of the insurgency, so much so that the Taliban can pay an average gunman $200 per month, whereas the average police officer is only paid ~$50 for similar work. Profits from the trade were estimated at $450–600 million between 2005 and 2008, which included income from: (1) levies on opium farmers; (2) protection fees from lab processing; (3) transit fees on drug convoys; and (4) taxation on import of chemical precursors (Bove, 2011). One in seven Afghans is estimated to be involved in some way in the drug trade. The estimated opium Gross Domestic Product (GDP) of Afghanistan is between $2.6 and $2.7 billion, which equates to 27% of the county's total GDP, both licit and illicit (Lacouture, 2014) (Fig. 16.3A–D). In the most recent data available from the UNODC (2014) it is clear that the year 2014 was another record at 224,000 ha, which produced 85% of the world's opium and 77% of its heroin (Economist, 2015). These are staggering figures that cannot be easily swept away or ignored in such a poor country, especially as one can see the direct relationship of this drug production and the resulting revenues that occur as a partial result of climate change, water diminution, and the necessity for people to have some means to earn a livelihood. Furthermore, this opium trade thrives on the lawlessness of the ongoing conflict in Afghanistan, where large-scale poppy-growers commonly use their political connections to keep prying government officials at bay. Eradication efforts thus have repeatedly targeted the poorest farmers because they are

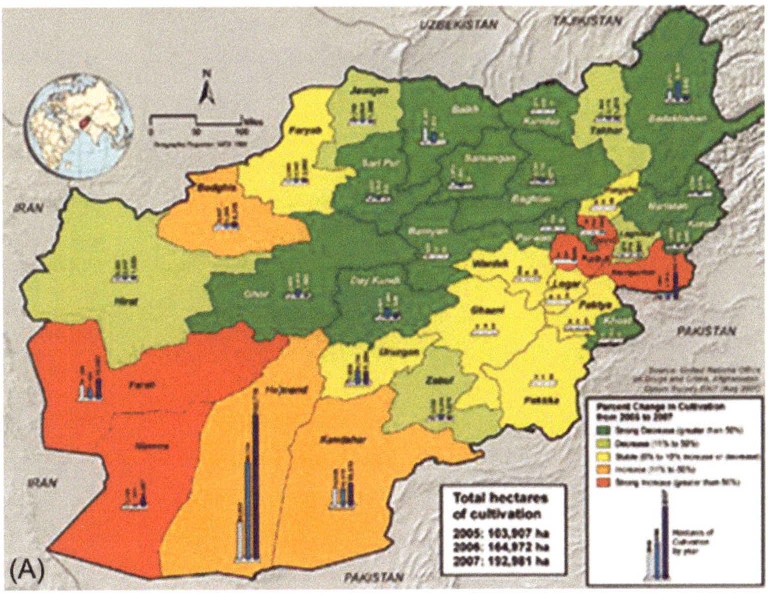


FIG. 16.3 Maps of opium-poppy cultivation in Afghanistan: (A) 2005–2007 with hectares (ha/year) as histograms in each province per year, from strong decrease *(dark green)* to strong increase *(red)*. (From United Nations Office on Drugs and Crime; Afghanistan Opium Survey 2007, August 2007.)

*(Continued)*

**Opium Cultivation in Afghanistan, 2008 (at province level)**

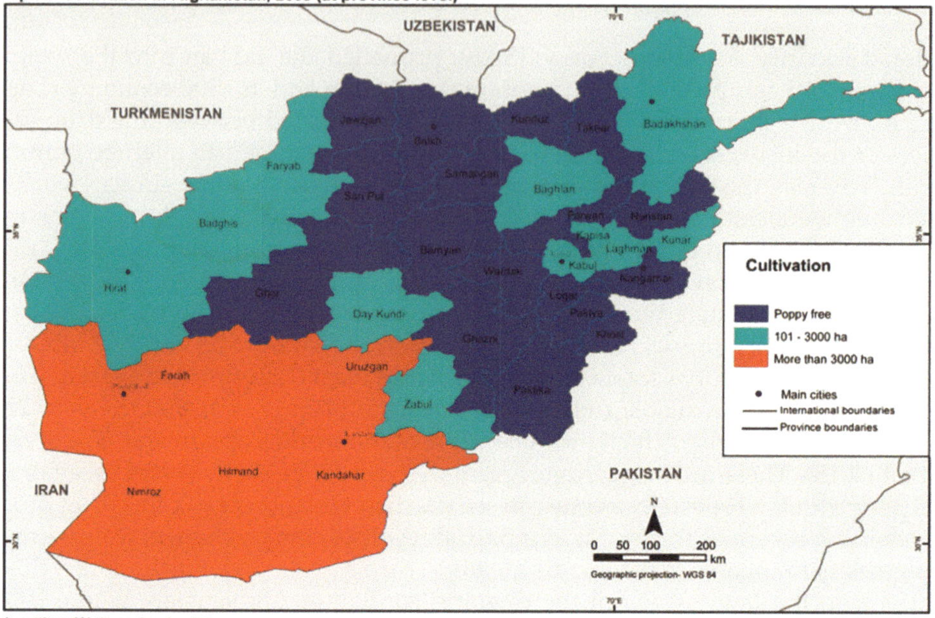


Source: MCN - UNODC Afghanistan Opium Survey 2008
Note: The boundaries and names shown and the designations used on this map do not imply official endorsement or acceptance by the United Nations

(B)

**Opium poppy cultivation in Afghanistan, 2009 (at province level)**

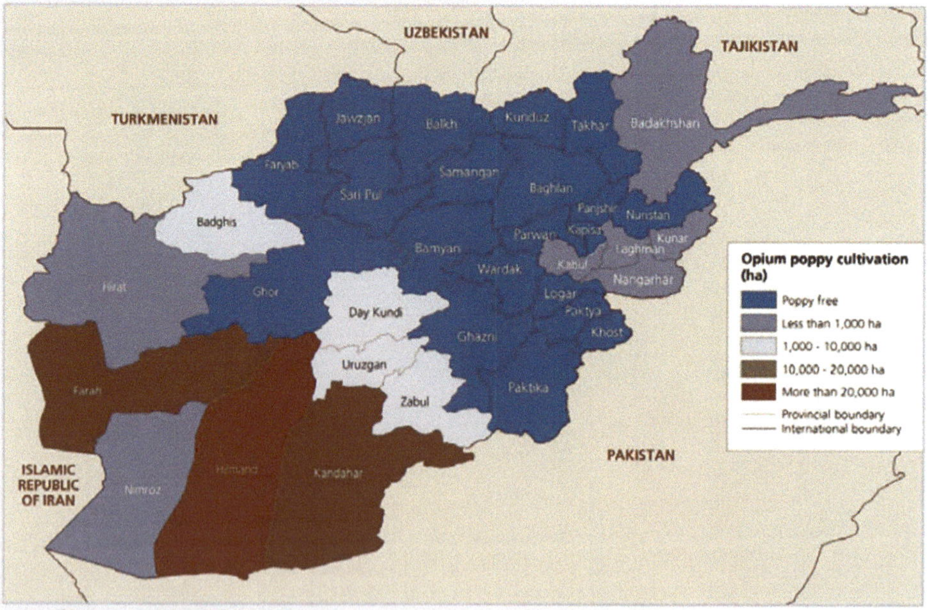


Source: Afghanistan opium survey, 2009, UNODC.

(C)

FIG. 16.3, CONT'D (B) 2008 with more than 3000 ha under cultivation *(red)*, to poppy free *(dark blue)*. (From Ministry of Narcotics—United Nations Office on Drugs and Crime; Afghanistan Opium Survey 2008.) (C) 2009 with more than 20,000 ha *(dark brown)*; 10,000–20,000 ha *(light brown)*; 1000–10,000 ha *(light blue)*; <1000 ha *(light gray)*; and poppy free *dark blue*. (From United Nations Office on Drugs and Crime; Afghanistan Opium Survey 2009.)

*(Continued)*

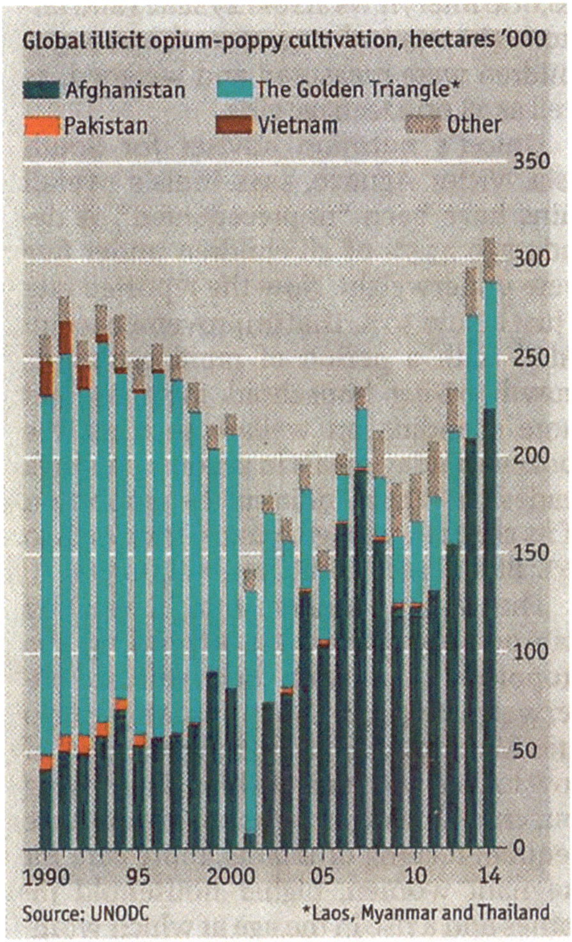


(D)

FIG. 16.3, CONT'D (D) Graph of global illicit opium-poppy cultivation in hectares '000. Afghanistan as the lowermost and darkest color on the graph clearly is the underpinning of the greatest production of such problematic material on Earth. *From UNODC data in The Economist, July 4, 2015.*

vulnerable, which makes the violence only worse; desperate opium farmers stripped of their livelihoods receive support from the Taliban (Economist, 2015).

## 16.2.1 Climate Change and the Flourishing Opium Trade

Hilmand (Helmand) Province is an exemplar of the complex connections and strong feedbacks between climate change, drying of the vital Hilmand River, cultivation of drought-resistant poppy, and the ongoing conflict in Afghanistan (Kovach, 2011). The weak central government and the ongoing civil conflict have led to years of environmental degradation (Shroder, 2012, 2014) and poor mechanisms for water management in the strong desert environment of the Seistan depression into which the Hilmand River flows. This neglect and lack of technical competence after three decades of war have also decreased the amount of water flowing into the province from the Hilmand River. On top of this are the compounded effects in recent years of on-going climate change, which has decreased precipitation and further lowered the levels of the Hilmand River. The result has been to leave the region in a prolonged and significant drought. As the drought has continued and made traditional agriculture for the growing of staple crops progressively more difficult, many farmers have had to turn to the drought-resistant, opium-poppy crop (Fig. 16.4A–C)

The resulting dramatic expansion of the opium harvest has made Helmand Province the leader in Afghanistan in that crop, which has in turn brought the country into the forefront of a resurgent international trade in opium in the world. This has contributed to the survival and resilience of the anti-government insurgents, in the forefront of which are the Taliban. In a systems analysis of the material flows in the environment and agricultural sectors, compared to the social and political sectors, the Helmand villagers are caught squarely between climate-change induced, water-management problems and the civil conflict such that they have little choice but to cultivate opium (Fig. 16.5).

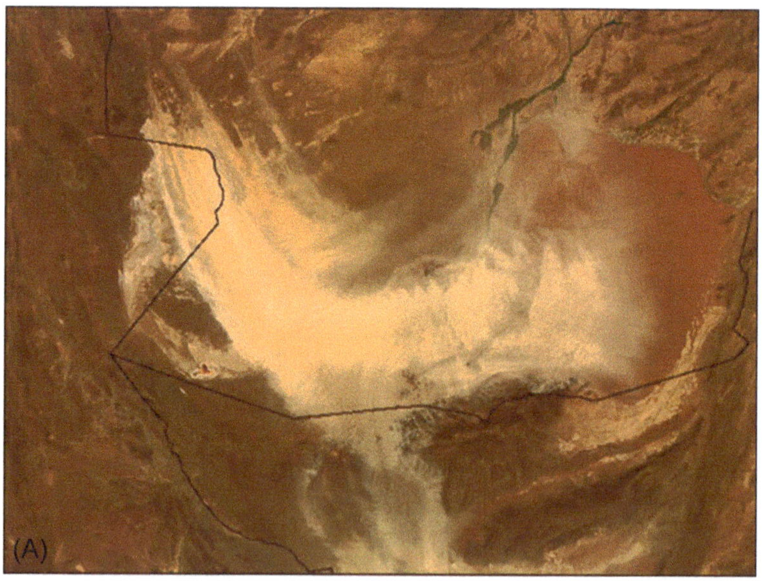


FIG. 16.4 Pictures that indicate the increasing aridity that has been developing in the Seistan Basin into which the Hilmand River empties: (A) Dust storm in MODIS image over the lower Seistan Basin on Sep. 23, 2003, imaged by the Terra satellite system. Such dust storms are very common in this region and reflect the high winds of the *badi-sado-bist-roz*, or wind of 120 days that blows strongly in the summer months out of the Central Asian high pressure into the low pressure of the Indian monsoon that approximately coincides with the Afghanistan–Pakistan border at that time. Also observable are the plentiful irrigation areas associated with the Hilmand River entering from the upper right (northeast), much of which is the area of the American-built Kajikai Dam and reservoir, as well as extensive areas that are now replete with major opium-growing agriculture.

*(Continued)*

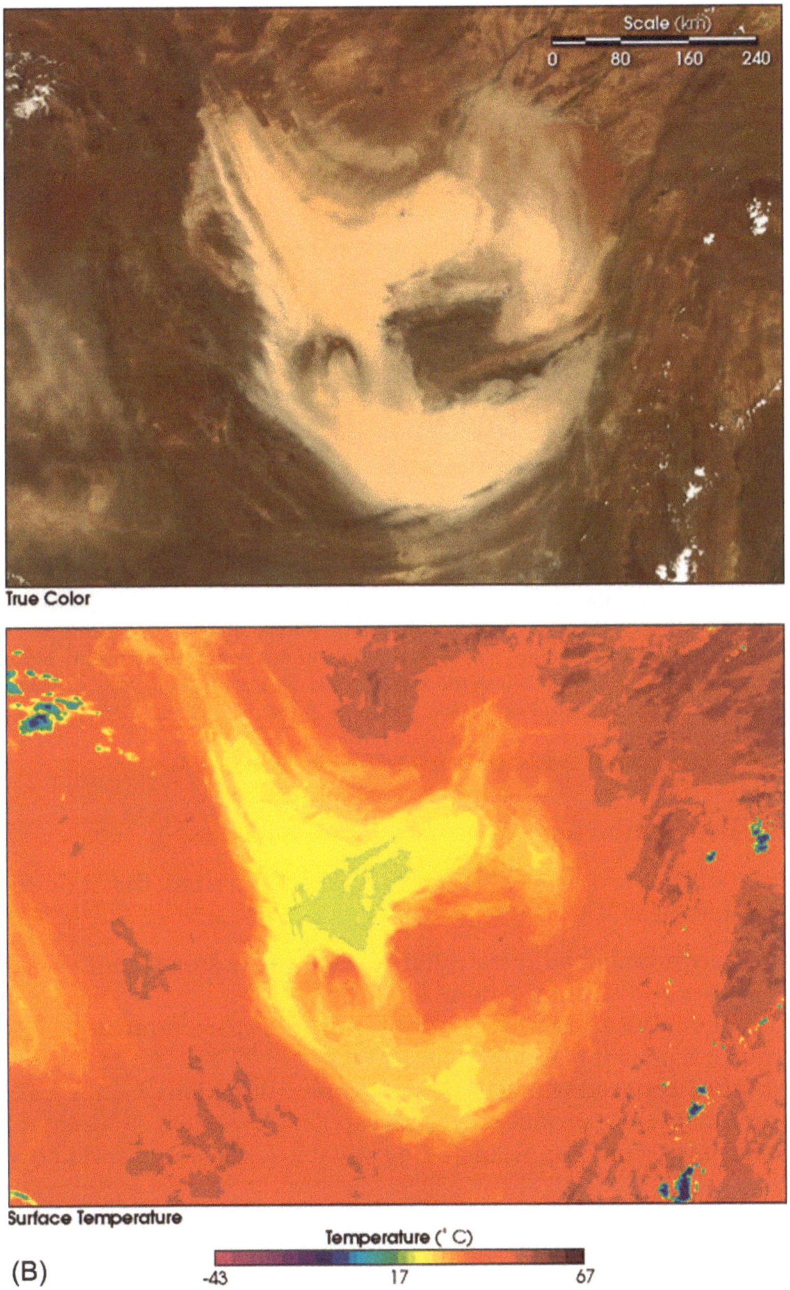


FIG. 16.4, CONT'D (B) Dust storm blowing over the lower Seistan Basin on Jun. 14, 2004 imaged by the Terra MODIS satellite system of the NASA Goddard Space Flight Center, and showing true color at the top and temperature of the cooler dust than the warmer ground surface at the bottom. In 2001, the Hilmand River dried up entirely in its lower reaches so that the fine sediments in the river bottom and hamoun lake beds on the Afghanistan–Iranian border provided plentiful dust to be mobilized in the high winds. This has continued in subsequent years, as this image shows.

*(Continued)*

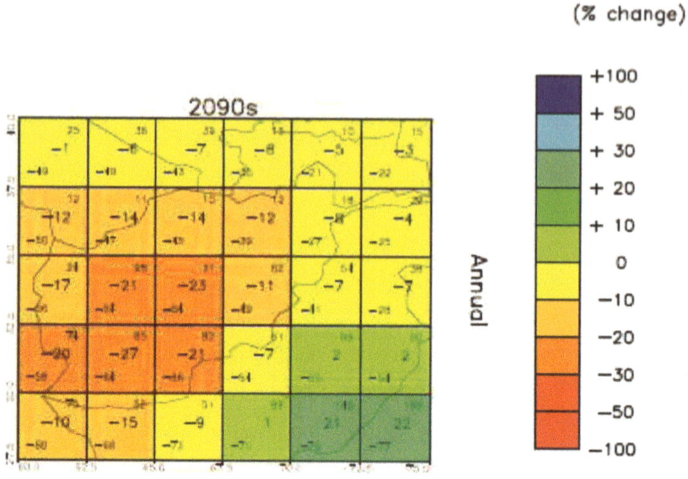


(C)

FIG. 16.4, CONT'D (C) Extensive different quantitative computer models of world climates allow reasonable projections out in subsequent years. A strong model based on different predictions of emissions of greenhouse gases by the IPCC (United Nations Intergovernmental Panel on Climate Change) allows projections into the future year by year. This one shows a −20% to −30% change in annual precipitation in the Seistan Basin into which the Hilmand River drains in the year 2090, 75 years hence, and is an example of one of the many negative quantitative predictions available of the future water availability in the region. *From Savage, M., Dougherty, B., Hamza, M., Buttefield, R., Bharwani, S., 2008. Socio-economic impacts of climate change in Afghanistan. DFID CNTR 08 8507. A report to the Department of International Development, Stockholm Environment Institute; 7 pages. http://www.necsi.edu/afghanistan/pdf_data/2007447_AfghanCC_ExS_09MAR09.pdf (accessed 02.07.15).*

## 16.2.2 Opium Production and Economic Addiction

The historical roots of the opium economy in Afghanistan show that it developed because of a number of factors (UNODC, 2003), including:

- lack of effective government administration (central government control ~30%, Taliban control ~10%, tribes and warlord control ~60%);
- degradation of agriculture and most economic infrastructure because of decades of war;
- a war economy and related black marketeering;
- a nexus established between war, crime, and opium cultivation;
- opium as a profitable crop, produced with cheap labor (women, children, refugees);
- abundant inputs for opium poppy, including suitable land, water, and know-how from itinerant labor;
- opium as a form of savings, a source of liquidity, and a collateral for credit;
- opium as an insurance against poverty and hunger; farmers sell future crops to narco-users for subsistence; and
- opium requires no marketing or storage, and it can be sold easily on the spot markets.

At the same time that opium cultivation seems to offer many positive results to poverty-stricken Afghan people, still the problems with local drug abuse are also extreme, in

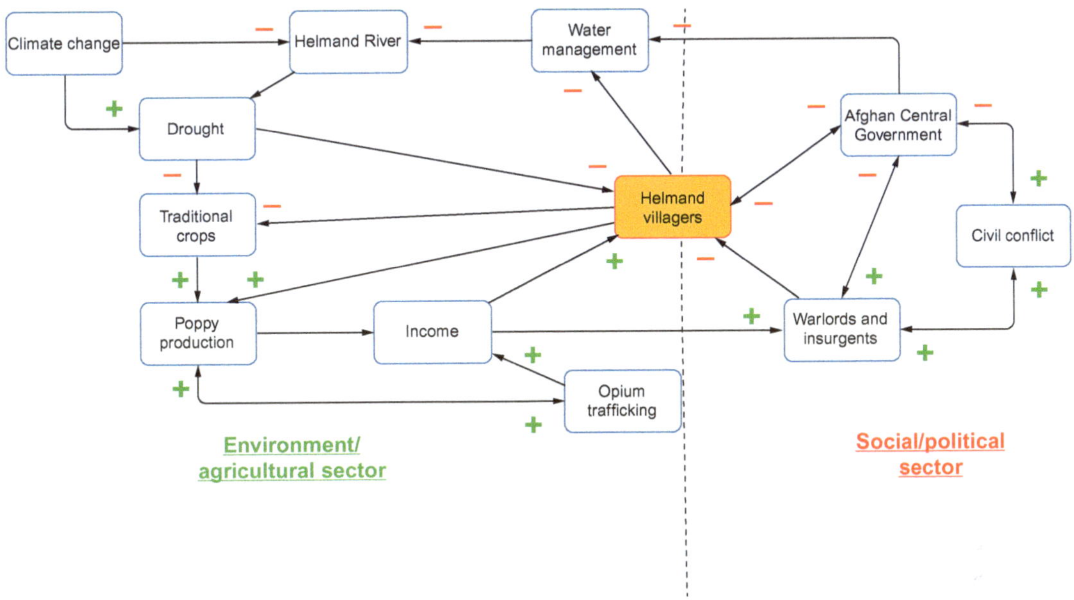


FIG. 16.5 Systems analysis of the environmental/agricultural sectors versus the social/political sectors in the Hilmand river basin. The connection between climate change, environmental degradation, violent civil conflict, and opium production in the Hilmand Province is dynamic and complex, with the flows from both sectors focuses on the villagers in Hilmand Province. Climate change and degradation of the environment are significant inputs into this operation, whereas civil conflict and instability are both inputs and outputs, indicated by the positive (+) and negative (−) signs. The result, of course, is open trafficking of opium and its derivatives. *From Kovach, T., 2011. Helmand Province, Afghanistan, Conflict and Climate Change. Inventory of Conflict and Environment (ICE) Case Studies. Number 261. American University, Washington, DC; 13 pages. http://www1.american.edu/ted/ICE/helmand.html (accessed 30.06.15).*

that misuse and addiction have continued to escalate dramatically in recent decades. About 1 million addicts have been estimated by the Afghanistan Ministry of Health, with some 15% being women and children (Associated Press, 2013), but the Asian Development Bank (Starkey, 2009) estimated twice as many at 2 million. Many people do not even understand the addictive and debilitating consequences of smoking or injecting opiates, and some families even give opium to children to quiet them or to treat minor afflictions.

This great and new addiction problem in Afghanistan has been due to the prolonged human deprivation and suffering there, the breakdown of social controls, the return of refugees who developed drug problems in refugee camps, and the almost completely unlimited availability of the various opiates in Afghanistan. The many war wounded who abound in Afghanistan for obvious reasons, became addicted as a consequence of the primitive first aid and large-scale use of opium, morphine, and heroin as painkillers. Nonetheless, it should also be noted (UNODC, 2003) that drug abuse in Afghanistan is still relatively low compared to the neighboring countries of Iran, Pakistan, and a number of Central Asian countries.

## 16.3 FAILURES OF OPIUM-CONTROL EFFORTS IN AFGHANISTAN

### 16.3.1 The National Problem

The country of Afghanistan has now come to rely upon an opium economy for much of the past three decades or so, with the result that the processes are now deeply embedded in Afghan society. Thus ending this economic addiction is likely to take as long or even longer to bring to any sort of resolution, if indeed, such is even possible. Over $6 billion (Rehabs.com, 2014) has been invested in such drug eradication in Afghanistan to date, unfortunately to quite limited effect. The military strategy of counter-narcotics in Afghanistan has clearly been not only ineffective, but even counterproductive, especially given the ever-growing levels of violence in the poppy-growing regions of the country. Both Afghan and US officials have freely admitted that the illegal poppy production has corrupted the government from bottom to top, with cabinet officials, provincial governors, warlords, local militias, and the Taliban insurgency all completely implicated (Waldman, 2004; Gates and Lektzia, 2004; van Ham and Kamminga, 2006–07).

In the latest report by the United Nations Office on Drugs and Crime (UNODC, 2014), almost all of the metrics on cultivation and sale of opium poppy have increased (Table 16.1), except for the price, which has gone down recently, probably because of the increasing oversupply. The problem is a complex combination of narcotics bazaars, special finances, profuse narco-users who ignore Islamic strictures, greed, warlords, and the illegal world opium trade. In fact, opium has become an "economic narcotic" for whole segments of Afghan society (UNODC, 2003): (1) as a commodity, opium is an income generator; (2) as a source of liquidity, it is a means of exchange; and (3) as a payment mechanism, opium is a way to store value and fund transactions. Traders in opium frequently act as narco-users or narcotics money lenders operating under varying degrees of risk, because: (1) opium serves as a means of *salaam*, or informal advance payments; and (2) money-lenders' capital

TABLE 16.1 Key Findings From the Afghanistan Opium Survey 2014 by the UNODC and the Islamic Republic of Afghanistan: Ministry of Counter Narcotics

| |
|---|
| 1. The total area under opium poppy cultivation estimated at 224,000 ha—7% higher than 2013 |
| 2. The vast majority (89%) of opium cultivation was in nine provinces in Southern and Western region, which include the most insecure areas in Afghanistan |
| 3. Hilmand remained Afghanistan's most major opium-cultivating province, followed by Kandahar, Farah, and Nangrahar |
| 4. Opium cultivation increased in most of the main poppy-cultivating provinces, but stabilized in Hilmand itself (+3%) |
| 5. Total eradication of opium poppy decreased by 63% in 2014, to 2692 ha |
| 6. Average opium yield was 28.7 kg/ha in 2014, which was 9% more than in 2013 (26.3 kg/ha) |
| 7. Potential opium production in 2014 was estimated at 6400 tons, an increase of 17% from 2013 (5500 tons). |
| 8. At US$ 0.85 billion—about 4% of Afghanistan estimated GDP, the farm-gate value of opium decreased by 13% in 2014 |
| 9. The price of opium decreased in all areas in Afghanistan, probably because of increased production supply |

*Note:* The key findings have been edited for clarity and brevity, and not all of the findings are included here.

can assist farmers with regenerated cashflows, by: (a) rapid turnover trade (low profit); (b) shipments to border regions (moderate profit); or (c) smuggling opiates internationally across borders (high profit).

Opium is also an ideal commodity for marketing, trade, and speculation, because:

- it is compact to transport;
- it is durable to store;
- it has a high intrinsic value ($350–$400 per kilogram);
- only a few licit agricultural commodities (truffles) go past this ($800/kg);
- high risk of interdiction at borders maintains high prices;
- opium is a commodity suitable for trafficking, especially in provinces where warlords extract protection taxes;
- traffickers in some regions recycle part of their profits into poor villages to gain respect and protective gratitude from local villagers;
- a clear nexus exists between drug trafficking and warlordism; and
- re-emergence of drug cultivation and recrudescence of violence in certain provinces are well-known recurrent phenomena.

## 16.3.2 Temporary Success

Combatting the opium trade in Afghanistan (Choi et al., 2009) was deemed as essential for the US by the 9/11 Commission, but efforts so far have been a near complete and utter failure (Blanchard, 2009; Pain, 2012). In fact, the late Richard Holbrooke, the US Special Representative for Afghanistan and Pakistan, noted in 2009 that the counter-narcotics effort in Afghanistan was "the most wasteful and ineffective program that [he] had seen in 40 years in and out of government." Five pillars had been put forward to solve the problem: (1) alternative livelihoods; (2) elimination and eradication; (3) interdiction; (4) judicial reform; and (5) public information. However, many problems with these "pillars" occurred anyway, particularly with an overemphasis on eradication, which instead had contributed to increased insecurity, ever-greater corruption, and more poverty in Afghanistan (Choi et al., 2009; Lacouture, 2014). In fact, all forms of eradication (beating with sticks, cutting down, aerial spraying with poisons, etc.) (Fig. 16.2) negatively impact the lives and health of farmers and wage laborers, their families, and their livestock (van Ham and Kamminga, 2006–07). Such strong-arm methods of eradication further alienate local populations and only add to associated violence and resistance to the government.

Peters (2009), after an extensive investigation in the field, argued that a complete and effective anti-drug strategy in Afghanistan would have to combine her revised nine pillars to be really effective. She noted especially that the following would likely produce useful results if fully implemented: (1) diplomatic initiatives for regional peace and free trade; (2) a properly equipped and implemented counterinsurgency strategy; (3) blended intelligence and law-enforcement efforts; (4) military strikes against the drug lords; (5) the creation of a farm-support network; (6) a public-relations campaign; (7) the isolation of drug and terror funds; (8) an alternative-livelihoods program; and (9) eradication only as a last resort (although that process generally came first on most

US and Afghanistan governments' programs). Targeting the criminals and not the farmers was seen as a logical and best way to proceed. The problem has been repeatedly that all of these pillars have been tried over and over in various combinations, with varying degrees of competence and organization in their implementation, but only mixed results have occurred.

Providing a viable, long-term plan for sustainable development has been viewed as essential to overcoming poppy production, with a special emphasis on alternative licit new crops (rose oil, saffron, nuts, raisins, various fruits, cotton, etc.; Van Ham and Kamminga, 2006–07), which would require simultaneous establishment of the resources and facilities required for their cultivation, processing, storage, and international market exportation. In addition, it has been expected that this would include improving irrigation and access to water, educating farmers on how to plant alternative licit crops, developing large-scale processing and warehousing facilities, and establishing networks linking farmers and producers to the buyers throughout the international markets. Further recommended was that these factors should take place parallel to the international and public-works campaigns designed to greatly improve Afghanistan's infrastructure.

These recommendations go hand in glove with other, perhaps stronger recommendations to form resource corridors in Afghanistan that would facilitate wholesale extraction of natural resources, refining, dramatic infrastructure enhancement, and international transportation (Shroder, 2014, 2015). All of this, if done properly with strong environmental and political controls, actually might give the country a chance to prosper beyond warlords and criminal gangs. The possibility exists, however of course, that the same criminal proclivities that have existed in that region for the past many decades will be able to subvert these important notions and plans for a better future and run away with the new opportunities. These possibilities led to Umar Sheraz's ideas of future possibilities in Afghanistan resulting in: (1) continuation of the Great Game; and, (2) other unrealistic expectations; as well as (3) an ongoing uneasy coexistence, along with all of the concomitant other problems attendant to these possible scenarios (Shroder, 2015, Figure 1). The scenario of the desired regional peace and prosperity, with political stability and socio-economic development are illusive goals that retreat in the face of all the other possibilities that are underway at the present time.

The continual deterioration of the Afghanistan state following the 2001 invasion appears to stem from the particularly effective way in which the narco-terrorist regime of the Taliban and their supporters have been able to profit from the chaos that they create (Lacouture, 2014). In order to combat this lethal synergy, strategies of counterinsurgency and counter narcotics must work cooperatively and effectively without obvious weakness. Any "center of gravity" or focus of attention on these issues must rest with the impoverished citizens of Afghanistan to be able to discern the validity or not of the US-vaunted war on drugs and the war on terrorism to make their own lives better. In order to feel economically and personally secure, the people of Afghanistan must view the government of President Ashraf Ghani as legitimate and the most likely political source to help improve their lives. In the absence of such views, the narco-terrorist regime that has become the default distributor of goods and services to many, will be able to continue to exploit the Afghan people, with the result that lawlessness and violence will not retreat (Fig. 16.6A and B).

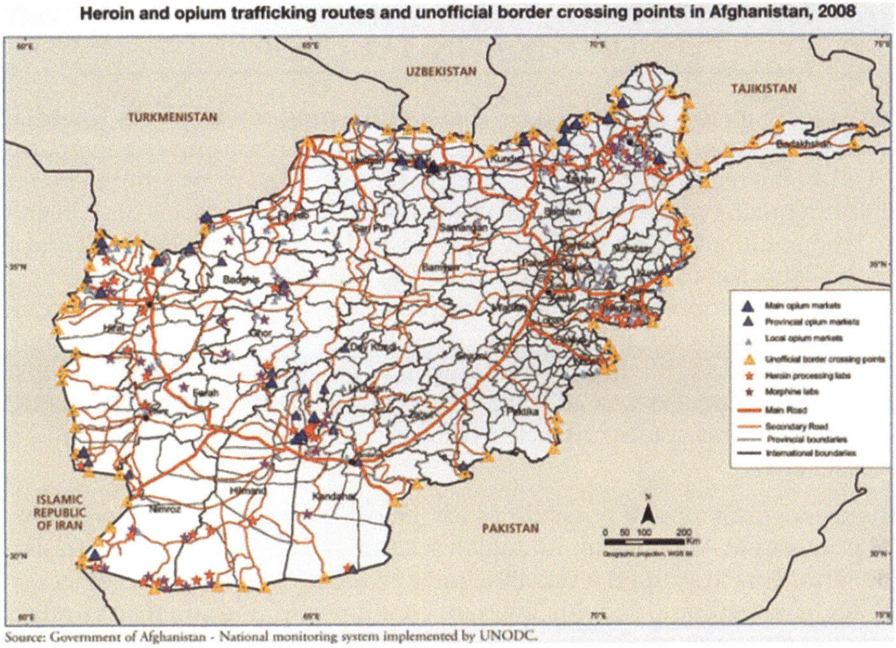


(A)

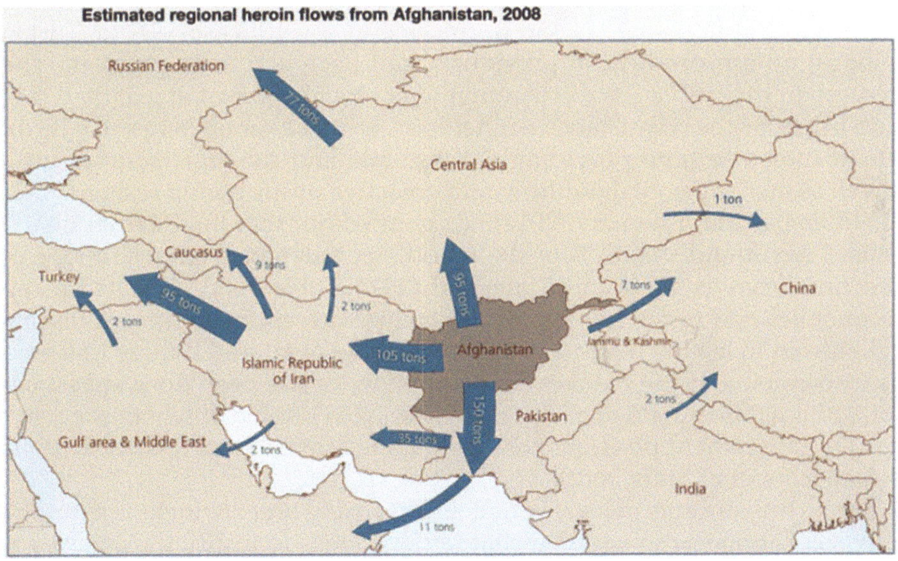


(B)

FIG. 16.6 Drug trafficking routes out of Afghanistan into other countries. (A) Heroin and opium tracking routes and unofficial border crossing points in Afghanistan, 2008. At least 75 unofficial border crossing points are indicated and more than that are known to exist so this is only a rough map of such phenomena. (From Government of Afghanistan—National monitoring system implicated by UNODC; http://www.unodc.org/images/afghanistan//Maps/ACImap_p61.jpg.) (B) Estimated regional heroin flows from Afghanistan, 2008. *From Addiction, Crime and Insurgency, UNODC, Vienna, October 2009; http://www.unodc.org/documents/afghanistan//Counter_Narcotics/Addiction_Crime_and_Insurgency_The_transnational_threat_of_Afghan_opium.pdf (accessed 01.07.15).*

## 16.4 LEGALIZATION ISSUES AND OFFICIAL PURCHASE OF THE ANNUAL CROP

In recent years, in the face of ever-rising opium-production levels (Nordland, 2015; Parenti, 2015), and the promise of new seeds to boost opium output (Khan and O'Donnell, 2015), as well as the recognition of the complete failure of the prior efforts to beat back the cultivation of so much opium in Afghanistan, quite alternate scenarios have been floated to attempt to find solutions. Accordingly, van Ham and Kamminga (2006–07) and the Senlis Council in UK and Kabul (now rebranded as the International Council on Security and Development) have proposed licensing poppy cultivation for the production of essential medicines, which would be an integrated counter-narcotics, development, and counter-insurgency model for Afghanistan (Anonymous, 2005; ICOS, 2007; McNeil, 2007; Aslan, 2008). Such a resolution also was adopted by the World Medical Association (2008), when they noted that an illegal opium crisis clearly exists in Afghanistan, and therefore the World Medical Association:

- supports the investigation of possibilities for the controlled production of opium for medical purposes in Afghanistan through a scientific pilot project in Afghanistan; and
- urges governments to support a scientific pilot project to investigate whether certain areas of Afghanistan could provide the right conditions for the strictly controlled production of morphine and diamorphine (heroin) for medical purposes.

The case has been made repeatedly for reforming the opium industry in Afghanistan with legalization and licensing, as had been done in several cases (India, Turkey) where rampant illegal opium production previously had been a serious problem. The British-based medical journal, *The Lancet* (Anonymous, 2005), editorialized that this idea of legalization might be the only chance Afghanistan would ever have to solve its drug problem, while at the same time providing a pragmatic and dynamic solution to its future peace, as well as meeting a vital public health objective of supplying essential medications in the developing world. Rahmaty (2014) has pointed out that the International Narcotics Control Board has found that 80% of the world's population faces a shortage of opium-based medicine, even as Byrd and Mansfield (2014; Table 16.2) indicated in opposition that the economics and necessary support of competitor countries are not appropriate to allow Afghanistan to legally manufacture inexpensive opiates. All these ideas are in need of serious discussion and not blanket condemnation, as has been done subsequently by a number of authors (Greenfield et al., 2009; Windle, 2011), some of whom seem to be trying to justify prior government positions that were quite oppositional to any ideas of legalization (U.S. Department of State, n.d.).

Opposition to the concepts of legalization and licensing seem to include a number of governmentally self-serving or even somewhat fatuous ideas to justify the contrary positions. These include: (1) illegal production would continue anyway; (2) diversion from the legal to illegal markets would be inevitable; (3) diversion would involve further corruption; (4) it was noted that there would likely be no market; and (5) Afghanistan lacks the institutional capacity to support a legal pharmaceutical industry. Little account has been made of the fact that prior problems with illegal opium production had occurred in Turkey and India and

TABLE 16.2 Reasons That Licensing Afghanistan Opium for Medical Use Would Not Be Likely to Work

1. With large increases in Afghan opium production in 2013 and 2014, risk exists that resulting frustration may give rise to a search for extreme but unproductive solutions
2. No easy solutions are available to solve the illegal narcotics problem in Afghanistan. The proposal the country should shift to licensed production of opium derivatives for pain medication will not work
3. Due to severe problems with governance, rule of law and security, opium licensing in Afghanistan would be subject to extremely high leakages. The comparative advantage that Afghanistan now has in supplying the illicit market means that it would be likely expand cultivation to meet demand in both markets
4. Afghanistan is a labor intensive, relatively high-cost producer of opium, and prices for licensed opium are much lower than on the illegal market, so profits might well be marginal or even negative
5. Existing producers of licensed opiates—Australia, Turkey, India, France, and others—would strongly oppose any move to let Afghanistan become a competitor on the licensed market
6. Even if a more liberalized market for opiates is envisioned, technological advances and modern techniques in other countries mean that Afghanistan could not be a competitive producer

*From Byrd, W.A., Mansfield, D., 2014. Licensing Afghan opium for medicinal work: why it won't work. U.S. Institute of Peace; Peace Brief 170; 4 pages. http://www.usip.org/sites/default/files/PB179-Licensing%20Afghan%20Opium%20for%20Medicinal%20Use.pdf (accessed 27.06.15).*

the problematic situations had been largely overcome by legalization and licensing in those countries. A tendency just to give up on anything that Afghanistan might try is apparent in much of the Western literature, and although that pessimistic viewpoint might be quite justified and practical, given many past failures, it does unfortunately foreclose on an any possible new solutions that the optimistic new government of President Ashraf Ghani might ever want to try.

### 16.4.1 Resistance From Legal Opium-Growing Counties

The fact that opium is grown legally in Turkey, India, Australia (Tasmania), France, and Spain for sale to the world market for pharmaceuticals means that those countries have a vested interest in opposing any such similar opium production under license in Afghanistan because it would undercut their legal sales. Furthermore, in spite of the efforts of some 1200 narcotics control officers, about 30% of India's legal crop seems to be diverted to the illegal market (McNeil, 2007), which has obvious ramifications for the Afghanistan situation where corruption is even more endemic. Also, India is the only country allowed to sell raw opium tar, similar to that gathered by hand in Afghanistan, so there may be handling and processing mechanisms developed in India that would be useful to consider in gaining greater control of at least some of the Afghan opium production.

Inasmuch as 80–90% of the world's illegal opium is already being grown in Afghanistan, it would seem logical that some attempts at legalization and licensing would begin to exert a measure of control on a quite unmanageable illegal situation. Nevertheless, the short-term thinking and self-serving attitudes from countries where legal opium production is already underway seem to have curtailed such alternative thinking, even in the face of such dominant high illegal production in Afghanistan.

## 16.4.2 Resistance From the US Government

Table 16.3 presents the negative position of the US government on any legalizing and licensing opium production in Afghanistan, which at first reading appears to be a rather overbearing, and in places even somewhat amateurish attempt to list all the reasons why this could not possibly work for this most difficult country of Afghanistan. British critics of the American position have noted that opposition to any change in the position of the US government on opium in Afghanistan was partially due to the power of the religious right-wing, as

TABLE 16.3 Supposed Myths and Facts for Fighting the Opium Trade in Afghanistan Proposed by the International Narcotic and Law Enforcement Affairs Division of the U.S. Department of State (n.d.)

*Myth No. 1:* Afghanistan's abundant poppy crop can be refined into biodiesel to serve as the country's principal agricultural product

*Fact:* Refining biodiesel from poppies sanctions an illicit activity, increases demand for poppy cultivation, and benefits narco-traffickers and insurgents.

- Cultivating, processing or possessing opium is a violation of Afghan law
- Dividing poppy cultivation into licit and illicit segments would complicate significantly the task of identifying illicit cultivation
- Alternative uses of opium poppy will raise its price and will encourage more illicit cultivation. The increased production will ultimately reduce the price, but increase the supply, making the country more dependent on poppy cultivation
- Alternative uses of poppy will reduce the narco-traffickers subsidy burden and allow them to save more money while ensuring a long term supply
- Countries that produce opium for licit purposes have strict controls, sophisticated law enforcement and licensing systems (all of which Afghanistan lacks)—and still admit to significant illegal diversion

*Myth No. 2*: Farmers plant poppy because they have been doing so for generations; it is part of their culture and they have no alternatives available

*Fact:* According to the UN Office of Drug Control (UNODC), only 6.4% of the Afghan population currently participates in opium poppy cultivation and opium is grown on less than 2% of Afghanistan's farmable land. Although some small fraction of the Afghan population has traditionally cultivated opium poppy, Afghanistan's opium boom began in the 1980s as drug traffickers capitalized on the chaos created by perpetual conflict

- According to the UNODC, 70% of the cultivation in Helmand, the nation's largest poppy-growing province, began in just the last 4 years, coincident with the Taliban's resurgence in the south

*Myth No. 3:* Legalizing poppy for licit medical use (morphine) would solve the drug problem in Afghanistan

*Fact:* There is no legitimate world demand for legal Afghan opium, as the current supply of licit opiate raw materials already exceeds world demand. Using Afghan poppy for legal opiates is neither a feasible nor desirable solution

- Countries that produce licit opium have strict controls, sophisticated law enforcement and licensing systems (all of which Afghanistan lacks)—and still admit to significant illegal diversion
- Historical experience from Bolivia and other countries show the often disastrous effects of legalizing drug production without the requisite market demand or law enforcement and control mechanisms. Afghanistan has far less regulation capacity than all of these countries, virtually guaranteeing that an unacceptable amount of government-subsidized opium would be diverted to the illegal drug market
- Legalization is ultimately counterproductive and dangerous, and would expand and entrench the drug trade, undermining ongoing efforts to bring security and sustainable economic development

*Myth No. 4:* Buying-out the poppy harvest would solve the drug problem in Afghanistan

TABLE 16.3 Supposed Myths and Facts for Fighting the Opium Trade in Afghanistan Proposed by the International Narcotic and Law Enforcement Affairs Division of the U.S. Department of State (n.d.)—cont'd

*Fact:* Afghanistan's National Drug Control Strategy and the US Strategy for Counternarcotics in Afghanistan reject "silver bullet" approaches like buying-out crops from farmers. The governments of Afghanistan and the United States are opposed to the legalization of opium in Afghanistan and to buying the crop, as are the relevant technical agencies of the United Nations

- There are no shortcuts to fighting opium production in Afghanistan. A buyout scheme would not be a one-time effort, would cost at least hundreds of millions of dollars annually, and would be more expensive than the international community could realistically undertake. The Government of the Islamic Republic of Afghanistan does not have the physical or financial capabilities to buy up the annual harvest, either. Poppy cultivation also violates Afghan law
- Moreover, the real effect of such a buyout could be a significant upsurge in cultivation, absent very robust law enforcement capabilities to ensure that growers sell only to the government. At present, only a fraction of Afghanistan's population is engaged in the drug trade and opium is grown on less than 2% of Afghanistan's farmable land. A buyout policy would likely create a perverse incentive, leading more Afghan farmers to cultivate, and increase the cost of a buyout dramatically. It would also disincline farmers to switch to sustainable licit alternatives

*Myth No. 5:* The United States thinks the best way to solve Afghanistan's drug problem is to simply eradicate the poppy fields

*Fact:* The Government of the Islamic Republic of Afghanistan has developed a comprehensive counternarcotics strategy widely supported by the international community, and the US government has a similarly balanced supporting strategy

- The GIRoA is implementing a comprehensive National Drug Control Strategy that includes institutional capacity-building, law enforcement and interdiction, demand reduction and treatment, criminal justice reform, public education, regional cooperation, agricultural development, and governor-led poppy eradication
- The United States has eliminated support for the Ministry of Interior's Poppy Eradication Force (PEF). The USG continues to support governor-led eradication on a case-by-case basis where licit alternatives exist

*Myth No. 6:* The United States is planning to execute an aerial spray eradication program in Afghanistan despite opposition from the sovereign Afghan government

*Fact:* The United States has never implemented aerial spraying in Afghanistan and does not plan to execute such a policy in the future

*Myth No. 7:* The narcotics trade and the Taliban-led insurgency are unrelated or only tangentially related. The counternarcotics effort wastes precious resources that would be better spent fighting the insurgency. The opium trade and the insurgency are closely related. Poppy cultivation and insurgent violence are correlated geographically, and opium now provides the Taliban with a significant portion of its revenues

- The UNODC and other major international stakeholders all acknowledge that the Taliban-led insurgency and the Afghan drugs trade are increasingly linked
- In 2009, after NATO defense ministers agreed that ISAF could play a role in counternarcotics in Afghanistan, NATO issued an order for ISAF to target drug traffickers and drug-producing labs where there is evidence that they provide "material support" for the Taliban-led insurgency
- The links between the insurgency and the drug trade are especially strong in contested areas of the south, where drug proceeds provide a substantial amount of insurgent revenue. As found elsewhere in the world, the illegal drug trade is strongest in areas where security and rule of law are weak. It sustains and promotes lawless activity, corrodes good governance, and threatens to undermine the legitimate Afghan government and our efforts to assist it in waging a counterinsurgency

*Myth No. 8:* The Government of the Islamic Republic of Afghanistan lacks the will or the capacity to take serious action against the drug trade

*(Continued)*

TABLE 16.3 Supposed Myths and Facts for Fighting the Opium Trade in Afghanistan Proposed by the International Narcotic and Law Enforcement Affairs Division of the U.S. Department of State (n.d.)—cont'd

*Fact:* Where the security situation allows, the Government of Islamic Republic of Afghanistan can and does take action on the drug trade; as the recent successes in dramatically reducing poppy cultivation in Balkh, Badakhshan, Nangarhar, and Helmand demonstrate

- Afghan government leadership in major poppy-producing provinces has shown that Afghanistan can and will reduce poppy cultivation, improve access to alternative livelihood programs, and extend the rule of law, given adequate security
- Even in secure areas, corrupt practices and other improper influences can and have had an effect on the ability to target and prosecute the most significant traffickers. It is important that GIRoA act decisively to reduce such activities that are unhelpful to countering the drug trade and advancing the rule of law

*Myth No. 9:* The US government can combat the opium trade by using mycoherbicides to eradicate opium poppy

*Fact:* The USG does not and has no plans to use mycoherbicides in Afghanistan

- Mycoherbicides are naturally occurring biological organisms that could theoretically attack illicit drug crops. Over the past decade and at congressional direction, INL has investigated the use of mycoherbicides for both coca and opium eradication. The USG continues to research mycoherbicides
- It is unknown whether mycoherbicides would jump species and thus destroy food crops
- The effect on humans is as of yet undetermined. Some reputable researchers have claimed evidence of carcinogenic effects

*Myth No. 10:* There is a single miracle crop that can easily replace poppy

*Fact:* It is incredibly difficult to match the value of opium poppy with a single legal crop, although it is critical that we provide sustainable agricultural alternatives to opium poppy in Afghanistan

- USAID has reviewed several alternative crop suggestions including soy, cotton, sunflower, jathropa, and others. There are isolated cases where, under certain conditions, the price of licit crops can equal that of poppy. However, the farm gate price of opium poppy can be manipulated to regulate supply by the small cartel of brokers, traders and narco-criminal networks that control opium processing and trafficking in Afghanistan
- Crops with the greatest potential are the high value horticultural crops in which Afghanistan has traditionally enjoyed competitive advantage. These crops include, but are not limited to, grapes, almonds, pistachios, pomegranates, apricots, and melons

*Myth No. 11:* Interdiction alone can solve the counternarcotics problem

*Fact:* While interdiction is a critical piece of the US government's counternarcotics strategy in Afghanistan, it will not alone solve the drug problem. The US government strategy provides a balanced, whole-of-government approach to targeting narcotics traffickers and drug lords, while enhancing our focus on agriculture, demand reduction, public information, interdiction, and rule of law

*Note:* This table is presented almost exactly as listed on-line, along with its errors, nonsequiturs, and failures to answer the appropriate questions, but with a few additions to enhance clarity. Answers to a number of the issues raised here are discussed in the text.

well as the law-enforcement groups and paramilitary contractors who had profited so much in the early eradication efforts. Their opposition to change was almost "theological," it was thought (McNeil, 2007). This situation of such strong opposition in the United States would seem to deserve some greater assessment in a more rigorous scientific environment to separate the more valid from the less robust points of view, insofar as that can be accomplished.

To be sure, weak government institutions, endemic corruption, and woefully limited experience with any sort of workable control mechanisms do, indeed, raise doubts about Afghanistan's ability to successfully implement such a strategy. Nevertheless, with 100% of the present huge opium crop going to the illicit market already, even a slight scientific testing of the legalization idea by the national government and its international government support toward a more rational approach would appear worth a try, to face what appears to be an uncontrollable force of agricultural reality in that country.

Some of the problems with the answers provided in Table 16.3 include: not bothering to answer whether or not opium tar could be converted to biofuel (Fig. 16.7); the fact that prices for legal pain killers elsewhere are artificially maintained higher than necessary; and the fact that legal inexpensive opium from Afghanistan for pharmaceuticals has not been scientifically studied or market tested, so no one really knows whether or not it would be feasible.

### 16.4.3 Cogent Resistance From the US Institute of Peace

On the other hand, analysis by Afghanistan economic experts Byrd and Mansfield (2014) offers more recent thoughtful and analytical assessments of the legalization situation that cannot be ignored because of their clear message. For example, they indicate that Afghanistan's clear advantage in the illicit drug trade is based largely on the illegality of the opium and the concomitant high prices that it can now command as a result. The country would not appear to be competitive any longer in a more liberated and regulated global regime for opiates

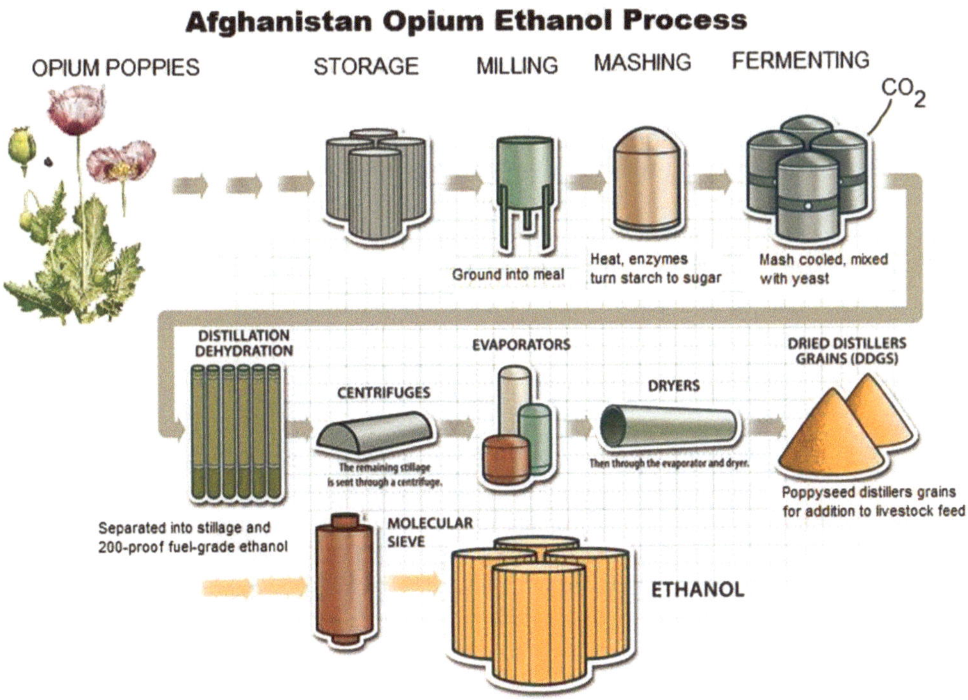


FIG. 16.7 A process for conversion of opium produced in Afghanistan into biodiesel or ethanol has been suggested to convert some of the crop into fuel. *From Mike Licht; https://commons.wikimedia.org/wiki/File:Afghanistan_Opium_Ethanol_Energy_-_04082008.jpg (accessed 01.07.15), with permission.*

(Table 16.2). Such an argument from the point of view of economists can carry a good deal more weight than other arguments, providing other unforeseen factors do not come into play (Rahmaty, 2014), such as more focused involvement of the GoA to get more directly involved in controlling costs and distributions.

### 16.4.4 Recognition of Climate Reality

Parenti (2015) has made a strong case that poppy cultivation is exactly how Afghan farmers are adapting to anthropogenically forced climate change, even if they don't understand the cause for their necessary change in agricultural practice. With the drought-tolerant crop to rely on, they have noted that all poppy really needs is a little water early on, and it is then fairly hardy without regular irrigation. A fair number of these farmers, including women, are getting out of debt with poppy as a source of funds, so it would likely be a hard sell to get them to change their farming behavior much. As Parenti (2015, p. 195) has noted:

> Trying to stamp [poppy] out or wish it away has failed, to continue on with such a fantasy will only make life worse for Afghan farmers and spread instability.

The conclusion seems inescapable, that cultivation of opium poppies in Afghanistan is not going away any time soon, that everything that has been tried in the past has not worked and is unlikely to work in the future, and that well-meaning governments should think of new ways of doing business entirely on these issues.

Perhaps the only solution to this seemingly intractable drug-production problem in Afghanistan will be the likely prospect, predicted recently in *The Economist* (2015), that synthetic opiates will soon be driving the traditional opiate smugglers out of business by home-brew opium clubs based in garages in the West where the huge and most lucrative illegal markets exist. Such a prospect is not at all heartening to drug officials in the West, of course, but it would offer some relief to those who trying to curtail the trade in Afghanistan. What the poor farmers in Afghanistan would do to support themselves otherwise in a future of increasingly water-starved people remains to be seen.

## 16.5 CONCLUSION

A number of quite different solutions have been proposed to overcome the high levels of opiate drug smuggling out of Afghanistan, which has more drug production than the rest of the world combined (Fig. 16.8). All of these solutions are so far quite to no avail, due to greater or lesser acceptabilities and possibilities (spraying, hand eradication, border interception, precursor chemicals, conversion to biodiesel, licit pain-killing drugs), and so forth. If the present-day economics between legal and illegal drug movements and sales prices indeed do show that Afghanistan will not be able to compete in a global licit marketplace, then the rotteness of illicit opiate production in Afghanistan will have to await the attention of later actions. A more attentive religious orthodoxy against such strong drug use in Islam would be a possible solution and a great hope that could be investigated with appropriate and sympathetic religious authorities. A less corrupt and more honest, technocratic government would

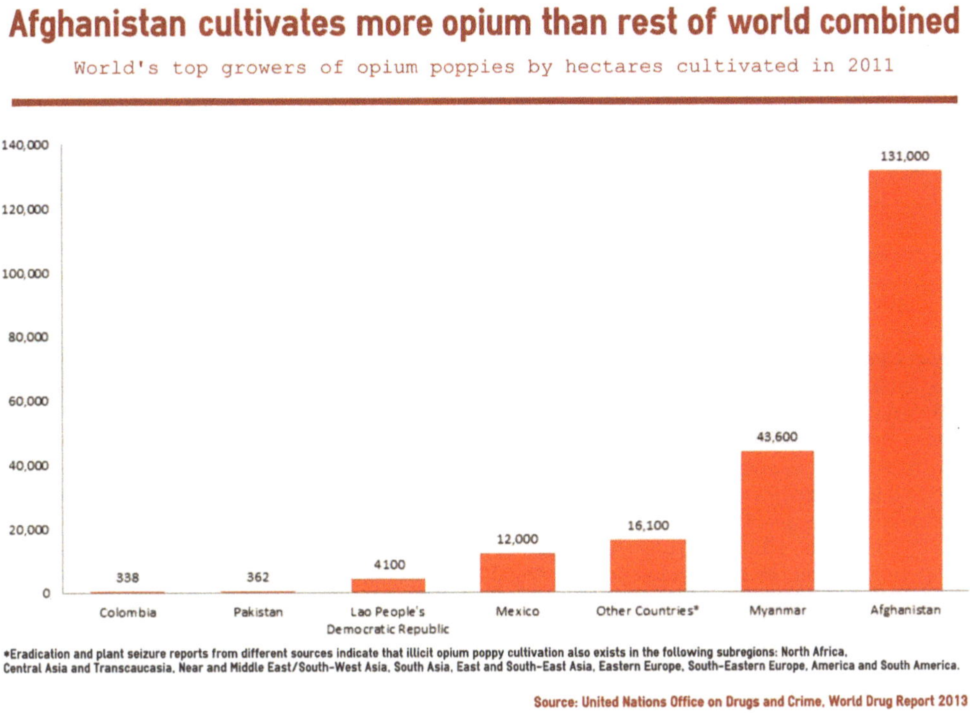


FIG. 16.8 Afghanistan cultivates more opium than the rest of the world combined, as shown in this graph by hectares in 2011, and these figures are quite out of date, with as much as 209,000 ha in 2013, 224,000 ha in 2014, and perhaps as much as 250,000 ha estimated in 2015. *From UNODC, World Drug Report, 2013; http://www.cnsnews.com/news/article/ali-meyer/after-12-yrs-us-occupation-afghanistan-sets-record-growing-opium (accessed 02.07.15).*

be necessary under President Ashraf Ghani that might be able to overhaul its internal administrative structure and associate corruption enough to deal appropriately between farmers, warlord, and criminal gangs. Reaching accommodations would be advisable between Taliban insurgents, and the "foreign" spoiler forces that have been encouraged directly and indirectly by surrounding countries that are interested in resource acquisition of one kind or another (Shroder, 2014). The future looks complex and messy, with many different competing agencies, all seeking some personal advantage as the climate changes and water availability are ever more doubtful.

## References

Anonymous, 2005. Pragmatism over opium production in Afghanistan. Lancet 366 (9491), 1052. 24 September. http://www.nytimes.com/2004/04/10/world/afghan-route-to-prosperity-grow-poppies.html (accessed 27.06.15).

Aslan, R., 2008. How opium can save Afghanistan. The Daily Beast. http://www.thedailybeast.com/articles/2008/12/19/how-opium-can-save-afghanistan.html (accessed 27.06.08).

Associated Press, 2013. Afghan opium production hits record high, U.N. says. CBS News. http://www.cbsnews.com/news/afghan-opium-production-hits-record-high-un-says/ (accessed 03.04.16.).

Blanchard, C.M., 2009. Afghanistan: narcotics and U.S. Policy. Congressional research service 7-5700; www.crs.gov, https://www.fas.org/sgp/crs/row/RL32686.pdf (accessed 26.06.15).

Bove, V., 2011. The political economy of opium in Afghanistan. Institute for Democracy & Conflict Resolution (IDCR); Briefing paper (IDCR-BP-10/11); 9 p. http://www.idcr.org.uk/wp-content/uploads/2010/09/10_11.pdf (accessed 26.06.15).

Byrd, W.A., Mansfield, D., 2014. Licensing Afghan opium for medicinal work: why it won't work. U.S. Institute of Peace; Peace Brief 170; 4 pages. http://www.usip.org/sites/default/files/PB179-Licensing%20Afghan%20Opium%20for%20Medicinal%20Use.pdf (accessed 27.06.15).

Choi, M., Elko, B., Krentel, J., Lander, N., Valenti, A., 2009. Combating the opium trade in Afghanistan. New Voices in Public Policy, 4; 34 pages; School of Public Policy; George Mason University.

Cogan, C.G., 1993. Partners in time: the CIA and Afghanistan since 1979. World Policy J. 10 (2), 73–82.

Economist, 2015. Afghan highs. The Economist, July 4, 33.

Farrell, G., Thorne, J., 2005. Where have all the flowers gone? evaluation of the Taliban crackdown against opium poppy cultivation in Afghanistan. Int. J. Drug Policy 16, 81–91.

Gates, S., Lektzia, D., 2004. Drugs, Governance and Civil Conflict. International Peace Research Institute, Oslo. http://www.isn.ethz.ch/Digital-Library/Publications/Detail/?lang=en&id=28525 (accessed 27.06.15).

Goodhand, J., 2005. Frontiers and wars: the opium economy in Afghanistan. J. Agrar. Chang. 59 (2), 191–216.

Greenfield, V.A., Paoli, L., Reuter, P.H., 2009. Is medicinal opium production Afghanistan's answer? lessons from India and the world market. J. Drug Policy Anal. 2 (1). Article 3. http://faculty.publicpolicy.umd.edu/sites/default/files/reuter/files/afghanistan_legal_opium.pdf (accessed 27.06.15).

ICOS (International Council on Security and Development), 2007. Poppy for medicine. http://www.icosgroup.net/static/reports/poppy_medicine_technical_dossier.pdf (accessed 27.06.15).

Khan, M., O'Donnell, L., 2015. Afghanistan's poppy farmers say new seeds will boost output. The World Post, Huffington Post and Berggruen Institution. http://www.huffingtonpost.com/2015/05/05/afghanistan-opium_n_7216756.html (accessed 26.06.15).

Kovach, T., 2011. Helmand Province, Afghanistan, Conflict and Climate Change. Inventory of Conflict and Environment (ICE) Case Studies. Number 261. American University, Washington, DC. 13 pages. http://www1.american.edu/ted/ICE/helmand.html (accessed 30.06.15).

Lacouture, M., 2014. Narco-terrorism in Afghanistan: counternarcotics and counterinsurgency. International Affairs review; The Elliott School of International Affairs at George Washington University. Fall 23 (1), 1–9. http://www.iar-gwu.org/node/39 (accessed 25.06.15).

McCoy, A., 1997. Drug fallout. Progressive Magazine, August; 3 p. http://www.thirdworldtraveler.com/CIA/CIAdrug_fallout.html (accessed 26.06.15).

McNeil, D.G., Jr., 2007. Could Afghan poppies be painkillers for the poor? The New York Times, 14 October; 3 pages.

Nordland, R., 2015. Afghan opium cultivation rises to record levels. New York Times, 12 November. http://www.nytimes.com/2014/11/13/world/asia/afghan-elections-cited-as-factor-in-record-levels-of-opium-production.html?_r=0 (accessed 26.06.15).

Pain, A., 2009. Livelihood Trajectories in Afghanistan. Case Study. AREU, Kabul. Draft Paper.

Pain, A., 2010. Livelihood Trajectories in Afghanistan: Evidence from Badakhshan. AREU, Kabul. http://www.areu.org.af/Uploads/EditionPdfs/1003E-Afghanistan%20Livelihood%20Trajectories%20-%20Evidence%20from%20Badakhshan%20CS%202010.pdf (accessed 25.06.15).

Pain, A., 2012. Afghanistan's Opium Poppy Economy. Middle East Institute, Washington, DC. http://www.mei.edu/content/afghanistans-opium-poppy-economy (accessed 25.06.15).

Parenti, C., 2015. Flower of war: an environmental history of opium poppy in Afghanistan. Johns Hopkins University Review of International Affairs. SAIS Rev. 35 (1), 183–200.

Peters, G., 2009. Seeds of Terror: How Heroin is Bankrolling the Taliban and Al Qaeda. Thomas Dunne Books, St. Martin's Press, New York.

Prados, J., 2002. Notes on the CIS's secret war in Afghanistan. J. Am. Hist. 89 (2), 466–471.

Rahmaty, S., 2014. A solution for Afghanistan's opium crisis? The Diplomat, 2 December; 3 pages. http://thediplomat.com/2014/12/a-solution-for-afghanistans-opium-crisis/ (accessed 26.06.15).

Rehabs.com, 2014. Afghanistan, the war, and a flood of heroin. http://www.rehabs.com/explore/afghan-opium/ (accessed 25.06.15).

Shroder Jr., J.F., 2012. Afghanistan: rich resource base and existing environmental despoliation. Environ. Earth Sci. 67 (7), 1971–1986. http://dx.doi.org/10.1007/s12665-012-1638-7.

Shroder, J., 2014. Natural Resources in Afghanistan: Geographic and Geologic Perspectives on Centuries of Conflict. Elsevier, San Diego, CA, USA.

Shroder, J.F., 2015. Progress with Afghanistan extractive industries: will the country know resource success or failure evermore? Extract. Indust. Soc. 2, 264–275.

Starkey, J., 2009. Afghanistan's forgotten heroin addicts. Development Asia, Analyzing Issues that Matter, Asian Development Bank, No. 5, October–December; 4 pages.

The Economist, 2015. Only connect: home-brewed heroin may soon be in the works. May 23, 67.

U.S. Department of State, International Narcotics and Law Enforcement Affairs, n.d. but post-2001. Myths & facts about fighting the opium trade in Afghanistan. http://www.state.gov/documents/organization/142643.pdf (accessed 24.06.15).

UNODC, 2003. The Opium Economy of Afghanistan: An International Problem. United Nations Office of Drugs and Crime, Vienna. 222 pages.

UNODC, 2014. Afghanistan Opium Survey 2014. United Nations Office on Drugs and Crime, Islamic Republic of Afghanistan, Ministry of Counter Narcotics. November, 67 pages. https://www.unodc.org/documents/crop-monitoring/Afghanistan/Afghan-opium-survey-2014.pdf (accessed 15.07.15).

Van Ham, P., Kamminga, J., 2006–07. Poppies for peace: reforming Afghanistan's opium industry. Wash. Q. 30 (1), 69–81.

Vicens, A.J., 2014. We spent $7.6 billion to crush the Afghan opium trade—and its doing better than ever. Mother Jones. http://www.motherjones.com/politics/2014/10/afghanistan-opium-poppy-heroin-record-levels (accessed 03.04.16.).

Waldman, A., 2004. Afghan route to prosperity: grow poppies. The New York Times, 10 April. http://www.nytimes.com/2004/04/10/world/afghan-route-to-prosperity-grow-poppies.html (accessed 27.06.15).

Windle, J., 2011. Poppies for medicine in Afghanistan: historical lessons from India and Turkey. J. Asian African Stud. 46 (6), 663–677. http://roar.uel.ac.uk/3486/1/Poppies%20for%20Medicine%20in%20Afghanistan%20-%20pre-print%20copy.pdf (accessed 27.06.15).

World Medical Association, 2008. WMA resolution on poppies for medicine project for Afghanistan. Adopted by the 59th WMA General Assembly, Seoul, Korea, October 2008. http://www.wma.net/en/30publications/10policies/p30/ (accessed 27.06.15).

CHAPTER

# 17

# Hydropolitics of Afghanistan and Its Future

*S.J. Ahmadzai*

**Center for Afghanistan Studies, University of Nebraska at Omaha, Omaha, NE, United States**

OUTLINE



## Abstract

Afghanistan is the source of many rivers flowing into its neighboring countries. The last four decades of instability since 1970 have pushed Afghanistan into isolation and left it out of the benefits of the water-management structures established by its neighbors. Post-9/11 Afghanistan has been re-establishing itself, not only at regional level but also in the international arena. This would also increase Afghanistan's involvement in efforts for transboundary water management that will impact the mechanisms previously established by others in the absence of participation by Afghanistan. Moreover, increasing demand for water, and development plans of Afghanistan in its basins will have effects on the downstream riparian states that could incite negative reaction from them. This chapter explores and analyzes the history of hydropolitics inside Afghanistan and in regards to its relationship with neighbors. The chapter also discusses the current status of any available water-management mechanisms that will be impacted by Afghanistan's comeback to the transboundary water arena and the uncharted field of hydropolitics in the region.

**Keywords:** Water resources, Diplomatic isolation, Transboundary water management, Harmon Doctrine, President Ghani, President Karzai, Dam building, New water policy, River basins, Water demand



http://dx.doi.org/10.1016/B978-0-12-801886-6.00017-3

## 17.1 INTRODUCTION

Understanding the context of a political debate regarding the use of water resources requires the study of the variables that contribute to making it contested, controversial, or disputed in geographic surroundings of a country or area. These variables, for example, could be the result of: (1) economic activities of nations benefiting from a river flowing in and out; (2) being a riparian or source country; (3) a military and economic power; (4) agricultural dependencies and food security; (5) advanced and expensive irrigation infrastructure that has already been set up in riparian states; (6) impractical or outdated water management mechanisms and accords, such as the 1929 Nile Treaty between Britain and its Egypt forbidding upstream nations from constructing anything on the Nile river that could reduce the water flow (Kendie, 1999); and most importantly; (7) climate change.

Countries around Afghanistan (with the exception of China) depend upon the water resources originating from Afghanistan, albeit this dependency varies in different countries. The population boom in these countries has resulted in increasing demand for water and water-related, socio-economic activities. Moreover, bad water-management practices and underdeveloped irrigation systems have even raised the problem to unprecedented levels. The demand for solutions makes policy makers search for ways to address the burgeoning needs of their populace. These efforts can be projected either by direct or indirect use of force, or joint efforts with their neighbors to manage and/or adapt to the changes affecting them. For example, in 1894 when Mexico contested the US use of waters of Rio Grande that had resulted in decline of water in the river after flowing into Mexico, both countries agreed on setting up a joint commission to address their problem (Akweenda, 2002). The approach was selected by the US government leadership despite the explanation (*Harmon Doctrine*) given to the then Secretary of State of the United States by Attorney General Harmon, who claimed that it was the sovereign right of the United States to use the water of Rio Grande, with no foreign obligation. As a result of such cooperation and understanding, the needs of the people on both sides of the border, both countries agreed on a mutually accepted, water-management mechanism and signed a treaty in 1906 to manage fairly the waters of Rio Grande and avoid any potential conflict that could arise (Akweenda, 2002).

In international arenas, states attempt to exert influence over efforts that are supposed to result in policies, mechanisms, and procedures that will impact their people and their interests or change behaviors. They use resources available to them, be they natural or otherwise, to influence decisions that affect them (Guess and Farnham, 2011). Theoretically, these efforts are, in fact, doing politics and when such efforts involve valued natural resource such as water to influence decisions or mold them in one's favor, and for the sake of presenting a simple argument, they can be called hydropolitics or as Anthony Turton states, "hydropolitics is seen as the authoritative allocation of values in society with respect to water" (Turton, 2002). At local levels within a country, this theory can explain the actions of local power holders in an area who try to use water to influence others. For example, in Samangan province of Afghanistan, Sexton (2012) noted that a warlord was using water from the Khulm and Balkhab river basin to gain more political clout and dominate over the residents of downstream areas.

President Mohammad Ashraf Ghani, on his trip to Iran in April 2015, implied that the ill-treatment of Afghan refugees in Iran could have adverse effects on the issues relating to the waters of Helmand River flowing into Iran. President Ghani's such implicit warning can be understood as a tool to persuade or force Iran to change its behavior and unfriendly policies towards Afghan refugees residing in Iran (1TVNews, 2015). Yet again, it is water used as a political tool to influence policies of another country that are impacting the citizens of Afghanistan.

Afghanistan's geographic location assigns it a unique position by putting it right between South and Central Asia, making it not only a crossroad for cultures but also for economic activities driven by a variety of resources. The neighboring six nations, as well as having four transboundary river basins that connect Afghanistan with its neighbors; many correlative factors occur that can make Afghanistan a place of contest among these nations and at the same time a place for cooperation.

As examples: (1) the Amu Darya River basin, which includes the Northern basin, connects Afghanistan with its three Central Asian neighbors of Tajikistan, Uzbekistan, and Turkmenistan; (2) the Hari Rud River basins connects Afghanistan with Iran and Turkmenistan in the northwest; (3) the Helmand River basins connects it with Iran in the southwest; and (4) the Kabul River basin connects Afghanistan with Pakistan in the east. Afghanistan's fifth river basin or a sub-basin; (5) the northern basin has rivers such as the Shirin Tagab, Balkh Aab, and Samangan that do not flow into Amu Darya but are counted as part of the greater Amu Darya basin rivers (Fig. 17.1).

In addition to many other factors, parts of the dynamics that define the relationship of Afghanistan's neighbors with Afghanistan have been transboundary waters. This statement can be applied to similar cases or countries in other parts of the world. For example, the shared-water resources in the Indus valley have been one of the dominant topics of discussion defining the relationship between India and Pakistan and their efforts to avoid potential conflicts. In fact, they have been relatively successful in managing their relations concerning the Indus River basin through the Indus Water Treaty.

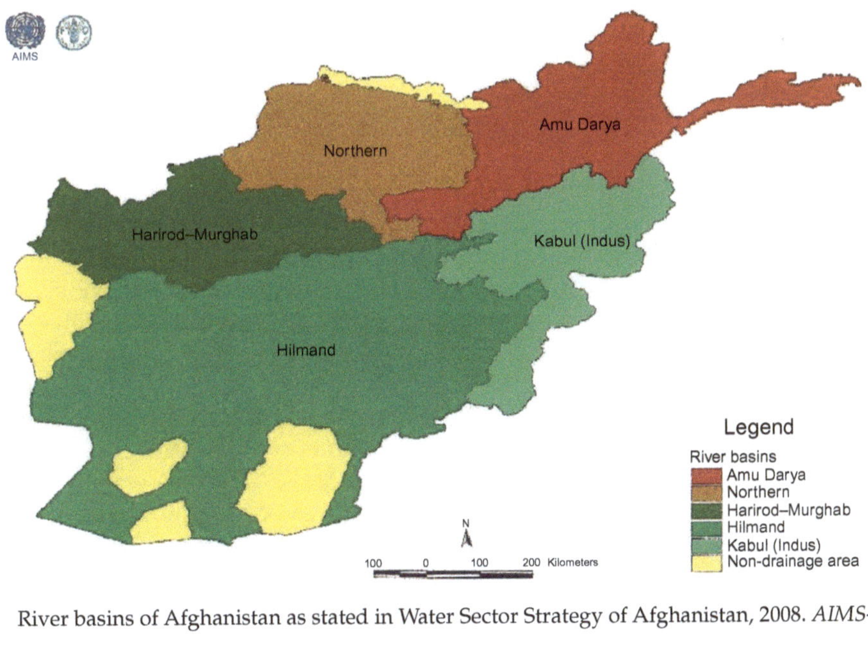


FIG. 17.1 River basins of Afghanistan as stated in Water Sector Strategy of Afghanistan, 2008. *AIMS-FAO.*

For Afghanistan, getting into political debate and negotiations with its neighbors regarding transboundary water resources is an uncharted field both internally and externally, because Afghanistan has historically been long ignored by the powers surrounding it and so far it has only an old water agreement on the Helmand River with but one of its six neighbors, Iran. Even that Helmand River treaty with Iran proved to be very controversial inside Afghanistan. After the communist coup of 1978, according to many informal sources, Musa Shafiq, the last Prime Minister of Afghanistan under King Zahir Shah was executed for not only having been a loyal supporter of the monarchy and the royal family, but also for being the architect of the last reforms in the government and the signatory of the Afghan–Iran water treaty on the Helmand River. This was done because to the communists, signing that treaty and agreeing to the unfavorable terms was committing treason against the country and the people. Prime Minister Shafiq's execution changed the narrative of transboundary water discussions in Afghanistan forever, and made any future efforts by any leader in the country very controversial because the memories of the Prime Minister's execution haunted every political leader who thought of negotiating Afghan waters with its neighbors.

In the summer of 2014, the author of this chapter visited Kabul, Afghanistan, to discuss a possible transboundary-water conference that would include water experts from the region to academically discuss regional water issues and present their opinions on the ways to tackle future transboundary waters-related problems. The idea was met with strict resistance from the officials at the Ministry of Water and Irrigation and Ministry of Foreign Affairs. In fact, the Deputy of Minister of Energy and Water, Mr. Shojaudin Ziaie, revealed to the author that any discussion about Afghanistan's transboundary waters could not be conducted without the approval of the National Security Council chaired by then President Hamid Karzai. The reason for such a strong restraining policy or muzzling was the lack of trained capacity in the Afghan government to successfully conduct transboundary water negotiations or discussions with any of the neighbors. Regional political-power imbalance was another reason discouraging Afghanistan from getting into any sort of dialogue with its neighbors regarding transboundary waters. The Afghan government's approach to transboundary water discussions reflects the lack of political will and power that has been caused by lack of capacity and the deficit of trust between Afghanistan and its neighbors. This factor directly relates water issues to Afghan national security, which can be understood since water scarcity and mismanagement can cause national security problems in the long run. Mr. Ziaie's mentioning of a famous phrase that "might makes right" explains the situation Afghan government is concerned about because Afghanistan at the present time lacks the might or the intellectual or background education to have fair and favorable negotiations with its neighbors.

According to a report published by the Office of the U.S. Special Inspector General for Afghanistan Reconstruction, "Afghanistan imports approximately 73% of its total electricity" from its neighbors and this is expected to rise in the future (SIGAR, 2015). Being altogether too much dependent on neighbors in such a politically and environmentally volatile and unstable region is not to the advantage of Afghanistan, and to mitigate its dependence on neighbors, Afghanistan plans to build hydropower dams and also to divert water for agricultural purposes. This will cause great anxiety among the neighbors of Afghanistan because they depend on unrestrained waters from Afghanistan. For example, construction of the Salma Dam on the Hari Rud River in Herat province had already caused Iran and Turkmenistan anxiety about the future allocation of water and its impact on their activities downstream. According to an ex-Afghan diplomat, whom the author cannot name here, in 2011 the Presidents of

Iran and Turkmenistan sent a joint letter to President Hamid Karzai of Afghanistan stating their displeasure and concerns about the Salma Dam. The letter was presented to President Karzai in a meeting by the Iranian Ambassador to Afghanistan, Mr. Fida Hussain Maliki (Anonymous, 2015). Moreover, if Afghanistan constructs a planned dam on the Kokcha River, a major tributary of Amu Darya in the north of the country, as had been planned long since by Soviet advisors, it will invite a negative reaction from downstream countries such as Uzbekistan and Turkmenistan. Uzbekistan had already expressed its displeasure with the USSR on another planned dam in Tajikistan, the Rogun Dam of 1970s. After the collapse of the USSR, this project faced strong opposition from Uzbekistan since it would impact water flow into Amu Darya and subsequently into the agricultural areas of Uzbekistan. As a result, the relationship between Tajikistan and Uzbekistan has been deteriorating and so far, major international organizations, such as the World Bank or Asian Development Bank, have not shown interest in funding this project (OOSKAnews, 2013).

During his 2015 visit to the United States, President Mohammad Ashraf Ghani, in his talk at the Center for Foreign Relations in New York City on March 26, 2015, outlined his national programs in three different sectors indicating his priorities. He marked harnessing waters of Afghanistan as a second opportunity or program from which Afghanistan can benefit (The Arg, 2015). As Afghanistan greatly needs to use its natural resources for the betterment of its people, its developmental programs and priorities will no doubt create further anxieties among the neighbors. These anxieties can lead to different interpretations and speculations about the future intents of the leadership of Afghanistan.

The last 40 years of turmoil and political instability have cost Afghanistan a tremendous loss of opportunities and capacity, whereas, in the meantime, the neighbors of Afghanistan have greatly benefited and had opportunities to build plentiful hydropower and irrigation infrastructures of their own. Moreover, they had not consulted with Afghanistan regarding their projects on transboundary waters, or perhaps there was no urge to contact Afghanistan because the various Afghan governments had been busy with internal affairs or conflicts and had no interest or chance in taking on external affairs such as transboundary waters. As a result, Afghanistan has been left out of commissions, consortiums, and water management bodies such as the International Fund for Saving the Aral Sea (IFAS) established in 1993. The IFAS is relevant to Afghanistan, because the waters of the Amu Darya end up in the Aral Sea and any development or administrative activity regarding the rivers contributing to the Aral Sea is directly of concern to Afghanistan.

The following information about the importance of Afghan basins for Afghanistan and its neighbors will give readers better ideas about why and how any developments upstream or downstream can impact the relationship between Afghanistan and other riparian states benefiting from the Afghan rivers. Such developments on Afghan rivers entail risks, as well as offering opportunities for transboundary cooperation among riparian states surrounding Afghanistan.

## 17.2 CENTRAL ASIAN STATES AND THE AMU DARYA BASIN

Two important rivers provide the largest sources of water in Central Asia, the Amu Darya and the Syr Darya. The Syr Darya originates in Kirghizstan, and is not a topic of discussion in our chapter but will be referred to on occasion. The Amu Darya starts as the Panj tributary in the northeastern mountains of Afghanistan in the Wakhan glaciers at an altitude of about 4900 m and flows roughly northwest for 2540 km, of which 1250 km are within the jurisdiction of Afghanistan or make the international boundary between Afghanistan and its northern neighbors (Ahmad and Wasiq, 2004). Out of the mountains where it originates and below its confluence with the Vakhsh River of Tajikistan, the Amu Darya flows westwards and constitutes the international boundary between Afghanistan and its three Central Asian neighbors, Tajikistan, Uzbekistan, and Turkmenistan. The major northern Amu Darya tributaries are the Vakhsh River, Kafiringan River, Surkhan Darya River, and Sherabaddarya Darya. To the south, on the Afghanistan side, more than one river contributes to the Amu Darya basin, such as Murghab, and Balkhab, but only the Kokcha, and Kunduz are the two rivers that flow into Amu Darya. As a result, most of the rivers flowing into the Amu Darya originate from Afghanistan and Tajikistan. The remaining rivers that contribute to the Amu Darya basin (but not to the river) dry up in Afghanistan, Uzbekistan, and Turkmenistan, and are considered blind rivers. Afghanistan does not have any treaty with any of the Amu Darya riparian states than regulate the share and usage of water in the Amu Darya basin. Most of the negotiations in the past have been about the Amu Darya as an international boundary between Afghanistan and the Soviet Union, not water usage (King and Sturtewagen, 2010) (Fig. 17.2).

The average annual flow of water in the Amu Darya is about 75 billion cubic meters (bcm), but this discharge can get as high as 108 bcm in the summer and as low as 47 bcm in the winter (Ahmad and Wasiq, 2004). In another paper, Shobair (2010) estimated that the average annual

FIG. 17.2 Amu Darya Basin. *UNEP.*

flow of water in Amu Darya is 80 bcm (8000 mcm) and Afghanistan uses only 7% of this, whereas it contributes 27.5% to the Amu Darya every year. Table 17.1 produced by Shobair (2010) can present a better view of annual water contribution and usage by each riparian state of the Amu Darya Basin.

Looking at the numbers listed in Table 17.1, a clear and wide disparity exists in the amount of water each state contributes to the Amu Darya and the amount of water they use. This shows how much each state is dependent up the Amu Darya. Moreover, the two major users of Amu Darya water are Uzbekistan with 14 million people benefiting from the Amu Darya, and 4 million people in Turkmenistan benefiting from the river as well (Wegerich, 2011). The total population of both countries, as of 2014 was 28 million for Uzbekistan and 5 million in Turkmenistan (CIA World Factbook, 2015). Both countries mainly use this water for agricultural purposes that make up a large share of their economies.

Afghanistan has plans to build at least three major dams that can impact the flow of water in the Amu Darya River. It plans to build irrigation and hydropower projects on the Kokcha River in Takhar province, the Kelagi Irrigation and Hydropower project on the Kunduz River in Baghlan province, and the Amu Irrigation and Hydropower project on the Amu Darya in Kunduz province. All projects combined will be using about 6000 million m$^3$ water by year 2020, showing a 1.4% increase of water usage in Afghanistan (Shobair, 2010), which if compared to the current annual flow of water from Afghanistan into Amu Darya is minimal and has little impact downstream. However, with the unpredictable pattern of weather in the region and the predicted changes in climate, such a small percentage of increased water usage in these Afghan projects can create significant anxiety in Uzbekistan and Turkmenistan, who among all Amu Darya riparian states depend the most on the water in the Amu Darya River. This will further make them apprehensive of any future plans Afghanistan might have in the long term.

However, if Tajikistan plans any other major irrigation or hydropower projects on Amu Darya tributaries, it will be received with anger by Uzbekistan and Turkmenistan, as has been seen with the case of Rogun power project in Tajikistan that has been a sore point in the

TABLE 17.1 The Water-Flow Shares of the Amu Darya River Basin by Country

| Riparian Country | Generated Average Annual Flow (million m$^3$) | Share of Total Average Annual Flow (%) | Average Annual Water Use (million m$^3$) | Share of Total Average Annual Water Use (%) |
|---|---|---|---|---|
| Tajikistan | 50,000 | 62.5 | 7500 | 11 |
| Afghanistan | 22,000 | 27.5 | 5000 | 7 |
| Uzbekistan | 5000 | 6.3 | 33,000 | 47 |
| Kyrgyzstan | 1500 | 1.9 | 1500 | 2 |
| Turkmenistan | 1500 | 1.9 | 23,000 | 33 |
| Total | 80,000 | 100 | 70,000[a] | 100 |

*Note:* Mr. Sayed Sharif Shobair, served as Chief Engineer and Coordinator of the Emergency Irrigation Rehabilitation Project of the Minister of Energy and Water of the Islamic Republic of Afghanistan.

[a]*The missing 10,000 million m$^3$ is probably "lost" along Amu Darya and in its delta at the Aral Depression (Shobair, 2010).*

*From Shobair, S.S., 2010. The Afghan Part of Amu Darya Basin: Impact of Irrigation in Northern Afghanistan on Water Use in the Amu Darya Basin. United Nations Economic Commission for Europe (UNECE). http://www.unece.org/fileadmin/DAM/SPECA/documents/ecf/2010/FAO_report_e.pdf (retrieved 20.05.15).*

relationship between Tajikistan and Uzbekistan. Moreover, if Afghanistan asserts its legal claim on the share of water it is producing into the Amu Darya and asks for the payment of the extra water that is being used by Uzbekistan and Turkmenistan, it will stir a not-so-positive interaction with them that can have obvious negative future implications.

Currently, Afghanistan has no formal functioning agreement with Central Asian states on the water usage and/or allocation of the Amu Darya. A 1977 interaction in Tashkent between Afghanistan and the USSR bore no fruit when Afghanistan's request of 9 $km^3$ was countered with an offer of only 6 $km^3$ and as result of that no agreement was reached (Rycroft and Wegerich, 2009). In fact, Afghanistan has been left out entirely of the past water management structures, mechanisms, and dialogues concerning the Amu Darya basin. That could be due to many reasons, such as instability in Afghanistan and Afghanistan's sometimes limited political weight in the region, and/or the policy and priority differences among Central Asian states regarding water management and allocation. But because Afghanistan is the second largest contributor of water into the Amu Darya after Tajikistan, and a post-9/11 Afghanistan is quite different from the Afghanistan of the pre-1980 era, it would be difficult for the Central Asian states to ignore Afghanistan in the future, which could mean more enhanced and cooperative relations with Afghanistan concerning the Amu Darya waters.

## 17.3 TURKMENISTAN, IRAN AND HARI RUD–MURGHAB AND HELMAND RIVER BASINS

Afghanistan's northwestern and western neighbors are connected with Afghanistan by two river basins, the Hari Rud–Murghab and the Helmand. In this section we will first talk about the Hari Rud river basin that is shared with Iran and Turkmenistan, and then we will focus on Helmand river basin that is shared only with Iran.

## 17.4 HARI RUD–MURGHAB RIVER BASIN

Named after its major rivers, this river basin is made up of the two major rivers of the Hari Rud and Murghab beside other seasonal rivers that originate in the Baba Mountains (Kohi Baba) of central Afghanistan, meandering westwards to Iran and Turkmenistan. At a point where the river flows north toward Turkmenistan, the Hari Rud River forms the northwestern section of the border between Afghanistan and Iran before entering Turkmenistan, where it is renamed the Tejen. The Murghab River flows north in Badghis province of Afghanistan. Flowing northwards to Turkmenistan, at its peak times, the Murghab river flows into the Kara Kum (Qara Qum) canal that also takes water from the Amu Darya and transports it to the city of Marv (Mary) of Turkmenistan and the areas beyond and around it for drinking and irrigation purposes. The Hari Rud River is the only source of water for Dosti Dam built jointly by Turkmenistan and Iran at the border area of Sarakhs. Construction of the dam was completed in 2004 and its operation started the following year (TAICE, 2015). The strategic importance of the dam can be understood from its being the major source of drinking water for the twin cities of Sarakhs on both sides of the border and the larger city of Mashhad in Iran. In addition, the Hari Rud River is also the source of water for agriculture beyond Dosti Dam in Turkmenistan.

Inside Afghanistan, in total 300,900 ha of land is cultivated in the Hari Rud–Murghab basin before its major rivers cross over into Iran and Turkmenistan, making 9% of the total irrigated land in Afghanistan (Rout, 2008). The number of people benefiting from the waters of the Hari Rud–Murghab river basin in Afghanistan is some 2,145,000 people (King and Sturtewagen, 2010) (Fig. 17.3).

Afghanistan has no water management or allocation agreements, either with Iran or Turkmenistan to regulate Hari Rud River. The Murghab River, on the other hand, is not considered to be the source of any transboundary conflict between Afghanistan and Turkmenistan because regional topography and water utilization capacity on the Afghanistan side limits the use of the water, thus causing no concerns for Turkmenistan (Hansaz, 2012). In addition, Turkmenistan's needs in that geographical area are more or less met by the Amu Darya water diverted through Qara Qum canal. However, the Hari Rud River can be, or is the cause of concern for Iran, because Iran happens to be the major benefactor of the waters of the Hari Rud River. The Salma Dam in Herat provinces was one of the other points of contention between Afghanistan and Iran, which invited the ire of Iranian officials on many occasions, complaining that Afghanistan did not consider the needs of Iran and Turkmenistan while building the Salma Dam. Worth mentioning also in this regard, is that Iran did not consult Afghanistan about building the Dosti Dam downstream on the Hari Rud River either. Such measures of ignoring each other on important projects with transboundary impacts can or will be reciprocated by Afghanistan in the future, if the environment of distrust continues to exist between both countries.

FIG. 17.3 River basins of Afghanistan showing the Hari Rud and Murghab river basins in the northwest of the country. *USAID-Afghanistan Energy Infrastructure Center.*

Occasionally, local officials in Herat and Farah provinces blame Iran for the insecurity in their areas, and in particular any attempts by insurgents to hamper the construction of the Salma Dam are attributed to Iran. Salma Dam construction had already been delayed a number of times, costing Afghanistan and India a lot in time and money. The construction of the dam was completed in 2015 and water collection has already started. The dam is built with the financial assistance from India and is considered to be one of the biggest landmarks of Indian assistance to Afghanistan. As a sign of friendship between India and Afghanistan, in August of 2015, the Afghan government decided to rename the Salma Dam to the Afghan–India Friendship Dam (Tolo News, 2015).

## 17.5 HELMAND RIVER BASIN

Afghanistan's largest river basin, the Helmand, plays a very significant role in the economy of Afghanistan and Iran's border provinces of Sistan and Baluchistan. The overall basin is fed by the waters of Helmand River and its many small and big tributaries such as the Arghandab River, Musa Qala River, and Arghistan River. Farah and Khash Rivers are the other two major rivers that contribute to the depression of Hamoun Lakes in the Seistan Depression into which the Helmand flows, but are not Helmand river tributaries and are other sources of water in the Helmand River or Seistan Basin. This depression is divided into four lakes or hamouns: Puzuak, Saberi, Shapour, and Helmand (Dehgan et al., 2014) Starting in the mountains of Central Afghanistan in the Paghman Mountains near Kabul Province, the Helmand River meanders southwest towards Iran and drains into the Hamoun Lakes to the west of the capital city of Nimroz province of Afghanistan. The Helmand River is of strategic importance to Afghanistan and Iran, and with which Afghanistan has its only water sharing agreement on its transboundary waters. On the Afghanistan side, residents of a number of the southwestern province, and on Iran's side of the border, residents of Seistan and Baluchistan provinces depend on the water of the Helmand River. From its starting point in the Hindukush Mountains in the north of Kabul to its drainage point in Hamoun Lakes, the total length of the river is approximately 1300 km or 800 miles (King and Sturtewagen, 2010). Before draining into the Hamoun Lakes that mark the boundary between Afghanistan and Iran, the river flows entirely inside Afghan territory. The total area cultivated by the Helmand River in Afghanistan is 1,376,000 ha, making it Afghanistan's largest cultivated area (Rout, 2008) with 5,800,000 people living in the basin (King and Sturtewagen, 2010).

The current boundary between Afghanistan and Iran at the southwestern corner of Afghanistan was marked in 1905 by British Colonel Henry McMahon. The reason to change boundary was due to the fact that Helmand River had changed its course (northwards) in 1891 and the water did not flow into Rud-i-Parian to the west of the Helmand River which was previously marked by the Sistan Arbitration Commission led by a British General Goldsmid in 1872 as the border between Afghanistan and Iran (Persia). The Second Sistan Arbitration Commission that marked the current boundary was headed by Col. Henry MaMahon (Sykes, 1915). It was all because of the water and a weaker Afghan government that had allowed the Iranian (Persian) government of that time to expand its territorial claims eastwards to include a number of the current water sources of the Hamoun Lakes. This portion of boundary poking into southwestern corner of Afghanistan is called the McMahon line (Fig. 17.4).

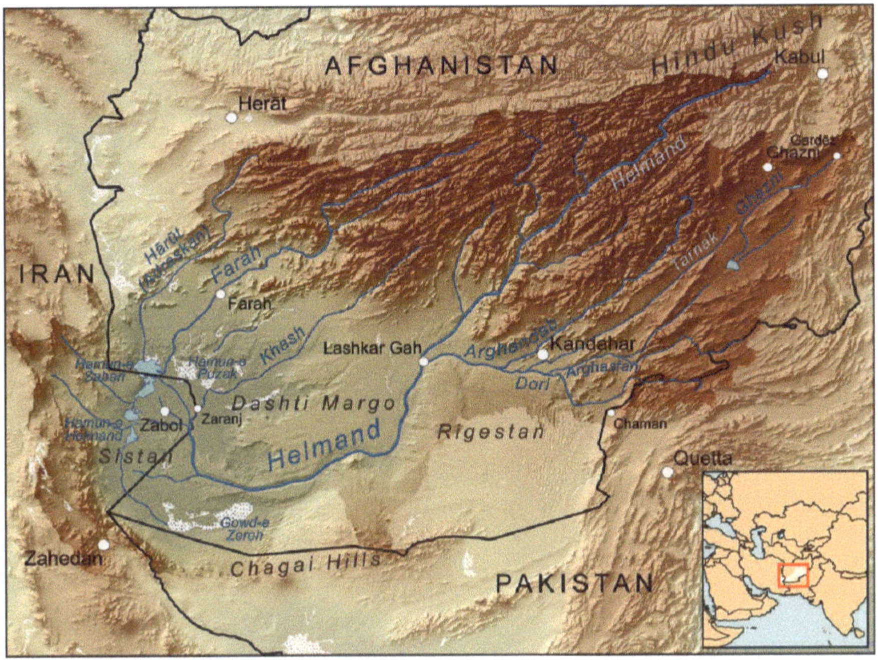


FIG. 17.4 Helmand River Basin. *Wikipedia.org*.

The Helmand basin is another basin where Afghanistan and Iran are at loggerheads with one another. In the past, the first point of contention was the Helmand Valley Project that included the construction of Kajakai Dam on the Helmand and Dahla Dam on the Arghandab River. Interactions between both countries resulted in Afghanistan's first treaty on water allotment and regulation in 1973. According to article 2 of the treaty, Iran will, on average, consume 22 cm water per second. In addition, Afghanistan will also allow Iran to use 4 cm/s of additional water as a goodwill gesture, thus making the total amount of water 26 cm/s. Article 5 of the treaty states that Afghanistan cannot deprive Iran of the amount of the water agreed upon in this treaty but has total sovereignty over the use of the remaining water inside Afghanistan (Governments of Afghanistan and Iran, 1973). Table 17.2 is the actual annual amount of water agreed by both for each month of the year in the treaty.

So far, due to instability in Afghanistan, the treaty has not been fully implemented and Afghanistan has not been compensated for the excess water flowing into Iran, as was originally agreed upon. Moreover, during the last 40 years, Iran has significantly increased its use of water from the Helmand River or the Hamoun Lakes without consulting Afghanistan on any of its water-infrastructure developments in the areas around the Hamoun Lakes.

Another point of contention is a dam planned in early 1970s and not yet finished, which is the Kamal Khan dam on the Helmand River in Nimroz Province of Afghanistan. Although it was to be regulated by the 1973 treaty, construction of this dam, just like Salma Dam in Herat, has invited Iranian ire at times when the people of Zaranj city in Nimroz are in a dire situation and have to import water and power from Iran, despite the fact that a lush river is flowing to

TABLE 17.2 Treaty of 1973 Between Afghanistan and Iran on the Disposition of the Waters of the Helmand River

| Month | Based on Average of 22 cm/s (as Formally Agreed) | Based on Average of 4 cm/s (as Goodwill) | Total Average of Formally Agreed and Goodwill Amount of Water per Second |
|---|---|---|---|
| Oct. | 4.23 | 0.77 | 5.00 |
| Nov. | 10.75 | 1.97 | 12.72 |
| Dec. | 19.48 | 3.56 | 23.04 |
| Jan. | 29.35 | 5.32 | 34.67 |
| Feb. | 66.12 | 12.04 | 78.16 |
| Mar. | 61.90 | 11.23 | 73.13 |
| Apr. | 26.30 | 4.81 | 31.11 |
| May | 7.64 | 1.39 | 9.03 |
| Jun. | 16.71 | 3.02 | 19.73 |
| Jul. | 11.61 | 2.11 | 13.72 |
| Aug. | 7.93 | 1.44 | 9.37 |
| Sep. | 1.98 | 0.34 | 2.32 |

the south of their city but they cannot benefit from it. Lack of a dam has caused the people of Nimroz, especially in the city of Zaranj, to have many social, economic, and health problems. Examining the satellite images of the areas around Zaranj city and comparing them with the agriculture fields on the other side of the border in Iran, it is easy to conclude that the people of Nimroz are in a worse condition than their neighbors to the west of them. Completion of the half-constructed Kamal Khan dam has been delayed for decades due to logistic and security reasons that are occasionally attributed to Iranian interference. If and when completed, according to a statement by the Minister of Water and Energy of Afghanistan, Mr. Ismail Khan, Kamal Khan Dam will have the capacity to produce 9 MW of electricity and irrigate around 80,000 ha of land in the province (Azadi Radio, 2011), which will drastically improve the lives of millions of people in the area. Better water-management systems, and electricity will also help in providing the people of Helmand River basin in Afghanistan with better alternative livelihoods to replace opium production, which has been one of the major financial source for terrorists and insurgents in Afghanistan (Chapter 16, herein). Moreover, Afghan opium has been the main source of heroin used in the streets of Iran, Russia, the UK, and many other countries in Europe. Like Afghanistan and Pakistan, with 1.5% of their adult population addicted to drugs mainly produced in Afghanistan, Iran is one the three countries with the highest number of addicts in the world (UNODC, 2014).

With increasing dependency on the waters of Hamoun lakes, any development on the Helmand River or the tributaries of the Hamoun Lakes will no doubt invite more Iranian criticism. Iran has the potential of destabilizing the western provinces of Afghanistan, or it could cooperate with Afghanistan in finding long-term peaceful solutions to the problems of both countries under international laws and agreements that do not undermine the long-term interests of Afghanistan. Iran's military, political, and technical powers can be used to either make the dire situation in Afghanistan worse, or they can help Afghanistan manage its waters for the benefit of all. As Sinaei (2012, p. 115) has stated in his article, "these resources provide an opportunity for a state to exercise power and change the conduct of other states." Iran's advanced infrastructure and the political fragility of its Sistan and Baluchistan provinces are several of the reasons to claim more water from Afghanistan, but it should not hinder any legitimately needed future or delayed developments on the Helmand River in Afghanistan.

## 17.6 PAKISTAN AND KABUL RIVER BASIN

Kabul River basin connects Afghanistan with Pakistan and is one of the most populated river basins of Afghanistan. With the total area of it being some 76,908 $km^2$, it is also the most densely populated basin of Afghanistan, with a population density of 93/$km^2$ and the total population of an estimated 7,184,974 people living in the basin (Kamal, 2004). The main river of the basin, Kabul River originates in the mountains of Paghman in the Sanglakh region and is the main source of irrigation and drinking water for the residents of Kabul city and its surrounding areas. The river flows eastwards to Laghman and Nangarhar provinces and enters the Federally Administered Tribal Areas of Pakistan in Khyber Pakhtunkhwa Province. The total length of the river to where it joins Indus river in the Khairabad area of Attock is about 700 km (435 miles), of which 560 km (350 miles) is within Afghanistan (King and Sturtewagen, 2010). Attock city is located about 80 km to the northwest of Islamabad and about 100 km to the southeast of Peshawar city. The main tributaries of the Kabul River in Afghanistan are, the Logar, Ghorband, Panjsher, Alingar, and Kunar Rivers (Fig. 17.5).

Afghanistan already has three water reservoirs, the Naghlu, Sarobi, and Darunta on the Kabul River and is planning to build a dam on Kabul River in Nangarhar Province. Among the existing three reservoirs, two of them, the Naghlu and Sarobi are mainly used for hydropower generation and only Darunta is used for both hydropower and irrigation purposes. The electricity generated from all three dams does not meet the demands of the population living in the Kabul basin area, thus Afghanistan has to import electricity from Central Asian States such as Uzbekistan and Tajikistan through transmission lines crossing the Hindukush Mountains that are always vulnerable to avalanches in the Salang area of the mountain range. In the winter of 2014–15, most of Kabul city remained in the dark for more than a month because avalanches had totally destroyed the transmission lines and the Afghan government did not have any backup system available to address the problem.

The Kabul River has the potential of producing more electricity to address the Afghanistan power shortage but any new dam will have its effects on the level of water in the Indus River in Pakistan. The total volume of water at Warsak Dam located in Kyber Pakhtunkhwa Province of Pakistan is 17 MAF (21,000 million $m^3$) where the Kama multipurpose dam, when built, will use 0.1 MAF (123 million $m^3$) water from the river in Nangharhar Province of

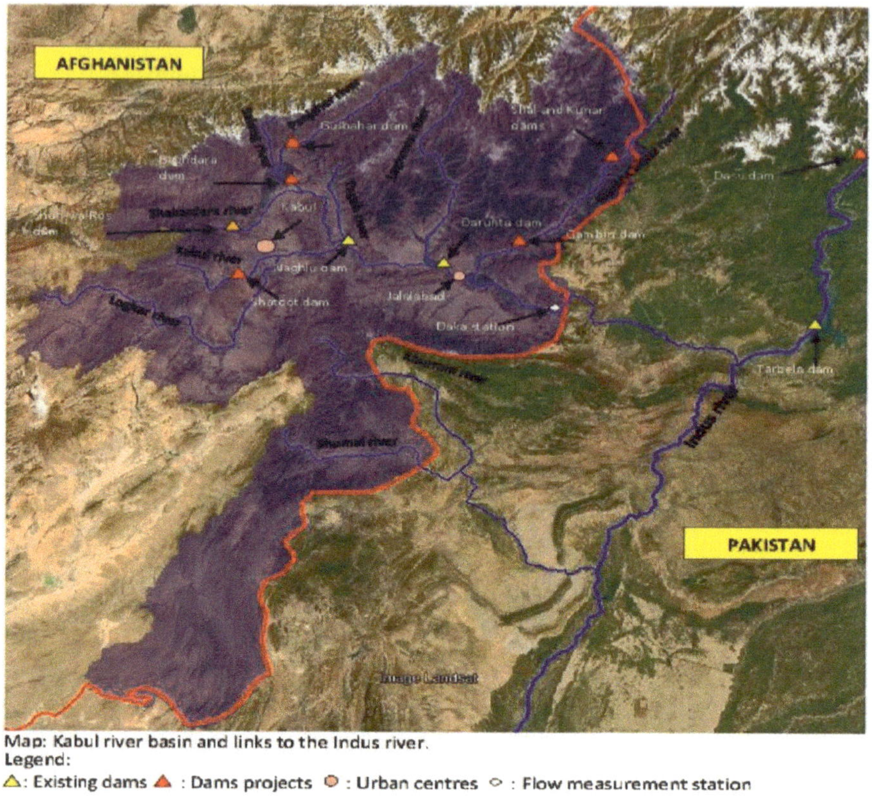


FIG. 17.5 Kabul River Basin. *http://www.thethirdpole.net*.

Afghanistan. This figure will increase to 350 million m$^3$ in the years following is construction, and is also expected to irrigate about 14,000 ha of land (Aziz, 2013).

Afghanistan and Pakistan do not have any treaty or water-sharing agreement to regulate their shares of Kabul River water and they are not expected to have one any time soon. Many reasons exist as to why it is difficult to have a water agreement between both countries. Trust deficit, and Pakistani interference in Afghan affairs since the 1970s, when the first anti-Afghan government fighters fled to Pakistan, are some of the reasons. The history of this antagonistic relationship between the two countries also traces back to the establishment of Pakistan as a country, half of which is claimed by Afghans as their lost territory that British India had forcefully annexed in 1893 by drawing the infamous Durand line (the current de facto boundary between both countries). The Durand line has not been recognized by any Afghan governments and agreeing to a water treaty would require formal identification of a point of entry for Kabul River into Pakistan. This, of course, would be interpreted as recognizing the Durand line as the international boundary with Pakistan, unless some different mechanism can be applied that does not mention the Durand line as an international border between both countries.

Despite all of the bilateral issues, in 2006, both governments showed willingness to initiate initial talks about a water-management treaty. The World Bank, with its history of facilitating the Indus River Treaty between India and Pakistan, was trusted to facilitate the dialogue on a possible Kabul Rive Treaty. The efforts unfortunately bore no fruit and the talks did not proceed further (Ahmad, 2013). The year 2006 was also the time when Taliban reemerged and the insurgency was fully back in the south and east of Afghanistan. Lack of capacity in Afghanistan, and instability are other reasons hindering establishment of a water treaty with Pakistan. During the last 30 years, Pakistan has increased its use of water from the Indus River, of which the Kabul River is a major tributary, and has not provided any financial support to Afghanistan to enhance management of the Kabul River (Hansasz, 2011). On the contrary, Afghan and international sources have accused Pakistan of doing the adverse by supporting the insurgency in Afghanistan (Shroder, 2014).

Recently, the Chinese government has expressed interest in helping Afghanistan to build the Kama hydropower dam. Given China's friendly relationship with Pakistan, it is expected that Pakistan will not object to the project as it did when India proposed building the dam. India and Pakistan are archrivals and have gone to war three times before. Construction of the dam will not only manage water flow during flood times, which brings havoc to the areas of Kyber Pakhtunkhwa of Pakistan every year, but it will also help Afghanistan to address its irrigation and power problems.

## 17.7 CONCLUSION

The people of South-Central Asia have been using their rivers for two major purposes of agriculture and generating hydropower. Afghanistan seems to be lacking stability, resources, infrastructure, and technical capacity to harness its water for either of the purposes; this is mostly due to wars since 1979 and financial incapacity prior to that, as most of the mega-agriculture or hydropower generating projects were made possible with foreign assistance such as that of the United States for Helmand and Arghandab River Valley project in the 1950s. This situation is not that different now. Afghanistan still lacks financial resources to take on such mega-projects. In fact, India has been funding the construction of Salma Dam on the Hari Rud River, which has been marred by many delays so far. In the last 14 years, the Afghan government and international community failed to initiate any large-scale irrigation projects that could have considerable impacts on the water flow in the Afghan rivers. The decrease in the amount of water in Afghan rivers has mostly been due to climate changes.

Despite all of its limitations, Afghanistan will have to come up with plans to address its burgeoning demand for water and power. To do so, the Afghan government has already started devising policies and plans, which will eventually be implemented by utilizing the current water resources that have been flowing un interruptedly across international boundaries into the neighboring countries. Adding more to the existing problems is climate change, which will affect water availability in the rivers of Afghanistan and the region. Scientific data show that the major source of water in Afghanistan, the glaciers in the north-eastern corner of the country, have been drastically shrinking. According to a report produced by the United Nations Environment Program, Afghan glaciers have receded more than 50% since the 1930s (United Nations Environment Program, 2007). This could be a major threat to the future availability of water in Afghanistan and in the region. Any Afghan effort to harness its waters will alarm its neighbors, who depend on the waters originating from Afghanistan.

On a grand level that involves all of Afghanistan's neighbors, all states will eventually have to recognize the looming danger of climate change and population increase and their needs. As part of the neighborhood, Afghanistan's legitimate needs cannot be ignored any more by its neighbors as has been the case for the last 40 years, where all water-related developments in the countries surrounding Afghanistan were done without any consultations with Afghanistan or taking Afghanistan's future needs into consideration. All politics, policies, and planning will have to adapt to conditions and realities that future generations will face, and will have to come up with plausible solutions that are rooted in not politics but realities. This can be challenging to achieve but not impossible.

## References

1TVNews, April 22, 2015. Retrieved May 1, 2015, from 1TV News Web site: http://www.1tvnews.af/en/news/afghanistan/16250-ghani-mentions-border-water-while-referring-to-refugees-problem-in-iran.

Ahmad, D.S., 2013. Towards Kabul Water Treaty: Water Cooperation for Managing Shared Water Resources-Policy Issues and Options. International Union for Conservation of Nature, Karachi.

Ahmad, M., Wasiq, M., 2004. Water Resource Development in Northern Afghanistan and Its Implications for Amu Darya Basin. The World Bank, Washington, D.C. (Retrieved 19.05.15).

Akweenda, S., 2002. From Harmon to Helsinki: the evolution of key principles in international water law. In: Turton, A., Henwood, R. (Eds.), Hydropolitics in the Developing World: A Southern African Perspective. African Water Issues Research Unit, Pretoria, pp. 97–111. www.internationalwaterlaw.org (Retrieved 27.04.15).

Anonymous, January 15, 2015. Ex-Diplomat (S.J. Ahmadzai, Interviewer).

Azadi Radio, August 19, 2011. (in Dari) Contract to Construct Kamal Khan dam has been signed with a Tajik company (in. Azadi Radio). Prague. http://da.azadiradio.org/content/article/24292699.html (retrieved 18.06.15).

Aziz, K., 2013. Need for a Pak-Afghan Treaty on management of joint water courses. Criterion Quarterly 2 (4). http://www.criterion-quarterly.com/need-for-a-pak-%E2%80%93-afghan-treaty-on-management-of-joint-water-courses/ (Retrieved 30.06.15).

CIA World Factbook, 2015. Central Intelligence Agency. Retrieved from https://www.cia.gov/index.html.

Dehgan, A., Palmer-Moloney, L., Mirzaee, M., 2014. Water security and scarcity: potential destabilization in western Afghanistan and Iranian Sistan and Bluchistan due to transboundary water conflicts. In: Weinthal, J.T., Nakayama, M. (Eds.), Water and Post-Conflict Peacebuilding. Taylor and Francis, London, pp. 305–326. http://environmentalpeacebuilding.org/assets/Documents/LibraryItem_000_Doc_937.pdf (retrieved 19.06.15).

Governments of Afghanistan and Iran, March 13, 1973. Helmand River Water Treaty. Retrieved 18.06.15.

Guess, G.M., Farnham, P.G., 2011. Cases in Public Policy Analysis. Georgetown University Press, Washington, D.C.

Hansasz, P., 2011. The Politics of Water Security in the Kabul River Basin. Strategic Analysis Paper. Future Directions International, West Perth.

Hansaz, P., 2012. The Politics of Water Security Between Afghanistan and Iran. Strategic Analysis Paper. Future Directions International, West Perth.

Kamal, G.M., 2004. River Basins and Watersheds of Afghanistan. Afghanistan Information Management Service (AIMS). Afghanistan Information Management Service (AIMS), Kabul. http://www.nzdl.org/gsdl/collect/areu/Upload/1710/Kamal_River%20basins%20and%20watersheds2004.pdf (retrieved 23.06.15).

Kendie, D., 1999. Egypt and the hydro-politics of the blue Nile river. Northeast. Afr. Stud. 6 (1–2), 141–169 (Retrieved 16.02.15).

King, M., Sturtewagen, B., 2010. Making the Most of Afghanistan's River Basins. East-West Institute, New York. http://www.ewi.info/idea/making-most-afghanistans-river-basins (retrieved 22.05.15).

OOSKAnews, February 22, 2013. OOSKAnews, Inc. Aquanow: https://www.ooskanews.com/story/2013/02/world-bank-says-tajikistan%E2%80%99s-roghun-dam-not-threat-regional-security_154476 (retrieved 20.06.15).
Rout, B., 2008. How the Water Flows: A Typology of Irrigation System in Afghanistan. Afghanistan Research and Evaluation Unit (AREU), Kabul. Retrieved from http://www.areu.org.af/Uploads/EditionPdfs/811E-Typology%20of%20Irrigation%20Systems.pdf.
Rycroft, D., Wegerich, K., 2009. The three blind spots of Afghanistan: water flow, irrigation development, and the impact of climate change. China Eurasia Forum Quart. 7 (4). http://www.researchgate.net/publication/41660030_The_Three_Blind_Spots_of_Afghanistan_Water_Flow_Irrigation_Development_and_the_Impact_of_Climate_Change (retrieved 22.05.15).
Sexton, R., 2012. Natural Resources and Conflict in Afghanistan: Seven Case Studies, Major Trends and Implications for the Transition. Afghanistan Watch, Kabul. Retrieved from http://www.watchafghanistan.org/files/Natural_Resources_and_Conflict_in_Afghanistan/Natural_Resources_and_Conflict_in_Afghanistan_Full_Report_English.pdf.
Shobair, S.S., 2010. Teh Afghan Part of Amu Darya Basin: Impact of Irrigation in Northern Afghanistan on Water Use in the Amu Darya Basin. United Nations Economic Commission for Europe (UNECE), Kabul. http://www.unece.org/fileadmin/DAM/SPECA/documents/ecf/2010/FAO_report_e.pdf (retrieved 20.05.15).
Shroder, J.F., 2014. Natural Resources in Afghanistan. Elsevier, San Diego, CA. Retrieved 04.08.15.
SIGAR, 2015. Quarterly Report to the United States Congress. Office of the Special Inspector General for Afghanistan Reconstruction. SIGAR, Arlington, TX. https://www.sigar.mil/pdf/quarterlyreports/2015-04-30qr.pdf (retrieved 30.06.15).
Sinaei, V., 2012. Hydropolitics and human security: water cooperation in relations between Iran, Afghanistan and Turkmenistan. Iran. Rev. Foreign Aff. 2 (3), 111–133. http://profdoc.um.ac.ir/articles/a/1027822.pdf (retrieved 23.06.15).
Sykes, L.-C.P., 1915. A History of Persia. MacMillan and Co., London. Retrieved 18.06.15.
TAICE, June 5, 2015. Tossab Consulting Engineers Company (TossAb International Consulting Engineers). Retrieved from TAIECE Engineers: http://www.taice.de/project1.htm.
The Arg, March 26, 2015. Office of the President of Afghanistan. Retrieved May 5, 2015, from Afghan President's Office Website: http://president.gov.af/en/news/transcript-of-the-conversation-by-president-ghani-in-the-council-for-foreign-relations
Tolo News, August 27, 2015. Retrieved from Tolo News Website: http://www.tolonews.com/en/afghanistan/21105-salma-dam-renamed-afghan-india-friendship-dam.
Turton, A., 2002. Hydropolitics: the concept and its limitations. In: Turton, A., Henwood, R. (Eds.), Hydropolitics in the Developing World: A Southern African Perspective. African Water Issus Research Unit, Pretoria, pp. 13–19. www.internationalwaterlaw.org (retrieved 27.04.15).
United Nations Environment Program, 2007. Global Outlook for Ice & Snow. UNEP/GRID-Arendal, Birkeland. http://www.unep.org/geo/geo_ice/PDF/full_report_LowRes.pdf (retrieved 05.05.15).
UNODC, 2014. World Drug Report 2014. United Nations, New York. http://www.unodc.org/documents/wdr2014/World_Drug_Report_2014_web.pdf (retrieved 26.06.15).
Wegerich, K., 2011. Water resources in Central Asia: regional stability or patchy make-up? Cent. Asian Surv. 30 (2), 275–290. http://dx.doi.org/10.1080/02634937.2011.565231.

CHAPTER

# 18

# Future of Water Management in Afghanistan: Capacity Building, Risk Assessment, Cooperation, and Good Water Governance

*J. Shroder*
University of Nebraska at Omaha, Omaha, NE, United States

OUTLINE



## Abstract

Water management in Afghanistan is beset by numerous problems, including difficulties of negotiating transboundary cooperation over water between Afghanistan and any single country, or with groups of the neighboring countries of Pakistan, Iran, Turkmenistan, and Uzbekistan, and Tajikistan into whose territory flow the rivers that head in Afghanistan. At the present time the rivers of Kabul, Helmand, Hari Rud, Murghab, and Amu Darya have no truly functioning water treaties about them. Risk assessment for water-cooperation negotiation involves a variety of cost-benefit assessments and risk-opportunity situations that any country must consider as they attempt to move toward a more comfortable position in water negotiation. Enhancing

*Transboundary Water Resources in Afghanistan*
http://dx.doi.org/10.1016/B978-0-12-801886-6.00018-5

cooperation over water is accomplished by paying greater attention to water management, water-capacity development, or hydro-education of government officials, and generally improved water governance overall. New President Ashraf Ghani could become an essential technocratic water champion for Afghanistan.

**Keywords:** Water governance, Water management, Hydro-education, Democracy principle, Voluntary principle, Autonomy principle, Universality principle, Equity principle, Water cooperation, Risk assessment, Water-cognizant capacity development

## 18.1 INTRODUCTION

The water-management situation in Afghanistan, which developed in this new millennium, once the Taliban government of incompetence and mismanagement had exited in the face of the invasion by the Western Coalition of NATO forces in late 2001, was one of early recognition of insufficient water-cognizant capacity in the country. As a new government was being organized in Kabul in the early 2000s, some of the main tasks were to recruit enough local technical experts to take on the huge number of bureaucratic tasks needed to run a government with reasonable competence. Such expertise was indeed in short supply in the country after three decades of a nearly destroyed educational system. Soon it became obvious to the foreign nongovernmental government organizations (NGOs) and their technical advisors, that the original, never robust technocracy of the government had been further decimated by the prior decades of negative events, and especially by the loss of those who knew so much about water issues.

In all probability, the water experts that existed in Afghanistan prior to the outbreak of the war in the late 1970s either left the country to escape the looming chaos, converted to other jobs, or have since passed away. Many of these hydro-experts or hydro-engineers found water-related, or other jobs elsewhere and never returned home, which left a technical deficit in hydro-capacity, that is only slowly being replaced again. In any case, it became necessary to regenerate a new population of water-cognizant personnel for the improvement of the water situation in Afghanistan. Only with a new cadre of water experts can the country engage in future years in meaningful negotiations of cooperation over water with the neighboring countries.

In any regional situation involving water, three possible steps can occur in the process of dealing with the problems; regional conflict, negotiation, or cooperation (Dinar et al., 2013). Obviously the regional-conflict step has been the prevailing situation in Afghanistan, and although it has seemingly not been about water in any substantive or overt way, nonetheless that fact must be assumed as a background motivation for at least some of the fighting. Negotiations leading to cooperation are a preferred option that is addressed in this chapter.

Bogardus (1964) has defined cooperation as a process through which individuals and groups may move up from one level of social development to a next, more prosperous one. The principles of such cooperation would include: (1) the *democracy principle* of managing the means of any cooperative arrangement; (2) the *voluntary principle* of joining into or leaving from any cooperative agreements; (3) the *autonomy principle* of self-sustainability of the arrangements; (4) the *universality principle* of setting up goals that are attributable to the participants; and (5) the *equity principle* of participating in, and sharing the benefits of the arrangements (Dinar et al., 2013). These five principles are obvious goals that any future cooperative deal with the neighboring countries would have to address, but a number of other incentives would have to be provided as well. In order to work in the real world, the cooperative arrangements have to give additional gains to the participants by addressing issues of justice, fairness, and the resulting benefits and joint costs of adhering to the cooperative agreements. Thus in simple terms, what incentive could possibly be in any deal to attract Afghanistan into any cooperative agreement on water derived from inside its borders? What clever mix of incentives could be worked out to make the people of Afghanistan, and their justifiably rather suspicious and conservative government bureaucracy pay attention to the possibility of negotiating cooperation agreements on transboundary water resources?

In general terms, some of the strategies for developing and promoting regional cooperation on transboundary waters coming from Afghanistan include overall strategic goals and related items specific to the goals (Table 18.1). These goals and specificities can give some incentives to stakeholders in Afghanistan that would make them want to engage in transboundary negotiations, but in general, some significant other rewards would be required to gain sufficient attention among the cognizant population of Afghanistan to achieve this result (Table 18.2). Otherwise, Afghan citizens would feel rather aggrieved that their best

TABLE 18.1 Strategies for Regional Cooperation in Transboundary Waters From Afghanistan

| Strategic Goal | Specific Items |
|---|---|
| Strengthen policy and regulation (both domestic and cross border) | 1. Build local capacity in Afghanistan<br>2. Improve data and information collection, analysis, and sharing<br>3. Increase awareness and involve stakeholders in decision-making |
| Support long-term national and regional programs | 1. Adopt basin-wide approach<br>2. Promote consensus on importance and urgency of transboundary water problems<br>3. Promote harmonization of water regulation among basin countries<br>4. Encourage participation of all subsectors |
| Evaluate basin-wide development alternatives | 1. Identify, modify, and promote successful regional shared-water agreements<br>2. Facilitate exchange between riparians<br>3. Set up hydrological data services—data and information sharing |
| Expand areas of interaction and joint interest | 1. Promote joint water-related and nonwater-related projects benefiting all basin countries<br>2. Establish dispute-resolution mechanisms<br>3. Include out-of-basin (foreign) expertise |

*Note:* Strategies can provide some incentives to Afghanistan but ultimately additional true incentives would have to be applied.
*From Dinar, A., Dinar, S., McCaffrey, S., McKinney, D., 2013. Bridges over Water: Understanding Transboundary Water Conflict, Negotiation and Cooperation. World Scientific, Hackensack, NJ, USA.*

TABLE 18.2 Types of Incentives to Gain Attention in Afghanistan in Working Toward a Goal of Transboundary Negotiations on Water

1. Financial rewards for water deliveries downstream (PES: payments for ecosystem services)
2. Offers of engineering water works designed by foreign engineers and built in Afghanistan
3. Provision of remote sensing data and up-to-date hydrologic interpretations
4. Transboundary guarantees of good behavior and reduction or elimination of cross-border violence
5. Linkage of dam building and irrigation projects in Afghanistan with overall resource-corridor growth based upon the newly developed extractive industries (Shroder, 2014a,b, 2015)
6. Linkage of university education programs on water with transboundary water conferences hosting faculty and senior-level/graduate students
7. Establishment of multiple university programs on surface and ground-water programs with faculty doubling as government water advisors

*Note:* Incentives can be provided by neighboring countries and foreign-aid agencies.

welfare had been ignored and disregarded in favor of some illegitimate explanations. The development of such situations in this region of South Asia, because of a past history of so many disagreements and outright violence in the past, is an invitation to immediate failure.

## 18.2 PROBLEMS WITH COOPERATION

Numerous difficulties can and do arise in any proposed system of cooperation for dealing with transboundary water resources. Afghanistan is a prime example of a country caught in a downward spiral of noncooperation because of many internal motivations and hindrances. Internal motivations mainly include the lack of hydro-expertise in the government, and a desire to not enter into any sort of agreements unless the hidden ramifications and complexities of transboundary water issues are completely understood. In general the hindrances can occur because of: (1) the technical complexity of cooperation projects; (2) the existence of differing goals between the countries concerned; (3) the ill-defined rights and responsibilities of each of the countries involved; (4) the existence of intractable neighboring countries who do not wish to negotiate about transboundary water under any circumstances, such as China, or whom might not negotiate in good faith such as Iran or Pakistan; and (5) the existence of additional considerations such as non-water-related political or economic circumstances. It has been observed (Dinar et al., 2013) that the so-called *grand coalitions* of countries, whereby all the riparian nations in a basin sign the various treaties, are actually rather rare and most characteristic of exceptionally friendly countries; far more common and successful, however, are the *partial coalitions* that are signed between only two countries, leaving several other countries in the same hydrological basin quite out of the negotiations.

In the case of Afghanistan, the number of surrounding countries (six), and the number of significant transboundary rivers (five), indicate the complexity and difficulty of Afghanistan for gaining agreements between: (1) China, India, and Pakistan over the Kabul River tributary to the Indus; (2) Iran over the Helmand; (3) Iran and Turkmenistan over the Hari Rud–Tenjen; (4) Turkmenistan over the Murghab; and (5) Turkmenistan, Uzbekistan, and Tajikistan over the Amu Darya. These large numbers of rivers and countries, some of whom, such as Pakistan and Iran, have been or are still belligerent with Afghanistan, indicate the improbability of achieving much in the way of major cooperative agreements in this region any time soon.

On the other hand, some would seek to explain an alternative point of view wherein a marginally operant water treaty was signed between Afghanistan and Iran in 1973, under the reigns of the Shah of Iran and the GoA that still seems to be at least partially or notionally in force. A case could be made, however, that the treaty is no longer valid at all because the Shah was driven out of power in 1978, and more than four, quite different governments have existed in Afghanistan over the past four decades since the signing of that treaty. In any case, the original treaty specified a certain amount of water to be provided to Iran from the lower Helmand River, for which a monetary fee would be paid by Iran to the GoA. In fact, however, no such monies were ever paid for water delivered, nor because of climate change and severe-drought diminution of Helmand River waters, has Afghanistan ever been able to deliver the required water on a regular basis. This has resulted in a number of cross-border incursions whereby Iranian forces have brought in bulldozers to divert part of the Helmand to their benefit, which has resulted in some violent confrontations with Afghanistan police and border guards. Thus in practical applications of water treaties in Afghanistan, no real treaties exist that seem to be in real force at the present time, even though a number of foreign agencies (World Bank, USAID, East-West Institute, etc.) and governments (Germany, United States, etc.) have been pressing the GoA recently to establish new water-treaty talks.

The position of the Afghanistan government in recent years has been to not negotiate over water under any circumstances until some future undetermined date when they might feel more comfortable that they understand all the many technical issues necessary. In fact this book is an effort to provide background understandings about the many issues dealing with water that apply in Afghanistan. This is done in order to provide the necessary water-cognizant capacity building for GoA officials so that they would feel more comfortable in attending scientific or political discussions about water issues dealing with their country. The modern position of GoA on the water that falls on, and flows within its borders is that it is grossly under-utilized at the present time, and far greater measures are necessary to dam rivers for production of hydropower generation and enhancement of irrigation in fallow lands that receive no water at present. This is a quite sensible government position, especially in view of the violent opposition to the GoA that occurs in the support for the Taliban insurgency maintained by neighboring countries, especially Pakistan, but also Iran to a certain extent.

## 18.3 WATER COOPERATION AND RISK ASSESSMENT

Countries have to find ways to compensate for, or to control risk, or they will not be able to enter into positive agreements with other countries. Perceived risk for any given country is the perception that an act of cooperation will expose the country to some sort of harm, will jeopardize something of value to the country, or will threaten the political future (or life) of individual policy makers (Subramanian et al., 2012). Five general categories of risk are generally perceived by decision-makers (Table 18.3). These risks exert an influence on decisions about cooperation. Sovereignty and equity risks, for example, are generally considered as core risks in that they are deep-seated and are known to tend to resurface in ongoing negotiations, even after deals have been made and benefits delivered. In most cases of risk assessment, to a greater or lesser extent, the dominant risk revolves around sovereignty and autonomy. The risk is mainly about a

TABLE 18.3 Categories of Risk to Countries in Any Possible Cooperation Over International Water

| |
|---|
| *Capacity and knowledge*—Confidence in ability to negotiate a fair deal; having enough and the correct information, data, and knowledge to accomplish this |
| *Accountability and voice*—Deliverability of benefits by the regional entity and co-riparians, commonly related to trust; having a say in decision-making in the governmental structures of the regional entity |
| *Sovereignty and autonomy*—Ability to act in the best interest of the country without constraints; making decisions independently |
| *Equity and access*—Fairness of (relative) benefits to country, including timing of benefits and costs and obtaining/retaining fair access to the river |
| *Stability and support*—Longevity potential of the agreement; in-country support of the agreement, including likelihood of ratification |

*From Subramanian, A., Brown, B., Wolf, A., 2012. Reaching Across the Waters: Facing the Risks of Cooperation in International Waters. Water Papers, Water Partnership Program. The World Bank. http://water.worldbank.org/sites/water.worldbank.org/files/publication/WaterWB-Reaching-Across-Waters.pdf (accessed 10.01.14).*

decision-maker who worries about the danger of intrusion into the country's authority to make sovereign decisions. This refers to the desire to maintain control over national development goals and related development of resources and infrastructure, as well as the right to make decisions independently, both of which are essential to Afghanistan at the present time.

In undergoing the risks of capacity and knowledge (Table 18.3), which are quite characteristic of Afghanistan, the country fears that it will be at a major disadvantage in several ways: (1) in having less negotiating capacity than co-riparians; and (2) in not having adequate or accurate data and information about its rivers and river basins. In general, decision-makers characteristically experience that which is known as accountability and voice risk (Table 18.3), when they fear the co-riparians, third parties, or regional institutions may not deliver the agreed-upon benefits. In this regard they may also be concerned that the country's interests are not adequately considered in joint decision-making processes. In addition, they may perceive a high probability that the regional institutional arrangements would not result in the flow of the benefits expected.

Countries are commonly also acutely concerned with the risks of equity and access. These include: (1) the amount of fairness in any deal regarding specified quantities (or qualities) of water, beneficial water flows, or project costs; and (2) entitlement to use the river. Some countries view prior appropriations or entitlements as an absolute right to continue with their historic uses. Other countries may want to gain access to a river running though, or originating in their territory, or perhaps they may want to attain benefits in proportion to the country's relative size in the basin or in the terms of percentage contribution of water to the basin. All of these factors may play into the various riverine situations in Afghanistan as the decision-makers contemplate negotiations with the surrounding countries. The risk of stability and support (Table 18.3) has direct national and personal implications in Afghanistan, especially because the country has so many diversified and powerful stakeholders, warlords, and other strong players who can interfere in so many ways with the negotiation process. The implementability of any agreement may depend entirely on the presence or absence of key stakeholder support, and the positive or negative public image of any particular decision-maker may be key as well.

## 18.4 ENHANCING WATER COOPERATION

Measures to address the categories of risk expressed in Table 18.4 are critical to moving toward cooperation. Risk reduction is an important process that needs to be addressed before a country such as Afghanistan could possibly progress to any sort of negotiated outcome. In a number of prior cases (Subramanian et al., 2012) reductions of risks can provide sufficient motivation for countries to reconsider a water-cooperative deal, or even to sign an agreement. These cases have included a number of categories of risk reduction that can be effective (Table 18.4).

In addition, political opportunity can be a critical factor in enhancing cooperation in a number of cases. At certain times, the perception of resulting national and regional political gains can even trump residual risk if conditions are right. In other words, cooperation with some risks still remaining can even be effected if sufficient political opportunity can be obtained as well. Subramanian et al. (2012) noted a number of examples wherein shifts in regional politics

TABLE 18.4 The Various Categories of Risk Reduction That Can Induce a Given Country to Come to the Negotiating Table

1. *Knowledge and skill expansion*—Training and studies to meet capacity and knowledge needs, and support for developing new skills
2. *Institutional design*—Tailoring the institutional arrangement to be a "fit for purpose" cooperative agreement for dialogue and action among the riparian nations downstream
3. *Agreement design*—Tailoring the agreement to the preference of political leaders involved in terms of its formality, scope, goals, and obligations, and so forth
4. *Program design*—Shaping the program to address Afghanistan's best interests and goals—sectoral linkages, long- versus short-term benefits, review and monitoring, and so forth
5. *Financing and/or guarantee*—Meeting financial needs and gaps identified, including third-party guarantees of financial obligations
6. *Facilitation* (by a third party)—Unbiased, third-party assistance in dialogues among riparian nations, including clarifications and interpretations
7. *Decision legitimacy*—Use of consultation, discussion and workshop meetings and other avenues for ensuring widespread domestic and regional-support decisions

*From Subramanian, A., Brown, B., Wolf, A., 2012. Reaching Across the Waters: Facing the Risks of Cooperation in International Waters. Water Papers, Water Partnership Program. The World Bank. http://water.worldbank.org/sites/water.worldbank.org/files/publication/WaterWB-Reaching-Across-Waters.pdf (accessed 10.01.14).*

enabled such progress, or the possibility that cooperation over water between countries could signal or lead to cooperation on other issues in regional and global meetings and discussions. Changes in national leadership can bring in new champions who are willing to take on new initiatives on cooperation. Such might be a special case for Afghanistan where the swearing in of the new president, Ashraf Ghani and his rival Abdullah Abdullah, in Kabul on Sep. 30, 2014 (Nordland and Walsh, 2014) certainly offers new hope of such a development. Ghani was a former official of the World Bank (1991–2001), which as an organization, has been pressing for transboundary water negotiations in Afghanistan for some time, so no doubt the new president will be quite familiar with the necessary processes to initiate new discussions.

In fact, third parties such as the World Bank and others of their development partners could play important roles in supporting Afghanistan with risk reduction. Such reduction in risk could include engaging with the country at appropriate scales, such as an entire basin, or sub-basins, or at the whole-country level with all drainage basins. Conduction of detailed and quite transparent risk assessments would be another method, as well as designing risk-reduction strategies, including financing and guarantees to target dominant risks, and periodically reassessing the risk situation, while employing new strategies as needed. The role of any possible partners that could be taken on with respect to enhancing political opportunity for Afghanistan is less clear, although much more research on that is needed, and potential partners such as Tajikistan, for example, could be encouraged. That country, or any others, would need to stay abreast of regional geopolitical developments in order to be ready, if and when opportunities might be presented, as well as continuing to implement appropriate risk-reduction strategies so that risks would have already been dealt with to whatever extent that is possible, if and when new opportunities were presented.

Key messages in risk reduction concerning transboundary water negotiations include at least 10 major points (Table 18.5). For example, the lack of detailed studies of such risk, while not exhaustive, does offer opportunities for important new assessments in Afghanistan and other countries that may become beneficial. Because risk is so critical in good

TABLE 18.5 Key Messages to Keep in Mind From the World Bank for Those Engaged on Cooperation Work on Transboundary Waters

1. Risk types are not much studied, but are critical in decision-making
2. Countries are not unitary actors, but instead several stakeholders are likely to be involved
3. Individual decision-makers matter, with champion people being a key
4. Solutions must be devised for special situations, and should map the needs of the particular country, most or all of which are unique
5. Risks will most likely require a diversity of interventions
6. Opportunities can outweigh residual risks
7. Politics are difficult to predict, so anticipation is critical
8. National, regional, and global events will affect opportunities
9. Long-term time commitment is needed
10. Deals are dynamic and will change so they need constant attention

*From Subramanian, A., Brown, B., Wolf, A., 2012. Reaching Across the Waters: Facing the Risks of Cooperation in International Waters. Water Papers, Water Partnership Program. The World Bank. http://water.worldbank.org/sites/water.worldbank.org/files/publication/WaterWB-Reaching-Across-Waters.pdf (accessed 10.01.14).*

decision-making, it is necessary to keep the pressure on to develop new understandings of possibilities. Special attention needs to be paid to the role of political economy in decision-making and the country's perception of risks. Similarly, close attention to the sensible desires of numerous different sorts of stakeholders in Afghanistan is necessary to aid in this process. Those desires would be capable of a reasonable questionnaire investigation in Kabul at the outset, for example. An additional factor might be for President Ghani to present soon an open-ended water manifesto that done cleverly enough could establish him as a water-authority champion, which would go a very long way toward helping the situation. A variety of well-timed solutions involving resource corridors, resource-extraction funding, education on resources and environmental control, desalination of water paid with profits, and other potentially highly beneficial activities could serve as part of major and rather unique solutions to problems in Afghanistan and would allow President Ghani's supporters to justifiably claim that technocratic solutions to raise the national income could lift Afghanistan out of its long nightmare (Shroder, 2012, 2013, 2014a,b).

The diversity of interventions that are possible in this resource-development scenario will allow a number of different actions to reduce risk and develop so many diverse opportunities that collectively these can outweigh residual risks sufficiently to make a difference. Basically enough new opportunities can come into existence with these changing developments that serious progress can be made. Furthermore, because opportunity can be such a powerful factor in determining the outcome of an offer for cooperation, these new possibilities must be monitored carefully by a newly efficient government, if such can be obtained. In the midst of these possibilities, politics are so essential that attention must be paid to any slight shifts of geopolitics so as to take advantage of any opportunity, should such arise. Additionally because national, regional, and global events can so much affect opportunities to change risk potentials, thinking needs to be developed to take advantage of any unforeseen opportunities. The recent recognition that the precipitous departure of the United States from Iraq under the President Obama administration did not work so well, and that a different mode is required in Afghanistan, are both examples of unforeseen opportunities to make new deals. Also in this dynamic mix is the fact that deals themselves can be quite changeable, so settling in for the long-term commitment to years of planning, facilitation, and confidence are quite likely to be required.

The result of cost-benefit analysis versus risk-opportunity assessment for any country involves plotting of where the country exists in a matrix of costs versus risks so that benefit and opportunity can be enhanced (Fig. 18.1). In the figure, the objective for any given country is to move to the upper right where the countries most likely to make a deal can exist (Fig. 18.2). Water cooperation involves working together towards a common goal, in a fashion that is mutually beneficial. The process refers to peaceful water management and the use of freshwater resources among various players operating at local, regional, and international levels. The absolutely vital nature of freshwater provides a powerful incentive for cooperation, which can compel stakeholders to reconcile over their differences, even with the most divergent views.

According to the United Nations (Anonymous, 2013), since 1948 only 37 incidents of acute conflict over water have occurred that involved violence, whereas some 295 peaceful international water agreements were signed instead. In fact, water poses both a threat as well as an opportunity to people anywhere. Three major linkages occur between conflict and water; these are access to adequate water supplies, water and livelihoods, and water management (Carius et al., 2004). Thus although water only rarely is responsible for starting a war between states, its scarcity can assume an increasingly contentious and violent role if, for example, such water-dependent sectors as irrigated agriculture are cut off from flow, which can lead

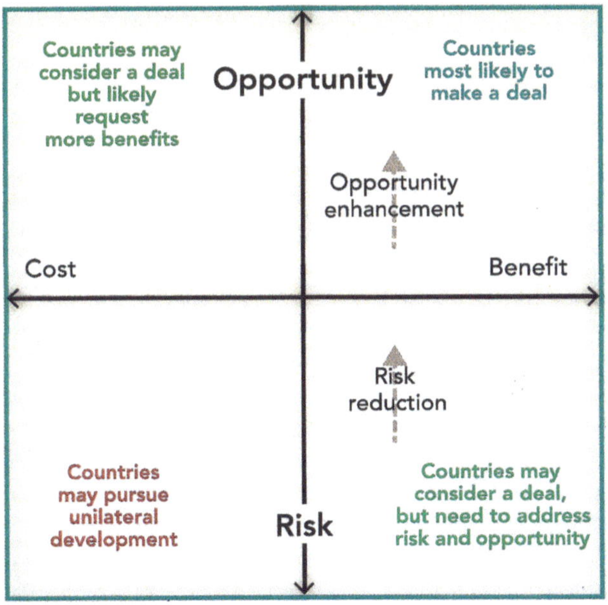


FIG. 18.1 Cost-benefit analysis to risks and opportunities on a road to cooperation in transboundary water resources. Countries that wish to consider possible cooperation in transboundary water resources assess their position somewhere along the horizontal axis in terms of net benefit (benefits less costs). The vertical axis is the net opportunity (opportunities less risk). Cost-benefits are economic considerations, whereas the risks to opportunities are political; once a country determines in which quadrant it exists, risk reduction and opportunity enhancement *(gray arrows)* can enable change for the better by moving more into the upper right where good deals are possible. *From Subramanian, A., Brown, B., Wolf, A., 2012. Reaching Across the Waters: Facing the Risks of Cooperation in International Waters. Water Papers, Water Partnership Program. The World Bank. http://water.worldbank.org/sites/water.worldbank.org/files/publication/WaterWB-Reaching-Across-Waters.pdf (accessed 10.01.14).*

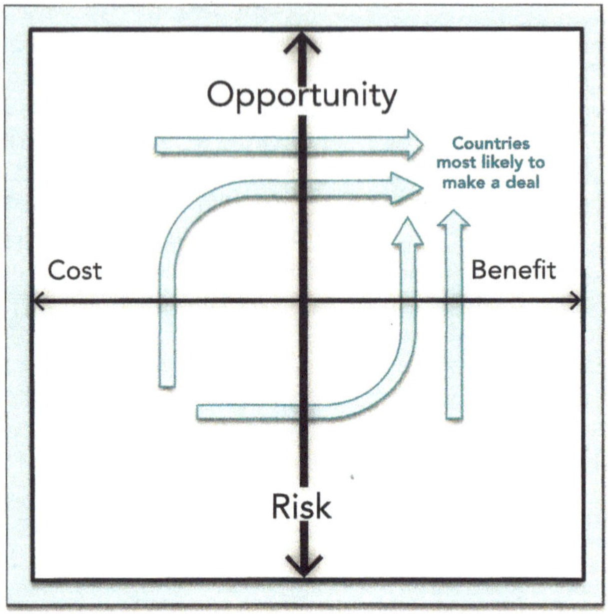


FIG. 18.2 Paths to cooperation in transboundary water management wherein the initial position of a country in any one of the four quadrants can be changed through various combinations of risk-reduction strategies combined with enhancing and seizing opportunities. *From Subramanian, A., Brown, B., Wolf, A., 2012. Reaching Across the Waters: Facing the Risks of Cooperation in International Waters. Water Papers, Water Partnership Program. The World Bank. http://water.worldbank.org/sites/water.worldbank.org/files/publication/WaterWB-Reaching-Across-Waters.pdf (accessed 10.01.14).*

to destabilizing migrations out of the affected zones. This has happened in the enormous droughts of 2001 in the lower Helmand, where thousands of people in both Afghanistan and Iran had to leave to seek a living elsewhere.

As Shrestha et al. (2013) have noted, approaches to water-resource management in the Hindu Kush–Himalaya have not fully accounted for the social, cultural, or political implications of water management and climate-change adaptation interventions that are required. Water management in South Asia has traditionally been handled at the state level rather than other models whereby drainage basins and transboundary consideration of the interests of the many others are considered instead. Typically rigid hierarchical management regimes in South Asia do not support flexible, cooperative approaches for coping with the ever-changing environment and sociopolitical landscapes. The changes related to transboundary waters and climate change that are so important to the future of Afghanistan will not be easy to deal with, given governmental rigidities, so the answer must be to find ways to meet local and regional needs more efficiently, lest an unresponsive government contributes further to the chaos that leads to more violence. In addition, highly regulated or restricted data and information-sharing protocols are counterproductive to good governance, and do not support informed decision-making. The unfortunate lack of cooperation in information exchange about water, or in the lack of sharing of appropriate technologies across borders, are serious impediments to adequate development of water resources and their management in the Afghanistan region.

Access to freshwater is a cross-cutting issue concerning cooperation problems that demands attention at all levels, as well as across sectors of societies. Issues concerned with water involve many stakeholders who generally have conflicting and competing needs that cross multiple physical, political and jurisdictional boundaries. Cooperation over this water is critical to the handling of such issues as decisions over allocation, upstream and downstream impacts of water pollution, water removal, and overexploitation, in addition to construction of water infrastructure and the decisions over financing the management of the water resources and the services associated with them (Anonymous, 2013).

In any negotiation over cooperation with transboundary water resources, the five reasons for seeking such water consilience include: (1) poverty reduction and equity; (2) economic benefits; (3) preserving water resources; (4) environmental protection; and (5) promotion of peace. Obviously the establishment of more inclusive governance over water and improved cooperation between different users can help overcome unequal distribution of the substance and increase access to it, which are essential to basic human needs. Industries increasingly face the uncertainty and challenges of water scarcity so that cooperation between the stakeholders is an obvious benefit to economies. Similarly, cooperation helps to exchange data and information that help find joint management strategies, as well as helping overcome cultural, political, and social tensions that can build trust between different groups, water communities, regions, and nation-states. In addition it is recognized that this attention to increasing cooperation with water resources is critical to meeting the eight major Millennium Development Goals of the United Nations (Anonymous, 2014). Other benefits to water consilience and cooperation include: (1) cost aversion; (2) promotion of more efficient techniques of water storage and distribution; (3) transboundary flood management; (4) spill-over associative effects leading to regional cooperation in collateral domains (politics, economics, environment, culture); (5) resource mobilization through cooperation between municipalities and private providers; (6) improved ecological benefits and environmental benefits; (7) enhancement to facing common external threats such as climate risks and water-borne disease vectors; and (8) increasing ability to deal with water-related natural disasters. At the same time, however, and especially in regard to Afghanistan where virtually any change in the status quo is met with resistance, numerous challenges against water cooperation can arise. Thus conflicts can occur between stakeholders at different levels on a range of issues that include multiple demands on the water, the qualities and quantities of the water, possible negative impacts on associated ecosystems and communities, the development of infrastructures, economic interests that change balances of power in the region, and difficulties of obtaining the massive quantities of finances required to adequately develop the water resource.

## 18.5 PROBLEMS WITH WATER MANAGEMENT

The management of water is highly complex and extremely political (Carius et al., 2004). The balancing of competing interests over water allocations, and the managing of water scarcity, require quite strong institutions, which are not known to occur much in Afghanistan. Many reasons occur as to why water management fails, which include inadequate administrative capacities, lack of transparency, ambiguous jurisdictions, overlapping administrative functions, fragmented institutional structures, and lack of necessary infrastructure. Dysfunctional bureaucracies are a characteristic of these sorts of failure. Access to reliable water databases is fundamental to understanding and management of water, which would include the absolutely fundamental tools of associated meteorological, hydrological, and socioeconomic data for deliberate and foresighted management of the water resources. Unfortunately, in the developing countries of Southwest Asia, such data are few and far between, with the result that adequate water management is most difficult. In addition, even where some data are available, the disparities among the various riparian entities hinders their capacities to generate, interpret, or legitimize data and leads to considerable mistrust and inabilities to engage in cooperative actions.

The management of water in a number of countries, and even within the United Nations itself, can be characterized as overlapping and competing bureaucratic management among the various governmental and bureaucratic agencies. Divergent management approaches and disaggregated decision-making serve contradictory objectives that lead to competing claims from different sectors (Carius et al., 2004). These factors are likely at least part of the reason for President Karzai's original restriction of any ministry from even discussing water issues below Presidential cabinet level. Such competing claims in a country with no formal or uncorrupted system of water-use permits, or where enforcement and monitoring are inaccurate, would be natural sources of disputes. The failure to take into account the local rights and practices without sufficient participation by local communities and water users would be a further characteristic of sources of problems, with the result that some system of discovery of these factors would have to be implemented to overcome difficulties. The alternative, of course, is violent confrontations about water, as has happened elsewhere in the world (Bolivia, China, India, etc.) on a number of occasions.

## 18.6 WATER-COGNIZANT CAPACITY DEVELOPMENT

Technical capacity related to water, the understandings in the government necessary to achieve water security in Afghanistan, are human, financial, institutional, scientific, technological, and service provisioning. This requires reaching toward a high level of education, training, and engagement in the new government of the country, which is a goal that will be most difficult to achieve, even in the best of times. Because water security has so many social, humanitarian, economic, legal, and environmental dimensions (Bigas et al., 2013), these multiple factors require an equally wide range of capacities and expertise that go quite a way beyond the immediate management of the available water resources. Obviously the development of capacity at the governmental institutional level in Afghanistan is critically important, because it sets up the entire framework for capitalizing on the human capacities in the government, and must be used to coordinate multi-sectoral policies. Institutional frameworks for trans-sectoral, decision-making and policy-construction in Afghanistan are characteristically poorly developed, which adds to the complexity and difficulty of reaching these objectives. As discussed at length in Chapter 13, hydro-cognizant capacity building in Afghanistan must include all the essential elements of water science and water management in order for the country to begin to prosper in terms of water knowledge and the ability to negotiate from positions of authority as the external diplomatic pressures about water increase on the country.

To achieve water security in Afghanistan thus will require inculcation of a wide range of water-knowledge capacities that must be built up in the country. This will require a multidisciplinary approach to address adequately the satisfaction of a large range of demands on the water. This range would include such diverse and sometimes somewhat incompatible components such as drinking water, irrigation water, hydropower generation, and ecosystem maintenance. When these demands meet seasonally related, uncontrolled surpluses (floods), shortages (droughts), and incompatibilities (clean water in urban settings, etc.), the inability to plan for the natural variations can be near fatal to an agency or a particular group of people.

Increased demand in certain places and times affects trade, commerce, economic growth, and development. These essential factors of any economy require increased skills and technical capacities to assess and monitor resources, as well as to understand the impacts of global change underway to: (1) better inform about adaptation and copy strategies; (2) prepare for, and recover from natural catastrophes; (3) increase cooperation over transboundary water resources; (4) integrate and incorporate recognition of the value of the ecosystem services into government policy (which requires educational means in the schools); (5) develop and support on-going dialogues across the various governmental sectors; and (6) bring on-the-job mutual learning to the decision-makers about these factors so that they will recognize their value, and without which the GoA will not prosper in its future endeavors.

The development of capacity throughout the GoA is needed at many levels within the government to ensure a reasonable future, and will require a cross-sectoral, enabling environment that will support water security. Capacity is needed at multiple levels, from the individual, to organizational, institutional, and educational complexes, in order to reduce vulnerability to water insecurity, where the poor capacity at various levels and across disciplines is generally cited as the major hurdle to the attainment of true water security (Tables 18.6–18.9). The

TABLE 18.6 List of Management Issues and Improvements Concerning Water Resources That Are Required in Afghanistan

1. Establish of permanent qualitative and quantitative national system of water resources monitoring and measurement
2. Develop and enhance meteorological and stream gauge network to estimate (forecast) water availability through each year
3. Develop a snow/glacier hydrology information system
4. Develop a permanent hydrological data information system and a database
5. Initiate water management systems even where problems are evident, regardless of data limitations
6. Establish firm national water quality standards (both primary and secondary), and/or, a legislated national water resources regulation act for both surface and ground water
7. Develop a water safety plan for the country that controls and assures quality
8. Establish a financial resource plan (resource-corridor approach; Shroder, 2015, etc.)
9. Encourage stakeholder participation
10. Mitigate imbalance between availability and supply of water
11. Water resources management, development, protection, and sustainability
12. Define water resources potential (availability), supply and demand
13. Define sustainability of water resources for socioeconomic development and environmental security

*From Saffi, M.H., Kohistani, A.J., 2013. Water Resources Potential, Quality Problems, Challenges and Solutions in Afghanistan: An Evaluation of the Past and Present Situation of Water Resources and Consequences on the Basis of Previous Water Resource Investigation and Present National Groundwater Monitoring Wells Network Data Management, Mapping, Visualization and Assessment. DACAAR. http://www.dacaar.org/Publications/6%20Research%20and%20Studies/WASH/Report%20on%20Water%20Resource%20Porential_opt.pdf (accessed 09.11.15).*

TABLE 18.7 Factors Required in an Enabling Environment in Afghanistan to Positively Improve the Water-Management Sector

1. Establish a coordination organization within the water sector and move towards an enabling environment to make policies, strategies, and legislation for integrated water resources management
2. Produce reform within the current institutional framework to implement the policies, strategies, and legislation
3. Set up the management tools (data management, evaluation, mapping, and visualization), which can help to make better policy-relevant decisions that are required by institutions to do their job

*From Saffi, M.H., Kohistani, A.J., 2013. Water Resources Potential, Quality Problems, Challenges and Solutions in Afghanistan: An Evaluation of the Past and Present Situation of Water Resources and Consequences on the Basis of Previous Water Resource Investigation and Present National Groundwater Monitoring Wells Network Data Management, Mapping, Visualization and Assessment. DACAAR. http://www.dacaar.org/Publications/6%20Research%20and%20Studies/WASH/Report%20on%20Water%20Resource%20Porential_opt.pdf (accessed 09.11.15).*

TABLE 18.8 Factors Required in Afghanistan to Initiate Technical or Conservation Measures With Water Resources

1. Recharge ground water by surface water during peak flow periods for aquifer recharge and recovery, by directing surface water into pits, trenches, boreholes, and infiltration basins
2. Harvest rainwater from precipitation collected from clean roofs
3. Construct infiltration ponds
4. Construct water storage infrastructure, including check dams, trenches, and capturing water by making storage reservoirs
5. Install desalination plants where the areas have saline water
6. Raise awareness on efficient usages of water
7. Construct water wells to high standards, and show that improper or nonstandard water-well construction actually promotes and facilitates contamination of the ground water
8. Promote water uses from nonconventional water resources, such as harvesting rainwater and treating waste water
9. Formulate regulations for water well drilling and removal of ground water

*From Saffi, M.H., Kohistani, A.J., 2013. Water Resources Potential, Quality Problems, Challenges and Solutions in Afghanistan: An Evaluation of the Past and Present Situation of Water Resources and Consequences on the Basis of Previous Water Resource Investigation and Present National Groundwater Monitoring Wells Network Data Management, Mapping, Visualization and Assessment. DACAAR. http://www.dacaar.org/Publications/6%20Research%20and%20Studies/WASH/Report%20on%20Water%20Resource%20Porential_opt.pdf (accessed 09.11.15).*

supporting policies, laws, and infrastructures will create the proper enabling environments that will allow for increased capacity that is based upon environmental potentials and limits, along with country-specific and regionally directed needs. The policies need to extend across disciplinary and institutional boundaries in order for the management of the water resources to take place in an integrated and sustainable fashion.

In order for the water-security challenges and gaps of Afghanistan to be met in any reasonable fashion, the policy makers of the country will need to identify these problems in a coherent fashion so that they can be addressed. These assessments will need to be undertaken at the appropriate scales so as to ensure that the proper institutions are involved, and the relevant levels of national to river-basin boundaries are adhered to. At the level of the individual, assessments of human capacities to provide water services, while also adapting to climate-change variations in fresh water are essential, whereas at the institutional level the assessment of general performance and capacities to respond to changing circumstances is also critical. In most regions in fact, increasing the water-related capacities depends upon increasing the number of adequately educated and well-trained water professionals and is quite necessary. Building up the institutions required to develop and maintain adequate water management is equally essential to ensure water security in future.

**TABLE 18.9** Factors Required in Afghanistan to Build Capacity and Knowledge About Water Resources

1. Invest in building of hydrogeological expertise
2. Increase the number of water-sector professionals
3. Increase investment in water-resources management, monitoring infrastructure, and research
4. Initiate practically oriented, scientific studies of surface and ground water that focus on water interactions, availabilities, recharge characteristics, water quality, sustainability, and management alternatives
5. Engage in policy relevant research and in alternative solutions
6. Water resource data management, assessment, mapping, evaluation, and visualization for future water-resources management, development, protection, and sustainability
7. Technology transfer and research development must take place to transfer new water knowledge on to the practical field level of restoring and reestablishing old practices
8. Place stress on a public awareness campaign, the role of women, and community participation in preserving ground water against depletion and pollution
9. Initiate water management where problems are evident, regardless of data limitations
10. Initiate practically oriented scientific surface and ground-water studies that focus on water interaction, availability, recharge characteristics, quality, sustainability, and management alternatives
11. Develop an aquifer classification system and address the problem of protecting vulnerable aquifers from contamination

*From Saffi, M.H., Kohistani, A.J., 2013. Water Resources Potential, Quality Problems, Challenges and Solutions in Afghanistan: An Evaluation of the Past and Present Situation of Water Resources and Consequences on the Basis of Previous Water Resource Investigation and Present National Groundwater Monitoring Wells Network Data Management, Mapping, Visualization and Assessment. DACAAR. http://www.dacaar.org/Publications/6%20Research%20and%20Studies/WASH/Report%20on%20Water%20Resource%20Porential_opt.pdf (accessed 09.11.15).*

As with most modern endeavors, any increase of water security requires ever finer degrees of availability of appropriate data so that the quantity and quality of water resources is better known, as well as the information about the financing necessary for implementation of interventions. Knowledge of available water resources and their expected dynamics, as well as its uses and users, depend upon the sound management of that resource. This management absolutely requires data of sufficient quality that can only be acquired with adequate human and financial resources. Higher quality data produced by close monitoring and reporting on the water resources leads to an increase in knowledge; this supports the ability to make informed decisions for enhancing water security through better water-resources management.

Development of water-capacity building in a country such as Afghanistan is a long-term process based upon leadership, good governance, knowledge management, and information transfer, all of which need to be continuously organized and adapted to the need of the stakeholders, along with feedback from them about its effectiveness. This is an organic learning or educational process based around agreed-upon norms, respect for value systems, and the fostering of self-esteem to enable retention of new understandings. For this process to be successful, it needs to integrate the external information and understandings into national priorities, processes, and systems. This can only happen by building upon existing capacities, which in Afghanistan are rather thin anyway, rather than creating new ones, so the capacity-building process can be somewhat arduous. The difficulty in doing this is caused by the problematic circumstances of the Taliban insurgency against the government and the real lack of accountability to the ultimate beneficiaries, the people of Afghanistan. True development of capacity really requires responsible educational management though the monitoring of outcomes and impacts from the point of view of the beneficiaries.

The means to accomplish all the necessary water-cognizant, capacity building in Afghanistan that is necessary will require a concerted and multi-phase effort through the good auspices of the major educational systems in the country. The best or most comprehensive of the NGOs who can still operate in Afghanistan in spite of the considerable danger to them are essential as well. Even then, however, success is not guaranteed in that difficult place. The offices of the United Nations (UN), the US Agency for International Development (USAID), the World Bank, US Institute of Peace (USIP), the World Bank, the Agha Khan Foundation, the East-West Foundation and a number of other water-cognizant NGOs with appropriate local experience (Germany, Norway, Switzerland, etc.) are the most logical donors. UN-Water has been noted recently (Bigas et al., 2013) as a logical agency to address gaps in capacity by: (1) emphasizing collaborations among agencies; (2) fostering trans-sectoral coordination at the national level; (3) establishing training facilities and educational means to support capacity development in institutions; and (4) setting up a knowledge community specific to water-security challenges. This capacity development of hydro-security in Afghanistan requires the means to apply cross-disciplinary, multi-level, multi-stakeholder, and gender awareness and gender-sensitive approaches to the problems. Such approaches can be brought about through incorporating new partners and donors from sub-regional and country networks into capacity development and demand centers that can scale up capacity at all relevant levels. Other factors that can be integrated into the process include capacity-building courses, educational materials, and examples based on experience and lessons learned throughout the country, as well as technical advisory assistance from local or foreign experts that is used to identify relevant expertise. This is asking for a great deal in backward and commonly rather intransigent Afghanistan, but the alternative is resigning to failure, which is not acceptable. Only a forward-looking approach is a reasonable alternative and concerted efforts toward capacity-building must be maintained in spite of long odds against success.

## 18.7 PROBLEMS WITH WATER GOVERNANCE

The provision of good governance in any region is obviously a prerequisite for achieving any sort of water security, as has long been recognized by the international community of experts on such matters (Bigas et al., 2013). Unfortunately, Afghanistan has never been recognized as any sort of paragon of such good governance. Instead, the weak institutional and legislative arrangements, poorly enforced laws, weak to non-existent accountability mechanisms, and extensive to endemic corruption produce quite poor governance mechanisms, which clearly hamper efforts to achieve very much water security. The governance mechanisms that are necessary to achieve true water security in Afghanistan include developing proper operating capacities, transparency in the government, participation in governance, accountability rather than anonymity, and access to legal recourse. These mechanisms and objectives (Table 18.6) can be best achieved through establishment of formal agreements and processes at the national and international levels, but whether or not this can be attained remains to be seen.

Good governance over water issues must rely upon well-designed institutions and competent people that are empowered to enact and enforce legislation and instruments of policy that are themselves quite conducive to attainment of predetermined goals with social, economic, and environmental importance to water security. Such competence in government can be expressed through different organizational structures and set up according to local conditions and capacities, and in harmony with agreed-upon domestic and international policy goals. The structures of such governance must be consilient with local power groups or the warlords typical of Afghanistan, and must fit into local arrangements when setting up management systems that are designed to improve water security in a sustainable fashion for the future of the country. A review of world water-governance systems shows a fair variety of water-allocation mechanisms, such as those that are market-based, or different water-allocation methods by a range of competent to corrupt officials (Le Quesne et al., 2007).

Water governance in the world is actually an ever-evolving process. This occurs because water governance is undergoing continual revision and refinement as the process responds to new data and information. This also occurs because of new problems and new experiences as the resource grows scarcer in the face of increasing demands upon it as populations increase exponentially and the climates of the world change. The achievement of water security in a country such as Afghanistan will require institutional and regulatory support, the general capacity of change, active and adaptive water-management structures, establishment of some new structures and new forms of relationships, and development of multi-layered models that are capable of integrating the complex natural and social dimensions characteristic of the Hindu Kush (Table 18.10). A number of new legal frameworks that are international and

TABLE 18.10 Key Issues That Are Critical for Establishing Improving Water Governance

1. Establish a hierarchy of river-basin and aquifer-basin systems, as appropriate, as the fundamental biogeographic unit for water management. Ultimately this will require coordination and cooperation across regional to international borders
2. Pursue efficiency gains and provide mechanisms for dispute resolution in order to enhance equity and flexibility in the allocation of water rights between competing uses
3. Reconcile the security of water rights with risk, the uncertainty of resource supply and availability, and sustainability by using methods that entail periodic reviews of permits, avoidance of monopolies, and transfer of negative externalities
4. Prioritize both the environment and vital human rights in water allocation policies, laws, and decision-making processes, including the requirements to assess and manage the flows within the environment
5. Integrate management of surface and underground waters with modern governance methods concerning land and biological resources
6. In areas where important river basin boundaries do not correspond with ground-water basins, special attention must be paid to ensure that unsustainable pumping does not exceed allowable limits, perhaps by establishing alternative mechanisms and incentives
7. Empower water users and stakeholders to assume additional responsibility, access relevant information, seek administrative and judicial remedies, and participate in decision-making processes with regard to water management and allocation
8. Account for traditional water-allocation systems, rights, and practices at the local level
9. Strengthen understanding and risk management of water-related natural hazards by setting up early warning systems
10. Protect freshwater ecosystems of high conservation value from depredations caused by infrastructure development by designating and managing protected areas

*Note:* These concepts can constitute useful benchmarks to assess how well the institutions and legislation of Afghanistan can respond to evolving socioeconomic and environmental conditions, as well as what kinds of reform might help achieve related water-security goals.

*From Bigas, H., Burchi, S., Steduto, P., van Beek, E., MacQuarrie, P., Earles, E., Jägerskog, A., Coates, D., Gumbo, T., Leeendertse, K., Donkor, S., Canales, C., Jouravlev, A., Lloyd, G., Liu, H., Wright, H., Zarkaria, S., Aureli, A., Demuth, S., deF Doria, M., Mishra, A., Stephen, R., Chociej, Z., Qadir, M., Ardakanin, R., Liebe, J., Briggs, M., Lifeng, L., Loures, F., Meng, J.-H., Symington, M., 2013. Water Security & the Global Water Agenda: A UN-Water Analytical Brief. United Nations University, Institute for Water, Environment & Health (UNU–INWEH), Hamilton, Ontario, Canada; 38 p. http://www.unwater.org/downloads/watersecurity_analyticalbrief.pdf (accessed 10.02.14).*

global for transboundary water management have been developed in the past two decades that offer some recourse (Bigas, et al., 2013) in the face of many new problems that have developed. Different developing regions in the world, with obsolete or non-existent, frameworks of water legality (Latin America, Caribbean, etc.), have been able recently to modernize and reform water legislation and water-management systems; it very much remains to be seen, however, whether or not Afghanistan will be able to do this any time in the near future.

The financing of improvement of water governance is essential, with the private sector in many countries poised to assist in many places. Global or local businesses in Afghanistan that are seeking to make profits in resource or other development in Afghanistan (Shroder, 2014a,b) would be wise to put water on their agendas in recognition of the importance of reliable water supplies and healthy living conditions to help ensure profits upon which they can rely for future growth. A next step has been suggested by Bigas et al. (2013) that corporate water stewardship would entail the adoption of a more consistent and transparent approach to the development and enforcement of modern legislation on water issues. Cooperation of the private sector with local universities, building new capacity in the governmental bureaucracy, generating new water data, and transferring water technology into the country would all be essential aspects designed to lift Afghanistan up from its present-day, rather diminished capabilities with water management.

## 18.8 CONCLUSION

The future of water management in Afghanistan, and the capacity building, risk assessment, cooperation, and good governance that will be required to make this all happen are a large order that will require many people and a great deal of robust funding to bring into reality. This volume on transboundary water resources is an early attempt to bring some measure of these efforts into being. Considerable major effort, of course, will be required to produce effective results. In all probability, partnering with the World Bank, USAID, and cognizant and interested NGOs and academics will be necessary to attract the attention necessary to do this work and achieve a strong measure of success.

### References

Anonymous, 2013. Water Cooperation: Information Brief. United Nations. Zaragoza (Spain) Conference 2012/2013. http://www.un.org/waterforlifedecade/water_cooperation_2013/pdf/info_brief_water_cooperation.pdf (accessed 27.09.14).

Anonymous, 2014. We Can End Poverty: Millennium Development Goals and Beyond 2015. United Nations, New York. http://www.un.org/millenniumgoals/ (accessed 27.09.14).

Bigas, H., Burchi, S., Steduto, P., van Beek, E., MacQuarrie, P., Earles, E., Jägerskog, A., Coates, D., Gumbo, T., Leeendertse, K., Donkor, S., Canales, C., Jouravlev, A., Lloyd, G., Liu, H., Wright, H., Zarkaria, S., Aureli, A., Demuth, S., deF Doria, M., Mishra, A., Stephen, R., Chociej, Z., Qadir, M., Ardakanin, R., Liebe, J., Briggs, M., Lifeng, L., Loures, F., Meng, J.-H., Symington, M., 2013. Water Security & the Global Water Agenda: A UN-Water Analytical Brief. United Nations University, Institute for Water, Environment & Health (UNU–INWEH), Hamilton, Ontario, Canada. 38 p. http://www.unwater.org/downloads/watersecurity_analyticalbrief.pdf (accessed 10.02.14).

Bogardus, E.S., 1964. Principles of Cooperation. Cooperative League of the U.S.A., Chicago, IL, USA.

Carius, A., Dabelko, G.D., Wolf, A.T., 2004. Water, conflict, and cooperation. Policy brief: The United Nations and Environmental Security. ECSP Report, Issue 10; http://www.wilsoncenter.org/sites/default/files/ecspr10_unf-caribelko.pdf (accessed 27.09.14).

Dinar, A., Dinar, S., McCaffrey, S., McKinney, D., 2013. Bridges over Water: Understanding Transboundary Water Conflict, Negotiation and Cooperation. World Scientific, Hackensack, NJ, USA.

Le Quesne, T., Pegram, G., von der Heyden, C., 2007. Allocating Scarce Water: A Primer on Water Allocation, Water Rights and Water Markets. Water Security Series 1. World Wildlife Fund, Godalming, Surrey, United Kingdom. http://assets.wwf.org.uk/downloads/scarce_water.pdf (accessed 10.02.14).

Nordland, R., Walsh, D., 2014. President is sworn in, and shares the stage. The New York Times International, Tuesday, September 30, 2014, A6.

Shrestha, A.B., Wahid, S.M., Vaidya, R.A., Shrestha, M.S., Molden, D.J., 2013. Regional water cooperation in the Hindu Kush Himalayan region. In: Griffiths, J., Lambert, R. (Eds.), Free Flow: Reaching Water Security Through Cooperation. United Nations Educational, Scientific and Cultural Organization, Paris. UNESCO Publishing; Tudor Rose: 65–69. http://unesdoc.unesco.org/images/0022/002228/222893e.pdf (accessed 10.01.14).

Shroder Jr., J.F., 2012. Afghanistan: rich resource base and existing environmental despoliation. Environ. Earth Sci. 67 (7), 1971–1986. http://dx.doi.org/10.1007/s12665-012-1638-7.

Shroder, J.F., 2013. Building resource corridors in Afghanistan: a solution to an interminable war? Earth September, 25–31.

Shroder Jr., J.F., 2014a. Developing Afghanistan resource corridors in an environmentally sound manner. Earth April, 8.

Shroder, J.F., 2014b. Natural Resources in Afghanistan: Geographic and Geologic Perspectives on Centuries of Conflict. Elsevier, San Diego, CA.

Shroder, J.F., 2015. Progress with Afghanistan extractive industries: will the country know resource success or failure evermore? Extract. Indust. Soc. 2, 264–275.

Subramanian, A., Brown, B., Wolf, A., 2012. Reaching Across the Waters: Facing the Risks of Cooperation in International Waters. Water Papers, Water Partnership Program. The World Bank. http://water.worldbank.org/sites/water.worldbank.org/files/publication/WaterWB-Reaching-Across-Waters.pdf (accessed 10.01.14).

PART III

# DISTANCE LEARNING IN CENTRAL AND SOUTH ASIA



Several of the authors of this book have taught at university level or in other educational capacities in Southwest and South Asia in the past and are well familiar with the "slings and arrows" of somewhat outrageous teaching fortunes that are too often visited upon the perhaps rather typically naïve university lecturer. Leaving cultural differences aside, however, and pausing for a moment to consider the stakes, teaching about the life-blood waters of the region to the more interested or intellectual inhabitants of the area would seem a generally advantageous endeavor. Accordingly, after decades in Southwest Asia spent watching all the local water hazards, flash floods, droughts, long lines to the pumps, and the disgusting pollution of sewage, plastic floaters, body parts floating by from incomplete cremations, and general water mayhem, any educational process that raises water awareness would seem beneficial. That is what this section of the book is about.

Several years ago, information coming from Afghanistan indicated a general lack of desire, or outright prohibitions about talking to foreigners about water, which seemed so glaringly at odds with common sense for addressing looming problems that we recognized an urgent need to develop a means for dialog. As a result, several of us in various agencies and universities sought ways to address the problems about water before they became even more acute than they were already. Production of distance-learning modules on water education to load on the web seemed like a good idea, providing that they could be translated into appropriate regional languages, and especially if enough band-width could be reasonably assured to present the detailed graphics. In addition, if needed, we thought it possible that the political and religious censors could be overcome or bypassed with physical transmission of educational materials on disk, or sent by courier or by email, and quality control could be maintained through rapid response to suggestions for modification, and with repeated iterations.

The DLMs are distillations of most of the topics discussed in this book, but some topics are far too complex to be understood in detail from only the bulleted presentations that we present in the DLMs. Nonetheless, the DLMs are loaded onto the website of the Afghanistan Studies Program of the University of Nebraska at Omaha at http://www.unomaha.edu/international-studies-and-programs/center-for-afghanistan-studies/academics/transboundary-water-research/index.php. Access to these materials will be by password, linked to a directory of approved personnel entitled to avail themselves of the information and questions that we have presented. In this fashion, we are seeking to raise the levels of understanding of major issues with water that will enable the newly risen Afghanistan nation state to enter into a new age of greater water-cognizance. We feel that only by this means will the needed knowledge on water issues become more universally available and amenable to improved understandings.

This book is an attempt to get many of the problems and the best ideas on water in the region into one place where they might be more easily used by others. Suggestions for improvement or requests for addition of new materials are encouraged to be sent to the senior author (email address elsewhere in the book). We are aware that some of these DLMs are rather too simplistic and do not really enter into any detailed expositions and explanations of the kind needed to address the many serious issues or problems that so bedevil the inhabitants of Central and South Asia. In fact, in Afghanistan especially a generation or more of good water engineers have been lost to the three decades of grievous violence, and it will take much time to train them back again. Thus this effort is designed merely to initiate the process of better understanding about water.

Our water-cognizance groups at the University of Nebraska, led in this instance by the six members of the Centers for International Studies and Afghanistan in Omaha, as well as the Daugherty Water for Food Institute based in Lincoln, seek to develop these DLMs to such a level that they can be expanded and augmented, first in the standard English language of science, and then translated into a variety of the local languages of Dari, Pashto, Urdu, and Tajiki to make the applications more locally available. Perhaps in this fashion the universality and applicability of the materials on water will be increased sufficiently to attract the attention necessary to edit the old and enable us to add new materials. Certainly the brevity and oversimplifications required to establish these materials in an electronic medium are recognized by us as a problem in need of further refinement. The needs in the acute situation of declining water availability in the midst of already reduced knowledge about water, however, was clear and thus we have established some 21 specific distance learning modules that are explained further in the following chapter.

CHAPTER

# 19

# Distance Learning Modules for Water

*J. Shroder*

**University of Nebraska at Omaha, Omaha, NE, United States**

OUTLINE



## Abstract

Distance learning modules (DLMs) are a means of transmitting information electronically to people able to assimilate the new knowledge. More than 21 DLMs have been constructed about the transboundary water resources of Afghanistan, with attention to the hydrologic cycle of the world, and its activity in Southwest Asia, especially in rivers and lakes. Measuring and managing the various elements of water figure prominently, whereas the implications of its diminishment or hazardous interactions with humanity through its catastrophic effects or through climate change offer the most serious impetus for seeking solutions to the most serious problems. Management of the vital resource is imperative, whereas adoption of market mechanisms to effect its distribution downstream seems a fair possibility, which new laws and treaties could help to make happen. Dams and irrigation works on all the major and minor tributaries will continue to provide solutions, even while adding to the complexity, along with the possibly rather severe hydropolitics that can happen within country and across borders. Underground water tends to be a great unknown in need of much further investigation and elucidation in terms of quantity and quality. Hydropolitics in Afghanistan are complex enough to demand close attention to major issues at hand. One of the essential issues is hydro-hegemony, or the manners in which nation-states manifest power and control over water resources and upstream-downstream distributions, is a newly recognized situation of potential great value to Afghanistan, but which is highly controversial. This book seeks to shed greater light on the topic of transboundary water resources from Afghanistan, with the hope of the authors that this will offer greater prospects for peacemaking in a contentious region long known for volatility and violence.



*Transboundary Water Resources in Afghanistan*
http://dx.doi.org/10.1016/B978-0-12-801886-6.00019-7

**Keywords:** Hydrologic cycle, Hydro-cycle in Southwest Asia, Rivers of Hindu Kush, Pamir, Hindu Raj, Afghanistan lakes, Measuring water, Climate change implications, Afghanistan transboundary resource management, Floods, Natural-hazard management, Market mechanisms for water sharing, Water law, Water treaties, River dams, Kabul-Panshir-Kunar river hydropower, Irrigation, Panj and Amu Darya hydropower and irrigation, Helmand hydropower, Helmand irrigation, Water in Islam, Afghanistan ground-water issues, Hydro-capacity building, Water information sharing, Upstream-downstream issues, Glacier inventory, High technology, Grant-money partnerships, Hydro-hegemony, Afghanistan hydropolitics

## 19.1 INTRODUCTION

This book was originally conceived by the senior author (JS) in an effort to continue the analysis of Afghanistan's resource problems that were unearthed in the effort to better comprehend all the enigmatic resource situations that had emerged in prior studies (Shroder, 2014a,b). The many issues of water in a chronically aqueously challenged and water-short place were discovered to be partly matters of multiple factors. Such factors include natural weather variation, coupled with long-term climate change, machinations by nearby foreign powers, multiple depredations caused by decades of civil war and invasion by outsiders, and no little water ignorance among the Afghan population at large, caused by the wholesale destruction of the educational system as a result of the near-ceaseless violence for much of the past four decades. Thus this book was organized to include a host of topics on water that would contribute some possible solutions to a number of difficulties in the generally fraught society of Afghanistan.

Hovering in the background of any discussion about water in Southwest and Central Asia are the arid winds of climate change that periodically blow over the nation with recurrent droughts that so threaten even simple existence. As these things go, generally climates do not suddenly switch from one condition of climatic equilibrium, to a new wetter or drier one, but instead various forms of a new drier, quasi-equilibrium seem to come and go on rather unpredictable schedules, with no one being able to see quite what will happen next. The environment of Afghanistan has long been threatened by all sorts of merciless natural and anthropogenic degradation (Shroder, 2012), with the result that the margins of toleration for further environmental diminishment through additional water shortage are generally quite small. Many people are thus seeking answers to how best to deal with water shortages as population pressures mount for ever greater access to the quite finite resource.

One partial solution that was seen by the authors of this volume includes education about the complex natures of the problem. We thought it likely, from our positions in decades of education about the geology and physical geography of Southwest Asia, that some of the water problems could be solved by clever engineering, others by careful education about how best to use, reuse, and not further diminish an already steadily declining resource. If people were fully aware of their precious water resources, perhaps they could see how better to handle the water in their lives, rather than taking it so much for granted. Indeed, because water is so essential to all aspects of even simple human existence anywhere, it is quite taken as a given that it must be more or less available nearly everywhere. Of course no one legitimately thinks that this is really true, as anyone living far from a water source or tap knows full well in their thirst, but still, the assumption exists that because water is such a common part of the Earth's crust, it will be available somewhere nearby in almost every situation. Helping people to deal with this as continual reality in their lives is, and has been, a driving theme for this written work.

When the first ideas were conceptualized about how this book would be organized and disseminated, we recognized that the information contained herein would be complex and not easily assimilated by a people whose educational background was limited necessarily by inattention in the midst of four decades of war. Nonetheless, we determined to seek ways to transmit good and reliable information into the war-torn and often volatile region in an attempt to help head off many knowledge gaps and misunderstandings about all things aqueous, even while we recognized that this would require a fairly large and likely rather expensive work, especially because so much effort would be necessary to accomplish anything of substance.

An additional effort was made by us to conceptualize exactly how we would get this information to the people who most needed it inside Afghanistan, but also into the nearby neighboring countries. The authors of this volume are all fluent in the English language, of course, but we directly recognized the necessity of providing translations into the regional languages of Dari, Pashto, Urdu, and Tajiki as likely necessities. A large problem with this, however, emerged in that, with the exception of the scientific Russian that most educated Tajikistanis speak and read as a matter of historical happenstance, several of the other languages lack most of the necessary scientific terminology to discuss the needed material on water issues. Generally in the experience of the senior author (JS) in his Fulbright lectures (Kabul University, 1977–78 and Peshawar University, 1983–84), the lack of technical vocabulary leads to the direct use of the untranslated foreign scientific terminology entire, which obviously decreases comprehension. Nonetheless, no particular solution to this terminology problem is apparent, other than teaching people what the scientific words mean. This, of course, also quickly leads to the need for introductions to the newest, or state-of-the-art technical capabilities, available in remote sensing, GIS, and data acquisition that have generally been denied to Southwest and Central Asia as a result of poverty of access to hard currency, the increasingly common but general unavailable high technology, and the lack of familiarity with aerospace exploration of high mountain areas.

We also noticed that we probably would have to develop a series of workshops to ask a great many questions about what government people would need or want to know in future. It was clear that the Afghanistan government ministries would need to become more fluent with water issues in order to deal better with the myriads of water details they would ultimately have to face, probably rather sooner than later. Because President Karzai had made such a big effort to squelch all the prior efforts made by the UN, the World Bank, and the US Agency for International Development (USAID) to jumpstart negotiations with the neighboring countries about water resources, it was clear to us that we needed to ascertain new ways to get water discourse on the table for discussion as soon as possible.

In fact, in this particular environment in Southwest Asia, we knew that practicalities would demand some sort of value-added procedure that would make it obviously worthwhile for the Government of Afghanistan (GoA) to pay attention to our requests to hold one or more academic workshops for government personnel and university professors and students. We saw this value-added procedure to be the explanation of a number of new ways of thinking about water in the Western universities and think tanks of which we were aware, as well as especially the great value of the many technologies coming from new generations of space-borne sensors in the modern-generation satellites. Many Pakistani students have availed themselves of education about these new technologies in the West, especially in universities in Europe and the USA, but because of historical animosities during the Cold War, and especially the past 40-odd years of war in Afghanistan, this has generally not been possible for students from Afghanistan or Tajikistan. Furthermore, a whole generation or more of hydrologically skilled water engineers has been lost from Afghanistan because of the long war and now must be retrained.

In the 1970s, UNO had been in the forefront of delivering high-quality, engineering education to Kabul University, as well as initial efforts at education in remote sensing, especially for its value for assessing glacier ice in the Hindu Kush and Afghan Pamir (Shroder, 1978, 1980, 1989). At that time also, the United Nations in Switzerland and the U.S. Geological Survey had asked the senior author (JS) to undertake an initial remote-sensing assessment of the glaciers of Afghanistan and Pakistan with early Landsat satellite imagery. This work was interrupted by the communist coup and takeover of the Afghanistan government in 1978 but continued anyway after successful removal of maps and imagery from Afghanistan through use of the diplomatic pouch.

In 1983–84, the senior author (JS) moved again to Pakistan to continue work with remote sensing of glaciers in the Western Himalaya, with Nanga Parbat and the Karakoram receiving the most attention. The Afghanistan region had to await renewed attention to the area in the 1990s and 2000s as UNO was asked by the USGS to become the Southwest Asia (Afghanistan and Pakistan) Regional Center for GLIMS (Global Land Ice Measurements from Space). The GLIMS project was a combined effort on the part of the USGS and NASA to use ASTER and other satellite imagery to assess the world ice mass as the climate changed. Much remote-sensing research resulted, as well as many publications (Bishop et al., 2014; Haritashya et al., 2009; Sarikaya et al., 2012; Shroder and Bishop, 2010a,b).

All of this work on the frozen phase of the $H_2O$ system in the mountains set standards for high-level assessments of this type, which is a remote-sensing methodology that has continued to evolve in useful ways that need to be transmitted to, and used effectively by, the scientists of Afghanistan, Pakistan, and Tajikistan. Thus our proposed 2015 workshops (see Chapter 20) in Islamabad, Pakistan, Kabul, Afghanistan, and Dushanbe, Tajikistan were designed to discover more of what the government officials and university personnel wanted to know about studying and understanding all aspects of the hydrologic cycle in their region with which we might be able to help them. Direct transfer of ASTER imagery to any of these countries assigned to our GLIMS Regional Center for Southwest Asia was not an easy task, partly because of rather simplistic security fears that such could be used for ill by certain unscrupulous persons might use the imagery or technology to support the Taliban or Al Qaeda in the region. In addition, the presently existing unfamiliarity of most scientists in Afghanistan with the appropriate methodologies to interpret such imagery beyond mere low-level "photo interpretation" was regarded by us as a detriment to be able to transfer any of these technologies. The workshops were thus designed in part to ascertain people and places with whom we might usefully work to develop new projects that would be mutually beneficial.

In addition to the conceptualization of our new projects on the hydrologic cycle in Afghanistan and several of the surrounding countries, we needed a modicum of funding to enable us to proceed. Accordingly, grant proposals to the Lounsbery Foundation and the U.S. Institute of Peace were developed successfully. These grants provided travel funding and office support as a kind of seed-money funding from which further funding work might be possible, once the possibilities were explained to the appropriate government and university personnel through the workshop medium. The funding also enabled production of the distance learning modules (DLMs) that could be used for capacity building through loading and use on our Transboundary Water Website, as explained in Chapter 13. These DLMs were the outgrowth of early plans for production of what were referred to initially as massive open online courses (MOOCs).

## 19.2 MASSIVE OPEN ONLINE COURSES

The idea for development of MOOC transmission of new information to legions of avid and eager students worldwide first occurred in 2011 (Selingo, 2014). The idea of exactly what was implied in a popular but free course taught by top educators was a surprise to many people in terms of the actual massiveness that could result. The idea was much ballyhooed and leapt into eagerly by many of the major universities. Other problems soon emerged, however, such as the high amounts of true physical labor required, lack of credibility that the MOOC idea would amount to anything serious, lack of adequate prerequisites on the part of the fervent but inadequate students, degree-credit cheating, laborious grading, and other difficulties (Pappano, 2012). MOOCs have rapidly evolved into various methods of presentation: from the so-called connectivist cMOOCs that focus on community and connections between people to help transmit the information, to the extended or xMOOCs that focus on scalability in size and self-motivation, so they have to be completely automated to be at all possible or even marginally effective. Every MOOC of whatever type has to developed according to some very negotiable elements that need to be worked out for every particular problem (Fig. 19.1; Table 19.1). The result in the case of the hydrocognizance we seek to develop in the many of the government agencies and universities in Afghanistan was a beginning set of DLMs that might be further refined later to accomplish some of the objectives.

## 19.3 WATER-AWARENESS DISTANCE LEARNING MODULES

The DLMs developed to increase the familiarity with water resources in Afghanistan were as eclectic a mixture of topics revolving around aspects of water in the country, as one might expect. This is because of all the other diversities and complex interactions that abound in the museum of ethnicities, extreme topography, and natural processes at work in Afghanistan already (Table 19.2). The 21 DLMs were selected fairly much at random, rather like an introductory course on the geography of water in Afghanistan that could be constructed; we did,

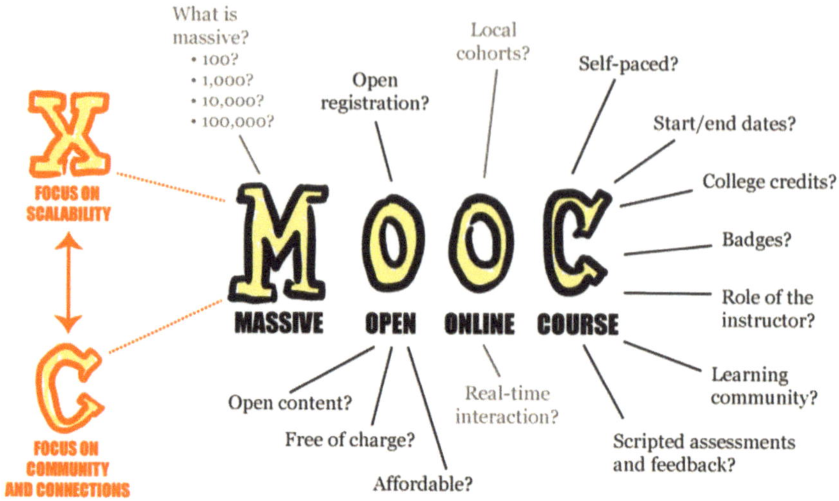


FIG. 19.1 Cartoon of the MOOC idea showing the highly fluid nature of exactly what might be needed in the construction of any similar instructional course situation. In the case of water awareness under consideration here, the development of greater knowledge base about water issues in Afghanistan was clearly necessary, with the result that greater familiarity learned about MOOC characteristics and presentation requirements shown here resulted in the more constrained, 21 distance learning modules (DLMs) that could be presented online efficiently, but which would not overtax the resources of the presenters.

TABLE 19.1 Problems With Development of Massive Open Online Courses (MOOCs) That Can Be Partially or Completely Overcome by Attention to Focus on Scalability (xMOOC) and Focus on Community and Connections (cMOOC) to Password-Protected, Distance Learning Modules That Have Been Developed Here (Table 19.2)

1. What is massive in a course (100, 1000, 10,000, 100,000 students)? Running a completely open course with no constraints can result in unmanageable situations if the course proves popular.
2. Should registration be open to anyone without restraint?
3. Should local cohorts to help manage the courses be retrained?
4. Should the courses be self-paced?
5. What would the start and end dates be set up and announced?
6. Should college credits be made available for each course?
7. Should badges or certificates of completion be made available to show the success of students?
8. What would the role of the instructor be and how involved with individual students would the instructor or local cohort be?
9. What is the nature of the learning community in terms of level of education, literacy, and engagement so as to advise in the construction of the courses?
10. Would any assessments of the students be made or any feedback for improvement of the courses be sought?
11. Would any real-time interaction between students and instructors or local cohorts be possible?
12. What would the expenses be to take such courses and would the courses be affordable to the students?
13. Or could the courses be made available free of charge?
14. Will the content of the courses be open to anyone to assess or should certain controversial or politically sensitive discussion be constrained?

TABLE 19.2 Distance Learning Modules (DLMs) for Water Issues in Southwest and Central Asia

| | |
|---|---|
| DLM 1 | Hydrologic Cycle |
| DLM 2 | Hydro-cycle in Southwest Asia |
| DLM 3 | Rivers of Hindu Kush, Pamir, Hindu Raj |
| DLM 4 | Afghanistan Lakes |
| DLM 5 | Measuring Water |
| DLM 6 | Climate Change Implications |
| DLM 7 | Afghanistan Transboundary Resource Management |
| DLM 8 | Floods and Natural-Hazard Management |
| DLM 9 | Market Mechanisms for Water Sharing |
| DLM 10 | Water Law and Water Treaties |
| DLM 11 | River Dams |
| DLM 12 | Kabul-Panshir-Kunar River Hydropower and Irrigation |
| DLM 13 | Panj and Amu Darya Hydropower and Irrigation |
| DLM 14 | Helmand and Hari Rud Hydropower and Irrigation |
| DLM 16 | Afghanistan Ground-Water Issues |
| DLM 17 | Hydro-capacity Building |
| DLM 19 | Water Information Sharing and Upstream-Downstream Issues |
| DLM 20 | Glacier Inventory |
| DLM 21 | High technology and Grant-Money Partnerships |
| DLM 22 | Hydro-hegemony |
| DLM 23 | Afghanistan Hydropolitics |

Each of these DLMs has a text for interested people to read and commit to memory, as well as a series of figures and tables so that questions can be answered later by those seeking to improve their understandings of the vital resource. The numbering of the DLMs is not completely consecutive (missing 15 and 18) because of the desire to leave openings for later appropriate additions. Several of the DLMs (Hydro-hegemony, Afghanistan Hydropolitics) are controversial and so must be revised extensively after external review as needed and restricted in use until they are more perfected.

however, remove a number of extra issues from immediate production as too controversial, or already well-known enough to overlook for now, or not quite necessary yet.

Ranging from pure physics, through regionally specific activities, to those topics known to be important but not yet well developed enough to present formally, the DLM lists represent our best thinking to date. We expect that many other DLM topics can and will be developed on an as-needed basis in the course of this work for the Lounsbery and USIP projects. Additional efforts that need a great deal more work are: (1) translations into the necessary languages of Dari, Pashto, Urdu, and Tajiki; (2) an investigation of band-width problems of presentation of the complex and graphics-laden DLMs online; (3) an assessment of review and censorship of the DLM products as being appropriate for access online inside Afghanistan; and (4) development of the password-protected Transboundary Water Website for Afghanistan.

First, the basic hydrologic cycle of the world is presented, with a modicum of understanding about temperature and pressure controls, transport of phases between locations, energy sources, phase changes, latent heats, and the locations of water types on the planet; all of these figure in detail about the basic science of water. Application of the hydrologic cycle to Southwest Asia discusses the sources of water and winds, together with information about precipitation, solid $H_2O$ in mountains, melting of snow and ice to run downstream, associated soil erosion, drainage systems, and climate controls. Glacier inventory is the measure of the storage aspects of ice resources in the high mountains. Afghanistan glaciers are discussed as some of the early inventory work enabled greater understanding of a commonly overlooked but essential part of the hydrological cycle in the country. The resulting major river systems that head in Afghanistan, the Kabul, Helmand, Hari Rud-Tenjen, and Murghab, and the Amu Darya are discussed in reasonable initial detail, before later more detailed treatments of certain special cases. Afghanistan lakes of many different types and sizes receive special attention.

In recognition of the inability of not being able to manage what you cannot measure, special attention has been paid to developing abilities to measure volumes of river water that Afghanistan has been sending down river without much thought to other countries for millennia. Such measurement is essential as climate change affects the volume of that water delivered, and makes life ever more difficult for both the upstream country of water origin and the downstream recipient of whatever is available. Transboundary water management is assisted greatly by access to advanced water accounting methods that enable accurate assessments of where the water is available, how it is used, and where it goes and in what phase state it exists before it is no longer accessible to humans. In its excess, water either floods or causes horrendous landslides and avalanches; in its dearth, damaging droughts develop. Greater availability of water in future to downstream nations may involve negotiated payments for upstream environmental protection or ecosystem services (PES) that preserves watersheds. This may be a new and quite benign method to change certain malignant behaviors into water-saving procedures of both the upstream riparian landowners and the downstream water recipients.

The complete lack of any viable or up-to-date water treaties of Afghanistan with any of its neighbors indicates that great room for improvement is certainly possible. Renewed negotiation is indicated as a necessity, once enough new water knowledge is transmitted. A discussion of existing dams and planned new irrigation constructions for all of the major drainages in Afghanistan helps bring an overview of that vital necessity, whereas discussions of hydropolitics and ground water add further depth to the DLM presentations. All of this in the paragraphs and chapters above in this book are the material about water that needs to be transmitted to the government and university personnel of Afghanistan as a first group of hydro-capacity-building DLMs.

DLMs can also include material on hydro-hegemony, or the exercise of national power over transboundary water, which because it also includes the means by which such effort can be levied against other nation-states as well, can be a highly contentious topic. The high technology aspects and grant-money partnerships offer a means to gather the funding and external expertise to study some particular aspect of water in one's country, particularly through the use of the plethora of new remote-sensing satellites. A DLM on information sharing and various issues between upstream and downstream riparian partners also offers new information and ways of thinking about water that may be advantageous to learn about. Certainly a variety of other DLM topics is also possible and should be considered if anyone has good ideas for development of other such topics in future.

## 19.4 CONCLUSION

The DLMs are complete at UNO and are being translated into the local languages of Dari, Pashto, Urdu, and Tajiki. They will be loaded online for anyone to use who has the password and access to the internet. We at UNO anticipate several things: (1) that additional ideas and materials will need future exposition; (2) that some mistakes may have occurred and will need to be corrected; (3) that some topics may be controversial and will draw the ire of one government or another; (4) that if enough controversy erupts, it may be necessary to modify or withhold certain statements to achieve political harmony; (5) that in spite of our desire to modify language so that it is acceptable to the maximum number, it may be necessary that the DLMs will need careful vetting by the different governments; (6) that if the DLMs are ignored by too many governments they will be seen as either not working or in need of serious revision; (7) that if the DLMs are found to be useful and are adopted by any of the governments, their value may enhanced by special material provided by the governments; (8) that Tajikistan's scientists are the first to exhibit considerable interest in the DLMs, as noted in the following Chapter 20; and (9) that Afghanistan's scientists, for whom the DLMs were initially developed the most, have not yet discussed the DLM idea in any concerted way but are expected to weigh in when they are granted access. The same may be true for the Pakistani scientists, although because Pakistan may feel as though it has the most to lose in any water discussion with Afghanistan, and because Pakistanis are fairly well educated at higher levels, feedback to UNO on the DLMs may be more limited from them.

The full implementation and later editorial revision or addition to the DLMs will require additional grant money to effect, because no budget from the taxpayers of the State of Nebraska can be expected to be used for this purpose. If the DLMs are discovered to be useful, however, the obtaining of additional funding is not viewed as any great hurdle to achieve.

## References

Bishop, M.P., Shroder Jr., J.F., Ali, G., Bush, A.B.G., Haritashya, U., Roohi, R., Akif Sarikaya, M., Weihs, B.J., 2014. Remote sensing of glaciers in Afghanistan and Pakistan. In: Kargel, J.S., Leonard, G.J., Bishop, M.P., Kääb, A., Raup, B. (Eds.), Global Land Ice Measurements From Space. Springer-Praxis, Heidelberg. ISBN-10: 354079817X | ISBN-13: 978-3540798170 (Chapter 23).

Haritashya, U.K., Bishop, M.P., Shroder, J.F., Bush, A.B.G., Bulley, H.N.N., 2009. Space-based assessment of glacier fluctuations in the Wakhan Pamir, Afghanistan. Clim. Chang. 94, 5–18.

Pappano, L., 2012. The year of the MOOC. The New York Times. 2 November; http://www.nytimes.com/2012/11/04/education/edlife/massive-open-online-courses-are-multiplying-at-a-rapid-pace.html?_r=0 (accessed 24.3.15).

Sarikaya, M.A., Bishop, M.P., Shroder, J.F., Olsenholler, J.A., 2012. Space-based observations of eastern Hindu Kush glaciers between 1976 and 2007, Afghanistan and Pakistan. Remote Sens. Lett. 3 (1), 77–84. http://dx.doi.org/10.1080/01431161.2010.536181.

Selingo, J.J., 2014. Demystifying the MOOC. New York Times; Education Life: 29 October. http://www.nytimes.com/2014/11/02/education/edlife/demystifying-the-mooc.html?_r=0 (accessed 24.03.15).

Shroder Jr., J.F., 1978. Remote sensing of Afghanistan. Afghan. J. 5, 123–128.

Shroder Jr., J.F., 1980. Special problems of glacial inventory in Afghanistan. In: World Glacier Inventory Proceedings Reideralp Workshop, September 1978 (IAHS-AISH) Publication No. 126, Hydrological Sciences Bulletin, pp. 142–147.

Shroder Jr., J.F., 1989. Glacierized areas of Afghanistan. In: Haeberli, W., Bosch, H., Scherler, K., Ostrem, G., Wallen, C.C. (Eds.), World Glacier Inventory, Status 1988. IAHS (ICSI)-UNEP-UNESCO, Teufen, Switzerland. pp. C39–C40, C346–C353.

Shroder Jr., J.F., 2012. Afghanistan: rich resource base and existing environmental despoliation. Environ. Earth Sci. 67 (7), 1971–1986. http://dx.doi.org/10.1007/s12665-012-1638-7.

Shroder, J., 2014a. Natural Resources in Afghanistan: Geographic and Geologic Perspectives on Centuries of Conflict. Elsevier, San Diego, CA, USA.

Shroder Jr., J.F., 2014b. Developing Afghanistan resource corridors in an environmentally sound manner. Earth 8.

Shroder Jr., J.F., Bishop, M.P., 2010a. Glaciers of Afghanistan. In: Williams, Jr. R.S., Ferrigno, J.G. (Eds.), Satellite Image Atlas of Glaciers of the World—Asia, pp. 167–199. U.S. Geological Survey Professional Paper 1386-F3. Washington, DC.

Shroder Jr., J.F., Bishop, M.P., 2010b. Glaciers of Pakistan. In: Williams, Jr. R.S., Ferrigno, J.G. (Eds.), Satellite Image Atlas of Glaciers of the World—Asia, pp. 201–257 U.S. Geological Survey Professional Paper 1386-F4. Washington, DC.

CHAPTER

# 20

# Concluding Transboundary Water Activities

*J. Shroder*

**University of Nebraska at Omaha, Omaha, NE, USA**

OUTLINE



*Transboundary Water Resources in Afghanistan*
http://dx.doi.org/10.1016/B978-0-12-801886-6.00020-3

## Abstract

*Transboundary Water in Afghanistan* is a book designed to include the latest information on the vital water supplies that emanate from the Hindu Kush to flow outward to the neighboring countries. Throughout the course of writing this book the authors have been involved with a variety of comprehensive scientific seminars, delivery of papers, and meetings with university and government agencies devoted to aspects of transboundary waters in this part of Asia, such as: (1) Central Asia Workshop Report, Istanbul—2014; (2) Dushanbe Forum of Mountain Countries—2015; (3) Seminar on Transboundary Water Resources of Pakistan and Afghanistan—Islamabad—2015; (4) Meeting with Afghanistan Ministry of Energy and Water—Kabul—2015; (5) Formulation of Integrated Comprehensive Transboundary Water Resource Center (TBWRC); (6) 5th International Hindu Kush Geosciences Conference (HKGC, Kabul—2015): (7) Mountains of our Future Earth International Mountain Conference, Perth, Scotland—2015; (8) Rational Use of Water Resources of Transboundary Rivers of Tajikistan and Afghanistan—Dushanbe—2015. The many engineering and other hydrological activities in Afghanistan increasingly planned or already underway are likely to be of considerable interest to surrounding countries; this book is a first attempt to put all these issues out for public discussion.


**Keywords:** Water symposia, Central Asian workshops, Dushanbe, Kabul, Islamabad, Ministry of Energy and Water, HinduKush Geosciences Conference, Mountains of our Future earth, Perth, Scotland, Transboundary Water Resource Center (TBWRC)

## 20.1 INTRODUCTION

This book, *Transboundary Water Resources in Afghanistan: Climate Change and Land-Use Implications*, is our effort to face up to highly contentious and commonly misunderstood matters of water in a part of the world where "highly contentious" and "misunderstood" are generally freighted with high levels of associated violence. It is obviously important to attempt to work out some of these problems as much as possible to help people's lives. The alternative is possibly yet another tipping point to an already failed state that wants so much to rise out of that benighted condition. In combination with the major program to develop resource corridors in Afghanistan (Shroder, 2013, 2014a,b, 2015), this information about the many different water issues in the country can help alleviate some of the current or looming problems by providing alternate thinking modes.

Even as transboundary water from Afghanistan flows from the highlands of the Hindu Kush into the surrounding neighboring countries, it offers much potential for development of new understandings about it. In the course of such flow, the concomitant educational transmittal of aqueous information should add to its importance for all. If this book succeeds at all in these endeavors, even in some small ways, it will have been a success because so little good fortune has ever been associated with the precious resource in this region. The people of Central and Southwest Asia certainly deserve some measure of advantage about this most vital of all the elements of life; perhaps some of the information imparted in this work can serve usefully in that endeavor. Water wisdom is as precious as the fluid itself; without it failure is certain. With it, the future beckons more benignly, rather than the dark uncertainty that lurks too much around the rather cryptic and uncertain transboundary water resources of Afghanistan.

In addition to ideas about water management in Afghanistan, of course, are the many relevant situations concerning water with all its factors of access, measurement, and use in the countries neighboring Afghanistan. The many water-related problems within, and between the countries of South and Central Asia has been noticed by a wide variety of foreign agencies and other governments, so much so that a variety of public and private conferences and symposia have been undertaken over the past few years to try to get issues on the table and discussed with greater transparency. As an example, the East-West Institute (EWI), International Water Association (IWA), and the International Union for Conservation of Nature (IUCN) hosted one such workshop in Istanbul in summer 2014 that exemplifies the issues with which Afghanistan and its neighboring countries must deal (Dalton, 2014).

## 20.1.1 Central Asia Workshop Report, Istanbul—2014

Thus the Amu Darya is the largest river in Central Asia and is part of the Aral Sea Basin linking Afghanistan, Kyrgyzstan, Tajikistan, Turkmenistan and Uzbekistan. These five countries are heavily dependent on the river and its tributaries for food, water and energy; however cooperation on the use of water resources in the Amu Darya Basin has long been quite limited unless forced by external power such as the Soviet Union while it existed. Two of these countries (Turkmenistan and Uzbekistan) utilize far more water than they produce, and so they are quite beholden to, but resentful of their debt to Tajikistan and Afghanistan upstream (Fig. 20.1; Table 20.1). As the regional population expands inexorably and economic demands in the basin increase, the riparian countries are most concerned they will need more

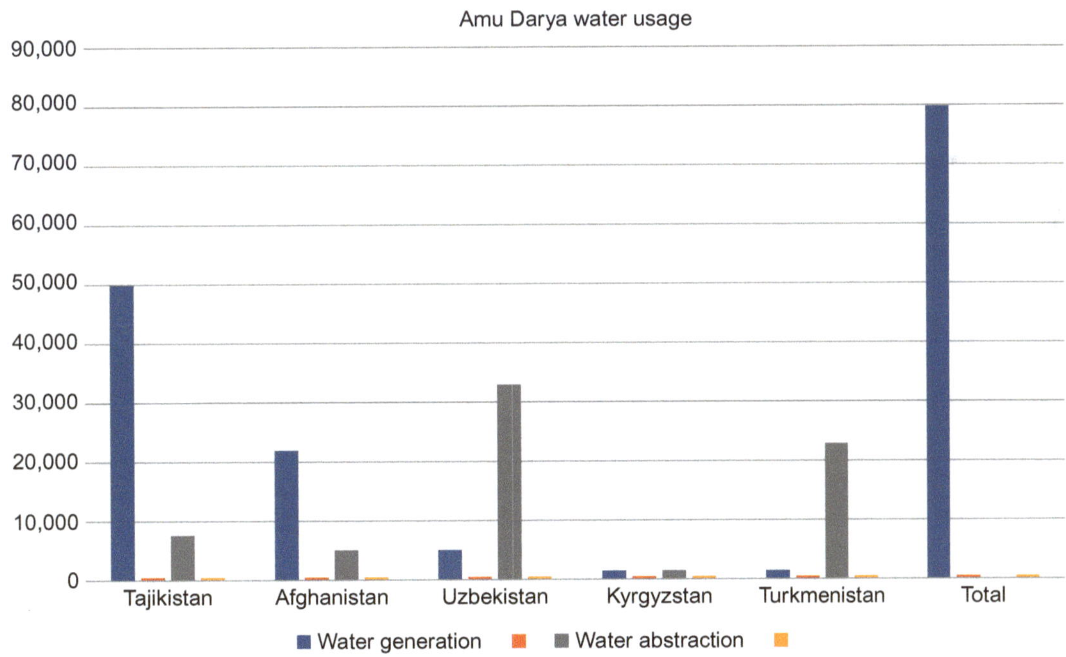


FIG. 20.1 Amu Darya water usage. Generation of the water is in *blue*, whereas that which is used is in *gray*. Tajikistan and Afghanistan clearly produce most of the water, whereas the other countries downstream that border Afghanistan use by far the largest share of the water.

TABLE 20.1 Amu Darya Water Flow and Usage

| Riparian Country | Generated Average Annual Flow in Million m³ | Share of Total Average Annual Flow in % | Average Annual Water Use in Million m³ | Share of Total Average Annual Water Use in % |
|---|---|---|---|---|
| Tajikistan | 50,000 | 62.50 | 7500 | 11 |
| Afghanistan | 22,000 | 27.50 | 5000 | 7 |
| Uzbekistan | 5000 | 6.30 | 33,000 | 47 |
| Kyrgyzstan | 1500 | 1.90 | 1500 | 2 |
| Turkmenistan | 1500 | 1.90 | 23,000 | 33 |
| Total | 80,000 | 100 | 70,000 | 100 |

water—but they may get less. Climate change, which has caused the river's seasonal flow to become erratic, has heightened concerns about diminishing flows.

The Istanbul workshop on the Amu Darya basin built on the accomplishments and lessons-learned from several prior international and local water initiatives addressing water challenges in the region. Using interactive sessions, the workshop sought to address the current policies, and institutional and technological barriers to sustainable and integrated resource use in the Amu Darya River Basin, as well as potential pathways for solutions. This workshop brought to the table a set of additional key actors without whose sectoral expertise and regional influence a dialogue on the closely inter-related agricultural, water, and energy challenges would have remained incomplete. Participants included UN officials and academics such as author J. Shroder and others, all with a mandate to address strong political sensitivities underpinning policy choices through diplomatic avenues.

The goal of the initiative was to foster an exchange among managers of water and energy, as well as agricultural resource managers and policy experts, both from the Central Asian region and internationally, on policy reforms and technological, institutional, and financing solutions in support of improved water management and mitigation of conflict over shared resources among antagonistic states. Essentially, the workshop supported participants to: (1) gain a better understanding of how to combine and apply the best practices and approaches on the water-energy-food nexus for solutions in river basins; (2) develop an integrated roadmap to plan for strategic investments in infrastructure solutions for water, food, and energy security, including new technologies such as efficiency-enhancing irrigation infrastructure, as well as renewable energy sources, whose potential remains vastly underexplored in the Central Asian region; (3) motivate participants to take practical steps toward implementing water-, energy-, and food-nexus planning and practices in the Amu Darya River Basin; and (4) explore new avenues for advancing hydro-diplomacy in Central Asia in the wake of the successful Conference on International Water Cooperation, held in Dushanbe in Aug. 2013 (and for which EWI served on the International Steering Committee), as well as with the advent of a profound transition in Afghanistan, whose agricultural and hydrocarbon resources were thought to play a key role in post-2014 economic reconstruction efforts (Shroder, 2014b; Dalton, 2014).

### 20.1.2 Dushanbe Forum of Mountain Countries—2015

The increasing recognition that serious water and related issues were already greatly affecting the countries of Central Asia, led directly to additional workshops. Thus in summer 2015 yet another effort got underway in Dushanbe, Tajikistan, to bring together >120 representatives of organizations from across Central Asia and the world whose work has focused on the concerns of mountain societies and environments. The forum was set up to investigate agriculture and food security under changing climates in the Central Asian mountains, as part of an effort to integrate watershed management as a viable solution for coping with, and adapting to coming changes (Batjargal, 2015).

The forum brought together government representatives at various levels, civil society groups, development agencies, and researchers to explore the links between mountains and water. In addition were included the contextual issues for sustainable mountain development including: (1) climate change and its impacts on mountain agriculture; (2) nutrition and health; (3) the role of women in mountain ecosystem stewardship; (4) integrated watershed management; and (5) disaster-risk mitigation.

Details of issues explored in the forum included: (1) food security, and creating conditions for entrepreneurship and better livelihoods, particularly among women; (2) methods for conflict-resolution about pasture management, water, and access to resources including frameworks for encouraging proactive participation by communities in the design of solutions; (3) the necessary governance mechanisms for achieving sustainable land-management practices; (4) the need for integrating disaster-risk mitigation into land-use management practices; (5) the design and delivery of knowledge-management solutions about water and mountains that offer not only technical solutions and platforms, but also the capabilities to help transform information into action on the ground, designed by and for the communities it is intended to benefit; and (6) multidisciplinary and cross-cutting approaches to the formulation of research problems and project delivery, and the requirement to strengthen communications between research and policy.

The Dushanbe conference concluded that priority actions should be to:

- diversify food systems by supporting mountain farmers and encouraging climate-smart agriculture;
- promote participatory governance of natural resources in watersheds;
- scale up sustainable land-management practices, including community-based, climate-risk mitigation plans;
- introduce equitable benefit sharing and compensation mechanisms to properly reflect the services provided by mountains for all;
- enable better knowledge exchange between practitioners, communities, and researchers;
- invest in decision-support tools and programs that integrate the concerns of mountain communities by working directly with them; and
- place sustainable mountain development concerns in development plans and use international meetings and conventions to emplace mountain issues in sustainable development goals and actions.

Thus, although the Dushanbe Mountain Forum was not only about many issues other than just water, the fact that climate change and water management were major features for discussion reflects the urgency needed to prepare for threatening developments in the region.

## 20.2 REMAINING WATER ISSUES

As this book was being conceived in 2014, during the final stages of completion of the previous book on natural resources in Afghanistan (Shroder, 2014b), a large number of additional water-related issues were recognized as needing additional discussion, as well as development for inclusion in the list of DLMs. The amount of information available to provide an intelligent and well-informed discussion on a variety of these issues was discovered, however, to be woefully limited and not well enough established in the open literature to enable adequate assessment. The result was the loss of inclusion of chapters on such concepts as: (1) the problems and possibilities of equitable division and apportionment of transboundary water resources; (2) the issues of ethnicity, gender, landowners, and landlords with regards to water; and (3) the strategic and security environments related to water issues in Afghanistan and surrounding countries. In the future, these topics should be assessed in detail by Afghans who are trained in the art and science of extracting and publishing relevant issues and data.

## 20.3 WATER SYMPOSIA IN ISLAMABAD, KABUL, PERTH, AND DUSHANBE—2015

As a result of early assessments of glaciers in Afghanistan at the behest of the US Geological Survey in the late 1970s (Shroder, 1980, 1989; Shroder and Bishop, 2010), we recognized early on that the water resources of Afghanistan and Pakistan were under considerable stress as the glaciers were observed to be mostly back- and down-wasting. Diminution of these frozen water resources meant that future melt-water supplies were bound to be reduced as the world's climate warmed through the inexorable consequence of increased greenhouse gases such as carbon dioxide ($CO_2$) and methane ($CH_4$) continuously pouring into the world's atmosphere by industrialization and human-driven agricultural needs. In combination with the devastating warfare and near constant insurgency in Afghanistan for the past four decades that has precluded proper scientific data collection and assessment of water resources, the result has been a great destruction of local ability to manage their water and any planning for the future.

In an effort to better understand water issues in South and Central Asia therefore, in the fall of 2015, three of us (S.J. Ahmadzai, P. Mcnamara, and J. Shroder) co-hosted transboundary water symposia at Quaid-i-Azam University in Islamabad, Pakistan, and the Tajikistan Academy of Science in Dushanbe, as well as with a meeting of Afghan water experts at the Afghanistan Ministry of Water and Energy in Kabul. In addition to the hosted symposia, also in Kabul was the *5th International Hindu Kush Geosciences Conference*, during which J. Shroder was asked to speak on water. Finally on the way back through Europe was to be a series of presentations by our group on water to an international physical- and social-science audience attending the world conference of *Mountains of our Future Earth* in Perth, Scotland in October. All of the extensive meetings and presentations mentioned above were funded by the Lounsbery Foundation, the US Institute of Peace, and the Water For Food Institute of the University of Nebraska.

### 20.3.1 Seminar on Transboundary Water Resources of Pakistan and Afghanistan—Islamabad—2015

In mid-September 2015, professors McNamara and Shroder were able to lead a seminar on transboundary water at the Department of Earth Sciences, Quaid-i-Azam University, Islamabad, Pakistan. About 150 faculty and students were in attendance to listen to the program, including 26 university professors and government bureaucrats, 15 doctoral students, 11 Masters of Philosophy students, 65 Masters of Science students, and 31 Bachelors students. Four speakers were set to raise the main issues at hand with transboundary waters.

First Dr. John (Jack) Shroder spoke for an hour on "Transboundary Water Resources of Pakistan and Afghanistan," which was a summation of this new book on *Transboundary Water Resources in Afghanistan*, which is to be published by Elsevier in 2016. Dr. Shroder had asked Dr. McNamara to mind the time and head Dr. Shroder off with a question-and-answer (Q&A) period after 1h. This approach worked very well and got the students engaged in detailed questions, during which we discovered that the students very much appreciated the direct engagement with foreign faculty such as those who were willing to come to the difficult environment of Pakistan.

Dr. McNamara used a portable microphone most effectively to elicit questions and comments from the students, which one of the administrators remarked on later was a technique that he had not seen before in a scientific meeting and he thought that it was quite effective, and that he would try it himself later. During the Q&A we discovered that the students were quite surprised to discover our openness in discussing such important topics as water in their region, especially insofar as their politicians seemed so inadequate at dealing with such things as new dam construction, hydroelectric production, or flood control. Dr. McNamara was especially good at getting the normally quite reticent female students to speak out, even those in full face-covering veils, some of whom spoke through less covered, more open female students and faculty who were willing to speak more freely in front of so many men. Several new female-faculty members also spoke well in the Q&A. A number of the female students expressed thoughts that seemed to indicate that students were too much ignored, or their thoughts and opinions were disregarded by the higher-ups, especially the politicians, who seemed to be universally reviled by the students, who periodically applauded enthusiastically.

At one point a female student asked in terms of the seminar topics, "But what can we really do?" Dr. Shroder quickly replied that she and the rest of the students were the future (of Pakistan), and that she and her other female associates were also the mothers of the future, so it was important to work hard on these most important water issues, especially as the climate was changing and the future of water in South Asia was so uncertain for so many millions of people.

Dr. Shroder's comprehensive and fact-filled talk was followed by a report from retired professor, Dr. Zulfiqar Ahmad, on, "Transboundary Water Resource Hydrograph of the Kabul and Indus Rivers in Relation to Earth Fill Dams." This Power-Point talk was an example of the best science work produced in Pakistan, and was a very comprehensive look at water issues in the country. For example, key water issues identified by Dr. Ahmad included:

- strong dependence on (snow and ice) melt water;
- Pakistan was naturally water stressed;

- a large irrigation system existed;
- transboundary water occurred (among Pakistan, Afghanistan, and India);
- a huge and growing population existed;
- many national and international activities occurred; and
- low coordination, collaboration, or sharing of knowledge and information were common.

Furthermore, in Pakistan the prospect of the predicted coming severe water scarcity and a huge increase in population was a most frightening prospect (Fig. 20.2).

The result is that great attention must be paid to dealing with these looming crises. The consequences were that the Water and Power Development Authority (WAPDA) of Pakistan had to attend to:

- construction of large and medium reservoirs;
- development of hydropower projects;
- water resources management (WRM);
- establishment of a Glacier Monitoring Research Centre within WAPDA and conduct studies related to WRM; and
- strengthening and upgradation of data-collection networks and management infrastructure.

Dr. Ahmad discussed in detail a number of other issues dealing with water in Pakistan in relation to water coming from Afghanistan and made several serious but somewhat enigmatic statements (in added **bold** below). Some of these are in opposition to much current observation and thought elsewhere and so could be mistaken as future projections, or government "spin," but that if true, in any case would necessarily indicate a high complexity of hydrologic processes that are in need of much additional study, as follows:

- Analysis of both the upper and lower belts show notable **increasing** trends in temperature and **precipitation** that will enhance the outflow runoff **downstream from Pakistan to Afghanistan** and will have a **positive** climate impact.

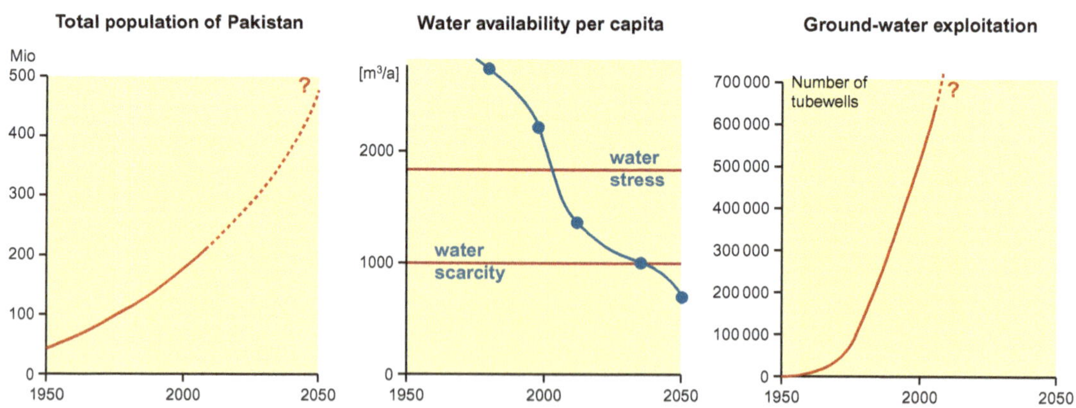


FIG. 20.2 Projections of Pakistan population, water availability per capita, and exploitation of ground water (from Zulfiqar Ahmad presentation). These projections indicate great problems and lack of sustainability in future, which are obviously not possible to endure with so that chaos and violence are highly probable as a result.

- Temperature and **rainfall** have shown recent **increasing** trends for the Chitral and Peshawar valleys (which are directly adjacent to Afghanistan and have interlinked transboundary drainage basins).
- The **increase in rainfall** will have a positive impact on water discharge for transboundary rivers. The increasing temperature will have impact on the glacier and snow reservoirs in Chitral mountains and it would generate the melt water.
- The retrieval of classified areas in remote sensing has shown that water declination causes the changing pattern of vegetation and invention of barren/reddened (false color infrared) land.
- Due to the decreased water supply from Afghanistan, the agriculture sector would face negative impacts in Khyber Pakhtunkhwa Province, especially in Peshawar valley.

These two lectures were followed by additional Q&A, and then following lunch, several additional talks were held. One by Dr. Shakeel Ahmad, who had studied the transboundary impact assessment of Indian dams on the Chenab River Basin from the perspective of the well-known Indus Water treaty, which in spite of numerous negative provocations and three wars between the Pakistan and Indian signatories, has held firm in much of its entirety for over half a century. His data do show that a decline in discharge on the Chenab has been noted in Pakistan downstream, which has been primarily caused by unannounced manipulation of dams upstream in the Himalayan foothills in India. Such changes cause great fears in Pakistan in the event of eruption of future hostilities. Dr. Ahmad's most striking recommendation, perhaps was that hydroelectricity produced in abundance by the many Indian dams could be imported by Pakistan, whereas Pakistan has been having so many political problems in getting very many, if any, new dams built on its own territory.

Dr. Arshad Ashraf then gave the last talk of the day, which was on the impact of climate change on water resources of the upper Indus Basin. His material from a subdivision of the Ministry of Agriculture was well known to Dr. Shroder, who had actually participated in the earlier collection of much of the data when he was working with Mrs. Rakhshan Roohi from the Ministry of Agriculture, who was originally responsible for the project in the 1990s and 2000s, as well as with Mr. Ghazanfar Ali when the two (Ali and Shroder) had a joint research project on glaciers and climate change in Pakistan funded by the USAID PEER program through the US National Academy of Sciences a few years ago (Bishop et al., 2014).

In the late afternoon of the conference the Q&A sessions continued with the students joining in most enthusiastically. Much discussion ensued over the idea that Pakistani and Afghan professors and students actually could travel to each other's countries and engage in discussions about their mutual water resources. The students exhibited considerable surprise that there would be interest and not fear in doing that. Dr. Shroder pointed out that science is about finding the truth of matters, and as with the weather which respects no national border boundaries, neither does the water, that it is a shared resource, not usefully subject to being kept as a secret between countries and that only by being discussed openly between the countries, could later good relations be developed and maintained on into the future. Open dialogues on these topics were seen as the scientifically and politically acceptable best means.

The ceremonies concluded with the awarding of certificates of achievement to a number of students, as well as the presentation of plaque awards to Drs. Shroder and McNamara as tokens of goodwill. The desire was in evidence to prolong and enhance further developments between Quaid-i-Azam University in Pakistan, and Kabul University and Kabul Polytechnic University in Afghanistan, with the University of Nebraska at Omaha (UNO) as a goodwill ambassador and broker of peaceful relations between the universities of the two nations. This occurred because of UNO's long four decades of such work throughout much of the latter half of the 20th century and the decade and a half of the new century. Several of the Quaid-i-Azam administrators and faculty noted that UNO's interest in continuing to do this work was well known and much appreciated, and we responded that we were hoping for similar good results in Kabul and that we would do our best to help improve relations between the two nations by continuing our work on this most vital of human resources, that of the absolutely essential water.

## 20.3.2 Meeting with Afghanistan Ministry of Energy and Water—Kabul—2015

In the meeting in Kabul in late September and early October, Mr. Farooq Qazizada, the Deputy Minister of Energy and Water, related at the outset that he was quite aware of the long association of the University of Nebraska at Omaha with his country. He mentioned the enhancement of the engineering college at Kabul University in the 1970s, the large UNO presence on Peshawar, Pakistan, during the Soviet-Afghan War in the 1980s, when UNO, as a kind of Ministry of Education in exile, helped Afghanistan with its primary and secondary education. The Deputy Minister noted that the many problematic issues in his country had included a lack of political will to address problems with water but now that UNO was back in his country, perhaps some of their difficulties with water could be addressed with the assistance of UNO.

The great lack of peace in Afghanistan for decades has meant equally great losses in all normal things, not the least of which is knowledge about water. Considerable amounts of training and development of new expertise are needed, and enhancement of capacity concerning water at the university level is essential. Sustainable capacity building appears to require development of a long-term strategy that would include a robust capability to engage in new research on water. This would require establishment of a new integrated and comprehensive Transboundary Water Resource Center (TBWRC) in Kabul that could host analytical hydrologists who had also been educated in the unusual hydropolitics of Afghanistan (see Chapter 17).

The well-known lack of water data in Afghanistan because of four decades of war is especially noteworthy and in need of serious remediation. Coupled with the lack of data sharing in the region, the fact that some water data are kept as state secrets in adjacent countries, and the lack of water infrastructure; these major water discontinuities are a striking absence in such arid countries replete with citizens who absolutely require plentiful water to maintain life. Knowledge gaps about water in Afghanistan are, of course, terrible impediments to any sort of negotiation with neighboring countries about water, and preclude any sort of joint water projects with them; as well as denying reasonable formulation of water policies. Development of new policy on water is now seen at the government level as an essential aspect for the country to undertake.

In terms of existing relations over water with neighboring countries, because the political will in Afghanistan has now changed favorably with the election of President Ashraf Ghani, the situation has changed greatly. Afghanistan and Pakistan are now seen to sometimes work well together on water discussions, but in recent requests by Afghanistan to Uzbekistan and Turkmenistan for discussions about water, no response has been given. Some limited discussion has occurred with Pakistan about development of water in the Kunar River tributary to the Kabul River. The British-built Durand-Line border between the two countries, however, is a bugaboo that no Afghan government is ever likely to recognize as a valid dividing line between them, because to do so would be to repudiate over a century of strong Afghan viewpoints to the contrary. Any negotiation over transboundary water between the two nations would require some sort of normal recognition of territorial sovereignty, and Afghanistan is unable politically to do that, thus an impasse. So at the outset in the development of transboundary-water policy, Afghanistan is seen as needing: (1) a new water research center; (2) capacity development about water; (3) good technical relations with neighboring countries; (4) water data sharing with neighboring countries; and (5) training in climate-water aspects and climate-change characteristics.

## 20.3.3 Integrated Comprehensive Transboundary Water Resource Center (TBWRC)

The Afghanistan Ministry of Energy and Water has specifically floated the idea that it would need help in developing a center for the express purpose of focusing upon a TBWRC. Perhaps this reflects initial planning for a major government request for foreign aid, but if so, it certainly seems to be a quite useful way to think about a heretofore insoluble problem. A possible way to begin to establish a TBWRC would be to develop one or more departments of such at Kabul University and Kabul Polytechnic University. This would begin to address the recurrent problem that once foreign aid comes into the country to establish a new program, all too often when the foreign aid comes to an end the continuity and sustainability of the project may terminate as well. If several TBWRCs were formed in the universities first and people were trained in maintaining them, then an external spinoff into a viable governmental organization might be possible at a later time.

Several additional ideas that were discussed were the formulation of a supreme council of land and water to establish the ability for bargaining and negotiation on a long-term basis, but this notion was not well developed and would probably overlap too much with existing environmental structures in the government. Finally, the involvement of public-private sectors in such things as the Kunar River water development projects might bring greater attention to water needs in Afghanistan.

## 20.3.4 Fifth International Hindu Kush Geosciences Conference (HKGC, Kabul—2015)

In late September at Kabul Polytechnic University the HKGC 2015 was sponsored by the Deutsche Gesellschaft für Nternationale Zussamenarbeit, Kabul University (KU), Kabul Polytechnic University (KPU), Marmaristan Consulting Services, Charles University, the Czech Republic, the Afghanistan Ministry of Mines and Petroleum, and the University of Nebraska at Omaha.

The conference was opened and introduced by the new Minister of Mines and Petroleum of Afghanistan, HE Dr. Daud Shah Saba, as well as the Minister of Higher Education,

HE Dr. Farida Momand. In addition, some words of welcome were extended by the university presidents, Dr. Ezatullah "Ahmed" (KPU) and Dr. Habibullah Habib (KU). Attended by some 30 geoscience faculty from Afghanistan, Europe, and USA, the large lecture hall was packed with students as well. Along with numerous papers on rocks, minerals, structures, and natural resource characteristics, otherwise papers were given on ground-water quality and quantity, as well as J. Shroder's paper on *Transboundary Water Resource Problems for Afghanistan and Solutions*. The 2-day conference went extremely well and allowed in-depth conversations with many old friends, as well as allowing access to the numerous students who came up for advice.

### 20.3.5 Mountains of our Future Earth International Mountain Conference, Perth, Scotland—2015

This conference in early October was hosted by Martin Price at the University of Perth and Greg Greenwood, Executive Director of the Mountain Research Institute (MRI) in Geneva, Switzerland. The conference involved a wide variety of physical and social-science issues in the mountains of the world for hundreds of scholars. This was a contribution to the global Future Earth program, a 10-year international research initiative that will develop the knowledge for responding effectively to the risks and opportunities of global environmental change and supporting transformation towards global sustainability in the coming decades. J. Shroder was asked to host two sessions at the conference: one on "*Challenges of land-use decision making in an increasingly nested and networked world*" and a second on "*Transboundary water issues*" that had five speakers on a variety of related topics about water that crossed boundaries in the world. Sher Jan Ahmadzai gave a paper on indigenous or traditional water management in Afghanistan, but he was able to include much new information on modern hydropolitics in Afghanistan as well.

In the course of meeting with the two sponsors, Price and Greenwood, it was determined that Greenwood would co-edit with Shroder a new book for Elsevier on *Mountain Ice and Water*, to be published in 2016 or 2017. The book will have about 10–15 chapters on mountain hydrology by a number of attendees at the conference from Africa, Asia, Europe, and North and South America. The contract to do this was signed with Elsevier by Greenwood and Shroder on Nov. 23, 2015.

### 20.3.6 Rational Use of Water Resources of Transboundary Rivers of Tajikistan and Afghanistan—Dushanbe—2015

In late Nov. 2015 a conference was held in Dushanbe, Tajikistan, that was physically attended by 25 participants from the highest levels of government (Table 20.2) in the Republic of Tajikistan (RT), while UNO participated electronically (Karimov, 2015). Conference objectives (Table 20.3) and existing problems (Table 20.4) were developed within a context of possible areas of cooperation between the two countries by using the DLMs developed by UNO. A working group of five doctoral members of Tajikistan's water-cognizant experts was formulated to work with UNO on these factors (Table 20.5).

Possible areas of cooperation include a number of new areas to be combined with the rejuvenation and effectuation of some older efforts that had been established earlier but which have been allowed to languish in certain cases. For example, further cooperation could

**TABLE 20.2** List of National Institutions and Organizations Who Had Representatives at the Meeting on Rational Use of Water Resources of Transboundary Rivers of Tajikistan and Afghanistan in Dushanbe, Tajikistan, in Nov. 2015

- Ministry of Energy and Water Resources of the Republic of Tajikistan (RT);
- Ministry of Education and Sciences of the RT;
- The Ministry of Economic Development and Trade;
- Committee for Environmental Protection (CEP) under the Government of the RT;
- Committee for Emergency Situations and Civil Defenses (CES&CD) under the Government of the RT;
- Research Center for Protection of Water Resources (RCPWR) CEP under the Government of the RT;
- Agency for Hydrometeorology of the CEP under the Government of the RT;
- Tajik Technical University named after M.S. Osimi;
- Tajik Agrarian University named after Sh. Shotemur;
- Institute of Water Problems, Hydropower and Ecology of the Academy of Sciences of the RT (AS RT);
- Institute of Economics and Demography of the Academy of Science;
- State Enterprise "Tajik Research Institute of Hydraulic Engineering and Land Reclamation" (SE "Tajik NIIGiM");
- Water Partnership of Tajikistan;
- Tajikistan National NGO Association
- Engineering Academy of the RT.

*From Karimov, S., 2015. Report on the conference devoted to "Rational use of Water Resources of Transboundary Rivers of Tajikistan and Afghanistan." Unpublished November 21, 2015 conference in Dushanbe, Tajikistan (S. Karimov, Director, Tajikistan National NGO Association; email to S.J. Ahmadzai, 14 December 2014; 9 page copy to J. Shroder).*

**TABLE 20.3** Objectives of the Conference Devoted to Rational Use of Water Resources of Transboundary Rivers of Tajikistan and Afghanistan, Nov. 2015, Dushanbe, Republic of Tajikistan

- To discuss issues related to possible cooperation between Tajik and Afghan scientists and leading experts in area of water management and rational use of water resources of transboundary rivers of Tajikistan and Afghanistan.
- To evaluate the current situation and problems existing in this particular area and identify mechanisms of better addressing such problems in cooperation with Afghan counterparts.
- To determine how DLMs as effective tools can be used to enhance level of ecological/environmental awareness of population in transboundary regions of Tajikistan and Afghanistan.

*From Karimov, S., 2015. Report on the conference devoted to "Rational use of Water Resources of Transboundary Rivers of Tajikistan and Afghanistan." Unpublished November 21, 2015 conference in Dushanbe, Tajikistan (S. Karimov, Director, Tajikistan National NGO Association; email to S.J. Ahmadzai, 14 December 2014; 9 page copy to J. Shroder).*

**TABLE 20.4** Problems That Exist With Transboundary Water Resources Between Tajikistan and Afghanistan

1. Absence of a permanent hydrological monitoring station along the Panj River.
2. Lack of information, limited access to existing data base, and low awareness of the border adjacent communities in Afghanistan and Tajikistan.
3. Lack of qualified personnel and mid-level professionals on hydrology, hydrometeorology, and rational usage of water resources.
4. Absence of a focal point for intergovernmental cooperation between specialists at different levels.
5. Insufficient use of modern communication systems in creation and establishment of scientific-educational institutions/organizations through applying distance learning modules.
6. Lack of environmental education programs focused on improvement of citizens' education and culture in the bordering areas as well as active involvement of the civil society organizations (CSOs) of Tajikistan and Afghanistan in addressing these problems.

TABLE 20.5 Working Group of Tajikistani Water Scientists Who Will Combine With UNO Counterparts in the Area of Water Management and Rational Use of Water Resources, As Well As Encourage the Creation of Necessary Structures and Documentation

1. Mr. Kobuliev Z.V.—Director of the Institute of Water Problems, Hydropower and Ecology of the Academy of Sciences of the RT, DSci, PhD, Professor, kobuliev@mail.ru.
2. Mr. Karimov Sh.T.—Head of Tajikistan National NGO Association, PhD, shamskarimov14@gmail.com.
3. Mr. Fazylov A.R.—Associate Professor of the Tajik Agrarian University named after Sh. Shotemur, PhD, alijon53@rambler.ru.
4. Mr. Boturov K.—Representative of the Agency for Hydrometeorology of the Committee for Environmental Protection under the Government of Tajikistan, PhD, boturov.kodir@mail.ru.
5. Mr. Saidov I.I.—Consultant of the Research Center of the Interstate Coordination Water Commission under the Ministry of Energy and Water Resources of the Republic of Tajikistan, Dsci, PhD, sibra@rambler.ru.

*From Karimov, S., 2015. Report on the conference devoted to "Rational use of Water Resources of Transboundary Rivers of Tajikistan and Afghanistan." Unpublished November 21, 2015 conference in Dushanbe, Tajikistan (S. Karimov, Director, Tajikistan National NGO Association; email to S.J. Ahmadzai, 14 December 2014; 9 page copy to J. Shroder).*

occur between Tajikistan and Afghanistan within the "Agreement between the Government of the Republic of Tajikistan and the Islamic Republic of Afghanistan on Cooperation for the Development and Management of Water Resources of the Panj-Amu Basins, dated 25 Oct. 2010." This is the first stage of cooperation between the two countries that is considered as the basis for implementation of other projects related to the Panj River Basin.

Joint efforts could be undertaken in reconstruction and rehabilitation of old hydrological stations and creation of new water-posts equipped with modern technology, which would be able to provide data not only on the level of water flows, but also on its quality (turbidity, sediment discharge, etc.). A good example of this could be the water-post of Ayvaj in the Panj River. Implementation of joint hydrological monitoring could be undertaken. Such an approach would ensure online hydrological data of the Panj River as well as enable recording of the actual flow or discharge of the river. Also extremely useful would be establishment of a mutually fruitful exchange networking on the hydrological data/information and experience about the Panj River with the use of modern communication systems, especially during possible emergency situations.

In the future, it would be much better and more efficient to focus on the implementation of not only the liquid effluents of water, but also on the management of solid runoff (sediments) as well. Also sometime in the future, it is recommended to have a joint designing and construction of structures to better regulate the use of watercourses of the transboundary rivers and protective constructions to ensure the normal operation of the water intake and other related facilities.

Improvement and adaptation of the legal framework concerning ongoing transboundary water-resource activities in accord with the legislation systems of the both countries would be a most valuable undertaking. One of the important tools of better addressing the problem of transboundary water use between the two countries is the educational component, including the use of DLMs and online learning modules. Some of this can be accomplished by design and development of thematic programs to increase the level of the environmental education of the population living on border areas of Afghanistan and Tajikistan with an active involvement of civil society organizations (CSOs).

Using the methods of distance education with the DLMs produced by UNO may address some of the problem, although care will be needed to bring up the level of education along the borders. Distance education is one of the ways to improve the level of awareness of almost any population on environmental management and ecology. It is thought to be one of the real solutions to the problems related to the management and rational use of the water resources of transboundary rivers as well.

Relevant training step by step is most important. The first step needs to be focused on retraining of water specialists of Tajikistan (representatives of the Ministry of Energy and Water Resources of the RT, the Agency for Land Reclamation and Irrigation, the Agency for Hydrometeorology of the Committee for Environmental Protection under the Government of RT, the structural units of the Academy of Sciences and other academic institutions of the RT, academicians, etc.), as well as the water specialists of Afghanistan. Such an approach would produce an effective product and would educate ordinary citizens in the future—the second step of the process.

Accordingly, the working group from Tajikistan (Karimov, 2015) has recommended a number of actions that will provide solutions, such as the following:

1. Taking measures to strengthen institutional capacity and tools of water and environment policy between Tajikistan and Afghanistan, including:
   - designing, developing, and adopting regulatory tools and mechanisms of cooperation on water and energy resources (creation of a legal framework and better legal environment for effective cooperation and activities between Tajikistan and Afghanistan);
   - developing and adopting scientific and educational programs and projects, including training and retraining of specialists, conducting researches, signing bilateral and multilateral agreements on: (a) systematic information exchange, (b) rational use of water and energy resources of the Panj River Basin, (c) training professional staff (improving capacity building); and
   - creating and establishing a proper information system (system for monitoring of water resources in the river basins, including the glaciers).
2. Preparing an analytical monograph on "*Review of the water and energy problems of the Panj River Basin: problems and solutions.*"
3. Assessing and evaluating the situation with existing capacity of Tajikistan and Afghanistan in using DLMs; exchanging mutually acceptable programs of training modules available in both countries.
4. Creating a single information system and database on water, energy, and environmental issues of the Panj River Basin.
5. Considering the water issues in a greater context integrated into the system of "water-energy-food-environment."
6. Designing and adopting a memorandum of understanding (MOU) between interested bodies (for instance, between the Institute of Water Problems, Hydropower and Ecology of the Academy of Sciences of RT and the University of Nebraska at Omaha, USA, and if it is necessary, with the relevant organizations of the IRA). In order to do this, Tajikistan must develop short, medium and long-term programs or design a "roadmap" with clear mechanisms of their implementation, etc.

7. Reviewing project proposals aimed to further intergovernmental development and rational use of water resources of the Panj River.
8. Developing consolidated criteria of analysis/assessment, evaluation and decision-making processes related to management of rational use and protection of water resources.
9. Establishing a scientific-educational center of DLMs under the Institute of Water Problems, Hydropower and Ecology of the Academy of Sciences of the RT, supported by the University of Nebraska at Omaha, NE, with the involvement of students from Afghanistan and Tajikistan.
10. Publishing textbooks and teaching materials related to appropriate profiles and relevant education programs in the area of water management and rational use of water resources with a special focus on transboundary rivers of Tajikistan and Afghanistan.

### 20.3.7 Orienting the Way Forward

In Jan. 16, 2016, after the entire text for this volume had been submitted by the December deadline to the Elsevier acquisition editors for publication, a request came from the US Institute of Peace to our group at UNO to work with Duran Research and Analysis in Kabul (http://duran.af). This group were actively attempting to get the Government of Afghanistan to do a better job with its efforts on water in the country, which is exactly what we had been working on ourselves. The result was access to many new ideas afloat to improve the transboundary water situation in Afghanistan and its neighboring countries. The preliminary ideas and results (Duran Research and Analysis, 2015) upon which we hope to be working on into the years 2016–17 are presented in Tables 20.6–20.10.

TABLE 20.6 Water Challenges in Afghanistan: Institutional Deficiencies in Approach and Coordination Concerning Water Resources

i. *Inadequate economic mechanisms*: An all-encompassing economic vision in the country is lacking, as is a shared vision for river-basin development, and such is required for success.
ii. *Disconnect with larger context*: Water-sector activities are largely detached from issues and debates about environmental and climate change, poverty reduction, population control, revenue generation, employment creation, urbanization, and refugee repatriation.
iii. *Inadequate governance mechanisms at national level*: New water law is inadequate concerning transboundary waters.
iv. *Dispersed institutional support and poor coordination*: Weak and unclear delineation of responsibilities exist on water between the ministries.
v. *Contested challenge of capacity*: Low knowledge or understanding of international water laws and regulations occurs.
vi. *Non-compliance to conventions and gap of regional and bilateral frameworks*: Only limited bilateral water agreements and no cooperation frameworks.
vii. *Lack of coordination and project-driven approach of international community*: Focus on emergency projects and short-term initiatives that are uncoordinated and unknowledgeable about other organization's foci are a recurrent problem
viii. *Absence of public-private partnerships*: Shared understandings about water are most limited between society, media, academia, and private sector, but should be augmented.

*From Duran Research and Analysis, 2015. Afghanistan's Trans-Boundary Waters: An Overview. Resources in Greater Central Asia, Kabul, Afghanistan. Heinrich Böll Stiftung. http://duran.af/downloads/ATBWs_Duran_hbs_final.pdf (accessed 16.01.16).*

**TABLE 20.7** Water Challenges in Afghanistan: Technical Deficiencies in Approach and Coordination With Afghanistan Water Resources

i. *Data gap*: Lack of hydrological, meteorological, geotechnical, and water data seriously impedes Afghanistan's ability to participate in regional dialogues.
ii. *Inadequate infrastructure*: Lack of infrastructure and equipment needed to efficiently conserve and utilize water resources.
iii. *Inadequate investment in ground-water recharge capacity*: Underground water is being extracted at a significantly dangerous rate and needs to be replaced.
iv. *Urbanization and water management*: Urbanization and high population growth rates are placing great stress on water resources and recharge of ground water.

*From Duran Research and Analysis, 2015. Afghanistan's Trans-Boundary Waters: An Overview. Resources in Greater Central Asia, Kabul, Afghanistan. Heinrich Böll Stiftung. http://duran.af/downloads/ATBWs_Duran_hbs_final.pdf (accessed 16.01.16).*

**TABLE 20.8** Water Recommendations in Afghanistan: Institutional Opportunities for Intervention and Development of Essential Water-Related Actions

i. *Develop a national economic development vision, and prioritize the water sector*: A national-level and cohesive cross-sectional economic vision for the nation is required that emphasizes transboundary water.
ii. *Clarify roles and improve coordination*: Uncoordinated and confused at present, water issues need to be systemized and coordinated through a strengthening of the new Supreme Council on Water in Afghanistan.
iii. *Establish and strengthen knowledge base on water-resource development and management*: The new Water Sector Strategy emphasizes needs for merging scientific theory with empirical and practical knowledge on water.
iv. *Address water within the broader context*: Climate changes (floods, droughts) is already adversely affecting lives with drastic consequences so development and management of water resources needs special attention and proper education.
v. *Improve national water governance mechanisms:* Legal and policy contexts do not address development and management of transboundary water resources, with the result that an adaptive and integrated management plan must focus on transboundary water issues.
vi. *Recognize the benefit of regional cooperation, hydro-diplomacy and compliance with international conventions:* Regional cooperation is absolutely essential to any effective management and development of water resources.
vii. *Promote long-term, region-program approach:* International support and investment are required in coming years because they are critical elements of future stability in the region.
viii. *Engage civil society, media, academia, and private sector*: Hydro-cognizant capacity building is a starting point to provide enough support from these sectors to help ensure some success in future.

*From Duran Research and Analysis, 2015. Afghanistan's Trans-Boundary Waters: An Overview. Resources in Greater Central Asia, Kabul, Afghanistan. Heinrich Böll Stiftung. http://duran.af/downloads/ATBWs_Duran_hbs_final.pdf (accessed 16.01.16).*

**TABLE 20.9** Water Recommendations in Afghanistan: Technical Opportunities for Intervention and Augmentation of Water Resources

i. *Address the data gap*: Gain use of innovative technologies (remote sensing, etc.) to collect, analyze, and present hydrological and meteorological data from within a new Resource Center on Transboundary Waters with plentiful international assistance.
ii. *Develop necessary infrastructure*: Improve water-storage capacity by building numerous small and less threatening dams and infrastructures to aid rural and urban water-management systems.
iii. *Improve ground-water conservation*: Ground-water recharge and rainwater harvesting are essential and must be developed.

*From Duran Research and Analysis, 2015. Afghanistan's Trans-Boundary Waters: An Overview. Resources in Greater Central Asia, Kabul, Afghanistan. Heinrich Böll Stiftung. http://duran.af/downloads/ATBWs_Duran_hbs_final.pdf (accessed 16.01.16).*

TABLE 20.10 Water Recommendations in Afghanistan: Civil Society Opportunities for Intervention and Advocation to Improve the Water Situation

i. Advocate for increased engagement of civil society, media, academia, and private sector in water development and management.
ii. Support and facilitate continuous indigenous Afghan research on all water issues.
iii. Advocate for improved legal and policy context of transboundary water resources.
iv. Expand Afghanistan's knowledge base on transboundary water resources.
v. Advocate for immediate collective review of UN international water conventions.
vi. Advocate for strengthened executive and coordination role of the Supreme Council on Water in Afghanistan.
vii. Advocate and strongly push for linkages with international and regional entities doing resource/database work concerning water.
viii. Advocate for Afghanistan government immediate action aimed at prioritizing the water sector.

*From Duran Research and Analysis, 2015. Afghanistan's Trans-Boundary Waters: An Overview. Resources in Greater Central Asia, Kabul, Afghanistan. Heinrich Böll Stiftung. http://duran.af/downloads/ATBWs_Duran_hbs_final.pdf (accessed 16.01.16).*

As the galley proofs of this volume were prepared and edited in the spring of 2016, word came from the Government of Afghanistan (GoA) through Duran and the US Institute of Peace that GoA was not yet ready to talk with the neighboring countries about water. This means that perhaps only further policy papers recommending certain water-related activities might work to provide assistance to GoA in its future endeavors with its own waters. With the unity government of President Ashraf Ghani and Chief Executive Abdullah Abdullah reportedly on the rocks in Kabul (Cunningham, 2016), and the calls for new elections in the face of ongoing government dysfunction in the face of continued confrontation with a robust Taliban insurgency, the probability that any water problems willed be solved anytime soon continues to be unlikely.

## 20.4 CONCLUSION

This book, the *Transboundary Water Resources in Afghanistan* is written from the point of view of the water that rises in the highlands of the Hindu Kush and the Afghan Pamir ranges, but that ultimately flows out over the borders to nurture the countries of Pakistan, Iran, Turkmenistan, Uzbekistan, and Tajikistan. As such, these as-yet unused waters from Afghanistan become a vital part of water use and water management across the international borders of the neighboring countries, so much so that any diminution of flow downstream is generally considered to be a hostile act made by upstream Afghanistan, even when it is entirely naturally caused. Thus all the engineering and other hydrological activities in Afghanistan that are increasingly planned or already underway in the country are likely to be of considerable interest to the neighbors; this book is a first attempt to put all these issues out for public discussion.

An essential aspect is our discovery that considerable eagerness exists in South and Central Asia for access to hydrological data and information, much of which can be provided by the distance learning modules that we have developed in English and translated into the regional languages of Dari, Pashto, and Urdu in Arabic script, and Tajikistani in the Cyrillic alphabet. Most strikingly in the course of doing this work was our discovery that these new educational ideas for additional DLMs and one or more transboundary water resource centers could be developed much further in order to greatly improve administrative capabilities in Afghanistan to improve water management in the country. In this fashion the looming threats of water diminution in the region could be at least partly alleviated, a fact that would greatly relieve all the governments of South and Central Asia who must deal with emerging unpleasant realities about water for a long time into the future.

## References


Ahmad, Z., 2014. Transboundary Water Resource Hydrograph of the Kabul and Indus Rivers in Relation to Earth Fill Dams. Power Point Presentation on 17/9/2014 at Quaid-i Azzam University at the Seminar of Transboundary Water Resources of Pakistan and Afghanistan.

Batjargal, Z.E., 2015. Dushanbe Forum of Mountain Countries—2015: Water and Mountains. Agriculture and Food Security Under Changing Climate in Central Asian Mountains: Integrated Watershed Management as a Viable Solution for Coping With and Adapting to Changes. Mountain Partnership Central Asia Hub; University of Central Asia; Bishkek, Kyrgyz Republic. http://www.mountainpartnership.org/fileadmin/user_upload/mountain_partnership/docs/Dushanbe%20Mountain%20Forum2015.pdf.

Bishop, M.P., Shroder Jr., J.F., Ali, G., Bush, A.B.G., Haritashya, U., Roohi, R., Sarikaya, M.A., Weihs, B.J., 2014. Remote sensing of glaciers in Afghanistan and Pakistan. In: Kargel, J.S., Leonard, G.J., Bishop, M.P., Kääb, A., Raup, B. (Eds.), Global Land Ice Measurements From Space. Springer-Praxis, Heidelberg. ISBN-10: 354079817X | ISBN-13: 978-3540798170 (Chapter 23).

Cunningham, E., 2016. A wave of political defections spells new trouble for Afghanistan. Wash. Post 29 March. https://www.washingtonpost.com/world/a-wave-of-political-defections-spells-new-trouble-for-afghanistan/2016/03/29/91c94682-f108-11e5-a2a3-d4e9697917d1_story.html (accessed 02.04.16).

Dalton, J., 2014. Central Asia: Workshop Report. Nexus Dialogue on Water Infrastructure Solutions; Istanbul, Turkey, 15–17 July 2014. East-West Institute; The International Water Association (IWA); International Union CN (IUCN). http://www.waternexussolutions.org/ContentSuite/upload/wns/all/Central%20Asia%20Nexus%20Dialogue%20Workshop%20Report%20V1.pdf.

Duran Research and Analysis, 2015. Afghanistan's Trans-Boundary Waters: An Overview. Resources in Greater Central Asia, Kabul, Afghanistan. Heinrich Böll Stiftung. http://duran.af/downloads/ATBWs_Duran_hbs_final.pdf (accessed 16.01.16).

Karimov, S., 2015. Report on the conference devoted to "Rational use of Water Resources of Transboundary Rivers of Tajikistan and Afghanistan." Unpublished November 21, 2015 conference in Dushanbe, Tajikistan (S. Karimov, Director, Tajikistan National NGO Association; email to S.J. Ahmadzai, 14 December 2014; 9 page copy to J. Shroder).

Shroder Jr., J.F., 1980. Special problems of glacial inventory in Afghanistan. In: World Glacier Inventory Proceedings Reideralp Workshop, September 1978 (IAHS-AISH) Publication No. 126, Hydrological Sciences Bulletin, pp. 142–147.

Shroder Jr., J.F., 1989. Glacierized areas of Afghanistan. In: Haeberli, W., Bosch, H., Scherler, K., Ostrem, G., Wallen, C.C. (Eds.), World Glacier Inventory, Status 1988. IAHS (ICSI)-UNEP-UNESCO, Teufen, Switzerland. pp. C39–C40, C346–C353.

Shroder Jr., J.F., Bishop, M.P., 2010. Glaciers of Afghanistan. In: Williams, R.S. Jr., Ferrigno, J.G. (Eds.), Satellite Image Atlas of Glaciers of the World—Asia, pp. 167–199. U.S. Geological Survey Professional Paper 1386-F3, Washington, DC.

Shroder Jr., J.F., 2014a. Developing Afghanistan resource corridors in an environmentally sound manner. Earth April: 8.

Shroder, J., 2014b. Natural Resources in Afghanistan: Geographic and Geologic Perspectives on Centuries of Conflict. Elsevier, San Diego, CA, USA.

Shroder, J.F., 2013. Building resource corridors in Afghanistan: a solution to an interminable war? Earth September: 25–31.

Shroder, J.F., 2015. Progress with Afghanistan extractive industries: will the country know resource success or failure evermore? Extractive Industries and Society 2, 264–275.

# Index

Note: Page numbers followed by *f* indicate figures, *t* indicate tables, and *np* indicate footnotes.

## D

## E

## L

## N

## U

Kolibri 360 GmbH
Ferdinand-Jühlke-Straße 7
99095 Erfurt, Deutschland
produktsicherheit@kolibri360.de